D0152245

Occupational Safety Management and Engineering

Occupational Safety Management and Engineering

Fifth Edition

Willie Hammer

Dennis Price

Prentice Hall
Upper Saddle River, New Jersey 07458

Library of Congress Cataloging-in-Publication Data

Hammer, Willie
Occupational safety management and engineering.
Willie Hammer and Dennis Price
p. cm.
Includes bibliographical references and index.
ISBN: 0-896515-3
1. Industrial safety. 2. Accidents Prevention I. Title.
CIP DATA AVAILABLE

Vice president and editorial director of ECS: **MARCIA HORTON**
Acquisitions editor: **LAURA CURLESS**
Production editor: **IRWIN ZUCKER**
Executive managing editor: **VINCE O'BRIEN**
Managing editor: **DAVID A. GEORGE**
Manufacturing manager: **TRUDY PISCIOTTI**
Manufacturing buyer: **PAT BROWN**
Marketing manager: **DANNY HOYT**
Copy editor: **ROBERT LENTZ**
Vice president and director of production and manufacturing, ESM: **DAVID W. RICCARDI**
Director of creative services: **PAUL BELFANTI**
Cover director: **JAYNE CONTE**

Prentice Hall, Inc.
Upper Saddle River, New Jersey 07458

Printed in the United States of America

10 9 8 7 6 5 4 3 2

ISBN 0-13-896515-3

Prentice Hall International (UK) Limited, London
Prentice Hall of Australia Pty. Limited, Sydney
Prentice Hall Canada Inc., Toronto
Prentice Hall Hispanoamericana, S.A., Mexico
Prentice Hall of India Private Limited, New Delhi
Prentice Hall of Japan, Inc., Tokyo
Pearson Education Asia Pte. Ltd.
Editora Prentice Hall do Brasil, Ltda., Rio de Janeiro

Contents

PREFACE

Willie Hammer noted in his Preface to the fourth edition:

> Occupational safety has changed since the first edition of this book came out in 1976. The United States is still the greatest industrial nation in the world, but many of its industries, its workers, the types of work they do, laws, public attitudes, and numerous other factors have changed. And so have the safety concerns of the workers, and their dependents, other relatives, neighbors, and the public in general.

He closed the opening paragraph of that preface saying, "This edition attempts to incorporate some of the most notable safety considerations that have taken place since earlier editions." Now, this edition makes that same attempt. The interim between the last edition and this edition was marked with some of the most dramatic changes in occupational history.

It is my privilege to make the additions and deletions that reflect some of the changes in occupational safety engineering and management since the last edition of this text about twelve years ago. One of the most dramatic changes to occupational safety comes from the proliferation of technology and the information revolution of this past decade. Software safety is now recognized as a part of occupational safety engineering and management. Software controls the energy of industry's machinery and products. This fifth edition reflects this industrial revolution by presenting the elements of a software hazard analysis program and software hazard analysis techniques. Severity-of-consequences hazard ratings, program size, and complexity are criteria used to determine the extent of software analysis to be employed for safety. The technical tools for software analysis, such as Code Walk-Throughs, Event Tree, Soft Tree, and Sneak Circuit Analysis, are applied to safety. Software safety analysis is an essential tool for the new millennium safety engineer.

The past decade has magnified the prevalence of computers in the workplace and the electronic office. Along with this has come the ubiquity of work-related musculoskeletal disorders. The repetitive motion injuries that sometimes result in these disorders involve various worker tasks, but cumulative trauma disorders to keyboard operators have drawn attention to this problem. A new chapter, Chapter 18, Work-Related Musculoskeletal Disorders, addresses this phenomenon. Evidence of the work-relatedness of musculoskeletal disorders, factors associated with them, the back belt controversy, and the steps to establish an ergonomics program to control these injuries are discussed.

During the past ten to fifteen years, another hazard has received special attention from researchers and regulators. It is the topic of a second new chapter, Chapter 26, Confined Space Entry. New confined space entry regulations now affect hundreds of thousands of work facilities and millions of workers. The hazards of confined spaces are described in Chapter 26, and guidelines for elements of a confined-spaces entry program are given briefly.

In addition to two new chapters, this edition reflects some significant changes in safety engineering and management since the last edition. Existing chapters have been revised to include these current topics, some arising out of new research, standards, and regulations. Discussions of workers with disabilities (Americans with Disabilities Act), workplace violence, older workers safety, and bloodborne pathogens (Bloodborne Pathogen Standard) are added to the chapter on Personnel, Chapter 9. In the past decade, behavior-based safety (BBS) programs have become a strong part of the safety movement. Chapter 10, Promoting Safe Practices, now includes a discussion of BBS. Chapter 15, Safety Analysis, includes the elements of a Process Safety Management Program and a discussion of What-if, Checklist, Hazard and Operability Study (HAZOP) and other analytic techniques now mentioned in the 29 Code of Federal Regulations. Nuclear waste, and various legal issues are new additions to other chapters. The book's contents have been revised to update topics, such as workers' compensation and workers' compensation fraud, fault tree analysis, hearing protection, environmental protection, fire protection, OSHA violation policy, the Emergency Planning and Community Right-to-Know Act, and system safety analysis. In many places, recent statistics now replace older data. Fifty-four references have been added.

The order of the chapters is changed. The first five chapters are on general introductory and administrative topics. Chapters six through fifteen are on subjects of concern to safety management and planning. The remaining chapters address safety engineering and program management of specific hazards.

These are some of the changes since the last edition: In the new millennium, workers participate more in their own protection than in the past. Managers are held more accountable for worker safety and health than before. Courts and lawyers have more influence in occupational safety than in the past. Communities are more involved in industrial safety than before. Safety engineering and management is more complicated.

My goal has been to maintain the basic no-nonsense, approach to safety that has characterized past editions. More information for managers of safety programs is given than in the past. Although much has changed, much has remained the same. The basic hazards (and preventative measures) from falls, mechanical injuries, heat and temperature, pressure, electricity, fires, explosions, toxic materials, radiation, and vibration and noise remain about the same. This revised edition retains and updates these topics and includes more details on some.

This edition is a small token of respect for Willie Hammer, whose dedication to the noble profession of safety engineering and management resulted in the first four editions of this text.

Dennis Price

CHAPTER 1

Accident Losses

When the Railway Safety Act was being considered in 1893, a railroad executive said it would cost less to bury a man killed in an accident than to put air brakes on a car. The railroad executive probably was not an evil or malicious man. In all probability he believed in God, was a good husband and loving father, and patted his dog when he came home. He would have done anything to avoid injury to his family or dog, but he considered the safety of other people only in monetary terms.

COSTS OF WELL-BEING

The struggle to provide safeguards to eliminate or reduce accidents and the injuries and damages that result has been influenced by two mutually opposing considerations: (1) costs of accident prevention, and (2) moral regard for human life and well-being. The moral consideration has developed because of massive numbers of accidental deaths and injuries. Influential persons who had witnessed or who knew of the effects of accidents called for corrective actions and new laws to safeguard workers and the public. Gradually compromises have come about between the benefits and the costs of accident prevention. Many of the larger companies have found the mutual consideration and compromise beneficial, and workers have found themselves safer than if no safeguards were provided. The result has been fewer walkouts and strikes and more efficient operations, as workers have less need to slow down to avoid or protect themselves against hazards. Another benefit to companies has been a reduction in costs of litigation and insurance premiums.

A report[1] prepared by the University of Michigan for the U.S. Department of Labor observed that "Chaos and disaster have traditionally been the harbinger of government programs intended to help the American worker. . . . In November, 1968, 78 workers died in a Farmington, West Virginia, mining tragedy. Less than a year later, the Coal Mine Health and Safety Act passed the Senate by a vote of 72 to 0." After the remainder of Congress and the President approved, the bill became law.

Unfortunately, unless accidental deaths and injuries are massive, bring about large or recurring lawsuits, or evoke unignorable comments from the news media, the federal government is averse to undertaking suitable corrective action. In the past, many owners of small businesses were intimately knowledgeable, concerned about, and personally responsible for their workers' safety. As large corporate entities emerged, their managers had more impersonal attitudes toward employees. They were chiefly concerned with operations at minimum costs and maximum profits, even at the cost of workers' well-being. During the 1990s large corporations were downsizing, trimming operations costs further to maximize the bottom line. In these environments, it is easy for executives to rationalize their satisfaction with underfunded and inadequate worker health and safety programs.

Because of the costs of accident prevention measures, early large corporations and other employers pressured governments and legislators against passage of any safety laws to protect workers and the general public. Over the years, this attitude has moderated somewhat, especially where a corporation finds safety is economically beneficial for business. Managers, however, are protected against personal liability claims for accidents in two ways—(1) each corporation is an impersonal entity, and (2) the corporations and thus the managers are covered by insurance—thus reducing their incentive to provide safe workplaces. As a result, workers have suffered deaths and injuries that might have been prevented.

Ever since 1908, when U.S. Steel began its first formal corporate safety program, large companies (and many smaller ones) have found that safety programs are beneficial in reducing the costs of doing business. They have found that the prime concern of almost every worker is his or her health and safety. If these are safeguarded in everyday activities, the worker is better motivated and more productive. The Railway Safety Act of 1893 not only reduced the number of trainmen killed and injured in train coupling accidents (see Fig. 1-1) but generated savings. Because there were fewer accidents, operations were faster and more efficient. Although the number of persons employed and the number of trainmen involved in accidents both increased substantially with the growth in railroads, the number killed or injured in coupling accidents dropped. The installation of a mechanical safety device undoubtedly was the principal factor in the reduction of coupling accidents.

Unfortunately, many small companies still believe erroneously that safety programs are nonproductive and unprofitable. Their safety efforts are minimal and their accident and injury rates are higher than those of large companies. Their chief complaint when any new law or mandatory standard concerning safety is considered or enacted is that the increased cost will put them out of business. To such organizations, apparently, monetary considerations are more important than the moral one to lessen accidental deaths and injuries.

THE INDUSTRIAL REVOLUTION AND ACCIDENTS

Concern about accidents and their avoidance has grown vastly since the start of the Industrial Revolution—in the United States especially with the advent of railroads and steamboats. Because of the benefits those new transportation industries brought about, the accidental deaths and injuries they caused were accepted. For a long time, almost daily, newspapers reported deaths and injuries of passengers, yet little corrective action was taken, except where it improved operations so that immediate costs were reduced. The problem of train coupling accidents grew so bad that the Rev.

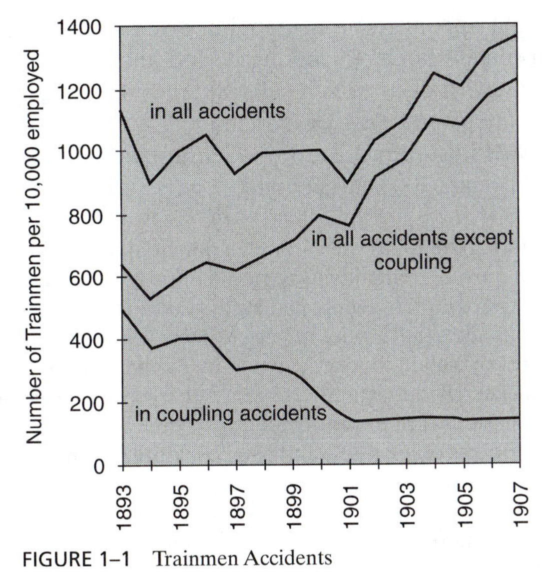

FIGURE 1–1 Trainmen Accidents

"Trainmen" includes enginemen, firemen, conductors, brakemen, and other trainmen.

Adapted from C. Eastman, *Work Accidents and the Law* (New York: New York Charities Publication Committee, The Russell Sage Foundation, 1910).

Lorenzo Coffin clamored for corrective laws. In 1893 the Railway Safety Act was finally passed at the behests of President Benjamin Harrison and his successor Grover Cleveland.

Over the years, because accidents were said to be inevitable, the public came to accept them, although many were easily preventable. However, industrial workers and the general public have become increasingly aware of hazards that could endanger not only the employees in industrial plants but outsiders as well—entire families and communities.

INCREASING HAZARDS

From the earliest times hazards have existed from which accidents have come about: falls from elevated places, impacts from falling trees and rocks, cuts from sharp objects, fires, drownings, and similar primordial events. Since the introduction of tools, people have been exposed more and more to increasingly complex hazards. For example, one of the leading categories of injury in the 1990s was cumulative trauma disorders (CTD) from repetitive motion in the workplace. A few decades ago these injuries were virtually unrecognized. Although back injuries were frequent, repetitive motion injury (RMI) was not an identified taxonomic rubric for categorizing some of them. Few people had even heard of carpal tunnel syndrome and other repetitive motion injuries. Changes in industrial technology brought this category of industrial injury to our notice. Another hazard arises from the fact that computers often control energy today.

Software might appear to be benign, but an error in software, virtually invisible to a worker, can remain latent for a long period and then suddenly unleash an unwanted flow of energy, with disastrous results. Sources of accidents are becoming increasingly complex and are often latent, or hidden.

In 1979, Hollister and Traut[2] made these comments, which might apply as much or more to the years 2000 and beyond:

> Safety during the 1980s will carry high potential cost. During the past century, we, as a society, have moved gradually away from concerns with hazards of the following types: hazards that were largely simple and easily recognized. Hazards that generally impacted only individuals and small groups of people in fairly immediate and recognizable ways. Hazards that were controlled to a large extent by the persons who might be harmed. Hazards that had associated benefits that were generally recognized and attained by individuals.
>
> We have in turn moved gradually toward a concern for hazards of this type: hazards that are often complex and hard to recognize or understand. Hazards that may impact thousands of people over long times. Hazards that are often controlled by persons other than those who might be harmed. Hazards that have benefits whose value to persons 'at risk' is only indirect and not easily measured.

Possibilities of accidents have been increasing not only in types but also in magnitudes, especially in industries like chemical plants, where they can affect entire areas and very large groups of people. In recent times accidents such as those in Bhopal, India, and the nuclear power plant at Chornobyl, Ukraine, have increased apprehension for the safety and well-being of workers and the public. These signaled a new era, in which the effects of industrial accidents can sometimes parallel and even exceed natural calamities.

Bhopal, India, December 23, 1984. At 11:30 p.m. workers detected a leak of water into a storage tank containing methyl isocyanate (MIC). About 40 tons of MIC poured from the tank for nearly two hours. The night winds carried the MIC into the city. Some estimate that 4,000 were killed, many in their sleep, and as many as 400,000 were injured or affected.

Chornobyl, Ukraine, April 26, 1986. At 1:23 a.m. an experiment in a nuclear power plant went wrong. This resulted in Reactor No. 4 exploding. The Russian reactor was built without containment shells. Because of the experiment, it was being operated at unstable levels, with its safety features disabled. The radioactive release covered hundreds of thousands of square miles. Ten years afterward, by some estimates, radioactivity from this event covered over 260,000 square miles. Ukraine officials estimate that 3 million people in that country suffered from the disaster. Thirty-two died in the accident or during efforts to put out the fire after the explosion. The overall death toll has been estimated by some to be over 10,000 persons.

PHYSICAL EFFECTS OF ACCIDENTS

The Division of Vital Statistics reports that accidents are the leading cause of death for persons from teenage up to age 45 (which are prime working years). The National Center for Health Statistics lists the three leading causes of death in 1995 to be heart disease, cancer, and stroke, followed by chronic obstructive pulmonary disease and accidents. In addition to the direct causes of deaths that Hollister pointed out as "simple and easy to recognize," accidents may result in long-term internal medical effects. Some heart and cardiovascular problems result from trauma and other stresses due to continued exposure to less easily recognizable unsafe conditions. Up to the age of 24, more deaths are due to accidents than to all other causes combined. In 1995 the death rate decreased substantially for those under 5 years of age, due primarily to decreases in fatal accidents for that age group.[3]

In the period since the Spanish-American War, battle deaths of military personnel have been far exceeded by fatalities from industrial accidents. Statistical data for accidental deaths are not available for 1898, when the Spanish-American War occurred. However, there were 385 deaths during that war compared to 520 fatalities for industry in Allegheny County, Pennsylvania, from July 1906 to June 1907. Evidently war was less dangerous than working in an industrial plant of the time.

War entails the intent of combatants to kill or injure; in industry there is no such intent, yet accidents kill people just as dead or wound them as badly. In a war the wounded, no matter how minor the injury, may be awarded Purple Heart medals. In industry many injuries, especially the lesser ones or those that are neither immediately apparent nor disabling, are never even reported. According to some estimates, the number of industrial accidents and injuries is actually ten times that cited by the National Safety Council. Even if they do report accidents, some industrial companies may downgrade their severity to lessen costs or indications of unsafe plants or operations. Nondisabling injuries are rarely reported; the injured person applies a Band-Aid or similar remedy and does little more. Other shortcomings in data on accident experience can include the following:

1. Unreported deaths and injuries may result from workers assimilating fine particles of toxic, debilitating, or radioactive substances after long-term exposure.
2. To maintain good safety records, some companies may transfer injured workers to less strenuous activities they are physically able to perform, reasoning that there have been no lost-time injuries to report.
3. Industries are not required to report injuries for certain employments—for example, those not covered by workers' compensation, such as some agricultural work or domestic service.
4. Certain losses are rarely or never reported. For example, railroads report damage that occurs to their equipment, track, and roadbed, and little more. No costs are included for injuries or damage occurring off the track or roadbed, freight losses, or evacuations because of ruptures to tank cars of toxic fluids.
5. Legal costs resulting from accidents are not reported.

In 1997 the National Safety Council (NSC) reported that during the previous year, 93,400 persons had been killed in accidents and 20,700,000 persons had suffered disabling injuries. Of these, 4,800 deaths and 3,900,000 disabling injuries resulted from accidents at work. Ten years earlier, the number of persons killed was just slightly higher, but the number of deaths from accidents at work was more than double the 1997 figure, and injuries were about one-half the 1997 figure. However, according to the NSC, the rate for worker unintentional-injury deaths has remained about the same from 1992 to 1996 at 4 per 100,000 workers.

NUMBERS OF ACCIDENTS VS. COSTS

Accidents and rates are sometimes measured and indicated by the numbers of occurrences, fatalities, or injuries. The noted jurist Learned Hand said the best measure of accidents is monetary losses. Over the years dollar values have become established as the principal means by which costs of accidents are estimated. Figure 1-2 indicates many of the economic factors involved because of accidents and attendant safety costs.

All of the items can be expressed as parts of an equation:

total safety costs = immediate losses due to accidents + rehabilitation and restoration
+ accident prevention costs + legal costs + insurance + welfare
+ other safety costs + immeasurables

The immediate costs are those due to work losses, damages to equipment and facilities that must later be repaired or replaced, medical, fire, and police personnel, and similar items. Welfare applies only to those who are put in need because of accidents. "Other safety costs" include the indirect effects beyond the immediate direct losses that have already taken place. These indirect effects may be increased due to apprehension regarding accident potential and added insurance premiums, added safeguards, or the cost of observance of new laws or the fines or penalties imposed because of nonobservance. In April 1986 the Occupational Safety and Health Administration ordered the Union Carbide Corporation to pay nearly $1.4 million for "conscious, overt and willful" safety violations at the Industry, West Virginia, pesticide plant where a huge toxic-gas leak had injured 135 people in August. In 1987 the assessed penalty against a group of construction companies reached $5 million because of one accident.

In 1991 a chicken plant in Hamlet, North Carolina, caught fire. Twenty-five employees were killed and 56 injured. Many of the injuries and fatalities allegedly occurred because some fire exits were blocked. North Carolina OSHA levied penalties of $808,000, and the company is defendant in at least ten civil suits filed by former employees and the families of those who were fatalities. The reaction of state OSHA officials was to initiate a more aggressive safety and health program. In the 1994–1995 fiscal year the North Carolina OSHA issued a total of more than $5.5 million in fines, compared to $3.1 million by the average state program, and $2.9 million by federal OSHA programs.

OSHA levied a $10 million fine after an explosion killed eight workers and injured 42 other workers and 70 members of a Louisiana community in 1991. The penalties were brought against a fertilizer manufacturer and natural gas supplier.

A corporation produced fertilizer at a plant in Lake Charles, Louisiana. July 28, 1992, the plant urea reactor exploded, destroying the facility and injuring three employees and four other persons. OSHA proposed a $50,000 penalty for each of the 87 employees exposed to the conditions that led to the single catastrophe. The total proposed penalty was $4.35 million.

In 1997 contractors performing demolition and salvage operations at the former Greater Pittsburgh International Airport received citations with proposed penalties of $1.65 million for allegedly violating asbestos and lead standards. Fines of over $800,000 were also proposed by OSHA in 1997 against the largest U. S. copper smelter for numerous health and safety violations, including life-threatening levels of airborne

A. Losses to or because of personnel:

1. Labor time for persons killed or injured.
2. Degradation of capabilities in activities such as employment, sports, or recreation.
3. Medical costs for immediate care of victims in an accident or for their extended care or rehabilitation.
4. Consortium of a spouse who has been killed or severely injured; awards to surviving children.
5. Labor time of fellow workers distracted by nearly being injured in a fire, collision, or other "near miss."
6. Time of managers and other supervisory personnel, clerical, public relations, other staff members whose activities were distracted or who were required to participate in corrective action.

B. Losses or damage can include accidents to:

7. Equipment, product, system, facility, plant or material damaged or destroyed.
8. Degraded operations.
9. Overhead costs while damaged or destroyed plants are not operating or productive.
10. Actions to correct deficiencies (redesign or retooling).
11. Disruption of operating schedules and lessened outputs because of accidents or unsafe conditions.
12. Recalls, including investigations, redesigns, corrections, public notifications, legal actions, and attendant problems.
13. Notifications, investigations, and liaisons with internal organizations, insurers, or municipal, state, or federal agencies.
14. Accident evaluation and reporting.
15. Obsolescence of a product and accompanying materials.
16. Loss of public confidence.
17. Loss of prestige.
18. Deterioration of morale of company personnel.

C. Hazards, accidents, or their possibilities create losses and costs for:

19. Insurance premiums for workers' compensation.
20. Automobile liability and physical damage liability and coverage against fires, marine, boiler, and machinery, and glass damage.
21. Amounts not covered by deductibles or because limits were exceeded.
22. Legal costs such as review of insurance, litigation initiated because of accidents and their defense.
23. Fines, penalties, or other punitive actions because of violation of public laws.

D. Costs to the government include:

24. Preparation and issuance of laws, standards, and regulations, plus administrative actions for their observance.
25. Payment for jurors, judges, and other court personnel.
26. Workers' compensation for governmental self-insurers, including costs for administering programs, paying for injuries, medical fees, rehabilitation, and bookkeeping costs.
27. Unemployment compensation for workers laid off because of accidents at plants where they were employed.
28. Agencies specifically designated for accident prevention, control, and minimization, such as road signs and their maintenance.
29. Those portions of police, fire, and other public departments concerned or involved with hazardous activities and emergencies.
30. Social security payments to persons injured in accidents but not adequately covered by other insurance.
31. Aid to dependent families of persons killed or disabled because of accidents, and payment of welfare (including need for food stamps) because injuries result in persons or families being below a minimum income level.

FIGURE 1–2 Losses Due to Accident and Other Safety Costs

lead and cadmium. It was alleged that, although the plant had spent $3 million to enclose its operations and treat air before it left the plant, provisions for ventilation inside the enclosed area were inadequate. This OSHA inspection resulted from an EPA referral. In 1997 OSHA cited a plastics company for willful violations of regulations, including 63 instances of record-keeping violations and one instance of failure to provide a machine guard. The fine assessed was $720, 700.

On November 29, 1995, the Assistant Secretary of Labor for Occupational Safety and Health stated during congressional testimony that in fiscal year 1995 "OSHA issued initial penalties of $100,000 or more against 122 employers compared with 68 such penalties in fiscal year 1994." Nevertheless, he pointed out, "Fatalities still occur at a rate of 17 workers a day, and another 16,000 workers are injured on the job every day." However, OSHA is now testing a new enforcement strategy, designed to reduce inspections and penalties for companies who have effective safety and health programs.

In 1997, a metal-forging plant agreed to pay $1.8 million in penalties for an accident at its Houston facility that killed eight workers and injured two others in December 1996. It was alleged that workers were replacing seals on a pressure vessel without full implementation of a lockout/tagout program and without specific procedures in place for the work they were doing.

If a state chooses, it may operate its own OSHA-approved health and safety program; however, federal regulations still apply when they are more stringent than state standards. Therefore, when an accident occurs, it is possible that there will be a joint state and federal investigation. This happened to an Oakland, California, plating company after a September 1993 fatality. Cal-OSHA inspected not only the plant where the accident occurred but also two other facilities operated by the company. Cal-OSHA issued seven willful/serious, 17 serious, one general/regulatory, four repeat general citations, and a failure-to-abate citation. These penalties totaled $552,000. Federal OSHA cited the company for two willful violations and seven serious violations, and a violation for failure to implement a program—all related to the confined-space standard. These federal penalties totaled $189,000. The combined total was $741,000.

An accident does not have to occur before an industry is assessed for nonobservance. In 1994 the leading OSHA General Industry Citations were $2.43 million for 15,586 violations of hazard communications regulations, $2.09 million for 5,776 violations of lockout/tagout regulations, and $1.6 million for 3,575 violations of machine-guarding regulations. Violations in these three areas were the most heavily fined that year.

Fines can be backbreakers for small companies. Relatively small companies accounted for 70 percent of the $132 million in penalties proposed by OSHA in 1995. They received 84 percent of the inspections and 81 percent of the 248,267 OSHA violations that year. Being small does not provide an excuse from safety responsibilities, nor an exemption from penalties for safety violations.

The National Safety Council (NSC) reported total safety losses of $121.0 billion in 1996 in the workplace. Their figure includes wage and productivity losses of $60.2 billion, medical costs of $19.0 billion, and administrative expenses of $25.6 billion. It includes employer costs of $11.3 billion for things such as the money value of time lost by workers other than those with disabling injuries, who are directly or indirectly involved in injuries, and the time required to investigate injuries, write up injury reports. It also includes damage to motor vehicles, work injuries of $1.6 billion, and fire losses of $3.3 billion. But total safety costs are indeterminable; for example, premium costs are not included, lawyers may not indicate amounts of settlements, and welfare costs because of accidents are not included. NSC estimated that the value of goods or services each worker must produce to offset the cost of work injuries in 1996 was $960. Year-to-year costs cannot be compared because of progressive changes, refinements, and advances in data gathering and reporting; however, it is conservative to say that costs are increasing annually.

At one time, the Bureau of Labor Standards considered that total accidents for a year were roughly four times the combined total of the workers' compensation payments plus medical costs. Note that this is predicated only on injury and involves no equipment, materials, or facilities loss. Ratios developed by other researchers indicate that monetary losses employers can suffer may vary from 1:1 (only workers' compensation costs) to 20:1 (twenty times the workers' compensation costs). The uncertainty is evident. Companies sometimes believe that insurance will protect them against monetary loss because of accidents. However, in almost every instance investigators have found that total losses far exceed the amounts reimbursable by insurance companies.

LESSENING SAFETY COSTS

To reduce the cost of insurance, some companies self-insure themselves or use deductibles. Those deductibles are amounts not covered by insurance. Many other safety costs are not covered by insurance, such as penalties. For example, in July 1971 the State of Ohio sued the Penn Central Railroad for more than $14 million in damages for "unsafe conditions" on its property in Ohio. A fine of $110,000 would also be levied for each day of alleged continued violation of state laws.

Companies and other organizations have resorted to the use of self-coverage for accident costs that managers believe will not impose disastrous financial loads. Costs of probable future losses are generally estimated at less than the cost of insurance premiums.

Coverage is obtained from an insurer for an amount greater than the deductible level. The insurer will pay any loss between the deductible and a maximum amount, if one has been stipulated. For example, in 1973, a munitions train blew up in Roseville, California, creating damages estimated at $10 million. The railroad was self-insured to $2.5 million; a commercial insurer carrier had to absorb the remaining loss of $7.5 million. In this case, the railroad had to pay 25 percent of the loss. If the total loss had been $5 million, the railroad would still have had to pay $2.5 million or 50 percent. Insurers would have had to pay nothing if this loss were less than $2.5 million. (Coverages of all different types and limitations exist and are available.) Twenty-four years later, unexploded munitions from that train were discovered and exhumed at the original accident site. Evidently, the liability for accidents such as this may persist long after the initial event, and an insurance company's settlement offer must be treated carefully.

ACCIDENT LOSSES FOR PERSONNEL VS. EQUIPMENT AND FACILITIES

In February 1973 an explosion took place in an empty liquefied natural-gas storage tank on Staten Island, New York. The collapse of the tank killed 37 workers and three safety inspectors. The total cost to insurers was only $2.8 million for the 40 persons killed and $40 for the tank. The Labor Department cited the company that owned the tank for: failing to develop and maintain an effective fire prevention program, not instructing employees regarding unsafe conditions, failing to use equipment by a nationally recognized testing laboratory, not promptly restoring fire-alarm signaling equipment to use after tests, and failure to provide a place of employment free of recognized hazards. Civil penalties totaling $250,000 were proposed against the company. In addition, a penalty of $1,600 was assessed against the company that provided the workers for the repair and cleaning. This case is not unique; a comparable accident is noted in Fig. 1-3.

Shortly after midnight on June 24, 1971, an explosion tore through a portion of a tunnel being constructed for the Los Angeles Metropolitan Water District (MWD) in Sylmar, California. Seventeen men, 16 employees of the Lockheed Shipbuilding and Construction Company (the contractor for the tunnel) and one employee of the MWD, were killed.

The explosion was evidently due to the ignition of flammable gas released as the tunnel advanced through an oil-bearing region. On the previous day, an explosion and flash fire in the tunnel had injured four men. Twelve hours later work resumed, after the company agreed to improve the ventilation system and to provide continuous testing for flammable gases. According to testimony before a committee of the State Assembly, digging was stopped as many as 30 to 40 times that night to allow the concentration of gas to drop to a safe level. During this time many electrical devices known to be sources of ignition were operated and the miners were permitted to smoke freely. The situation was highly conducive to an accident. Nine miners were killed outright, the others were suffocated in the smoke and dust.

The company and three of its employees were charged with 60 counts of violation of state safety codes, resulting in the deaths of the 16 employees. The resultant criminal trial lasted six weeks. The Lockheed Shipbuilding and Construction Company was found guilty of 16 counts of gross negligence and of 9 safety code violations and faced a maximum fine of $106,250. It was acquitted on 6 counts of safety code violation. The tunnel project manager was first found guilty of 16 counts of gross negligence and of 9 safety code violations. He was initially sentenced to 20 years and 6 months in jail, but this was then reduced to 5 years in jail and 10 years of probation. The verdict was overturned and a new trial was to be conducted, but the project manager pleaded no contest to one count each of gross negligence and safety code violation. He was fined $6,875 and put on probation for 3 years.

The work crew supervisor was found not guilty on the 4 counts of safety code violation with which he had been charged. The safety engineer was initially found guilty of 3 safety code violations, for which he was sentenced to 18 months in jail (modified to 6 months in jail and 10 years of probation). He was acquitted on 8 counts of gross negligence. At a retrial, pleaded guilty to one safety code violation and was fined $625 and placed on one year's probation.

Some of the features of this trial were unique. The case was the first in which a corporation has been held liable for gross negligence. It was the first time anyone was prosecuted on criminal charges for violation of safety provisions of the state's labor code. This provision of the code had been written in 1913. It was the first time in California that anyone was convicted of criminal liability where there was no showing of guilty intent.

On July 13, 1973, the Governor of California signed into law legislation aimed at making possible more severe punishment for employers whose failure to provide safe working conditions caused the death of an employee. The act replaced the former labor code provision, which specified misdemeanor penalties for such offenses, with one that permitted prosecution under manslaughter provisions of the penal code.

The trial involved determination of guilt under criminal charges. In addition, the dependents of the Lockheed men killed instituted a civil suit against the MWD, State of California, and the Mine Safety Appliance Company (which made the gas testing equipment) which was settled out of court for $9 million. Because of workers' compensation laws, they could not sue Lockheed. The family of the MWD man sued Lockheed but could not sue the MWD for the same reason.

FIGURE 1–3 Sylmar Tunnel Disaster

In one of the few published comparisons of costs, Ludwig[4] pointed out: "One plant's accident/injury performance was so good that insurance-covered costs were only $435 rather than the carrier's estimated cost of $30,000. With new projects, such potential savings may be made evident by having the carrier quote what costs would be under other circumstances." Any comment that the cost of accident prevention is economically unjustifiable is belied by Ludwig's comparison.

INCREASING MAGNITUDE OF ACCIDENT LOSSES

Not only are hazards becoming more complex and the possibilities and extent of fatalities and injuries increasing, but so are potential monetary losses. Figure 1-4 shows the growth in accident losses that took place in the chemical and allied industries since the first listing was made by the American Insurance Association (A.I.A.) in 1960 and was made every five years, until discontinued.[5] (*Note:* 1979 is the date of the last Technical Survey No. 3, published by the A.I.A. and from which the authors prepared the curves shown.)

The annual loss in the United States and western European countries increased from $11.9 million in 1960 to $280 million in 1978. The total loss due to the accident at the nuclear power plant at Three Mile Island in Pennsylvania, which brought about revisions to provide added safeguards in all similar plants, has been estimated at over $4 billion. This accident occurred because a cooling pump failed, and a pressure relief valve opened. When the pressure returned to normal, the valve remained open. About a million gallons of contaminated water escaped through the valve onto the basement floor and into tanks in auxiliary buildings. The affected unit, TMI2, is in a permanently shut down and defueled state. In 1990 the licensee, General Public Utilities Nuclear

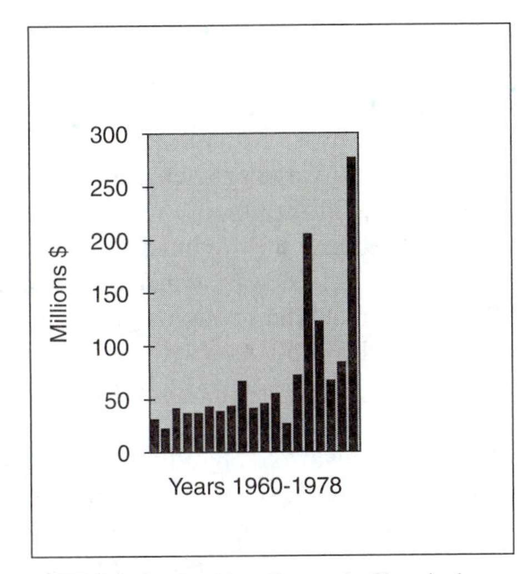

FIGURE 1–4 Accident Losses in Chemical Plants and Related Industries

Corporation, submitted a plan stating that it would deposit $249.9 million (1994 dollars) in an escrow account for decontamination of the unit.

In 1960 only two accidents in the chemical or allied industries involved a loss of $1 million each; by 1977 there were 15. Until 1968 there were no accident losses of $10 million or more; two of that magnitude took place in 1968, and in 1976 and 1977 there were three each.

AWARDS FOR INJURIES

The increase in monetary losses has been indicated by the sizes of awards made as the result of litigation because of accidents. The vulnerability of industry to high costs from accident-related court awards was demonstrated in 1997 in a product-design liability case, when a plaintiff family was awarded $262 million from a major automobile manufacturer. A 6-year-old was thrown from a car involved in a traffic accident. The claim alleged the manufacturer of the car was negligent in providing a latch design on a van's rear door that flew open during the accident. We see that the legal award from one accident today can approach the annual accident loss reported in 1978 for the United States and western Europe.

Multimillion-dollar awards are not unusual. To the cost of the loss awarded by a court must be added that of any defense. Not every plaintiff in an accident suit has a successful case, but the percentage is substantial. The United States has become the most litigious country in the world, and a great part of this litigation is due to suits generated by accidents.

In 1970 a jury made the highest award in legal history to that date ($3,650,000) for an injury. A young diesel mechanic was injured when a bolt snapped so that a large piece of electrical equipment fell and crushed part of his skull. He survived, but in a completely helpless state. He was to receive $3,000,000 for himself (half for the care he would need over the 40 years he was expected to live), $500,000 for his wife, and $50,000 for each of his three children.

In 1973 even that amount was exceeded when a 27-year-old man and his wife were awarded $4,735,996 in Los Angeles to compensate for a severe injury suffered by the man when he was struck by a 630-pound pipe.

Lesser but still massive awards are being made. In 1973 a farmer on top of a grain bin was pushing grain with a 20-foot metal pole. The pole touched an overhead 12,000-volt line. The shock knocked him off the bin and he fractured his back so badly that he became a quadriplegic. The farmer was awarded $1.8 million.

Also in 1973, a 69-year-old man who lost his right arm when removing scrap from a punch press was awarded $1,750,000 (with interest the total would be over $2 million). The claim alleged the manufacturer had failed to equip the press with a proper safety device.

In 1993 a Virginia jury awarded $8 million to a worker who received back injuries when a metal door on an electrical cabinet containing railroad switch gear fell onto him. The cabinet had been washed out by a storm and had fallen over beside railroad tracks. The worker was trying to remove and replace components from the cabinet, while it was tipped over. The injuries did not involve paralysis or amputation. The claim alleged the railroad company was negligent because they sent a worker into an emergency situation who was not trained for emergency response. While this amount was reduced as part of the appeal process, it demonstrates the dollar exposure that accidents can bring when the legal system is involved.

Three workers sued a computer equipment manufacturer for musculoskeletal disorders (MSDs) they alleged were caused by keyboard entry activities, using keyboards manufactured by this company. The workers were awarded almost $5.8 million in 1997. Repetitive stress injuries (RSIs), resulting in MSDs, have increased greatly since 1987, adding another third-party-liability dimension to work-injury costs.

In 1997 it became evident that legal costs of accidents and injuries can also be incurred because of a company's treatment of injured workers long after the accident is over. A major clothing manufacturer was found guilty of discriminating against five employees who had job-related injuries.[6] The company was ordered to pay $10.6 million in compensatory and punitive damages for allegedly requiring the workers to participate in a job reentry program, which forced them to work at lower pay at jobs found on a supplied list. More than 100 additional employees have filed suit against the manufacturer, and the company is appealing the judgment. This court action underscores the importance of a company's program for the rehabilitation of an injured employee, including job reentry policies and actions.

Effective accident prevention programs in the United States can reduce inflation and drastically reduce suffering from injuries and accidental death. Since most companies that undertake accident prevention programs are forced to do so by legal and economic requirements, it is advisable to review in the next few chapters some of the legal aspects of liabilities for accidents.

BIBLIOGRAPHY

[1] University of Michigan, Survey Research Center, *Survey of Working Conditions,* prepared for the U.S. Dept. of Labor, Employment Standards Administration, Washington, DC, Aug. 1971.

[2] H. Hollister and C. A. Traut, Jr., *On the Role of System Safety in Maintaining "Affordable" Safety in the 1980s,* SAND 79-1671 C (at the Fourth International System Safety Conference, San Francisco, CA, 1979).

[3] R. N. Anderson, K. Kochanek, and S. L. Murphy, "Report of Final Mortality Statistics, 1995," *Monthly Vital Statistics Report,* Vol. 45, No. 11, Supplement 2, (PHS) 97-1120, 1997.

[4] E. E. Ludwig, "Designing Process Plants to Meet OSHA Standards," *Chemical Engineering,* Sept. 3, 1973.

[5] *Hazards Survey of the Chemical and Allied Industries* (Technical Survey No. 3), American Insurance Association, New York, 1979.

[6] CTDNews, "News Briefs: Workers Win $10.6 Million in Lawsuit," *LRP Publications,* Vol. 6, No 10, Oct. 1997.

EXERCISES

1.1. Is the most important reason for accident prevention economic, legal, or moral?

1.2. List ten types of economic losses that can result from accidents or their possibilities.

1.3. What is the primary concern of each worker?

1.4. Are accidents a major or a minor cause of all deaths at various ages?

1.5. Have there always been similar types of hazards and accidents? Did the number of accidents increase with the Industrial Revolution? Why? Are the types of accidents increasing or decreasing? How have they changed?

1.6. Do data of the Department of Labor Statistics and the National Safety Council indicate all types of accidents? Why are some accidental injuries never reported?

1.7. Are safety programs nonproductive or a waste of funds, as some small managers claim? Do such programs save money for employers?

1.8. Could an accident such as that which took place in Bhopal, India, occur in your vicinity? Why?

1.9. Are the cost magnitudes of accidents increasing or decreasing? How about awards for lawsuits for seriously injured persons? Are juries becoming more or less sympathetic toward plaintiffs who may have been seriously injured in accidents?

1.10. Is litigation for accident injuries increasing or decreasing?

CHAPTER 2

Liabilities and Safety Legislation

In the United States, as in Britain where they originated, there are three types of laws: statute, common, and administrative. Statute laws, which probably have existed since before the Code of Hammurabi (1750 B.C.), are promulgated by royal edict or a ruling body, the highest governing power in any jurisdiction. Common law was used first in England and later in all English-speaking countries, with regional variations. Common law is not established by statutes but follows precedents set through judicial decisions. Administrative law is established by an executive in prescribing the criteria under which any statute or desired control will be carried out.

The operation of economic laws often leads to the creation of statues, common law, or administrative regulations. A few important aspects of these laws, both economic and legal, regarding accidents are presented in this book. Because so many laws oriented toward accidents and safety now exist, we can only touch on a few of the most noteworthy milestones.

All engineers should know something of the legal ramifications of their work: their responsibilities and liabilities under the laws; how the laws came about; and how they affect employers, engineers, fellow employees, and all other persons.

STATUTE LAW

Early law treated as a criminal any person who for any reason caused injury to another, as though the act had been done intentionally. (The "eye for an eye" dictum of Hammurabi was part of his Code—a statute.) The person who injured another was considered to have committed a crime against the peace of the community. Action by a community or its ruler was intended to lessen the tendency of the family of the person injured or killed to revenge the hurt and thereby to start a blood feud. Action for revenge could constitute a threat to peace that the local community or rulers could not permit. As late as the second century in certain areas of Greece, judgments were even made on inanimate objects that caused fatalities; found guilty, they were to be cast beyond the borders of the realm.

Offenses against the ruler, state, official religion, or community in general were crimes for which penalties were to be levied. Much later, misdeeds against individuals that were not considered criminal acts became known as civil wrongs, to be adjudicated between the parties concerned, with the ruler acting only as referee. Gradually a body of law came into being which indicated the rights and duties of individuals and the care they must exercise toward each other in their relationships (especially in the fulfillment of contracts).

Initially, customs and usage, tribal laws, and edicts of rulers or governing bodies were accepted only within local areas. The growth of trade and contact between different groups led to confusion, owing to new conditions and differing interpretations of the same laws. To clarify matters these laws, written and unwritten, were assembled and reviewed by contemporary authorities. Those that appeared to be most suitable were compiled into documents that were then approved by a ruling body or person.

Provisions of such a document, generally called a "code," became the current law and superseded all others within its jurisdiction. In addition to that of Hammurabi there were the Codes of the Twelve Tables (Roman), of Theodosius and Justinian (both Eastern Roman), and, much later, of Napoleon (French). Such codes constituted the "statute laws" of their respective times and jurisdictions. A royal decree or action by a legislative body (of which there were few in those days) was required to change or eliminate a code. This system, originated and still used in continental Europe, is generally not as flexible as the common law system that developed in England and is used there and in most other countries with an English heritage. Much British (and American) law is in the form of statutes, but an even greater mass exists in the common law.

COMMON LAW

Development of English common law began after the Norman Conquest when limited, local feudal laws were gradually replaced by the more widely applicable decisions of courts appointed to be the king's representatives. The royal courts' derivation of the common law was based on judges' decisions in disputes brought to them for resolution. It became a practice for judges to record judgments in cases as they were decided. These precedents provided a measure of guidance for future judges on how they might rule in cases involving similar circumstances.

Over the years, there have come to be courts for initial cases, courts for appeals, and supreme courts where final decisions are rendered. In many instances, a common law judge may not be the final arbiter of a dispute. A common law judge can reverse another's decision if he or she believes an error was made initially or if the circumstances on which the original decision was based have changed. That reversal in turn can be set aside by another judge with higher authority. Legislative action creates statutes which are superior to the common law. A statute can be used to overcome any common law precedent that might be unpopular or undesirable. Or, by a change in membership of a court, the minority may become the majority and the original decision negated.

MISADVENTURE (ACCIDENT)

As late as the middle of the thirteenth century, even though fatal injury had been an unavoidable accident (misadventure) with no intent to injure another, that fact could be pointed out only during a plea for the king's mercy to pardon the defendant. Pardons granted were often predicated on the defendant's making peace with the kin and friends of the victim. In many such cases, the victims' families found it advantageous to accept pecuniary settlements as restitution. Under the blossoming laws in England of the fourteenth through sixteenth centuries, more and more it became the custom that some persons causing unintentional harm to another could be forgiven by suitable monetary payments. Generally, in very early English law, forgiveness was extended only to the wealthy and landed.

STRICT LIABILITY

Some occurrences involving misadventures or accidents were regarded as so adverse and contrary to the public well-being that any perpetrator was considered "strictly liable" for any loss. The concept of "strict liability" at the time included the expectation that all persons had obligations to prevent certain damages, such as by controlling any fire, even in a hearth, where the risk existed that the fire might spread to burn down a house or entire neighborhood. Strict liability also extended to the duty of common carriers, innkeepers, and other persons to safeguard passengers and travelers and their belongings.

Not until the seventeenth century did a common person charged with causing an injury have a chance of being exonerated from blame for an accident. It had to be shown the malefactor had been completely without fault and the misadventure had been entirely unintentional. Some of the first claims citing absence of fault were due to injuries caused by the unintended discharge of firearms. Problems concerning misadventures, and decisions and discussions regarding them, continued throughout the seventeenth and eighteenth centuries.

In the eighteenth century legal decisions were accepted distinguishing between acts that were willfully inflicted and those that were unintentional. The willfully inflicted act against another person in violation of a legislated law was (and still remains) a crime. A violation of a duty not covered by statute became a tort, for which damage could be sought in a civil, as opposed to a criminal, court. (Figure 2-1 presents a few of the legal terms often used in the common law.) In 1682, Sir Thomas Raymond (in *Bessey v Olliott*) had said: "In all civil acts the law doth not so much regard the intent of the action, as the loss and damage of the party suffering." However, this concept changed later. The question of whether an accident had been inadvertent gained great importance with the Industrial Revolution, an era of individualism and intensive growth of commerce and industry, when common and statute laws and their judicial interpretations were oriented toward the wealthy.

Master, servant , and stranger : in law, usage for common persons now referred to as employer, employee or worker, and visitor (any third party such as a guest, customer, visitor or member of the general public).
Care and great care: the high degree of care that a very prudent and cautious person would exercise for the safety of others. Common carriers, such as airlines, bus companies, and railroads, must exercise a high degree of care.
Reasonable care: that degree of care exercised by a prudent man in observance of legal duties toward others.
Slight care: that degree of care less than that which a prudent man would exercise.
Exercise of due care: Every person has a legal duty to exercise due care for the safety of others and avoid injury to others if possible.
Standard of reasonable prudence: a person who owes a legal duty must exercise the same care that a reasonably prudent man would observe under similar circumstances.
Proximate cause: relationship between the plaintiff's injuries and the defendant's failure to exercise a legal duty, such as reasonable care. Example: *A* playfully pushes *B* in a crowded space, so *C* is hit by *B*, loses his balance, falls and is injured. *A*'s push is the proximate cause of *C*'s accident.
Liability: an obligation to rectify or recompense any injury or damage for which the liable person has been held responsible.
Strict liability: the growing concept that a manufacturer of a product is liable for injuries due to defects without a necessity for plaintiff to show negligence or fault.
Negligence: failure to exercise a reasonable amount of care or to carry out a legal duty with the consequence that injury or property damage occurs to another.
Negligence per se: violation of a law the courts have adopted as an absolute standard of conduct. Violation results in liability unless the violator has a legal excuse.
Master-servant relationship applied to a guest: a master is liable for any negligence of his servant acting within the scope of his employment.
Gross negligence: conduct involving failure to use even slight care, a complete lack of regard for the safety of others, or intentional failure to perform a required and apparent duty regardless of the severity of the consequences of his act.
Willful or reckless conduct: outrageous and reckless disregard for other's rights or well being and for possible consequences. Indicates not only a complete lack of care, such as could be considered gross negligence, but an intention to exercise no care at all.
Contributory negligence: an injured person's care for his own safety was less than that reasonable for a prudent man under existing conditions, he is considered negligent and the defendant will not be held liable.
Assumption of risk: A person who is aware of a danger and its extent, and knowingly exposes himself, assumes all risks and cannot recover damages even though he is injured through no fault of his own.
Tort: a wrongful act or failure to exercise due care for which civil legal action may be taken. Commission of a tort carries no right to recovery unless the tort is shown to be the proximate cause.
Joint tort liabilities: when an injury is caused by two or more persons, each of whom failed in a legal duty, they are in joint tort. The degree of suffering from each injury may then be assessed against the plaintiffs individually in case of successive injuries, or jointly in concurrent injury.
Damages, compensatory: compensation to an injured person for the loss he has suffered and may continue to suffer.
Damages, punitive: awards made in some states to compensate an injured party for intentional, malicious, or outrageous misconduct by the defendant, usually made so that the latter will not repeat the offense.
Safe premises: the duty of an owner to user of land to keep his premises in a condition that is reasonably safe for those lawfully there.
Dangerous instrumentality: a person who keeps, maintains, transports or stores a dangerous creature, device or substance is liable for injury or damage, regardless of fault, even when he exercises due care.
Defective design: more dangerous than an ordinary consumer would expect when used in an intended or reasonably foreseeable manner, or if the benefits fail to outweigh the risk inherent in the design.

FIGURE 2–1 Some Fundamental Legal Terms and Principles

Ultrahazardous operations: any person who conducts an operation that involves abnormal risk to others is strictly liable for injury or damage that results, even if these operations were conducted with care.
Foreseeability: a man may be held liable for actions that result in injury or damage only when he was able to foresee dangers and risks that could be reasonably anticipated.
Foreseeability applied to premises: an owner, lessor, or other party responsible for premises must anticipate to foresee all possible risks to which the public might be subjected during their use, to maintain them in a safe condition, and to provide warnings of any unsafe or unusual condition.
Foreseeability applied to rescue: any foreseeable act that places a rescuer in the same danger as in an injured person the rescuer is attempting to aid is considered negligence by the person who committed the initial act.
Fright without physical contact: at one time it was considered that a plaintiff could not recover damages unless injury was due to physical contact. This principle has been modified or repudiated over the years by many states to permit a plaintiff to collect damages for neurological or emotional disturbances that occurred without physical injury.

FIGURE 2–1 Continued

INDUSTRIAL REVOLUTION

Over the years, judges in England shifted their concern more and more from the very wealthy landed owners to the increasingly powerful industrialist classes. The leaders of the new enterprises could obtain great economic returns only with low-paid workers and the hazardous conditions of the developing factories, mines, and overcrowded urban areas. The accepted philosophy of the industrialists of the time was laissez-faire—to let natural laws operate by imposing no restrictions.

Just to exist and eat, workers had to accept work under extremely hazardous conditions. Deaths and injuries increased tremendously because of unsafe and unhealthful working and living conditions. Accidental tragedies became normal occurrences. Worker protection was woefully inadequate against the risks that arose with new, powerful, and increasingly complicated machinery that had few safeguards.

In Britain, judges made decisions in which they tended to withhold liability for accidents of dangerous enterprises even where a fault was obvious (Fig. 2-2), imposing new rulings that employers could use as legal defenses. Instead of concern for the injured person in the concept of strict liability, it was contended that imposition of liability would be harmful to the growing enterprises.

NEGLIGENCE

Common law interpretations and judicial decisions in the United States followed many of those originated in Britain. The decisions reflected the times, favoring industrialists and other entrepreneurs, and only after many years were they changed in favor of the workers and general public.

The principle of strict liability, which had been intended for the compensation of victims, went into temporary decline in both Britain and America. It gave way to the

New precedents regarding liability are being set almost daily by courts throughout the United States. A few of these judicial concepts have been or are especially significant to workers who may have been injured.

Assumption of Risk. This idea was based on an ancient law of contracts so old that the contention may have been originated by Homer, proceeded through Aristotle, the Romans, and later into the common law. Therefore no first decision can be cited. Under this concept, any worker could decline to enter into an employment in which he might reasonably consider there could be an injury to himself. A similar common law principle was established in the United States in Massachusetts in 1842.

The Fellow-Servant Rule. A decision regarding the nonliability of employers was enunciated in 1837 (*Priestly v Fowler*) in England. A butcher's boy was directed to take some goods to a certain place. Another servant was the driver of the van. As a result of being overloaded, the van broke down, and overturned, the boy was injured, and he sued the butcher. The court ruled the employer, the butcher, was not liable because the boy and driver were fellow servants. In the United States, its use as a defense for railroads was first abolished in Georgia in 1856, followed by Iowa in 1862.

Rules of Privity. The concept of privity was also based on contract law. Privity exists between a seller and buyer who enter into contract. If *A* contracts to sell an item to *B*, privity exists between them. In 1842, in the case of *Winterbottom v Wright*, a Lord Abinger ruled that unless privity exists between two parties, one cannot sue the other. Winterbottom drove a mail coach for the Postmaster General of England, to which it had been sold by Wright, who had built it. The coach overturned because of a defective wheel and Winterbottom was injured. He sued Wright. Lord Abinger ruled that because privity had not existed between the two contending men, Winterbottom could not sue Wright. The Postmaster General, who had constituted one side of the contract, had neither been injured nor filed suit. Wright had sold nothing to Winterbottom and was therefore not liable for damages.

The decision affected not only Winterbottom, but practically all workers in industrial plants who were injured by unsafe machinery. The machinery in all the plants was bought by employers, owners, or corporations, who were rarely injured. The injured workers had no contracts with the machinery manufacturers. The employers might be sued, but for a while even that possibility was ruled out. (Later, except under conditions, under workers' compensation employees usually gave up the right to sue employers for injuries).

Minor and limited exceptions were made to the privity rule after a few years, but the first major renunciation of that doctrine in the United States (it still holds in Britain) came in 1916 in an opinion by Justice Cardozo of New York in *MacPherson v Buick Motor Company.* Buick sold a car to a dealer, who resold it to MacPherson. While MacPherson was driving the car, a wheel collapsed and he was injured. He sued Buick for negligence in failing to inspect the wheel and detect the defect before the car was released for sale. Buick contended that because MacPherson had bought the car from a dealer, privity did not exist and MacPherson could not sue Buick. Justice Cardozo rejected that argument and began the general abandonment of the privity concept in matters of product liability.

That is also of interest in the case of *Elmore v American Motors. Mrs.* Elmore was driving an American Motors car when something on its underside failed, causing it to swerve and hit the car driven by Mrs. Waters. Mrs. Waters was killed. Her family and Mrs. Elmore sued American Motors. The company claimed that because Mrs. Waters was not a user or operator of the car, the company was not liable for damages. In 1969, a court not only ruled against the car company but did so in terms which extended coverage to all such affected persons, or bystanders, and killed the rule of privity forever in product liability cases. Thus, in industrial accidents, persons injured by defective equipment can sue the equipment manufacturers, even though they were not operators but only non participating bystanders.

Implied Warranty of Safety. In another case involving automobiles, Hennigsen bought a Plymouth from Bloomfield Motors, who obtained it from Chrysler Motors. While Mrs. Hennigsen was driving it a few days later, the steering mechanism failed, the car hit a brick wall and she was injured. The car was so badly damaged it was impossible to prove the steering mechanism had been negligently manufactured. The Hennigsens brought suit, but Chrysler claimed the warranty in the sales contract

FIGURE 2–2 A Few Important Early Liabilities and Accidents Legal Decisions

limited its liability only to replacement of any defective parts. The court ruled that any product being offered for sale carries with it an implied warranty that it was reasonably safe. An additional item of note regarding this case was that the Hennigsens won without having to prove there had been negligence on the part of Chrysler or Bloomfield Motors.

Contributory Negligence. In 1850, a court ruled in *Brown v Kendall* that to impose liability for accidental injury the "the plaintiff must come prepared with evidence to show the intention was unlawful, or that the defendant was in fault, and the conduct of the defendant was free from blame, or he will not be liable."

Thus if an industrial worker was injured by a piece of equipment, he had to show its manufacture had been negligent in some way. (Generally, most successful claims were based on failure to warn of a hazard or is a poor production process.) If the injured worker had been negligent to even the slightest degree, the defendant was considered free from fault. The fact that the plaintiff's negligence had contributed to the accident and injury rendered him unable to collect any damages. With the acceptance of the concept of comparative negligence, liability is assigned to the degree that contending parties are considered responsible

"Deep Pocket" Awards. A suit may involve a number of defendants, not all at whom may have large amounts of monetary resources or property. If some of the defendants cannot pay their share of any award, it must be paid by those who can. As a result, many suits for injury include any large organizations with resources, although the person thought to be the cause of the fault may not have resources. For this reason too, attorneys include in suits any and all possible defendants.

Resurgence of Strict Liability. Fault-based (tort) liability involves the legal inquiry whether or not the charged party acted or failed to act as a reasonably prudent person would or would not act under all of the circumstances. Such a civil wrong may result from a willful or deliberate act, knowing failure to act, or inadvertent or negligent failure to exercise reasonable care. A major decisional change occurred in 1963 with the adoption of the concept of strict liability by the California Supreme Court in the case of *Greenman v Yuba Power Products.* Tort need not be shown under strict liability cases.

Greenman was injured by a power tool made by Yuba, which his wife had bought him as a present. The court pointed out that "a manufacturer is strictly liable in tort when an article he places on the market, knowing it will be sold without inspection for defects, proved to have a defect that causes injury to a human being." In other parts of its opinion the court indicated there had been a defect in design which made the product unsafe for use and, in effect, that there was no difference between a design and a manufacturing defect when strict liability is imposed. Subsequent rulings, especially in other states, distinguished between the two but still indicated that strict liability applied to either one. In addition, the strict liability concept was ruled to apply to facilities as well as products.

Under the doctrine of strict liability the plaintiff need not prove negligence on the part of the manufacturer, but only that any injury was attributable to a defect or an unreasonably dangerous condition that existed in the product when it left the control of the manufacturer. Almost all states have now adopted the concept, and the need to prove negligence has practically been destroyed as a basis for suits in product liability cases. There has therefore been a great increase in the number of cases of industrial injuries where equipment manufacturers are being sued. The chief defense by manufacturers in court is to show that the equipment was not "unreasonably" dangerous. Because the dividing line between "unreasonable" and "reasonable" is rather tenuous and varies from jury to jury, the results of such trials are often unpredictable.

FIGURE 2–2 Continued

concept of negligence through fault, which also developed in the nineteenth century. Proof of negligence on the part of the defendant became an essential requirement for recovery of damages by the plaintiff. Until first enunciated by the courts, common law had no doctrine of negligence. Negligence quickly became a separate tort.

The new concept of need to prove negligence or other fault was adopted so rapidly, that by the middle of the century it appeared always to have existed. By 1850 it was said "the plaintiff must come prepared with evidence to show that the intention was unlawful, or the defendant was in fault; for if the injury was unavoidable, and the conduct of the defendant was free from blame, he will not be liable." Supposedly, having to prove negligence did not abandon the aim of compensating victims and of eliminating dangerous conduct, but now aimed only to eliminate conduct considered unreasonably dangerous.

What was considered reasonable and what unreasonable when applied to dangers, risks, or activities has never been defined definitively. If the economic benefits were attractive, managers of enterprises considered the presence of most hazards "reasonable." Because their interpretations were extremely biased, employers declared many highly dangerous activities not to be so. When accidents and injuries occurred, employers contended, through lawyers and the courts, that any hazards and risks had only been reasonable. This attitude endangered not only workers but also the general public. Working conditions in many plants continued to grow even more abominable.

Management in the twenty-first century will find the courts applying two important tests to determine their responsibility in hazard identification and mitigation: foreseeability and prudence. Both must be executed reasonably. In logic, there are two kinds of reason: inductive and deductive. The safety professional has both inductive and deductive tools, or techniques, to apply to both hazard recognition and evaluation. The reasonableness of accepting risk is determined in part by two factors that interact in the risk equation: probability or frequency of occurrence and severity of the undesired event. In a tort case, the legal inquiry is whether or not the party charged acted or failed to act as a "reasonably prudent person" would or would not act under all of the circumstances. "Reasonableness" is the issue. Management must realize that "reasonableness" in foreseeing hazards and responding prudently to risk is now not entirely without definitional insight.[7]

AMERICAN LAWS

After the Revolution, the statutes and common laws derived from England changed. Westward growth of the country continued, and economic and social differences between the two countries grew wider. Attitudes, philosophies, and stresses of the rapidly expanding United States induced tremendous legal alterations.

STEAM ENGINES, BOATS, AND LOCOMOTIVES

The advent of steam engines, boats, and locomotives, especially the last two, brought a tremendous number of accidents and, with them, increasing concern for the public's safety, and litigation. In the United States, in 1833, President Andrew Jackson thought the number of boiler explosions and other steamboat accidents such a problem that he asked Congress for legislation to control the "criminal negligence" involved. Accidents with railroads, which had been invented only a few years earlier, soon created a wider-ranging and greater safety problem, regarding which the federal government did nothing until 1893.

Railroad and ship explosions and other accidents were not limited to the United States. Transportation accidents were becoming common in all the European countries, where laws were being passed to help lessen the hazards and the losses in lives.

THE FOURTEENTH AMENDMENT AND SAFETY

After the Civil War the Fourteenth Amendment to the Constitution produced unforeseen safety and labor problems. ("No state shall make or enforce any law which shall abridge the privileges or immunities of citizens of the United States, nor shall any state deprive any person of life, liberty, or property, without the due process of law; nor deny to any person within its jurisdiction the equal protection of the laws.")

Passed to guarantee rights to the recently freed Negroes, the Fourteenth Amendment unexpectedly became a barrier to enactment of industrial labor and accident prevention laws. The due process clause was interpreted in such a way that it became an obstacle to passage of state laws in eliminating unsafe working conditions. Such laws supposedly violated "the liberty of the individual in adopting and pursuing such calling as he may choose." Employers considered such laws "meddlesome interference with the rights of the individual."

Also, any worker supposedly was free to enter into any contract desired in accordance with the concept of assumption of risk. If the employee was willing to accept an unhealthful or unsafe condition or one with a great risk of injury, there should be no legislative interference with the liberty of contract. Because of economic necessity individuals were often willing to accept poor working conditions. Very little was done by employers to provide safe workplaces. Their common belief was that each worker had a personal responsibility and concern to avoid any accident. Any accident that did occur usually was attributed to worker irresponsibility and negligence.

Social-minded citizens were generally unsuccessful in attempts to correct these inequities. Successful corrective measures to lessen workplace accidents and otherwise improve working conditions were initiated only with the growth of labor unions. Workers themselves came to realize that collective action was often the principal means of forcing corrective changes on an employer. Many industrial and sweatshop workers were immigrants from countries where they had grown familiar with various forms and benefits of workers' compensation laws.

LIABILITY LAWS

Some states introduced a few ineffective industrial codes in the 1880s. Alabama in 1885 was the first state to have an employers' liability law (copied substantially from one in Britain). Massachusetts followed with a similar one in 1887, and a few other states followed suit. Such codes varied widely from state to state.

These laws were intended to reduce employers' defenses against liability for accidents by eliminating the common law concept of assumption of risk. However, they did nothing to diminish the employers' defenses of contributory negligence by injured employees or fellow employees. It was on these latter two concepts that insuring companies of the employers generally and consistently won their cases.

Bowers[8] said of the period: "Employer's liability, under the influence of common law doctrine, was interpreted to mean exemptibility, rather than responsibility." With very few exceptions employers did little or nothing to alleviate the steadily growing injury and death rates.

In 1910 a New York commission issued a report[9] which listed five principal objections to the employer's liability system:

1. Insufficient compensation because of accidents.
2. Wastefulness, because it was too costly to employers, of no help to injured persons or their families, and of benefit only to insurance companies and the participating attorneys.
3. Delay, in that on the average it took six months to six years in that state for decisions. In other states the average time was even longer.
4. Employee-employer antagonism: although the two often had good relations, the interposition of insurance companies destroyed this relationship. The insurers were interested only in defeating any claims and provided little sympathy to the injured. Workers' resentment was thus directed to the employers, with which they were more familiar, than to the insurer.
5. Inconsistency of any settlements or awards for any accident proved to have been due solely to the employer's negligence.

WORKERS' COMPENSATION LAWS

Demands for correction of the liability laws were made throughout the country. A series of state laws for workers' compensation was first enacted in the United States about 1908 to 1910. Under President Theodore Roosevelt the first such laws for federal and state workers in Montana were enacted for protection of workers in very hazardous industries. This was 70 years after the first such law had been enacted in Prussia. Even tsarist Russia, at that time a very backward country, had its first law 56 years before the United States had. (Additional information on workers' compensation is provided in Chapter 3.)

Workers' compensation involves a form of "no-fault" insurance. In 1916, Justice Louis Brandeis stated [in *New York Central Ry. v Winfield* (1917), 244 U.S. 147, 37 Sup. Ct. 546, 554, 61 L. Ed. 1045]:

> In probably a majority of cases of injury, there was no assignable fault; and in many more it must be impossible of proof. It was urged: attention should be directed, not at the employer's fault, but to the employee's misfortune. Compensation should be general, not sporadic; certain, not conjectural; speedy, not delayed; definite as to amount and time of payment; and so distributed over long periods as to insure actual protection against lost or

> lessened earning capacity. To a system making such provision, and not to wasteful litigation, dependent for success upon the coincidence of fault and the ability to prove it, society, as well as the individual employee and his dependents, must look for adequate protection. Society needs such a protection as much as the individual: because ultimately society must bear the burden, financial and otherwise, of the many losses which accidents entail.

Worker's compensation laws are generally characterized by: strict (or no-fault) liability, compulsory insurance, administrative hearings, limited recovery for injuries (not including "incidental" damages such as pain and suffering), and limited attorney's fees. The fundamental principle of workers' compensation laws is to recompense workers for any loss in income-producing capability. However, changes have occurred over the years, so that a worker can be awarded damages for permanent injuries that do not cause loss of income. A very common example (see Chapter 28) is the hearing loss a person may suffer from working in a noisy environment. In many states hearing loss is a compensable injury for which workers can obtain compensation and yet continue at skilled labor. In past years, payments for medical and rehabilitation costs for workers injured at work have grown even more desirable and expensive.

LATER ACTIONS

Between the times of Grover Cleveland and Franklin Delano Roosevelt little federal action was taken for accident prevention to safeguard workers and the public. In one rare case, in 1912, Congress imposed a tax on white phosphorus matches. The phosphorus was attacking the mouths of workers so badly that horrible decays and other afflictions were occurring to the teeth, jaws, and lips, as women wet brushes laden with the damaging material to shape them.

Theodore Roosevelt's Federal Employer's Liability Acts were major advances. Additional laws were the Merchant Marine Act (1920) and the Longshoremen's and Harbor Workers' Compensation Act (1927). Meanwhile, as widespread needs arose, some safety measures were enacted by individual states or issued by private organizations. Commentators on industrial plant operations wrote that many accidents were due to poor equipment designs, with inadequate safeguards.

Little more was done nationally to improve workplace safety until the advent of President Franklin Roosevelt in 1932. Among his chief advisors were sociologists, labor officials, workers, and legislators who agreed on the need for accident prevention. Improvements were made, in spite of the opposition of business interests that generally fought them (and Roosevelt).

RESURGENCE OF STRICT LIABILITY

Safety improved somewhat because of changes in both statutes and the common law. A major milestone was reached with judicial decisions and rulings on strict liability in some states. Such a ruling made in California in 1963 was quickly followed by similar decisions in other states. A number of states had preceded California, but California's prestige caused many more to follow. Observance of the strict liability concept did more to eliminate defective designs than did all engineering school teachings, registration of engineers, and workers' compensation laws.

The doctrine of strict liability covered only certain states, but it applied to many accidents, because defectively designed industrial products were widely distributed. (See Fig. 2-2.) Even with the resurgence of strict liability, however, negligence is still a prime claim on which suits for liability are based. Of one type of negligence, often and widely claimed is "failure to warn" of any hazard. For example, failure to warn has become a vast source of litigation based on occurrences that formerly were commonly accepted by persons injured in falls from ladders. There have been so many such suits that ladders are now being manufactured bearing decals with warnings on their sides, tops, and steps.

Strict liability is one of several alternatives to the tort system. The tort system suffers from the deficiencies of high cost, long delays, proof of liability, identification of the responsible party, available defenses, and immunity from judgment. Today, alternative systems, in addition to strict liability, include limited liability (caps on recovery), "compensation" systems, excused negligence (Good Samaritan laws), and compulsory insurance. Each offers some form of relief from the deficiencies of the tort system.

The next significant milestone in industrial accident prevention was the passage of a federal law: the Occupational Safety and Health Act (OSHA) of 1970. This has done much to improve the safety of workplaces, although it has serious deficiencies. Even more deficient, although it does do some good, is the Consumers' Product Safety Act of 1972.

Today there are thousands of pages of safety-related regulations in the United States and other countries. Some are obscurely written. Some provide performance standards which are too broad; some provide specification standards which are too narrow. The relevance of such regulations to the actual reduction of accidents and the improvement of health and safety is debated by many. Compliance with regulations is not an absolute defense against liability. Noncompliance may not mean negligence per se. Many regulations are simply minimum standards of safety. Others are outdated and not state-of-the-art. Some, no doubt, are simply mistaken and erroneous.

In recent years the federal government has done little to improve safety in the workplace or for the general public. Federal administration executive actions have lessened the imposition of new safety measures and lessened attempts to apply existing standards.

At the time of this writing, bills are pending in both the U.S. Senate and the House of Representatives to reform OSHAct of 1970. The 1997 reforms proposed would:

1. Exempt employers from inspection merely for their having an average safety and health record.
2. Codify appropriations riders prohibiting the OSH Administration from conducting inspections of certain small employers.
3. Prohibit the OSH Administration from conducting most inspections based on employee complaints unless the employees have first brought their complaints to the attention of their employer and can prove that the employer has not taken action to correct them.

4. Slash penalties amounts and prohibit the OSH Administration from assessing penalties for first-time violations if employers correct violations within a "reasonable time."
5. Prohibit the issuance of criminal penalties.
6. Eliminate the National Institute for Occupational Safety and Health and the Mine Safety and Health Administration.
7. Amend the OSH Administration's record-keeping requirements so that only injuries involving lost time need to be recorded.
8. Amend the OSH Administration's standard-setting procedures, putting greater emphasis on cost-benefit analysis.
9. Establish a voluntary protection program award.

LIMITED LIABILITY

Recently there has been considerable interest in legislation that limits the amount of damages which might be awarded by the courts. There have been attempts to put caps on such things as medical malpractice. A history of restrictions on recoverable damages goes back to marine laws and limits on recovery for losses at sea. How this interest affects the health and safety of the workplace is yet to be determined.

EXCUSED NEGLIGENCE

There are legal protections (exculpation) for certain negligence acts. The purpose of these so-called "Good Samaritan" laws is to protect those performing essential duties for society. Persons of good faith coming to the aid of highway accident victims enjoy this protection, as do emergency responders, in many states. This protection has been extended to shippers of hazardous materials, and some regard the "Superfund" legislation of the Comprehensive Environmental Recovery and Clean-up Act (CERCLA) as providing this protection in order to have a means to restore the environment from past bad actions.

Legislation in the past has brought about some improvements in safety and accident prevention. However, the principal reasons for the improvements have been economic ones: increased awards for liabilities for accidents, continuing increases in workers' compensation premiums, and growing apprehension of massive industrial accidents that might affect the nearby public. The use of economic incentives for safety is now under review, and this powerful tool may not be available in the future.

BIBLIOGRAPHY

[7] D. L. Price, "Risky Business: Creating a Safe Environment," *Personnel,* Nov. 1986, American Management Association publication, pp. 62–67.

[8] E. L. Bowers, *Is It Safe to Work?* (Boston: Houghton Mifflin Company, 1930), p. 170.

[9] New York Employers' Liability Commission, *First Report,* 1910, vol. 1, pp. 19*ff.*

EXERCISES

2.1. State four legal milestones in the history of safety.

2.2. What are the differences between statute, common, and administrative laws? How do these relate to and affect economic laws in regard to accidents and safety? Which types of laws have precedence?

2.3. Define a "tort," "strict liability," and "negligence." What happened to the concept of strict liability?

2.4. What did employees have to do to be compensated for any accident for which they believed the employer was at fault?

2.5. What three defenses, originated early in the Industrial Revolution, did the employer use against any claim for liability for an injury? Do employers still have such protections?

2.6. Did all states have industrial employer liability laws in the nineteenth century? Of what did those laws consist? What were the principal objections to them? Were such laws effective?

2.7. Where and when were workers' compensation laws originated? When were they introduced in the United States? Were workers' compensation laws in the United States issued nationally or by state?

2.8. Describe safety measures for which at least three presidents worked.

2.9. State briefly toward the safety ends toward which the following were beneficial:

Workers' compensation laws.

Occupational Safety and Health Act (OSHAct).

Strict liability laws.

Product Safety Act.

2.10. Many persons who suffer injuries in accidents use trial lawyers on a contingency basis to prepare and plead their cases and receive as fees a percentage of any award. Discuss the advantages and disadvantages of such a system.

CHAPTER 3

Workers' Compensation

Common law obligations between a master (employer) and servant (employee) were based on the presumption that the worker was free to choose his or her employment, knowledgeable of hazards that might be present on the employer's premises and involved in any operations, and free to accept or reject its incumbent risks. (A stranger or guest had no such knowledge or right of elections.)

OBLIGATIONS TO EMPLOYEES

Historically, under common law an employer has been obligated to provide employees with:

1. A safe place to work.
2. Safe tools with which to perform the work.
3. Knowledge of any hazards that were not immediately apparent but which might be encountered during performance of the work.
4. Competent fellow employees and supervisors.
5. Rules by which all could perform safely and the means to ensure the rules were observed.

Prior to enactment of workers' compensation laws, only one means existed by which a worker could obtain indemnity for an injury resulting from his or her employment. This means was to prove in court that the employer's negligence had been the sole cause of the worker's injury.

Proving the employer had been negligent was generally a difficult, time-consuming, and costly procedure. An injured person usually lacked the resources to pay medical bills and personal living costs during the time the case was being brought to court and adjudicated, even when the employer's negligence was obvious. For example, it was noted[10] that "one Iowa case took 6 years, 4 injuries, and 4 appeals to the Supreme Court to determine a carpenter's right to indemnity. Another claim reached a final decision 12 years after the accident. A New York Commission noted it took 6 months to 6 years for cases to go through the courts." In many instances of temporary disability the person injured would not even institute a claim against the employer, fearing that after recovery he or she would not be permitted to return to the job or to find a similar one with another employer. Fellow employees who were knowledgeable of the employer's negligence refused to testify for fear they would lose their own jobs. And some courts were heavily biased toward the employers.

Even where the courts were objective, under the laws of the time employers could also defend themselves successfully, and generally did, by showing that any one of three conditions had existed which disqualified the injured worker from being indemnified:

1. There had been contributory negligence. Injured workers could expect no damages if they had been negligent to any extent, no matter how minor.
2. A fellow employee had been negligent. Injured workers could not expect damages if the injury resulted from the negligence of a fellow employee.
3. There had been an assumption of risk. Injured workers could not expect damages if they had become aware of the hazards of their work and accepted them.[10]

Examples of cases using each type of defense were legion. (Only in the past few decades has the defense of citing contributory negligence even to a minor degree been superseded. In some states, where employers are not entirely guilty, the employee's contribution to the accident lessens any award to the degree the worker was responsible.) Examples of use of the other defenses can be cited here.

Eastman[11] illustrates the second case in point. Suppose a yardmaster in Philadelphia, by reputation a reasonably careful man, puts a car of dynamite at the end of a train of cars instead of in the middle, as the rule of the company requires. Because of this carelessness the dynamite is blown up in a collision many miles from Philadelphia. A cow browsing in a field near the track and a station agent keeping his lonely post in a small country station next to the field are both blown to pieces. Now, in such a case the farmer could recover for the loss of his cow. But the station agent's widow could not recover for the loss of her husband because he was a fellow servant of the man whose supposed mistake or carelessness had caused the accident. Yet the station agent had had no more to do with the fellow servant's act or with his employment than had the farmer's cow.

Gagliardo[12] cites a case where a girl contracted tuberculosis while working for a candy company in 1924. She had worked in a cellar where there were no windows to provide light, the walls were wet, a cesspool backed up liquids that wet the floor, there were dead rats left about, it was drafty, and there were terrible odors. She was denied damages when the judge ruled: "We think that the plaintiff, as a matter of law, assumed the risk attendant upon her remaining in the employment" (*Wager v White Star Candy Company,* 217 N.Y. Supp. 173).

WORKERS' COMPENSATION LAWS

The first workers' compensation law, which applied to railroad employees only, was adopted in Prussia in 1838. Laws with additional and broader applicabilities were passed for all European workers: Sweden, Norway, Britain, Germany, Austria-

Hungary, and others. Tsarist Russia had a workers' compensation law for railroaders in 1852 and one for ship workers in 1878.

Not until 1908 did the U.S. Congress, prodded by President Theodore Roosevelt, pass a law that covered certain civil service employees in what were considered extra-hazardous occupations. The first state to pass a compulsory workers' compensation law was Montana in 1909, but it applied only to miners. The law was considered unconstitutional by the courts and thrown out. Another compulsory law, which applied to eight especially dangerous occupations, was passed in 1910 in New York. This law was declared unconstitutional the following year by the New York Court of Appeals in *Ives v South Buffalo Railway Co.*, 201 N.Y. 271 (1911). The New York court declared the new law violated the due process clauses of the federal (Fourteenth Amendment) and thus state constitution. In the same year, the tragic deaths in the Triangle Shirtwaist Factory Fire (Fig. 3-1) resulted in increased demands for workers' compensation laws.

The prestige of the New York court influenced legislators in many of the other states to believe that any compulsory law also would be held unconstitutional. Nine states passed workers' compensation laws which supposedly were noncompulsory in that employers could "elect" to come under the law. If not, the employer's common law defenses were curtailed or voided. Seven more states amended their constitutions to

Just before the end of the workday on March 25, 1911, a fire on the eighth floor of the building that housed the Triangle Shirtwaist Factory resulted in the deaths of 146 workers and serious injuries to 70 more. The disaster was one of the factors that led to the passage and adoption of workers' compensation laws.

As the 600 garment workers in the ten-story building started to leave for home at the start of the weekend, the fire started. Its cause was never determined but may have been a cigarette or match tossed carelessly onto some lint or cloth fragment. The fire spread quickly as attempts to extinguish it failed. The workers panicked and rushed for the exits. The exit that usually had to be used on each floor was a passageway only 20 inches wide. The employers had restricted it so inspections could be made as the workers left to prevent material from being stolen. A wider exit located on each floor had been locked to ensure the workers used the 20 inch passageway. At both points on all floors, the crush restricted persons from getting out. The two small elevators in the building could carry only a few people. A narrow circular stairway also quickly jammed.

A few persons managed to get down an available narrow staircase. Finally the wider exit on the eighth floor was smashed open. Less than twenty persons managed to go down the steep wrought iron fire escape before it collapsed, carrying with it those workers still on it. Some of the tenth-floor workers climbed to the roof. There a small number were able to cross to the next building over two ladders brought and held by students from a nearby university.

Because of the flames and smoke, some of the employees threw themselves down the elevator shaft. Scores of women and girls climbed through windows and onto ledges of the building and jumped as the flames neared them. The life nets brought by fireman were of little use and failed to stop the workers as in desperation they leaped in groups of three or four.

The owners were found not guilty of manslaughter charges. New fire codes were put into effect in New York. Another result was passage of the state's workers' compensation law. A year earlier a similar law there had been declared unconstitutional by the Court of Appeals.

The fire is still remembered for the horror of the victims' deaths. On March 25, 1986, a 75th-year memorial was held in New York, where the building still stands. A few of the surviving 1911 workers were present.

FIGURE 3–1 The Triangle Shirtwaist Factory Fire

legalize workers' compensation laws within limits, as they interpreted them, of the New York Court of Appeals' decision. One state, Washington, decided the New York decision was incorrect and passed a compulsory law, contending it was a proper exercise of the state's police powers. Its disapproval of the New York court's decision reflected common legal and judicial opinion throughout the United States. Other compulsory laws soon followed, all of course subject to the possibility they might be declared unconstitutional. However, demands for suitable action had been growing.

Not until 1917 did the U.S. Supreme Court declare affirmatively the validity of workers' compensation laws. Washington State's law, an "elective" law passed by Iowa in 1911, and a revised compulsory law passed by New York in 1913 were all upheld in separate decisions.

Since then almost all of the states that have had "elective" laws have replaced them with mandatory laws. If the New York Court of Appeals had ruled favorably on the first New York law, there probably would have been no need for any of the electives. By July 1986 only three states had such laws.

Employers finally accepted workers' compensation laws, reluctantly and belatedly, fighting them all the way. The U.S. Chamber of Commerce, no part of the federal government, indicated the aims of workers' compensation, long after the decision by Justice Louis Brandeis (see Fig. 3-2).

PROBLEMS OF NONUNIFORMITY

There are 57 workers' compensation laws. The 50 states, the District of Columbia, Guam, Puerto Rico, and the Virgin Islands each have one. The federal government has three. Each law is different and its interpretation is different. There are differences between the states in legal provisions and coverages, penalties, and benefits. In recent years these differences have grown greater as the various states (and the federal

In 1983 the U.S. Chamber of Commerce published an Analysis of Workers' Compensation Laws which stated: "Six objectives underlie workers' compensation laws.
They are:

1. To provide sure, prompt, and reasonable income and medical benefits to work-accident victims, or income benefits to their dependents, regardless of fault.
2. Provide a single remedy to reduce court delays, costs, and work loads arising out of personal-injury litigation.
3. Relieve public and private charities of financial drains incident to uncompensated industrial accidents.
4. Eliminate payment of fees to lawyers and witnesses as well as time-consuming trials and appeals.
5. Encourage maximum employer interest in safety and rehabilitation through an experience-rating mechanism.
6. Promote frank study of causes of accidents (rather than concealment of fault), reducing preventable accidents and human suffering."

FIGURE 3–2 From U.S. Chamber of Commerce

government) modified their provisions. The lack of uniformity has numerous disadvantages, and recommendations have been made to establish a federal law to cover all workers uniformly. Even after many years, such a uniform national law has not come about, nor does it appear imminent, because there is too much opposition.

When President Clinton's administration proposed the Health Care Reform Act, Title X of that act called for the health care portion of workers' compensation to be part of a new national health care system. This proposal featured 24-hour coverage independent of whether an injury or illness was occupationally related, medical-transaction cost controls, employees' choice of physician, and medical case managers. There was considerable opposition to these proposals. So, nonuniformity continues in workers' compensation laws.

Consider the coverage of a person who travels for an employer. Assume the person is a bona fide employee, not an agent or independent salesman, whose duties require him to travel to another city on the employer's business and stay there overnight.

In one New York case (*Johnson v Smith,* 188 N.E. 140 New York), a man contracted typhoid fever from food he had eaten in a restaurant while on his trip. The court ruled, in effect, that if the employer does not designate or provide a place for his employees to stay while on an overnight trip, injuries suffered at times outside those hours during which the employees were actually engaged in their employer's business do not fall within the scope of the workers' compensation laws.

In California a ruling for a comparable situation was different [*Wiseman v LAC,* 20 Cal. Comp. 306]. "As a general rule, a commercial traveler is regarded as acting within the course of his employment during the entire period of his travel upon his employer's business. His acts in traveling, procuring food and shelter, are all incidents of the employment, and where injuries are sustained during the course of such activities, the Workmen's Compensation Act applies."

This case involves an employee who met and started drinking with a girl while he was on an overnight business trip. They adjourned to his room, where eventually both burned to death in a fire ignited by a cigarette falling on their bedclothes. His widow sued for benefits under the Workmen's Compensation Act, and the California Supreme Court ruled in her favor. A case very similar to this one regarding a traveler being burned to death also occurred in England, but the traveler was from Michigan. Although his amatory misdeed and death took place overseas, his family received workers' compensation from Michigan.

Because the man from California (or Michigan) had been hired and was normally working there, he would have been covered by the provisions of that state's law even if he had been injured or killed in New York or any other state where he had been working temporarily, no matter how extended the time. To avoid problems because of the many different provisions in the states, some national companies use the following procedure. When a person hired by a national company in one state is transferred to work for a long time in another state, often the company has the worker resign. At the same time the person is hired for and put on the rolls of the branch of the company in the new state. This procedure lessens any legal problem that might arise because a worker hired in one state is working in another. (This procedure is rarely used when the worker's assignment is not for a long period of time.)

COVERAGES

Approximately 80 percent of all workers in this country are covered by workers' compensation laws. (*Note:* Changes in workers' compensation laws take place almost daily. To obtain current information, the laws and benefits of the states involved should be consulted.)

By February, 2000 only two states had elective laws for industry and five for agriculture. Most jurisdictions have compulsory laws. In addition, an employer may "volunteer" to accept coverage under workers' compensation laws even where not obligated to do so. For example, some states do not require certain employments or employers with fewer than a stipulated number of employees to provide coverage. The categories of persons not covered in some states are shown in Fig. 3-3.

WORKERS' COMPENSATION INSURANCE

Employers pay for the costs of workers' compensation. Benefits to injured employees may have to be paid for a long time. The employer may suffer financial reverses and go out of business. To ensure that benefits will be paid when and for as long as required, the laws call for an employer either to have insurance coverage for payments that may have to be made or to show the capability of carrying any foreseeable financial burden that could be imposed (self-insurance).

Some states require employers to obtain insurance from state funds set up for the purpose, either on an "exclusive" or "competitive" basis. Employers must insure their risks with state funds under the exclusive basis. Under a competitive situation, the employers may insure with the state fund if there is one, elect to use a private insurance company, or self-insure. In Texas, private insurers only were used until 1991; there was no state fund, and self-insurance was not permitted. The result was that by 1987 Texas "had the nation's highest workers' compensation rates, lowest benefits for workers and the worst record of workplace deaths and injuries."[13] This resulted in the creation by the Texas Legislature of the Texas Workers' Compensation Insurance Fund in 1991 to stabilize the competitive marketplace by selling workers' compensation insurance in Texas, and to serve as the insurer of last resort.

To self-insure, an employer must meet specific requirements of financial responsibility and capability, set by law, such as posting of bonds or securities as reserves for possible future obligation. The arrangements must be approved by the state's workers' compensation agency. Only very large companies and government agencies have been capable of self-insurance. At one time, many small units such as small cities found it more economical to self-insure. However, some accidents, generally involving motor vehicles, have resulted in lawsuits so great that some municipalities have been bankrupted or subjected to premium costs they found impracticable to pay.

COSTS OF WORKERS' COMPENSATION INSURANCE

Soon after the enactment of workers' compensation laws, employers began to object to the costs they had to pay, although the rates at the time were comparatively low. Since then the premiums to be paid have risen as coverage by workers' compensation

The categories of persons not covered under workers' compensation in some states include

1. Agriculture workers: At one time none of these workers were covered; now 38 states, Puerto Rico, and the Virgin Islands have some laws that provide at least some coverage. Fifteen more permit voluntary coverage by employers.
2. Domestic workers: All states and Puerto Rico have coverage for workers in domestic service, but some states have minimum requirements for hours worked or earnings.
3. Casual employment: In general, this is occasional, incidental, and not considered employment occurring at regular intervals. It has sometimes been defined by the number of days of labor, cost of labor, or is not in the course of the employer's business.
4. Hazardous employment: Some states indicate that the workers' compensation laws cover only employment listed as "hazardous" or "extra hazardous." However, the lists have broadened in many of these states, so that many occupations not ordinarily considered especially hazardous are now included.
5. Employees of charitable or religious organizations: Persons whose work is irregular, for short-term periods, or temporarily outside the scope of the regular activities of the employer are covered in only ten states and under the Longshoremen's and Harbour Workers' Compensation Acts.
6. Small organizations: In 14 states, workers' compensation laws do not apply where there are fewer than a stipulated number of employees. The remaining states have no such requirement.
7. Railroad and maritime workers: Most of these workers are not covered by state workers' compensation laws. Railroad workers in interstate commerce and maritime workers are covered by the Federal Employer's Liability Act. This is not a workers' compensation law, but it denies the employer the right to plead any of the common law defenses in any negligence action an employee may institute because of an injury.
8. Contractors and subcontractors: Independent contractors are not covered by the insurance of any one for whom they are performing services. An independent contractor is one who agrees to do a specific piece of work in accordance with his own capabilities and for which he will be paid an agreed reimbursement. The contractor will do the work without being subjected to another's orders. If the contractor is not truly independent, and the party for whom the services are being provided actually does direct the work, the contractor may not be independent, but an employee.
9. Minors are entitled to the same workers' compensation coverage other legally employed persons would receive under similar circumstances. In a few states, compensation where minors are illegally employed is increased if they are injured. In some states the amount received for a permanent injury is based on considerations of loss of the youth's future earning capacity.
10. Extraterritoriality: If a contract of hire is made within one state, and if the employee's residence or hiring was within another state, its workers' compensation laws still apply even if the injury occurs while serving a long temporary duty in another state. If the temporary duty is a very long one in another state, a national company will generally have the employee sign an agreement for the new state at the time of resigning from the first one.

FIGURE 3–3 Limitations on Workers

insurance (and insurance of all types) has increased tremendously. For this reason alone, employers have cut back on the number of personnel employed or moved their operations to foreign companies, where wages are lower, safety requirements are less stringent, and insurance and liability costs are less. Costs in different states differ, because not only are safety code requirements different, but so are the benefits that must be paid workers, hospitals, and medical personnel in the event of any accidents. The probability that accidents and injuries will occur and payments will have to be made also varies with the stringency of policing by the state, the employer, and the insurer. Figures cited for costs of insurance therefore will not apply between any two plants or states.

Except where efforts are made to hold or to reduce insurance costs, the costs of workers' compensation insurance have increased and will probably continue to do so. This is evident because the number of persons covered and the costs of benefits to be paid have also escalated. In 1960, 45 million workers were covered by compensation insurance; by 1978, coverage had increased to 76 million, or by 69 percent. Annual benefits paid during that time skyrocketed from $1.295 to $9.72 billion, an increase of 750 percent. (The chief reason costs have not risen further is that much high-risk industry has moved to overseas operations.)

In California there had been no increase in workers' compensation benefits since 1972. In 1983, rates for compensation insurance increased more than 15 percent. Where the minimum weekly benefit for temporary disability had previously been $49, in 1983 it increased to $84, in 1984 to $112. Similar maxima for total disability went from $175 per week to $196 and then to $224. The increases in benefits costs from 1986 to 1996 continued in California and were exacerbated not only by more claims, but also by a dramatic increase in fraudulent claims. The California problem with workers' compensation fraud received national attention and coverage by television investigative reporting. Some authorities say that 20 to 30 percent of claims filed are fraudulent. The result of these and other factors was that workers' compensation premiums rose from 30 percent of the payroll to 70 percent or more.

Fraudulent claims have become a nationwide problem, certainly not limited to California. In 1997, the Nevada State Industrial Insurance System listed the following indicators that a claim might be fraudulent.

1. *Monday Morning.* The alleged injury occurs late on a Friday afternoon but is not reported until Monday.
2. *Employment Change.* A reported accident occurs just before or after a strike, job termination, layoff, or end of a big project or of seasonal work.
3. *No Witnesses.* No one sees the accident, and the employee's own description does not logically support the cause of the injury.
4. *Suspicious Claims History.* A record of numerous, suspicious, or litigated claims.
5. *Treatment Is Refused.* The claimant refuses diagnostic procedures to confirm an injury.
6. *Late Reporting.* The employee delays reporting the claim.
7. *History of Changes.* The claimant has a history of frequently changing physicians, addresses, and employment. Beware if the allegedly disabled claimant is hard to contact at home.
8. *Group Claims.* Watch for the use of the same doctor and lawyer by several claimants.

The National Insurance Crime Bureau says that worker's compensation fraud costs the insurance industry $5 billion each year. The Insurance Fraud Bureau of Massachusetts studied workers' compensation fraud costs over a period of two and a half years. Concluded in 1993, the study showed that, in that state, 40 percent of fraudulent claims involved claimants working while collecting benefits, 16 percent were malingerers, and 15 percent were staged accidents. Workers compensation reforms were enacted in Massachusetts in 1992. Until then, there had been only one fraud conviction in 83 years of workers' compensation; in the five years that followed the reforms there were 6,800 cases of suspected fraud, 32 indictments, and 10 convictions. The Insurance Fraud Bureau of Massachusetts reported that almost simultaneously (i.e., from 1991 to 1994) the number of claims filed dropped 35 percent.

Other changes also have occurred. Claims for "cumulative trauma" injuries, injuries sustained over long durations of time, increased considerably nationwide. In 1993, about 48 percent of all California workers' compensation claims were for carpal tunnel syndrome. In 1997 it was reported that as much as one-third of the $60 billion annual workers' compensation cost paid by private-sector employees was due to musculoskeletal disorders claims. Sometimes these occurred during times when workers were being laid off. Workers generally have a year to file a claim after they become aware of a work-related injury. Those persons who were laid off or believed they would be (and were still eligible) would therefore file claims for benefits just before the end of the time limit. Claims were filed in record numbers. Firestone Tire and Rubber Company received more than 15,000 claims at two plants in 1982. General Motors received 750 claims for "cumulative trauma" at its Fremont plant alone.

In March 1988, newspapers reported that workers' compensation claims for benefits because of mental stress suffered had become an "insatiable beast" that devoured billions of dollars each year. Stress-related cases had grown 431 percent—from 1,282 to 6,812—between 1980 and 1986. From 1986 to 1997, claims arising from stress rose over 700 percent in California.

Rates charged for insurance coverage depend on the number of employees, the types of work involved, and the accident experience of the company covered. Premiums are usually adjusted for current claim experience, fluctuating with the losses the insurer suffers or believes it will suffer because of accidents at the employers' plants. Increases in premium costs have been so great in some cases that companies have been forced out of business. Each employer will therefore attempt to keep down the premiums that must be paid to the insurers.

The cost of any type of insurance is dependent on the risk the insurer feels it is taking in providing coverage. Premiums are based not only on past experiences and losses but on possible future losses, plus an amount for overhead, profits, other incidental costs, estimates by actuaries, and the employer's perceived degree of success in accident avoidance. Therefore, figures cited for costs of insurance for similar companies or plants will rarely be the same.

INSURANCE RATING SYSTEMS

This section describes private insurers' methods of arriving at the prices they charge for workers' compensation insurance. These methods are listed in Fig. 3-4.

KEEPING WORKERS' COMPENSATION COSTS DOWN

In the late '90s, the American Express Small Business Services gave eight cost-saving moves to help keep workers' compensation insurance costs under control.[14] These are shown in Fig. 3-5.

Various methods of determining the premiums to be charged include

1. Schedule rating: Safety conditions the average organization should meet are set as standards. Credits or debits are then assigned for better or worse conditions, and the insurance rates adjust from the standard. An incentive is therefore provided to improve working conditions.

2. Manual rating: Business and certain categories of workers are classified according to a manual which cites rates for the different classifications, depending on degree and class of hazards involved. For example, an industrial plant might have, among others, foundrymen, machinists, office personnel, and janitors, for which each rating classification is different. The number of $100 of payroll in each class is then multiplied by the applicable rate and the products totaled. This sets a premium for the entire facility.

3. Experience rating: The premiums are predicted on estimates of expected losses for the average employer in approximately the same manual classification. The premium is then adjusted for the loss experienced over the three previous years. A good record would reduce the premium; a poor one would increase it.

4. Retrospective rating: Retrospective rating relates premiums to experience in the current policy period. The employer being insured pays the expected premium at the beginning of the period for which coverage is being provided. An adjustment is then made at the end of the period which will reflect any injury-loss during that time.

5. Premium discounting: Overhead and administrative costs are comparatively less for servicing large companies which pay high premiums for their insurance. A plan was therefore developed under which these larger companies receive discounts reflecting the proportionately lowered expenses.

6. Combinations: Some features of these rating methods can be included by insurers to offer other premium rates.

FIGURE 3–4 Methods of Determining Insurance Premium Charges for Worker's Compensation

Ways to keep workers' compensation costs down

1. *Make sure your classification is correct.* This involves both the classification of a company and of its employees. If one worker is misclassified into a high-hazard job, the increase in premium costs can be significant. Most states use "The Scopes of Basic Manual Classifications" published by the National Council on Compensation Insurance of Boca Raton, Florida, as the basis for their classification schedules.
2. *Get out of the assigned-risk pool.* Determine why your company was placed there. Take steps to resolve the problems through appropriate actions such as the institution of targeted safety programs to reduce accidents or the use of a self-insured workers' comp purchasing pool.
3. *Conduct a payroll audit.* Overtime pay can be deducted to straight time in most states. Payroll figures are usually a basis for workers' compensation premiums.
4. *Use a deductible.* Premiums can be reduced by as much as 25 percentage this way. Deductibles range from $100 to $1,000 usually.
5. *Make sure your experience rating is correct.* The National Council on Compensation Insurance of Boca Raton, Florida, has worksheets available to help check this on your own.
6. *Use managed care to reduce your medical costs.* Over half of the states now allow managed-care providers to handle workers' compensation.
7. *Get injured workers back to work.* Early-return-to-work programs, when properly administered, are mutually beneficial to the worker and to the employer.
8. *Institute safety programs.* A safety program to recognize, evaluate, and control hazards can reduce the number of claims. These programs include safety audits, training, and preventive measures.

(American Express Company, 1997)

FIGURE 3–5 Ways to Keep Workers' Compensation Costs Down

WORKERS' COMPENSATION REFORM

The decade of the 1980s saw increased costs and reduced benefits explode in many of the workers' compensation programs across the United States. A flurry of reforms was instituted. By the early 1990s states such as California, Oklahoma, Oregon, Massachusetts, New York, North Carolina, and Texas had passed legislation to effect major cost savings and improvements. California repealed the workers' compensation minimum rate law and replaced it with a free-market, open rating system. They mandated an immediate drop of 7 percent in premiums and provided incentive for employers to promote a safe work environment by freeing insurance companies to offer lower rates for companies with fewer claims against them. They instituted antifraud measures and established new standards for stress claims. Their workers' compensation insurance rates dropped from $11.5 billion to $8.3 billion between 1992 and 1997.

Oklahoma reforms included reducing health care costs through managed care, requiring workplace safety programs, creating a more politically independent judiciary, strengthening fraud investigation and prosecution, increasing the use of independent medical examiners, and "making benefit payments more fair for both employers and employees." The results include a 4.5 percent reduction in workers' compensation costs in 1996, a decrease in claims filings of 8 percent from 1994 to 1997, an increase in fraud charges of 37 percent in 1996, and an increase of independent medical examiners of 42 percent between 1995 and 1997.

Reforms should be designed to increase the health and safety of the workplace. For example, Massachusetts reforms resulted in a dramatic reduction in accidents reported from 77,000 in 1990 to about 58,000 in 1993. New York 1996 reforms included the repeal of a unique unlimited liability exposure that added over $300 million per year in workers' compensation insurance costs. They instituted a standard that permits third-party liability action in very limited circumstances. They expanded managed care and allowed preferred provider organizations to offer workers' compensation medical services. They allowed employers to enter into nonbinding compensation arrangements with injured workers without admission of liability and established credits for safety and loss prevention programs and good safety records. More importantly, New York requires employers whose payroll is in excess of $800,000 and whose experience rating exceeds 1.2 to undergo a workplace safety and loss prevention consultation and evaluation by either the New York State Department of Labor or a private safety consultant.

Texas reforms include the creation of the Texas Workers' Compensation Commission, the Texas Workers' Compensation Insurance Fund (referred to above), and the Texas Workers' Compensation Research Center. While these changes to date have reportedly controlled costs and reduced rates, more importantly workplace injuries have dropped from 7.7 to 7.3 per 100 workers, while the national injury incidence rate increased from 8.4 to 8.9 per 100 workers.

Whether or not the workers' compensation reforms introduced in the 1990s will result in increased safety will have to be determined over a reasonable period of time. It is too early in this movement to evaluate the effects on costs, benefits, or safety.

REQUIREMENTS FOR BENEFITS

Although all the workmen's compensation acts are different, three fundamental conditions must be satisfied in any case for an injured worker or surviving dependents to receive benefits: (1) the injury must have resulted from an accident; (2) it must have arisen out of the worker's employment; and (3) it must have occurred during the course of the employment. One recourse insurers have to avoid paying a claim is to demonstrate that one of the three conditions was not fulfilled.

Injuries are not considered resulting from an accident where (1) injury was caused by the intoxication of the injured employee (although an employee will be entitled to compensation benefits if the accident in which the worker was injured was due to a cause other than his or her intoxication), (2) injury was intentionally self-inflicted (not common, but not unknown), (3) the employee has willfully and deliberately caused his or her own death, and (4) the injury arose out of an altercation in which the injured party was the initial physical aggressor.

One other condition exists that may vary with states: a statute of limitations indicating the time by which a claim must be filed after the accident took place.

In one case the exact time when an injury had first been reported became the governing consideration as to whether an employer should be held liable. On January 4, 1934, George Metesky of Waterbury, Connecticut, filed a claim in New York for workers' compensation. He claimed he was injured on September 5, 1931, when, while a generator wiper at a company power plant, he had been knocked down in a boiler room by a backdraft of hot gases that had affected his lungs. As a result he could no longer hold a job. The claim was disallowed because it had not been filed within a year of the accident. The disallowance was repeated each of three additional times the claim was submitted, the last in 1936. Because he claimed the accident had incapacitated him he kept trying to obtain compensation benefits from the power company. With each failure he grew more embittered. From 1940 to 1956 he planted and exploded 37 homemade bombs in public places (no deaths, few serious injuries) to publicize his grievance. In 1957 he was finally captured as the "Mad Bomber." After examination he was sentenced as a criminal patient to a mental hospital from which he was finally released in 1974.

DISAGREEMENTS

Disagreements between injured workers and their employers' insurers are generally based on differences of opinion regarding whether all stipulated conditions were met to receive benefits. In the event any disagreement can't be settled by the two, the matter is usually brought before the state's compensation commission for resolution. Lawyers are not required. Because some workers are unfamiliar with the existence of such commissions or disagree with their findings, they may hire a lawyer and take the case to court.

As far back as 1916, Justice L J. Wrenbury observed [in *Herbert v Fox* (1916) A.C. 405]:

> No recent act has provoked a larger share of litigation than the Workmen's Compensation Act. The few and seemingly simple words, "arising out of and in the course of the

employment," have been the fruitful (or fruitless) source of a mass of decisions turning upon nice distinctions and supported by refinements so subtle as to leave the mind of the reader in a maze of confusion. From their number counsel can, in most cases, cite what seems to be an authority for resolving his favor, on whatever side he may be, the question in dispute.

INJURY RESULTING FROM AN ACCIDENT

At one time the definition of "an accident" was "a sudden, adverse, unexpected event." However, judges and referees have ruled that an accidental injury can take place without occurring "suddenly." For example, many states now indicate exposures to harmful substances over lengthy periods of time can cause injury and death. There may have been no "sudden" onset of the injury, but the person was injured, an accident had therefore taken place, and the victim was entitled to compensation under the law.

[While the courts have been simplifying the definition of an "accident," other persons have been complicating them. Witness this definition from a statistical study on accident causative factors: "An accident, with or without injury, is in the main a morbid phenomenon resulting from the integration of a dynamic variable constellation of forces and occurs as a sudden, unplanned and uncontrolled event."]

Courts have also held that under workers' compensation laws an injury could be a psychological and not necessarily a physical condition. Increasingly, benefits have been awarded workers who have suffered mentally damaging stresses. Dependents of persons who died of heart attacks have been awarded compensation benefits when it was shown the attacks were the results of unusual mental stress from the work the victims undertook for their employers.

Conversely, it has been claimed a plant guard had hypertension so that he suffered a heart attack while making his usual rounds at night. The duties of the guard were not considered so stressful as to have caused the hypertension and attack. His death would not be considered one compensable under the workers' compensation laws.

INJURY ARISING OUT OF EMPLOYMENT

There is generally no question in this regard if the employee was doing work assigned by a supervisor, or carrying on any other activity normally expected of an employee.

An employee eats at a company cafeteria during his regular lunch period. He swallows a piece of glass that had inadvertently been dropped into the food and suffers an internal wound. It would probably be considered that the injury arose out of his employment. He would be entitled to workers' compensation. On the other hand, if the glass had been in a sandwich he had brought from home, he would not be covered because the hazard was not one to which he had been exposed by his employment.

Claims for workers' compensation are often made when the injury took place on the employer's premises but did not arise out of the worker's employment. A common illustration is when a man uses company equipment for personal reasons. A man is blinded in one eye by a steel chip from a piece of equipment he is using in his company's machine shop during his lunch period to repair a part for his boy's bicycle. His injury would not be compensable, because the activity did not arise out of his employment.

A question that often arises is whether or not the injured person was actually an employee. A man who agrees to paint the roof of an industrial plant is not an employee of the company if he does so under contract, according to an agreement as to when and how it will be done, using his own equipment, or without being subject to constant supervision from company personnel on how the details will be accomplished.

Generally, driving to and from work is not considered to occur during the course of employment unless there is a specific agreement that the employer will provide transportation. John normally drives to the warehouse area of the contractor for whom he works. The contractor then provides transportation to the actual site. John is covered during the time he is being driven to and from the work site. Also, the following might occur: John's boss calls him one night and says that on the following morning Sam will pick up John and take him directly to the work site. If John is injured en route, he will be entitled to compensation, because the employer normally provided transportation, and this was when John was injured.

In some states, rulings similar to that which resulted from the following situation indicated employees were entitled to benefits. John normally finishes work early and makes his way home over freeways that have little traffic at that time. However, his employer calls an unexpected conference which delays John's return until the freeways are extremely crowded. As a result, John is injured in an accident. The exposure to an abnormal hazard makes him eligible for workers' compensation.

Again there may be mitigating circumstances, so that injuries that occur during travel of employees to and from their homes are compensable. For example, an employer knows, or should know, employees who customarily take work home; any injury during such travel comes under the workers' compensation laws. Examples include an employee injured while taking home tapes from a recording session to work on them, and a claims adjuster who was killed en route home with case files on which he intended to work.

An employee in New York was so happy about the work he was doing and the friendly people with whom he was working, he started to dance a jig. Unfortunately, he fell and broke his leg. The Court of Appeals [in *Bletter v Harcourt* 250 N.E. 2d 572 (1969)] ruled that the good feelings of the injured party were work-connected and the injury was compensable.

In 1970, after he was fired, a Detroit automobile worker returned to his plant and shot and killed three supervisors. He was declared insane and committed to the state's mental hospital. In March 1973 the State of Michigan ruled he should be paid worker's compensation retroactive to the day he was fired. The ruling was predicated on the premise that working conditions had aggravated a preexisting but nondisabling tendency toward schizophrenia and paranoia until there was an acute, psychotic break with reality. Psychiatric treatment was required for this job-related disability; therefore the worker was entitled to compensation benefits.

Numerous similar related cases exist regarding degradation of physical capabilities because of employment, as a result of which compensation must be paid. Probably the most common example is the loss of hearing. Thus, a person's hearing is usually tested to determine both whether it is normal and whether any loss is greater than the usual degradation due to aging or attributable to a noisy environment in the workplace.

TYPES OF DISABILITIES

Compensation laws classify injuries in four or five categories, based on degree and permanence of disability. We list these below, from least to the most severe. It must be pointed out, however, that with hundreds of changes in benefits being made yearly, benefit lists are quickly out of date.

Temporary-Partial Disability

In most states a temporary-partial disability need not be reported. In some instances the worker may be injured, but the employer considers the injury so minor he or she does not report the event because it would affect the insurance premiums. Often when a worker suffers a mishap the supervisor tells him or her to go home for the day and files no report. Failures to report more severe mishaps have occurred. In 1986, the Union Carbide Company was fined $810,000 because of 81 failures to report on-the-job injuries from 1983 to January 1986. A temporary-partial disability may not preclude the person from performing full work duties. He or she may suffer some loss in time or wages, but full recovery is anticipated.

Temporary-Total Disability

The employee is incapable of any work for a limited time; however, full recovery is anticipated. The majority of compensation cases fall in this category.

Permanent-Partial Disability

The employee suffers an injury from which he or she will not recover totally. The worker will be able to perform some work, so earning capability is only partially affected. A worker who cannot return to the former job might be trained to do other work. This type of disability can be classified further into schedule or nonschedule injury. A *schedule* injury is usually stipulated in the compensation acts, with the number of weeks over which payments will be made, the weekly payment, and the total payment. Schedule injuries include, but are not limited to, loss of an arm, hand, finger, foot, toe, eye, and ear or ears. A *nonschedule* injury is of a less specific nature, such as a disfigurement or injury to the head. Awards for nonschedule permanent-partial injuries are determined in terms of the difference in wages the worker could earn before and after the injury, a percentage of permanent-total disability, or a relationship to schedule injuries.

Permanent-Total Disability

The employee suffers an injury that will prevent him or her from working at all or in any regular, common employment in the foreseeable future. Here again the laws differ regarding what constitutes permanent-total disability and the compensations persons so disabled will receive. Certain injuries are presumed to be permanent-total in most states: loss of sight of both eyes, loss of both hands or feet, or loss of one hand and a foot. Slightly more than half the compensation laws, including those of the federal government, require that benefits for permanent-total disabilities be paid for the entire period a disability may last, even if this is the balance of the injured person's lifetime.

The remaining laws either set a maximum total payment or limit the number of weeks during which payments will be made.

MONETARY DISABILITY BENEFITS

The benefits different states stipulate for identical injuries are inconsistent. As an example, a few of the maximum benefits for permanent partial disabilities can be cited (these amounts may have been changed).

At one point in time, loss of a hand in Colorado called for payment of $8,736, while Puerto Rico called for $9,000, Arkansas $24,332, Iowa $107,160, and Pennsylvania $116,245. Payment for the loss of an arm in Hawaii ($72,956) was more than for the loss of an arm at the shoulder in Georgia ($39,375) and many other southern states. Payment for the loss of a leg in Wyoming ($31,772) or in Ohio ($36,500) was less than for the loss of a foot in Iowa ($84,600) or Vermont ($81,375).

Because provisions differ, persons interested in a benefit amount would do well to consult state agencies.

DEATH BENEFITS

Payments are made for the support of the families or other dependents of fatally injured workers. In most cases payments will be made to a widow for life or until remarriage, and to children until they are 18 (longer if they are incapable of self-support). In the remaining states, benefits to survivors are limited either in duration of payments or in maximum amount. In addition, all states except Oklahoma grant a sum for funeral expenses.

EXTENT OF MEDICAL BENEFITS

Workers' compensation was originally designed to replace most of the monetary loss a worker might suffer as a result of an accidental injury. However, what in some instances has become more desirable to the employee and more costly to the employer are the medical costs for care, treatment, and rehabilitation.

Medical Care

All compensation laws provide that medical benefits will be paid. Eighty percent require full payment for all medical and hospital costs. The remaining 20 percent either limit the time the worker may receive benefits, the total amount, or both. For some accidents, medical care may be the largest benefit cost involved.

The provisions of the law indicate the surgical and medical services the injured party will receive. In some states the worker can select his or her own physician; in others, the employer provides a list of physicians from which the employee can choose. In a few states the employer selects a physician; however the injured worker can request a change if and when he or she desires.

Rehabilitation

In addition to surgical and medical services, some states require services for rehabilitation of workers. This includes prosthetic devices, crutches, physical therapy, and training to alleviate the effects of the accident and to help restore the worker's original physical condition and capabilities as much as possible or to permit the injured worker to follow a new occupation.

INJURY AND CLAIM NOTICES

When a serious injury takes place, notices must be provided to a number of organizations. A serious injury is the result of any accident in which a worker suffers a fatality, loss of a limb, or serious disfigurement or has to be hospitalized for more than 24 hours for other than medical observation.

The employer must immediately notify not only the insurance company but also the state agency which has jurisdiction for industrial safety and the federal Occupational Safety and Health Administration (see Chapter 4) or the state's counterpart. Generally, each insurer provides forms necessary for such reporting with instructions that indicate how they must be completed.

In most states an injured worker (or dependents) must wait for a prescribed period (from three days to a week) before monetary benefits begin. If the disability continues longer, compensation payments are usually retroactive to the date of the injury. No waiting period is required for in-patient hospital care, which may be provided from the time of the injury.

As mentioned earlier, statutes of limitations make it necessary that any claim for workers' compensation benefits be filed with the administrative agency within the time stipulated by law. This claim is in addition to the notice of injury and must be made unless excused by the concerned board or commission. Claim notice limits may range from six months to one year after death. In some instances there are special provisions for occupational diseases, because the exact date of injury may have been indeterminable.

HEARINGS

Most workers' compensation cases are settled routinely, without contest or disagreement. The insurer will satisfy itself that an accident occurred which had arisen out of and during the course of employment. Once the claim is filed, arrangements will be made for its settlement by one of three methods:

1. *Direct settlement system:* payments are begun by the insurer, subject to review by the administrative agency that the amount and duration are proper.
2. *Agreement system:* the employer and the employee reach an agreement before the latter can receive benefits. The agreement must then be approved by the state. If not approved, the employee is paid according to a temporary initial agreement, but a revised agreement later must be reached. If a revised agreement

cannot be arrived at, a decision will be made after a hearing by a workers' compensation board.

3. *Public hearing:* a commission hears the facts pertinent to a claim submitted by the injured person or survivors and determines and sets the benefits to be received.

A survey in California brought out that in two-thirds of the cases, benefit payments began within two weeks after notification of claims, and less than 20 percent took 29 days or more.

An injured worker or survivor who feels unjustly treated can request a hearing, for which no attorney is required. If the decision is still adverse to the worker or survivor, action can be taken through the courts. For this, an attorney may be used.

Because workers often do not realize the necessity for timely submission of claims or how poorly they may be compensated for a disabling injury, they may come to believe they have been unfairly treated by the employer, insurer, or hearing agency. Such uninformed workers may hire attorneys to bring suits for additional compensation. (All states except Nevada regulate the fees that can be charged by workers' compensation attorneys.) It is therefore advisable that employers endeavor to keep all workers informed of what to do after an injury occurs.

In July 1973, such an uninformed worker entered the office of his company's workers' compensation insurer and shot and killed two claims adjusters, critically wounded a secretary, and then committed suicide. He had been angered that the insurance company would make payments for his injury in small weekly amounts rather than in the one lump sum he wanted and believed was rightfully his.

In 1978 a man held four hostages for more than ten hours at the World Trade Center in New York. He claimed he had a bomb and had taken the hostages to protest the inadequacy of the workers' compensation payments he was receiving. His "bomb," which he threatened to use if his demands were not met, turned out to be four loaves of black bread in a paper sack.

ACTION AGAINST A THIRD PARTY

Under the workers' compensation laws, the employee gives up most of his or her rights to sue an employer for negligence. The employee does not give up the right to sue a third party. A few examples of the many types of third parties who could be (and at times have been) sued include: the manufacturer of the equipment that caused the injury; the driver of the vehicle that rammed the van in which the injured employee was making a delivery; the architect or engineer who designed the structure that collapsed and caused the injury; an inspection agency representative who certified as safe a pressure vessel that later burst.

For many years, because of the old rule of privity (see Chapter 2), workers injured by equipment their employers had purchased could not sue the manufacturer even if the equipment was badly designed or was hazardous in any way so that an accident resulted. Almost all awards won for injured persons were won by showing the manufacturer had been negligent or faulty in some way, generally because of either improper production or of improperly failing to warn the worker of the hazard involved in the item's use.

Although claims of negligence are still widely made, with the acceptance of the doctrine of strict liability by most states, suits against manufacturers by injured workers and other persons have increased rapidly. The Insurance Services Office points out that "workers injured on the job are involved in 11 percent of the product liability incidents resulting in claim payments. However, these incidents account for 42 percent of total bodily injury payments." More than 70 percent of the product liability claims and cases greater than $100,000 involve injuries to industrial workers. One hundred thousand dollars is approximately the average payment in an industrial product liability case; it is much higher than the overall average (about $14,000).

In 1986, because of the great number of claims, cases of litigation, and increased awards and premium costs, the federal government was asked to help with corrective action. A few states had enacted minor measures to control the liability problem, but the proposed action of the federal government was more pervasive. The first and principal proposal was for a "cap" for limiting the size of awards that could be made. Other proposals included limitations on punitive damages, on contributory damages to be paid for "pain and suffering," and on contingency fees for attorneys for the plaintiffs.

A young man in California lost eight fingers in a press that had a defective safety switch. Under the workers' compensation law current at that time in that state, a person who lost all his fingers in an industrial accident would receive slightly more than $40,000 plus a lifetime disability pension. As the outcome of the product liability suit in 1977, the young man was awarded $1.1 million to be paid by the press manufacturer's insurer.

In most states an injured employee covered by the workers' compensation act or the employee's survivors may not sue a fellow employee who caused the injury or death if the fellow employee was acting within the scope of his or her employment. In those states that do not specifically prohibit suits against fellow employees, the other employee can be sued if the death or injury was proximately caused by the willful and unprovoked act of aggression by the other employee; the intoxication of such other employee; or any act of such other employee which indicated a reckless disregard for the safety of the injured employee or a calculated and conscious willingness to permit another employee's injury or death.

In addition, a few states (Connecticut, Maryland, Massachusetts, Nebraska, and possibly Ohio) permit a fellow employee to be sued like any other third party for alleged responsibility for injury or death. Many outcomes might result in a hypothetical situation where a worker is injured or killed by a machine he or she was operating. Some of the complex legal ramifications involved might include (but not be limited to) the following points:

- The machine failed because of a negligently designed or manufactured component. The manufacturer could be held liable.
- On the other hand, the machine might have had a dangerous characteristic which the employee was aware of but did not report. In spite of the hazard the

worker continued to use the machine until an accident finally occurred, causing an injury. The manufacturer might not be held liable (assumption of risk). An attorney might then claim there was a failure to warn of the hazard in the use of machine.

- Or the employee may have reported the dangerous characteristic to a supervisor, who directed him or her to continue use of the machine because a production schedule had to be met. An accident resulted in which the employee was injured or killed. The manufacturer would generally not be held liable, but in those states that permitted it, the supervisor could be sued as a third party.
- Or the machine had a dangerous characteristic which the employee reported to the supervisor. The supervisor reported the problem to the plant engineering department, which added a safety device. However, in doing so they drilled holes in the machine which weakened it so that a failure occurred, injuring the employee. The machine manufacturer would not be held liable (unless it made the modification or provided faulty instructions on how it was to be done), nor would the supervisor. However, the plant engineer or the person who made the modification might be liable. (*Note:* Often, lawyers for plaintiffs will include in a suit everyone and all organizations that might even remotely be involved. To be exonerated, each defendant would have to show freedom from fault.)

If an accident occurred because of and during the course of employment, the workers' compensation insurer will make payments to the injured worker (or dependents) even if action is initiated against a third party. The worker or survivor may believe that a suit might bring more in an award than would workers' compensation benefits. In such a case the employee may transfer (subrogate) to the employer's insurer the right to sue the third party.

If the worker is awarded as much as or more than under workers' compensation, the insurer is reimbursed for payments already made. If the worker receives less than under workers' compensation, the insurer will continue to pay the worker's compensation.

When filing a suit, the worker or the worker's lawyer must notify the appropriate compensation agency, the employer, and the insurer that action has been instituted. The compensation insurer especially will probably assist the worker in the claim. This insurer will notify the defendant it has a lien on any amount the injured worker may be awarded until the sum it has paid has been reimbursed. Because of its interest, the insurer generally must agree to any out-of-court settlement the defendant may have attempted to make with the injured worker.

In an actual instance of a third-party suit, N. H. Siegel[15] described a case in which the driver for a company was killed in Connecticut. Connecticut is a state that permits fellow employees to be sued even when the company is covered by workers' compensation laws. The driver's widow sued three supervisors (who were also former friends of her husband). The company's workers' compensation insurer (and therefore the company) had to support her suit (to recover any losses under workers' compensation) against the employees the company was normally supposed to defend. In states with laws permitting suits against fellow employees, this anomalous situation would not be unusual.

INADEQUACY OF WORKERS' COMPENSATION

The University of Michigan reported in 1971:[15]

> . . . in the neighborhood of one-half million families a year are now left without income due to disabling occupational accidents, and those who are compensated receive far less in relation to their income than was the case 40 years ago.
>
> Because of an information system which leans heavily upon dramatic and publicized disasters to stimulate governmental progress, workmen's compensation in America continues to be a national disgrace.

Berkowitz and Burton[16] made a study of incomes disabled persons would receive. They pointed out that in total disability cases:

> In thirty-eight of the fifty-one states [includes District of Columbia], the maximum 1968 benefits (including any allowance for dependents) do not meet the poverty standard of living for the four person family. . . . It should be emphasized that the benefit levels are maximums, and many injured workers receive less. In no state would the minimum benefit level meet the poverty standard. . . . From 1968, the average state had a maximum benefit equal to only 54 percent of the state's average wage.

The National Commission of State Workmen's Compensation Laws[17] stated that in Louisiana a person who suffered a temporary-total disability would receive only 28.42 percent of the wages lost.

Also from Berkowitz and Burton: A 28-year-old man working as a "craftsman, foreman or kindred worker" would earn about $157,581 by the age of 64. If, however, he was injured so that he was permanently-totally disabled, his total income would be far less than that. In Pennsylvania, where he would qualify for a lifetime pension, he would receive $59,725. In New Jersey he would receive $85,607.

The situation could be even worse if he was partially disabled with a 50 percent disability. Theoretically, he could earn $78,791 from age 28 to 64. For his 50 percent disability he would receive only $13,565 in Pennsylvania and $9,753 in New Jersey.

This comparison was based on maximum benefits, and many workers or their families do not receive the maximum. Also, Pennsylvania and New Jersey are fairly liberal states. The explosion of the natural gas tank in New York (see Chapter 1) exemplified further the inadequacy of income benefits. An insurance industry paper[18] pointed out that each surviving widow of the 40 men killed would receive a maximum of $48 a week and a child $36. However, the maximum for a widow and child under New York law was $80 total. And further:

> Many states with inadequate compensation schedules plod along and continue to make nickel and dime changes in their benefit plans. New York, fortunately for those who lost a husband on Staten Island, is not one of those states. Benefits in New York State are among the most liberal in the country. . . . Occasionally though, we still think of the dependents of 29 workers who lost their lives in an explosion at the Thiokol Chemical Corp. plant in Georgia two years ago. For the most part, the wives and children of workers killed in that industrial accident must subsist on a maximum benefit level of just more than $40 a week for 400 weeks, or a total benefit of $17,000.[19]

It was pointed out[20] by one of the speakers at Park City, Utah, in 1981:

> The Workers' Compensation system was recognized as outdated and unresponsive. Other compensation systems, such as Social Security, are inappropriately bearing a significant cost of occupational disease. Benefits are woefully inadequate. The Workers' Compensation system does not provide to industry the necessary economic incentives to prevent occupational disease. . . . The prevention of occupational disease and injury has become increasingly important for industry. . . . The "hidden costs" of occupational disease and injury are well recognized as enormous, and are ultimately borne by all society.

The workers' compensation laws have helped somewhat to cover economic losses resulting from injuries and medical costs. However, they have failed in the original concept that the costs would lessen the numbers of accidents. To a great extent increased workers' compensation costs are simply passed on to the general public in the form of higher prices, and much of any improvement in workplace safety has been due to decisions in the common law, liabilities and fears of future ones, and enactment and execution of more stringent statute laws.

BIBLIOGRAPHY

[10] H. M. Somers and A. R. Somers, *Workmen's Compensation* (New York: John Wiley & Sons, Inc., 1945).

[11] C. Eastman, *Work Accidents and the Law* (New York: New York Charities Publication Committee, 1910), p. 1783.

[12] D. Gagliardo, *American Social Science* (New York: Harper & Row, Inc., 1949).

[13] Texas Sunset Advisory Commission, "Texas Workers' Compensation System Report," 6 December 1996 revision, 6 pp.

[14] American Express Company, "Keeping Worker's Compensation Costs Down," 1997 (http://www.aexp.com/smallbusiness/resources/managing/workcomp.shtml).

[15] N. H. Seigel, "The Fellow-Employee Suit: A Study in Legal Futility and Misguided Legislation," *Risk Management Today* (New York: American Management Association, 1960).

[16] Monroe Berkowitz and J. F. Burton, Jr., "The Income Maintenance Objective in Workmen's Compensation," *Industrial and Labor Relations Review,* Oct. 1970, pp. 3 ff.

[17] The National Commission on State Workmen's Compensation Laws in *Compendium on Workmen's Compensation,* Washington, DC, 173, p. 103.

[18] *Business Insurance,* Feb. 26, 1973, p. 1, col. 4.

[19] H. W. Heinrich, *Industrial Accident Prevention,* 3d ed. (New York: McGraw-Hill Book Company, 1950), p. 18.

[20] J. S. Lee and W. N. Rom, eds., *Legal and Ethical Dilemmas in Occupational Health* (Ann Arbor, Mich.: Ann Arbor Science Publishers, 1982).

EXERCISES

3.1. Discuss each of the six stated objectives expressed by the U.S. Chamber of Commerce and how well each has been met in actual practice.

3.2. What are the desirable characteristics that a compensatory system, such as workers' compensation, should have?

3.3. What three types of workers' compensation are used? Which is used in your state?

3.4. What are the three fundamental requirements for benefits under workers' compensation?

3.5. What is the legal definition of an accident? When would an injury not be considered an accident and an industrial worker not be covered by workers' compensation?

3.6. Describe the various types of disability.

3.7. Describe the benefits an injured worker can obtain under workers' compensation. Do those benefits appear to be adequate in your state?

3.8. Cite two examples of inequities between two states in benefits that workers could receive for the same permanent-partial disability.

3.9. How could an action against a third party arise?

3.10. How can an employer be sued if there is coverage under workers' compensation?

3.11. Describe some efforts to reform workers' compensation.

3.12. Describe some ways to keep workers' compensation costs down.

3.13. How are insurance company rates for workers' compensation coverage computed?

3.14. What are the inadequacies of workers' compensation? How can the inadequacies be eliminated?

CHAPTER 4

OSHAct and Its Administration

In most states, industrial safety laws, regulations, and codes and their administration had at best been spotty and inadequate to prevent great numbers of accidents. Because of the increasing costs of premiums for workers' compensation and litigation, some companies became more safety conscious; but after a few years the accident rate eased declining and then began rising again.

The inadequacy of administration of the industrial safety laws in California—one of the more stringent states—is shown by the Sylmar disaster in 1971. (See Fig. 1-3.) The State Assembly appointed a Select Committee on Industrial Safety to investigate the accident. Some of the committee's findings were reported in the *California Journal* (Feb. 1972). The following are a few extracts:

> The principal fact brought out in the hearings and related investigations is that the Division of Industrial Safety has increasingly sought "voluntary compliance" on the part of construction contractors, emphasizing "education" rather than strict enforcement of its safety orders. Although the legislative investigators pointed out repeatedly the division's first responsibility is the protection of workers, its enforcement policies in the construction section appear to have become strongly influenced by a desire to inconvenience contractors as little as possible. It had become the division's policy, in fact, that safety inspectors in the field were expected to issue letters to contractors listing any violations of state safety orders, rather than to put a "red tag" which might stop part of a project until the violation was corrected. In many cases this gave the contractors several weeks or a month before they had to worry about complying with the orders, as the inspectors were unlikely to return sooner. Similarly, the inspectors were expected not to use red tags which shut down the whole project. The guiding principle in the division, according to the testimony received by the Assembly committee, is that the contractors will take all necessary precautions for the safety of employees as they are informed of these precautions.
>
> This view is not widely shared by the individual safety engineers in the field, however. Experience has taught them that many contractors give little attention to safety precautions and will too often ignore state safety regulations if they are permitted to do so through lax enforcement.

> As a result of investigations conducted by the committee staff, it was brought out that there were a number of major construction projects in recent years in which contractors apparently had been permitted to ignore state safety regulations with impunity, with the result that there were a number of disabling injuries and fatalities which might otherwise have been avoided.

NEW CONCEPTS OF ACCIDENT AVOIDANCE

There were many ideas on the reasons why accidents had occurred, who was responsible for having caused them, and the corrective measures that should be taken. For example, until a few years ago it was considered that if a person was involved in an accident while he or she was exercising his or her responsibilities, it was probably the person's fault. Heinrich[19] had estimated that 88 percent of 75,000 accident cases reviewed were due to unsafe acts of persons. Similarly, when aircraft were involved in accidents, too often they were blamed on pilot error although the fundamental cause lay elsewhere. Development of the new concept of accident causes and what to do about them came from an unlikely source: ballistic missiles.

When ballistic missiles built for the Air Force were launched and then crashed with no pilot aboard, it was evident there was no operator error and the cause of any failure lay elsewhere. The fault must have been in design, manufacturing, or some other cause. A new engineering discipline, system safety, grew out of the idea. The concept was that accidents could be caused by an error not only of the operator but of personnel such as the designer, manager, manufacturing worker, maintenance man, or anyone else connected with the system. There might also have been an equipment or environmental deficiency. Not only did all potential hazards that might cause accidents have to be studied before the product was built or the system put into operation, but it was necessary to eliminate the hazards or control those that couldn't be eliminated. Thus not only was it necessary to ensure at the proper time and stage that the design was a safe one, but there had to be proper manufacture under good quality control, workers and operators had to be suitably trained, and maintenance had to be performed by knowledgeable persons when required. If Heinrich's data were reexamined in accordance with these newer ideas, it might be found that 88 percent of the accidents were probably due to causes other than error by persons immediately involved. Another new contention is that accidents are caused by unsafe conditions.

Gradually these new concepts of hazards and accident causes brought forward the idea of preventing injuries and damage by better prevention measures, especially through use of more safely designed and manufactured equipment. This idea was adopted in areas other than missiles and aircraft but, unfortunately, has seen only limited use, because entrenched ideas are difficult to discard even when they are wrong.

One result in the industrial and occupational safety area was the OSHAct, an attempt to reduce the number and severity of accidents by making equipment and procedures safer by mandatory means. Fundamentally, the OSHAct and the standards that were developed as a result attempt to ensure that employers observe the old common law obligations to their employees by providing a safe place to work, safe tools with which to work, and suitable work rules. In addition, the OSHAct requires information on hazards; safe designs for equipment and facilities; indicates penalties for nonobservance by employers of the standards; and includes other stipulations, guidance, and information. The fundamental aim of the OSHAct is to ensure "so far as possible every working man healthful working conditions and to preserve our human resources" (see Fig. 4-1).

Aims of the Occupational Safety and Health Act of 1970 can be achieved:

(1) by encouraging employers and employees in their efforts to reduce the number of occupational safety and health hazards at their places of employment, by stimulating employers and employees to institute new and to perfect existing programs for providing safe and healthful working conditions;

(2) by providing that employers and employees have separate but dependent responsibilities and rights with respect to achieving safe and healthful working conditions;

(3) by authorizing the Secretary of Labor to set mandatory occupational safety and health standards applicable to businesses affecting interstate commerce, and by creating an Occupational Safety and Health Review Commission for carrying out adjudicatory functions under the Act;

(4) by building upon advances already made through employer and employee initiative for providing safe and healthful working conditions;

(5) by providing for research in the field of occupational safety and health, including the psychological factors involved, and by developing innovative methods, techniques, and approaches for dealing with occupational safety and health problems;

(6) by exploring ways to discover latent diseases, establishing causal connections between diseases and work in environmental conditions, and conducting other research relating to health problems, in recognition of the fact that occupational health standards present problems often different from those involved in occupational safety;

(7) by providing medical criteria which will assure insofar as practicable that no employee will suffer diminished health, functional capacity, or life expectancy as a result of his work experience;

(8) by providing for training programs to increase the number and competence of personnel engaged in the field of occupational safety and health;

(9) by providing for the development and promulgation of occupational safety and health standards;

(10) by providing an effective enforcement program which shall include a prohibition against giving advance notice of any inspection and sanctions for any individual violating this prohibition;

(11) by encouraging the States to assume the fullest responsibility for the administration and enforcement of their occupational safety and health laws by providing grants for these States to assist in identifying their needs and responsibilities in the area of occupational safety and health, to develop plans in accordance with the provisions of this Act, to improve the administration and enforcement of State occupational safety and health laws, and to conduct experimental and demonstration projects in connection therewith;

(12) by providing for appropriate reporting procedures with respect to occupational safety and health, which procedures will help achieve the objectives of this Act and accurately describe the nature of the occupational safety and health problem;

(13) by encouraging joint labor-management efforts to reduce injuries and disease arising out of employment.

FIGURE 4–1 Aims of OSHAct of 1970

ENACTMENT OF THE OCCUPATIONAL SAFETY AND HEALTH ACT (OSHAct)

In December 1970 Congress enacted and President Richard Nixon signed the Williams-Steiger Occupational Safety and Health Act (OSHAct), which became effective on April 28, 1971. It applied to more than 5 million businesses involved in or affecting interstate commerce and 60 million workers. Any organizations covered by another federal agency such as the Atomic Energy Act of 1969 and the Coal Mine Safety Act of 1969 were exempted.

As a result of the new act, existing federal organizations were enlarged or transferred or new ones were created. Such a new one was the Occupational Safety and Health Administration (OSHA), which was put into the Department of Labor to set safety and health standards for nongovernmental employers. On the other hand, government agencies such as the Department of Defense and the Post Office Department became covered by OSHA after being directed by executive order that the provisions and standards of the OSHAct would be observed. The observance of all industrial plants, facilities, and related structures, with the operations they include, was to be ensured by visits by inspectors (OSHA Compliance Officers) located throughout the country. Where there was a lack of compliance in any plant, a citation would be issued which would indicate the standard not properly observed, the time within which corrections were to be made, and the penalty recommended.

OTHER ORGANIZATIONS

Another organization was the Occupational Safety and Health Review Committee (OSHRC), a quasi-judicial board composed of three members appointed by the President. Functions of the OSHRC are to hear and review alleged violations and, where warranted, to issue penalties and corrective orders. An accused employer can contest any citation, proposed penalty, or time stipulated to eliminate or abate a hazard.

The OSHRC reviews the case and decides its validity. Where elimination or abatement of a cited hazard cannot be accomplished within the allotted time, a reasonable extension may be granted. The employer must indicate that an effort has been made in good faith but that factors causing the delay are beyond the employer's control.

A third organization was the National Institute for Occupational Safety and Health (NIOSH), which is within the Department of Health and Human Services (DHHS). Elevated from what was once a lower-graded bureau, NIOSH carries out research and education functions. It is authorized to conduct research to permit determination of tolerance levels of hazardous substances and conditions that can be present in workplaces without deleterious effects within indicated times. NIOSH prepares recommendations to the Secretaries of DHHS and Labor, which, if adopted, would become provisions of the OSHA standards.

To assist in making these studies and developing any findings and recommendations, NIOSH may issue regulations requiring employers to measure, record, and make reports on employees' exposures to substances or physical agents that endanger health or safety. Financial assistance may be provided to employers who incur any added expense in carrying out these requirements. Costs of medical examinations and tests required for such programs can also be defrayed by NIOSH.

NIOSH will annually publish lists of substances with which toxic effects are associated; when requested by an employer or employees' representative, determine whether any substance found in the workplace has potentially adverse effects; conduct studies on industrial processes, substances, and stresses to see their potential effects on personnel, and publish these studies.

STATE INDUSTRIAL SAFETY PROGRAMS

As mentioned previously, one reason the OSHAct was passed was the inadequacy of existing state laws or their enforcement. Although the new law was a federal one, it permitted individual states to regain sole authority to police occupational safety if specific conditions were met. A diagrammatic relationship between OSHA, workers' compensation, and product liability is shown in Fig. 4-2.

A state that wants to regain control over industries within its borders that are covered by the OSHAct has to submit a proposed plan indicating how it intends to carry out a program at least as effective as the federal one. The plan has to be approved by the Secretary of Labor. The approved states agree to pay half the cost of its program and the federal government the other half. The Department of Labor must maintain surveillance over the states for three years to ensure that their safety responsibilities are carried out. Of the fifty states, half passed legislation with such plans and programs.

The enabling legislation and standards of the individual states sometimes turned out to be more stringent than those of the federal government. The reason may have been previous accident experience, types of activities in the workplace, or a desire to provide greater safety for their citizens than do the OSHA standards. The only consistency prevails where state programs were not approved and so the provisions of the federal OSHA remained in effect. State programs that were put into effect differ between states. Although those differences are now not great, standards have to be checked for specific requirements.

For example, in California the Sylmar disaster gave a decided impetus to the later approval of a state law, leading to Cal-OSHA. The standards and penalties in Cal-OSHA were based on years of experience with the state's safety codes. As a result, in the event of a serious accident an employer could be jailed under criminal law. Cal-OSHA also makes violation of a safety or health standard an offense applicable to any "employee having direction, management, control, or custody of any employment, place of employment, or other employee."

Some state provisions, such as those originating with Cal-OSHA, for example, that have resulted in better worker safeguards have included:

1. A stronger "general duty" clause than that of the federal OSHA standards, with quicker responses to any report of violations.
2. Higher maximum civil penalties that can be imposed on employers in the event of noncompliance, with more prosecutions.
3. More frequent plant inspections.

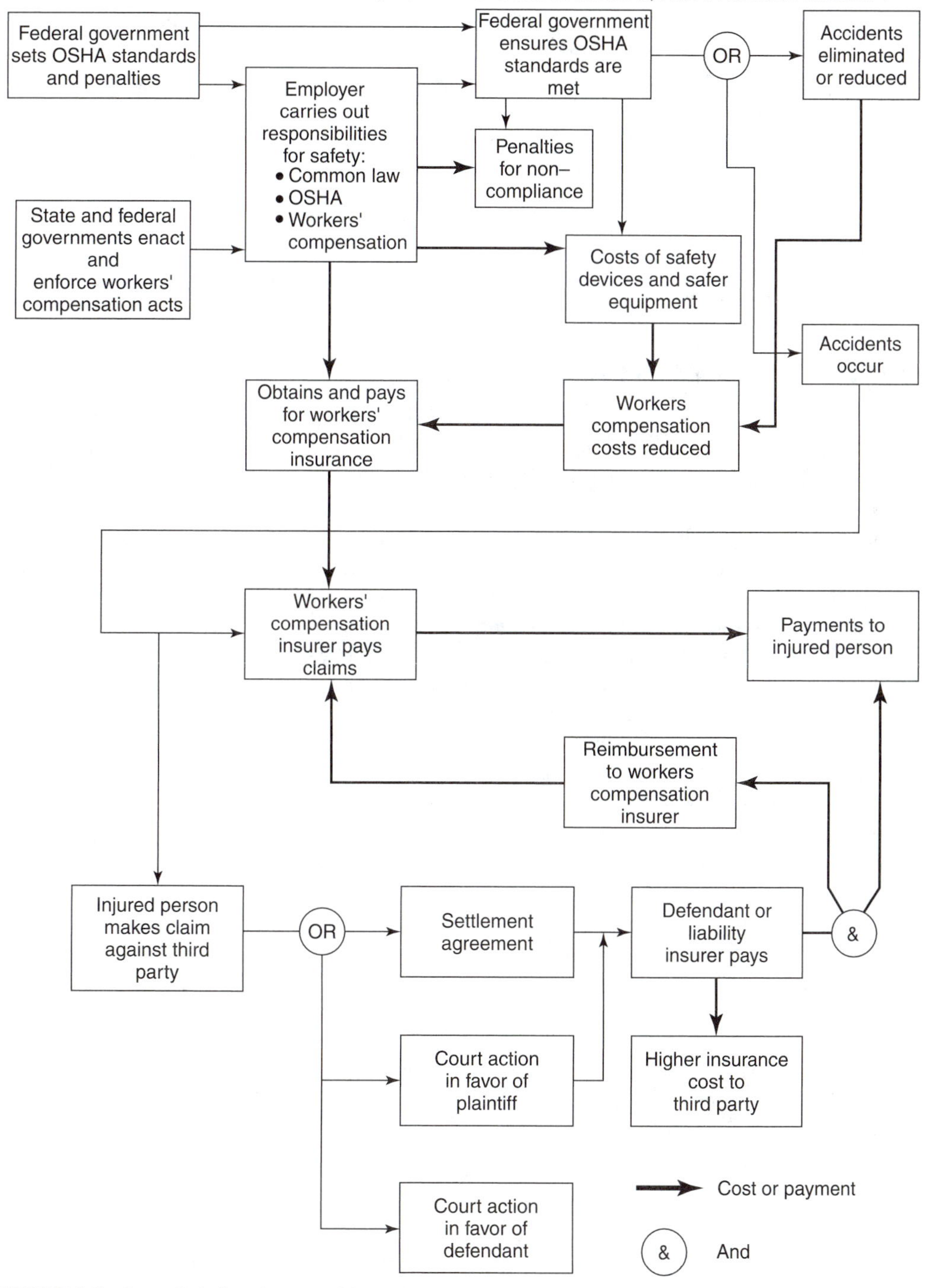

FIGURE 4–2 Some Relations between OSHA, Workers' Compensation, and Product Liability

4. The state has the authority (which the federal law has not) to order an employer to stop using a piece of equipment it has determined to be an imminent hazard.
5. State codes include requirements not present in the federal standard.
6. The state provides for employer consulting services and assistance that the federal government does not.

Other types of cooperation are maintained with other federal including the Nuclear Regulatory Commission, the Environmental Protection Agency (EPA), and the Consumer Product Agency. For example, a state will control problems of atmospheres inside a plant; the EPA of those outside. Because the requirements of the 24 states that have local laws differ, provisions of the federal law will be used.

RESPONSIBILITIES OF EMPLOYERS AND EMPLOYEES

Both employers and employees have responsibilities they must carry out under the act, although only employers can be penalized for failure to comply as or where necessary. The fundamental requirement of the employer is similar to that under common law: to furnish each employee a place of employment that is free from recognized hazards that are causing or likely to cause death or serious physical harm. In addition, the employer must comply with standards for injury avoidance; keep records of work-related injuries, illnesses, and deaths; and keep records of exposure of employees to toxic materials and harmful physical agents.

Employers must also notify employees of the provisions of the law, their protections and obligations. They must keep employees informed on matters of safety and health and on accidents and alleged safety violations in the place of employment. The employer must refrain from discriminating against employees who file complaints regarding hazardous working conditions.

The employee is obligated to "comply with occupational safety and health standards and all rules, regulations, and orders issued pursuant to this act which are applicable to his own actions and standards." In addition, the employee or representative may file complaints with the Department of Labor or with a Compliance Officer (see below), who may inspect the establishment. An employee may accompany the Compliance Officer during an inspection and may submit recommendations on new standards to be promulgated.

In the event an employee fails to observe the prescribed safety and health standards, the employer may be cited for a violation. For example, a sheet metal company was cited and fined for violations by workers which included: not wearing protective goggles or shields, poor housekeeping, oxygen cylinders lying near gas-filled cylinders, gas cylinders whose valves were not closed while not in use, and improper equipment grounding.

INSPECTIONS

To ensure compliance with provisions of OSHAct (or state codes that replace them), inspections originally could be made without prior notice by the Compliance Officer. Compliance Officers must be admitted and have the right to inspect safety records.

Generally, there was little problem in gaining access to a facility, but where an employer did deny permission, Compliance Officers could ask for a suitable court order enabling them to enter. Inspections were made where and when considered advisable by the responsible agency. In general, the following priorities were to be observed in making inspections:

1. An inspection would follow any accidental death or mishap in which five or more workers were injured. Under the law, such accidents must be reported within 72 hours of occurrence. In some states, a death or multiple serious injuries indicated the possibility of an imminent hazard.
2. A plant will be inspected to investigate a report of an imminent hazard, and to make certain that imminent hazards which have been noted have been eliminated.
3. Industries which are themselves considered especially hazardous will be inspected at frequencies and times determined by the cognizant OSHA office. The especially hazardous industries, those with high injury-frequency rates, include longshoring, lumber and wood products, and mobile homes and other transportation equipment.
4. Other industries will also be inspected on schedules to be established by OSHA offices.

VIOLATIONS AND PENALTIES

A sevenfold increase in the maximum limits for OSHA civil monetary penalties was stipulated in the Budget Reconciliation Act passed by the 101st Congress. It became effective after March, 1991. The maximum allowable penalty is $70,000 for each willful or repeated violation, and $7,000 for each serious or other-than-serious violation as well as $7,000 for each day beyond a stated abatement date for failure to correct a violation.

These amounts are ceilings—not floors. However, in order to ensure that the most flagrant violators are in fact fined at an effective level, a minimum penalty of $5,000 for a willful violation of the OSHAct was adopted.

The policy also applies to those states with OSHA-approved state occupational safety and health programs, under the congressional direction that these state plans must be "at least as effective" as the national plan. The participating states are being given a reasonable period to implement the new penalty structure, taking into account the states' legislative calendars.

The basic penalty process follows the criteria set forth in the Occupational Safety and Health Act, which is to determine penalties based on the gravity of the violation and the size, good faith, and history of the employer. Gravity determines the base amount; the other factors determine appropriate reductions.

All penalty amounts are proposed penalties issued with the citation. The employer may contest the penalty amount as well as the citation within the statutory 15-day contest period. Thereafter, the penalty may be adjudicated by the independent Occupational Safety and Health Review Commission, or OSHA may negotiate with the employer to settle for a reduced penalty amount if this will lead to speedy abatement of the hazard. The bases for proposing a penalty are described in Fig. 4-3.

Here is how the system for proposing penalties operates.

ADJUSTMENT FACTORS:

The size adjustment factor is as follows: For an employer with only one to 25 workers, the penalty will be reduced 60 percent: 26 to 100 workers, the reduction will be 40 percent; 101 to 250 workers, a 20 percent reduction; and more than 250 workers, there will be no reduction in the penalty.

There may be up to an additional 25 percent reduction for evidence that the employer is making a good faith effort to provide good workplace safety and health, and an additional 10 percent reduction if the employer has not been cited by OSHA for any serious, willful or repeat violations in the past three years.

In order to qualify for the full 25 percent "good faith" reduction, an employer must have a written and implemented safety and health program such as given in OSHA's voluntary "Safety and Health Management Guidelines" (Federal Register, Vol. 54, No. 16, Jan 26, 1989, pp. 3904–3916) and that includes programs required under the OSHA standards, such as Hazard Communication, Lockout/ Tagout or safety and health programs for construction required in CFR 29 1926.20.

SERIOUS VIOLATIONS:

The typical range of proposed penalties for serious violations, before adjustment factors are applied, will be $1,500 to $5,000, although the Regional Administrator may propose up to $7,000 for a serious violation when warranted.

A serious violation is defined as one in which there is substantial probability that death or serious physical harm could result, and the employer knew or should have known of the hazard.

Serious violations will be categorized in terms of severity — high, medium or low — and the probability of an injury or illness occurring — greater or lesser.

Base penalties for serious violations will be assessed as follows:

Severity	Probability	Penalty
High	Greater	$5,000
Medium	Greater	$3,500
Low	Greater	$2,500
High	Lesser	$2,500
Medium	Lesser	$2,000
Low	Lesser	$1,500

Penalties for serious violations that are classified as both high in severity and greater in probability will only be adjusted for size and history.

OTHER-THAN-SERIOUS VIOLATIONS:

If an employer is cited for an other-than-serious violation which has a low probability of resulting in an injury or illness, there will be no proposed penalty. However, the violation must still be corrected. If the other-than-serious violation has a greater probability of resulting in an injury or illness, then a base penalty of $1,000 will be used, to which appropriate adjustment factors will be applied.

The OSHA Regional Administrator may use a base penalty of up to $7,000 if circumstances warrant.

FIGURE 4–3 OSHA Penalty Policy

REGULATORY VIOLATIONS:

Regulatory violations involve violations of posting, injury and illness reporting and record keeping requirements, and not telling employees about advance notice of an inspection. OSHA will be applying adjustments only for the size and history of the establishments.

Here are the base penalties, before adjustments, to be proposed for posting requirement violations: OSHA notice, $1,000; annual summary, $1,000; and failure to post citations, $3,000.

Base reporting and record keeping penalties are as follows: Failure to maintain OSHA 200 and OSHA 101 forms, $1,000; failure to report a fatality or catastrophe within 48 hours, $5,000 (with a provision that the OSHA Regional Administrator could adjust that up to $7,000, in exceptional circumstances); denying access to records, $1,000; and not telling employees about advance notice of an inspection, $2,000.

WILLFUL VIOLATIONS:

In the case of willful serious violations, the initial proposed penalty has to be between $5,000 and $70,000. OSHA calculates the penalty for the underlying serious violation, adjusts it for size and history and multiplies it by 7. The multiplier of 7 can be adjusted upward or down at the OSHA Regional Administrator's discretion, if circumstances warrant. The minimum willful serious penalty is $5,000.

Willful violations are those committed with an intentional disregard of, or plain indifference to, the requirements of the OSH Act and regulations.

REPEAT VIOLATIONS:

A repeat violation is a violation of any standard, regulation, rule or order where, upon reinspection, a substantially similar violation is found.

Repeat violations will only be adjusted for size, and the adjusted penalties will then be multiplied by 2, 5, or 10. The multiplier for small employers — 250 employees or fewer — is 2 for the first instance of a repeat violation, and 5 for the second repeat. However, the OSHA Regional Administrator has the authority to use a multiplication factor of up to 10 on a case involving a repeat violation by a small employer to achieve the necessary deterrent effect.

The multiplier for large employers — 250 or more employees — is 5 for the first instance of repeat violation, and 10 for the second repeat.

If the initial violation was other-than-serious, without a penalty being assessed, then the penalty will be $200 for the first repetition of that violation, $500 for the second repeat, and $1,000 for the third repeat.

FAILURE TO ABATE:

Failure to correct a prior violation within the prescribed abatement period could result in a penalty for each day the violation continues beyond the abatement date.

In these failure to abate cases the daily penalty will be equal to the amount of the initial penalty (up to $7,000) with an adjustment for size only.

This failure to abate penalty may be assessed for a maximum of 30 days by the OSHA Area Office. In cases of partial abatement of the violation, the OSHA Regional Administrator has authority to reduce the penalty by 25 percent to 75 percent.

If the failure to abate is more than 30 days, it may be referred to the OSHA national office in Washington where a determination may be made to assess a daily penalty beyond the initial 30 days.

FIGURE 4–3 Continued

STANDARDS

The OSHAct empowered the Secretary of Labor to set safety and health standards. Until April 1973 the Secretary could stipulate unilaterally any standard it deemed advisable, including making mandatory any standard produced by an existing organization on a voluntary basis. After that date, open hearings had to be held on the issuance of proposed new or revised standards or on those already existing.

To meet the April 1973 cutoff date, the first version was therefore a hastily prepared assembly of standards taken to a great extent from those upgraded from the existing consensus standards plus any new provisions the new OSHA safety personnel considered advisable. Consensus standards included those issued by such organizations as the American National Standards Institute (ANSI), the National Fire Protection Association (NFPA), American Society of Mechanical Engineers (ASME), and many others. These standards were prepared as recommended guidelines to be observed voluntarily as deemed necessary by industry. Some of these had become legally mandatory within a political jurisdiction when a state or city required that the provisions of certain ones (for example, the ASME Code for Pressure Vessels) be observed. Many of these had also become economically mandatory where insurance companies raised their rates prohibitively where there was nonobservance. Other standards or requirements, such as those called out in the Walsh-Healy Act, had already been mandatory. (The Walsh-Healy Act requirements were imposed on any company that received a federal contract valued at $10,000 or more.)

The consensus standards developed and adopted by industry contained some provisions not looked on favorably by everyone. Some companies believed they were inadequate while they elected to ignore some of them. Inclusion of controversial provisions in the OSHA standards caused some persons in industry who did not like them to protest. As a result, out of an estimated 5,000 provisions, in October 1978 OSHA eliminated 928 rules branded as "nitpicking." Modifying the OSHA standards has been a continuing process to eliminate unduly costly and restrictive provisions, political pressures, or requirements outmoded by modern technology. Changes will also occur as experience indicates they are needed.

RECORD KEEPING

The OSHA standards require the keeping of numerous records by each employer. The cost required for preparation of these records was one of the most strenuous objections against the adoption of the OSHAct. Most records are maintained at the place of employment, where they should be available for inspection and use by the Compliance Officers or inspectors during any visits they might make there. (There are penalties for failure to comply.)

IMPACT OF OSHA

Some argue that the passage of the Occupational Safety and Health Act of 1970 has had minimal impact on safety. They contend that the trend to fewer accidents that occurred after 1970 was well established before then. The argument is supported by

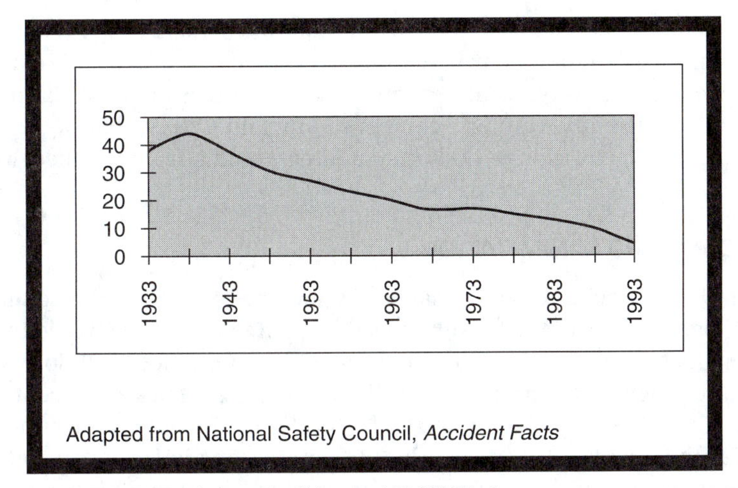

FIGURE 4–4 Workplace Fatalities per 100,000 Workers

the workplace fatality rates as shown in Fig. 4-4. This graph becomes less persuasive when we realize that in 1995, 41.2 percent of workplace fatalities were caused by transportation incidents and 20.3 percent by assaults and violent acts. OSHA has only recently begun to address workplace violence, and much of transportation safety is not their jurisdiction.

Bureau of Labor Statistics data indicates that accidents and injury rates are much higher for small companies than for large ones. The greatest number of complaints have come from companies having the highest injury rates—companies having between 20 and 250 employees. They have contended that the requirements imposed to meet the standards and to maintain the required records are economically unbearable and have petitioned Congress for relief. In addition, these small plants had courts issue orders restraining OSHA personnel from entering their plants without court orders. The contention was that if a Compliance Officer enters a plant without the owner's permission, rights under the Fourth Amendment against unreasonable searches are being violated. On the other hand, some companies welcomed inspections because they were reviews that might point out deficiencies missed by their own safety personnel.

Practically, no matter what the legal contention, it is because of the costs that companies, especially the smaller ones, have objected to the OSHAct and OSHA standards. Bringing equipment up to the levels required involves expenditures these employers do not feel are economically justifiable. Costs can be for (1) modifying or replacing existing equipment; (2) new equipment or added equipment; (3) record keeping; and (4) fines for violations for failures to properly observe standards. The first item is a first-time cost. The second forces employers to buy only approved equipment and facilities built that meet the OSHA standards. Thus it forces equipment manufacturers to incorporate certain safety features if employers are to buy and use them. They will have no desire either to pay to upgrade equipment or to buy new equipment for which they will find themselves in violation of the OSHA standards. Equipment manufacturers and their designers must therefore be familiar with and meet the OSHA standards if they expect to find markets in this country. The responsibility for the unsatisfactory equipment is the plant owner's; in Europe the onus for adequate and safe equipment is on its manufacturer. Thus, there is an interrelation between a manufacturer's liability, workers' compensation, and OSHA. Figure 4-2 indicates how the successful application of OSHA standards would affect the other two considerations.

OSHA AND HAZARD MINIMIZATION

As OSHA went into effect, magazines and newspapers contained numerous articles on the penalties that might result. It was thought that not only would the magnitude of the penalties have the effect of providing safer workplaces, but so would the possibility that employers themselves might be fined. (In most cases, except as noted later, the penalties proved to be minimal.) In addition, a few of the provisions of the OSHAct (not the standard) were overlooked, or the employers were uncertain about their interpretation or the degree or depth to which the standards would be applied.

The extent to which the OSHAct would be used to minimize hazards that had formerly been accepted in the workplace was early illustrated in 1972 when an employer was charged with a violation under the "general duty" requirement of the Act. Under that requirement, each employer must provide a workplace free of recognized hazards that are likely to cause serious physical harm or death. The company was said to have exposed employees to airborne inorganic lead dust in amounts in excess of those stipulated in the standard.

The company had required employees who might be exposed to the hazard to use respirators and to have blood and urine tests periodically. The tests would indicate whether the safeguards to prevent inhalation and absorption of the lead were adequate. When a test showed that an employee had an excess of lead in the blood or urine, he or she was transferred to a job where there was no exposure to the hazard. However, although the company had prescribed the use of the respirators, tests had shown there was an inadequacy, either in the equipment itself or in the workers' failure to use the inconvenient air masks. The OSHA Compliance Officer cited the company for failing to provide a safe workplace because of the hazard in the environment.

The company appealed the citation, but the appeal was denied. According to the findings, the presence of lead in the blood indicated that the company had not effectively eliminated the hazard. The need for respirators had put the burden for accident prevention on the employees rather than on the employer. The company was fined $600 and required to adopt engineering controls to reduce lead levels to an acceptable safe level within six months. The case was an early indicator of the government's attitude that hazards should be prevented because penalties would be increasingly costly (see Fig. 4-5).

ANTAGONISM TOWARD OSHA

Unhappily, progress formerly being achieved in worker safety by the OSHA has been retarded. In 1981, President Reagan issued an executive order requiring that no regulatory action be "undertaken unless potential benefits to society from the regulation

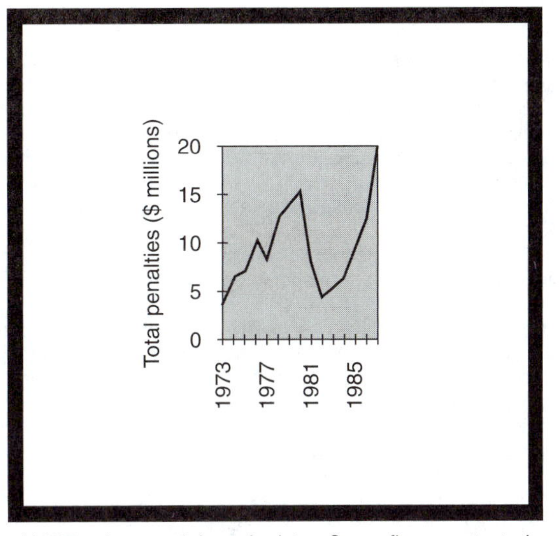

*1987 only a partial year's data. Some fines contested.

FIGURE 4–5 OSHA Fines Levied, 1973–1987

outweigh the potential costs to society." The action of the Administration to revoke, modify, or ignore provisions of the OSHA standards was based on the premise that they are inflationary. This action indicated a lack of understanding of the billions lost because of accidents (see Fig. 1-2). Opponents of the government policy pointed out that the persons most affected would be the lower and middle class of the public: workers and their families. The policy would result in the loss of safety in industrial plants and other facilities and activities.

Because of the executive order, cost-benefit or cost-effectiveness analyses became necessary before any new standards could be promulgated or existing ones modified (see Fig. 4-6). The result did much to stop the initiation of further safety and health standards. However, in 1987, the Supreme Court ruled that considerations of safety benefits outweighed savings in costs. Wherever possible, reduction in possible injury or death would govern any instance.

For many reasons, company estimates of such cost-benefit analyses have almost invariably been wrong. It will be pointed out in a later chapter that few engineers have ever had any education in accident prevention, because it is not taught in most engineering schools. Therefore they know too little about safety to be able to make proper determinations of the factors involved in any such analysis. Accordingly, costs of proposed safety actions are generally estimated by employers to be far too high and benefits far lower than they actually should be. Such estimates are usually highly biased, inaccurate, unreliable, and highly questionable. For example, when the OSHA standard recommended by NIOSH for reduction in the amount of vinyl choride permissible in the air was being reviewed years ago, spokespersons for the plastics industry estimated that the cost for compliance would be $90 billion. In spite of that, the new standard was issued under a prior president. Seven years later, the plastics industry

Problems with safety have come a long way since Hammurabi decreed "an eye for an eye" where accidents caused injuries. Over the many years since, it has gradually developed that persons injured have been indemnified by money payments rather than payments in kind. Generally, most safety considerations are evaluated in financial terms. All economic analyses include considerations and comparisons among four aspects:

Costs: the economic quantification of safety and of actual or potential accident losses, which generally include outlays required to produce safer equipment or operations.

Risks: judgments between economic cost evaluations of two or more alternatives, such as whether an accident preventive measure should be taken or which of several preventive measures should be chosen.

Benefits: (a) indicates the excess between how much an employer might be liable and how much might be gained or lost if no action or a different action were taken; (b) for both workers and employers, the lessened amounts in accidents, injuries, deaths, or other considerations, due to reductions in mishaps and savings in injuries, deaths, and monetary losses.

Effectivity: the ability to achieve a specific goal at the lowest cost.

Generally, with increases in well-designed and managed accident preventive measures, benefits and effectiveness will increase while ultimate costs and risks will go down. On the other hand, with no improvements where required or with decreases in effectiveness of accident prevention, risks and benefits will be adversely affected. Even with constant numbers of risks and accidents, monetary losses will increase because of the inflationary growth of costs of injuries, deaths, damages, and related monetary considerations.

FIGURE 4–6 Costs/Risks/Benefits/Effectivity

admitted that total compliance costs had been only $300 million. Evidently the estimated cost of compliance had been greatly exaggerated and highly inaccurate.

The Reagan-Bush Administration also attempted to abandon engineering controls as OSHA's preferred method of providing safety in the workplace, compared to supposedly cheaper but less desirable methods requiring workers to act safely or to observe stipulated safe procedures or warnings as the safeguards. Such was the situation concerning the case that arose in court regarding the use of respirators in cotton mills to eliminate lung problems of workers. There, too, as had happened with the inhalation of lead dust, the respirators had proved to be an unsatisfactory means of controlling the hazard. The Court of Appeals decided that evidence existed that respiratory equipment was inadequate compared with the far greater superiority of engineering controls (safe designs) in eliminating hazards. The U.S. Dept. of Labor found that the costs of the disease [byssinosos caused by cotton dust]—$7.51 billion—were more than 10 times the estimated cost of the cleanup, which was $655 million.

Catastrophes such as that in Bhopal and lesser ones in the United States made the public apprehensive of industrial safety. Because of the deterioration in lack of presidential support and the increase in workplace accidents, the number of such accidents or failures of companies to observe standards and reporting requirements increased. OSHA began imposing more massive fines as a result.

In April 1986, the largest fine to date imposed by OSHA, $1,377,000, was ordered against the Union Carbide Company (owner of the plant in Bhopal, India, where more than 200 persons had been killed by a leak in 1984). The fine was for 221 "conscious, overt, and willful" violations at the company's facility in Institute, West Virginia.

Among the deficiencies listed were 81 failures to list worker injuries serious enough to result in time off the job between 1983 and January 1985. In addition there were 91 other citations included among the total number of safety violations indicated. (Union Carbide finally agreed to pay $48,500.)

That fine was soon exceeded in 1987 by one of $2.6 million against IBC (meat packing). Similar large fines were imposed on the Chrysler Corporation ($1.5 million) and a number of other major companies for failures to properly report to OSHA the numbers of worker injuries in their plants.

All of these fines, however, were topped in October 1987 when a group of construction companies was assessed $5 million after 28 workers were killed and 10 injured when a building under construction collapsed. Reasons given for the collapse were "sloppy construction practices and obvious design deficiencies." Two previous failures because of the design deficiency had taken place elsewhere in the United States, but no one had been killed and no correction had been made. The great increase in fines levied by OSHA 1973–1987 can be seen in Fig. 4-5.

In September 1987 newspapers claimed OSHA:

- Was notably weak in fines it charged violators. (It was recommended that costs of serious violations be increased fivefold.)
- Allowed employers long delays before fines were paid.
- Frequently failed to verify that hazards had been corrected.
- Was deficient in writing rules and regulations required to produce safer and healthier workplaces.
- Failed to have the Justice Department take any criminal action against any organization or person where a loss of life resulted because of failure to observe its standards.

Under the Reagan-Bush Administration, OSHA in 1982 had begun exempting plants from inspections if their inspection logs showed that injury rates were below the average for the industry. This not only permitted many plants to avoid inspections, but also led to an increase in cheating on reporting. Many plants failed to report accidents. Finally, in March 1988, it was ordered that Compliance Officers would themselves examine hazardous areas instead of having to rely on records of job injuries and illnesses.

Opposition to plant inspections continued until, in the spring of 1997, Representative Joel Hefley introduced the OSHA Reform Act of 1997 to the 105th Congress. The bill called for the repeal of sections of the 1970 Act which give the Occupational Safety and Health Administration the authority to inspect employers and issue citations. Part of the argument against OSHA inspections is that they are ineffective in reducing accidents and injuries. The data graphed in Fig. 4-7 seem to support this belief.

THE NEW MILLENNIUM

President Clinton's Administration extended their campaign theme of "reinventing government" to the development of the "New OSHA" to meet rising criticism of the Agency and the Act. In May 1995 OSHA's Computerized Information System

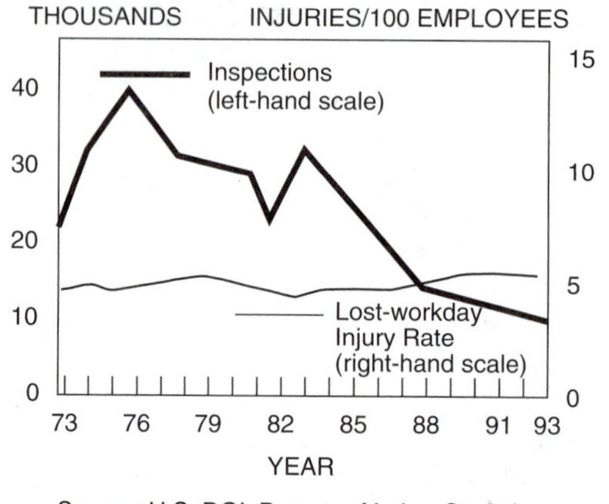

FIGURE 4–7 Manufacturing Sector Inspections and Lost-Workday Injury Rate

Source: U.S. DOL Bureau of Labor Statistics and OSHA Office of Statistics.

published "The New OSHA—Reinventing WORKER SAFETY AND HEALTH." This article states that OSHA's pride in its accomplishments is tempered by two realities. First, every year over 6,000 die from workplace injuries, about 50,000 die from illnesses caused by workplace chemical exposures, and 6 million suffer nonfatal workplace injuries. Second, the public views OSHA as driven by numbers and rules, not by smart enforcement and results. The report states that "Many people see OSHA as an agency so enmeshed in its own red tape that it has lost sight of its own mission."

These "realities" influenced the Clinton-Gore Administration to seek reform of the OSHAct of 1970 and the OSH Administration. Their goal was to **increase** worker protection, while **decreasing** red tape and paperwork. President Clinton said, "We have to recognize that, done right, regulation protects our workers from injury, and that when we fail, it can have disastrous consequences. I believe we can bring back common sense and reduce hassle without stripping away safeguards for our children, our workers, our families" (February 21, 1995).

THE NEW OSHA

Under the Clinton-Gore Administration, OSHA took steps to treat employers who have aggressive health and safety programs differently from employers who lack them. The intent is to offer employers a choice of how they will be regulated by OSHA. OSHA began to look for programs with management commitment, meaningful participation of employees, a systematic effort to find safety and health hazards whether they are covered by existing standards or not, documentation that the identified hazards are fixed, training for employees and supervisors, and ultimately a reduction in injuries and illnesses.

Employers with such programs are offered a partnership with OSHA. The "truly exceptional" employers in eliminating hazards and reducing injuries and illnesses are offered: the lowest priority for enforcement inspections (i.e., inspections will be rare), the highest priority for assistance, appropriate regulatory relief, and penalty reductions of up to 100 percent. For employers who are "well intentioned" but "have room for improvement," OSHA offers a sliding scale of incentives depending upon the degree to which the employer demonstrates real effort to find and fix hazards.

For firms that do not implement strong and effective health and safety programs, OSHA provides "traditional OSHA enforcement." "In short, for those who have a history of endangering their employees and are unwilling to change, OSHA rigorously enforces the law without compromise to assure that there are serious consequences for serious violators."

In the late 1990s the effort continued to develop readable regulations, based upon common sense. OSHA has used focus-group methods to assess problems with the readability and format of its rules. It has established a four-point strategy to develop standards that make sense: (1) identify clear and sensible priorities, (2) focus on key building-block rules, (3) eliminate or fix out-of-date or confusing standards, and (4) emphasize interaction with business and labor in the development of rules.

Working with "stakeholders" in business, labor, professional associations, and state government, OSHA is identifying the most pressing new priorities of agency action. Occupational asthma, reproductive hazards, metalworking fluids, asphalt vapors, commercial diving, welding hazards, workplace violence, and motor vehicle accidents are among the issues suggested for priority attention.

The agency is putting together a logical framework of basic building blocks for employer safety and health programs. The basic building blocks are scattered throughout the standards in an uncoordinated fashion. Among the elements to be included in this consolidation are: training programs, maintenance of records, monitoring of risk exposure, and medical surveillance.

OSHA is emphasizing hazard communication in the workplace, seeking new approaches to new hazards, as well as involvement in nontraditional sectors of industry. Hazard communication will focus on the most serious hazards. The new leading causes of suffering and disability in the workplace are musculoskeletal disorders, such as tendonitis, carpal tunnel syndrome, and low back pain. The average cost to business is $29,000 per musculoskeletal disorder. Six out of ten new occupational illnesses in 1992 were disorders associated with repeated trauma. The science of ergonomics can be used to address these hazards.

The service industries comprise a rapidly growing, nontraditional sector that in 1993 accounted for six of the ten industries with the largest numbers of workplace injuries. OSHA is increasing its involvement in health care, fast food, temporary service companies, and automobile repair facilities, because these experience injury and illness rates equaling or exceeding manufacturing industries such as steel, textiles, and paper.

One of the more interesting initiatives of OSHA is directed at compliance assistance through information technology. The OSHA System for Compliance Assistance and Referral (OSCAR) is available to promote easy public access to OSHA standards and interpretations. Electronic access to regulatory information and services is provided through the Internet. In 1995, OSHA directed 14.4 percent of its budget to compliance assistance; in 1996, 22.2 percent.

The OSHA Strategic Plan FY 1997–FY 2002 established three strategic goals: (1) improve workplace safety and health for all workers, as evidenced by fewer hazards, reduced exposures, and fewer injuries, illnesses, and fatalities; (2) change workplace culture to increase employer and worker awareness of, commitment to, and involvement in safety and health; and (3) secure public confidence through excellence in the development and delivery of OSHA's programs and services.

To accomplish these goals, OSHA plans to:

- Reduce three of the most prevalent types of workplace injuries and illnesses by 15 percent by focusing on those industries and occupations that cause the most injuries/illnesses and pose the greatest risk to workers.
- Reduce injuries/illnesses in at least five high-hazard industries by 15 percent, by focusing on those workplaces with the highest injuries and illnesses.
- Decrease fatalities in the construction industry by 15 percent, by focusing on the four leading causes of fatalities (falls, struck-by, crushed-by, and electrocutions and electrical injuries).
- Effect at least a 20 percent reduction in injuries/illnesses in at least 100,000 workplaces where the agency initiates a major intervention.
- Within four years of the effective date of significant final rules, achieve a 20 percent reduction in fatalities, injuries, or illnesses, or, for program rules or revisions, a 20 percent or greater increase in the rate of current industry compliance.
- Initiate inspection of fatalities and catastrophes within 1 working day of notification for 95 percent of occurrences to prevent further injuries or deaths.
- Initiate investigation of worker complaints within 1 working day or conduct an on-site inspection within 5 working days, so that, by FY 2000, 8 percent of all worker complaints that require on-site inspection are resolved within an average of 20 working days of notification to the employer.
- Evaluate and, if necessary, revise the "whistleblower" program.
- By FY 2000, resolve 75 percent of all "whistleblower" cases within 90 days.
- Establish Cooperative Compliance Programs in all federal enforcement states, by the end of FY 1998, to increase the number of firms with high injury and illness rates which implement effective safety and health programs.
- By FY 2002 ensure that 50 percent of the employers in general industry, who are targeted for or request an OSHA intervention, will have either implemented an effective safety and health program or have significantly improved their existing program.
- Make all standards, regulations, and reference material available in a user-friendly manner from the OSHA Home Page on the Internet by FY 2000.
- By FY2000, establish a referral clearinghouse for dissemination of occupational safety and health information.

- By FY 2002, respond to 95 percent of requests for information from the clearinghouse within 3 working days.
- Increase the number of plain-language standards from one in FY 1998 to five per year by the end of FY 2000, with 75 percent of employers and workers rating OSHA plain-language standards as readable and understandable.
- Ensure that 90 percent of small business employers and workers receiving OSHA's assistance rate their experience as useful.
- By FY 2002, ensure that 95 percent of stakeholders and partners rate their involvement in OSHA's stakeholder/partnership process as positive.
- By FY 2002, fully implement the information systems necessary to collect agency performance data and develop the capacity to analyze OSHA's performance.
- By FY 1999, implement a revised performance appraisal system which links competencies and performance to agency outcomes through the agency's Performance Plan.
- By FY 2000, ensure that 80 percent of employers and workers interacting with OSHA will rate OSHA's staff's professionalism, competence, and knowledge as satisfactory.
- Complete the redesign of all OSHA field offices by FY 1999 to better impact worker injury, illness, and death.
- Reduce the paperwork burden associated with OSHA regulations by 25 percent by the end of FY 1998, and thereafter by 5 percent each year, through FY 2001.

OSHAct reform is difficult to achieve. After five years of oversight hearings Senator Edward Kennedy and Representative William Ford introduced the "Comprehensive OSHA Reform Act" to the 102nd and 103rd Congresses (1991–1994). The 104th Congress (1995–1996), 105th Congress (1997–1998), and the 106th Congress (1999–2000) received various legislative bills for OSHAct reform from Senator Nancy Kassenbaum, Representative Cass Ballenger, Senator Mike Enzi, and Representative Jim Talent. From these several bills, only two became law.

On July 16, 1998 the OSHAct of 1970 was modestly amended by Public Law 105–198 to forbid the Secretary of Labor from using enforcement activities, such as the number of citations issued or the penalties assessed for the purpose of evaluating employees directly involved in OSHAct enforcement activities. Also, the Secretary was not to impose quotas or goals for such activities.

Public Law 105–197, the Occupational Safety and Health Administration Compliance Assistance Authorization Act of 1998 required the Secretary to establish and support cooperative agreements with the States. Under this act, also passed into law on July 16, 1998, employers may consult with state officials about compliance with occupational safety and health requirements and about voluntary efforts that employers may undertake to establish and maintain safe and healthful employment. OSHA will exempt employers from certain inspections if they:

1. Request and undergo on-site consultative visits and correct identified hazards.
2. Agree to request subsequent visits if major changes in working conditions or processes occur which introduce new hazards.

3. Implement procedures for regularly identifying and preventing hazards and maintain appropriate safety training for management and nonmanagement employees.

Changing the OSHAct requires persistence and patience for all concerned.

EXERCISES

4.1. Each state has had an industrial safety code for many years, so why was it necessary to enact the OSHAct? What is the difference between the OSHAct and OSHA?

4.2. Indicate the aims of the OSHAct and what the aims are intended to accomplish.

4.3. What are the functions of the OSHRC and NIOSH?

4.4. How many states have agreed to share part of the costs of programs and standards that are in accord with those of the federal OSHA?

4.5. What is an OSHA Compliance Officer? Can the Compliance Officer be refused access to an industrial plant to make an inspection? Who may accompany workers on such a visit?

4.6. What degrees of violations are there according to federal standards?

4.7. What can an employer do if there is a notice of a violation or an attendant penalty?

4.8. Approximately what order of priorities are OSHA officers to observe in making inspections?

4.9. How severe can penalties be for serious violations of standards that result in deaths?

4.10. From what principal sources were OSHA standards derived?

4.11. Why can it be expected that the standards will be modified over the years? How is it done?

4.12. Have the OSHAct and the necessity to observe the standards had a beneficial, adverse, or little effect in lessening accidents?

4.13. What have been some economic effects of the OSHA standards on equipment manufacturers?

4.14. What happens if adequate presidential support is not given to OSHA?

4.15. What are the changes in the "New OSHA"?

4.16. What are the goals of OSHA through 2002 and how are they being achieved?

CHAPTER 5

Standards, Codes, and Other Safety Documents

Hammurabi's Code, about 1750 B.C., contains probably the first known written admonition regarding the need for accident prevention. It states the penalties an incompetent builder would suffer. ("If a builder constructs a house for a man and does not make it firm and the house collapses and causes the death of its owner, the builder shall be put to death.") Penalties are also prescribed for lesser results of accidents. Since the day when Hammurabi's law was pressed in cuneiform wedges and baked, the penalties have become less severe, but the need for care has not only remained but increased, especially with the growth of technology.

In 1979, Hollister and Traut[2] pointed out that federal and state laws, standards, codes, and related documents on energy alone regarding environment, safety, and health laws require a working knowledge of an estimated 8 million pages of requirements. To these must be added local requirements to obtain an estimate of the immensity of the problem in only this single aspect of accident causation and prevention. OSHA requirements for industry alone refer to approximately 20 other sources,[21] all of which would combine to form a stack of paper several feet high. The types of documents related to accident prevention are indicated roughly in Fig. 5-1, while Fig. 5-2 indicates some sources of safety data, standards, and codes in the chemical industry. In recent decades the pages of laws, standards, codes, and other documents related to health and safety have multiplied until they are now a great challenge to understand and master.

Standards are produced by military, governmental, professional, and other technical organizations. There are national and international standards. Two of the most prominent standards organizations are the American National Standards Institute (ANSI) and the International Standards Organization (ISO).

To clarify a number of terms often used in accident control, the following indicates some common usages as they relate to safety matters. That the delineation is not fixed can be seen in the *Federal Register*, Vol. 37, which is titled *Occupational Safety and Health Standards*, whereas Vol. 36, with similar requirements, is titled *Safety and Health Regulations for Construction.*

1. A criterion is any rule or set of rules that might be used for control, guidance, or judgment.
2. A standard is a set of criteria, requirements, or principles.
3. A code is a collection of laws, standards, or criteria relating to a particular subject, such as the National Electric Code or the California Vehicle Code.
4. A regulation is a set of orders issued to control the conduct of persons within the jurisdiction of the regulatory authority.
5. A specification is a detailed description of requirements, usually technical.
6. A practice is a series of recommended methods, rules, or designs, generally on a single subject.
7. Design handbooks, guides, or manuals contain nonmandatory practices, general concepts, and examples to assist a designer or operator.

Wordings have to be carefully scrutinized when these are prepared and adopted to ensure they are proper. In the United States an omission, improper inclusion, or deviation may in the event of an accident be the basis for litigation.

FIGURE 5–1 Standards, Codes and Other Safety Criteria

USES FOR STANDARDS AND CODES

A famous author wrote: "The American has always reacted to the setting of standards the way Count Dracula responds to a clove of garlic or a crucifix." That is the way many corporate personnel and engineers feel, so any new safety requirement is usually met with varying degrees of opposition. This is especially true when observance of the standard means increased care, inconvenience, or cost; loss of freedom to do as one wishes; and possibly litigation.

On the other hand, criteria, which range from work rules to design standards, can be extremely beneficial in accident avoidance. Standards (to use the term as a general designation) often contain much useful technical information and promote consistency so there will be a basic level of safety in similar equipment, material, or operations. They tend to eliminate the need for the designer to hunt for information, decide whether or not any proposed design is safe or taboo, or make decisions regarding safeguards to be provided. They help lessen contentions caused by differences in opinion between designers, manufacturers, managers, and other persons about levels of safety, types of equipment, precautionary measures to be observed, and safeguards to be incorporated. They provide excellent sources of information on hazards and methods for their elimination or control. The criteria or requirements they contain were originally generated to avoid recurrence of accidents, past problems, or situations that careful consideration showed could develop into problems unless suitable precautions were taken. In effect, they indicate to designers what should not be done.

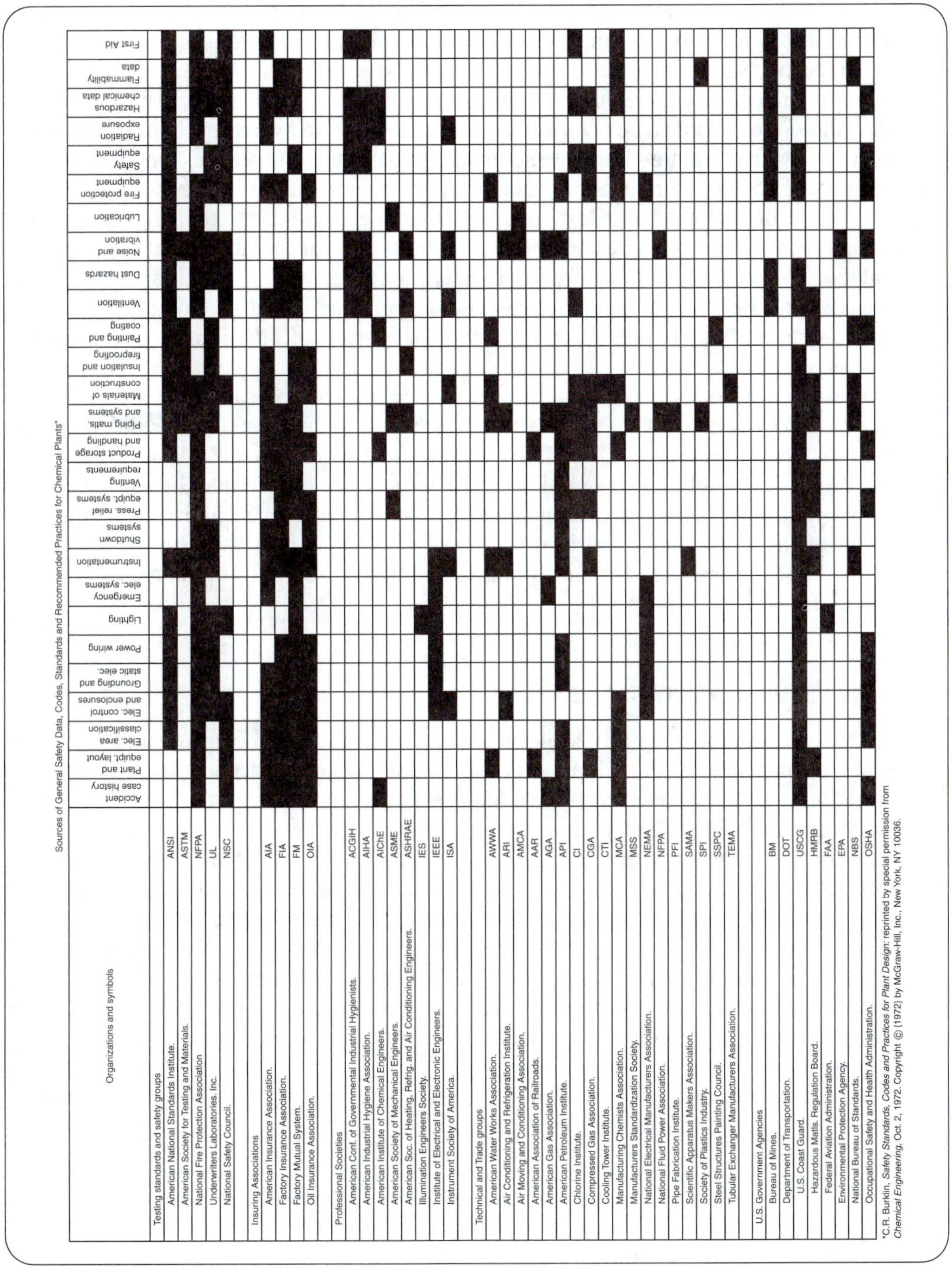

Sources of General Safety Data, Codes, Standards and Recommended Practices for Chemical Plants*

Organizations and symbols		Accident case history	Plant and equipt. layout	Elec. area classification	Elec. control and enclosures	Grounding and static elec.	Power wiring	Lighting	Emergency elec. systems	Instrumentation	Shutdown systems	Press. relief equipt. systems	Venting requirements	Product storage and handling	Piping matls. and systems	Materials of construction	Insulation and fireproofing	Painting and coating	Ventilation	Dust hazards	Noise and vibration	Lubrication	Fire protection equipment	Safety equipment	Radiation exposure	Hazardous chemical data	Flammability data	First Aid
Testing standards and safety groups																												
American National Standards Institute.	ANSI			■	■	■	■	■		■				■	■	■	■	■	■	■	■	■	■	■	■	■	■	■
American Society for Testing and Materials.	ASTM									■					■	■	■	■			■							
National Fire Protection Association	NFPA	■	■	■	■	■	■	■	■	■	■	■	■	■	■	■			■	■	■	■	■		■	■	■	■
Underwriters Laboratories. Inc.	UL				■	■	■	■		■	■				■	■	■	■		■	■		■	■		■	■	
National Safety Council.	NSC	■	■	■	■			■								■			■	■	■		■	■	■	■	■	■
Insuring Associations																												
American Insurance Association.	AIA	■	■	■	■	■		■	■	■	■		■	■	■	■	■		■				■		■	■		■
Factory Insurance Association.	FIA	■	■	■	■	■	■			■	■	■	■	■	■				■	■			■			■	■	
Factory Mutual System.	FM	■	■	■	■	■	■	■	■	■		■	■	■	■	■	■		■	■				■		■	■	
Oil Insurance Association.	OIA	■	■	■	■	■	■			■	■	■		■		■	■						■					
Professional Societies																												
American Conf. of Governmental Industrial Hygienists.	ACGIH																		■	■	■			■	■	■		■
American Industrial Hygiene Association.	AIHA																		■	■	■			■	■	■		■
American Institute of Chemical Engineers.	AIChE	■												■				■							■	■		
American Society of Mechanical Engineers.	ASME											■			■							■						
American Soc. of Heating, Refrig. and Air Conditioning Engineers.	ASHRAE														■		■		■		■							
Illumination Engineers Society.	IES							■																				
Institute of Electrical and Electronic Engineers.	IEEE				■	■	■	■	■	■																		
Instrument Society of America.	ISA				■					■						■			■		■				■			
Technical and Trade groups																												
American Water Works Association.	AWWA		■							■					■	■		■					■					
Air Conditioning and Refrigeration Institute.	ARI				■					■					■						■							
Air Moving and Conditioning Association.	AMCA																				■	■						
American Association of Railroads.	AAR		■											■														
American Gas Association.	AGA	■							■						■						■							
American Petroleum Institute.	API	■	■	■		■	■			■	■	■	■	■	■						■		■					
Chlorine Institute.	CI											■		■	■	■			■					■		■		■
Compressed Gas Association.	CGA		■							■		■		■	■	■							■	■		■		
Cooling Tower Institute.	CTI															■												
Manufacturing Chemists Association.	MCA	■	■	■	■									■		■							■	■		■	■	■
Manufacturers Standardization Society.	MSS														■													
National Electrical Manufacturers Association.	NEMA				■	■	■	■	■														■					
National Fluid Power Association.	NFPA														■						■							
Pipe Fabrication Institute.	PFI														■													
Scientific Apparatus Makers Association.	SAMA									■																		
Society of Plastics Industry.	SPI														■												■	
Steel Structures Painting Council.	SSPC																	■										
Tubular Exchanger Manufacturers Association.	TEMA															■												
U.S. Government Agencies																												
Bureau of Mines.	BM																		■	■			■	■	■	■	■	■
Department of Transportation.	DOT																											
U.S. Coast Guard.	USCG	■	■	■	■	■	■	■	■	■	■	■	■	■	■	■	■		■	■			■	■		■	■	■
Hazardous Matls. Regulation Board.	HMRB		■							■		■	■		■	■		■	■									
Federal Aviation Administration.	FAA							■																				
Environmental Protection Agency.	EPA																				■							
National Bureau of Standards.	NBS									■					■	■		■									■	
Occupational Safety and Health Administration.	OSHA	■		■	■	■	■					■		■				■			■		■	■		■		

*C.R. Burklin, *Safety Standards, Codes and Practices for Plant Design:* reprinted by special permission from *Chemical Engineering*, Oct. 2, 1972. Copyright © (1972) by McGraw-Hill, Inc., New York, NY 10036.

FIGURE 5–2 Sources of General Safety Data, Codes, Standards, and Recommended Practices for Chemical Plants

Standards generally indicate safety measures to be taken, although in many instances they may contain an inadequacy, incompleteness, or failure to properly deal with a hazard. Safety engineers can use them as bases on which to prepare safety programs by indicating deficiencies in compliance. They can use them to indicate to managers the safety norms to be achieved for any operation and whether those norms have been achieved. Obtaining improvements in equipment or procedures may be easier if it can be shown that standards or other criteria recommend or require their use.

Risks of accidents are less in plants that meet standards than for those that in some way do not. As a result, costs for liability insurance will be higher for plants that fail to meet prescribed or recommended standards. In addition, losses from lawsuits and awards and penalties against a company would be greater if it could be shown there was failure to comply with provisions of any applicable standard's provisions.

Standards, codes, and other criteria can also provide a basis for preparing checklists for use by engineers, safety personnel, or others to audit or study facilities, equipment, or operations for hazards. Each requirement can be made into an item to be checked by the safety professional, supervisor, or other person making the review or analysis. A list of pertinent items can be prepared, either using each provision as stated or converting the statement into a question. Figure 5-3 not only lists abbreviated statements that could be used for an investigation but also refers to the pertinent OSHA standards.

MANDATORY VS. VOLUNTARY STANDARDS

The OSHA standards state in Section 1910.2(f): "'Standard' means a standard which requires conditions, or the adoption or use of one or more practices, means, methods, operations, or processes, reasonably necessary or appropriate to provide safe or healthful employment and places of employment." The word "requires" makes any applicable provisions in the OSHA standard mandatory. Most industrial standards, especially those prepared prior to passage of the OSHAct, consisted of nonmandatory criteria unless they were specifically contained in a federal, state, or municipal requirement. Standards issued by organizations such as the American National Standards Institute (ANSI), the American Society of Mechanical Engineers (ASME), National Electric Code (NEC), or National Fire Protection Association (NFPA), to cite only a few, are voluntary or "consensus" standards adopted by agreement with participating members.

Thus, none of the 12,000 standards issued by the ANSI is in itself mandatory, since each contains the following statement:

> An American National Standard implies a consensus of those substantially concerned with its scope and provisions. A National Standard is intended as a guide to aid the manufacturer, the consumer, and the general public. The existence of an American National Standard does not in any respect preclude anyone, whether he has approved the standard or not, from manufacturing, marketing, purchasing, or using products, processes, or procedures not conforming to the standard.

OSHA SAFETY CHECK SHEET

Company _______________________________________ Date of Check __________

Check Findings

Yes No N.A.

1. Log of occupational injuries (OSHA #100) maintained. (1904.2)
( ) ( ) ( )
2. Record of occupational injuries of illnesses (OSHA #101) up to date. (1904.4)
( ) ( ) ( )
3. Summary of occupational injuries or illnesses posted (OSHA #102). (1904.5)
( ) ( ) ( )
4. Permanent passageways and aisles appropriately marked. (1910.22)
( ) ( ) ( )
5. Passageways, storerooms, services rooms, and work areas clean and aisles clear. (1910.22)
( ) ( ) ( )
6. All floors clean and dry. (1910.22)
( ) ( ) ( )
7. Floors free of protruding nails, splinters, holes, and loose boards. (1910.22)
( ) ( ) ( )
8. Where storage is on a second floor, balcony, or other overhead areas—load limits are marked. (1910.23)
( ) ( ) ( )
9. Floor openings of any kind have standard railings or floor hole cover constructed as required. (1910.23)
( ) ( ) ( )
10. Temporary floor openings have standard railings or are constantly attended by someone.
( ) ( ) ( )
11. Every openings, floor or platform 4 feet or more above ground level has a standard rail and toe boards guarding. (1910.23)
( ) ( ) ( )
12. All stairs with four or more risers have standard hand rails. (1910.23)
( ) ( ) ( )
13. Portable step ladders are in good condition. (1910.25 and 1910.26)
( ) ( ) ( )
14. Portable rung ladders have safety feet securely bolted or fastened. (1910.25)
( ) ( ) ( )
15. Wood ladders are stored in shelter or out of elements. (1910.25)
( ) ( ) ( )
16. Metal ladders are equipped with nonslip material on rungs. (1910.25)
( ) ( ) ( )
17. Metal ladders are not used in areas where exposed to electric circuits. (1910.26)
( ) ( ) ( )
18. Exits are visible and clearly marked and illuminated by a reliable light source. (1910.36)
( ) ( ) ( )
19. Work areas have at least two exits. (1910.36)
( ) ( ) ( )
20. Where exits are not clearly visible from work areas, signs pointing to exits are posted. (1910.37)
( ) ( ) ( )
21. Doors or openings which are not a means of egress are marked "Not an Exit." (1910.37)
( ) ( ) ( )

FIGURE 5–3 OSHA Safety Check Sheet Continued on page 78

22. Permissible noise exposure levels have been checked in areas where necessary and personal protective equipment provided as needed. (1910.95)
( ) ( ) ( )
23. Approved containers for flammable materials are available. (1910.106)
( ) ( ) ( )
24. Closed containers are provided for oily-rag disposal. (1910.106)
( ) ( ) ( )
25. Hard hats are provided and used in areas where impact or falling-objects hazards exits. (19010.136)
( ) ( ) ( )
26. Employees handling heavy objects are provided and use protective footwear. (1910.136)
( ) ( ) ( )
27. Separate toilet facilities are provided for each sex, and facilities are clean. (1910.141)
( ) ( ) ( )
28. Toilet facilities are adequate for the number of employees. (1910.141)
( ) ( ) ( )
29. Lunch room is adequate for the maximum number of employees who may use it. (1910.141)
( ) ( ) ( )
30. Covered receptacles are provided in lunch room and are emptied not less than once daily. (1910.141)
( ) ( ) ( )
31. A positive "lockout system" is provided to render those machines operated by electric motors inoperative while repairs or adjustments are being made. (1910.147)
( ) ( ) ( )
32. "No Smoking" signs are posted in areas where conditions require them. (1910.145)
( ) ( ) ( )
33. Physician-approved first aid kit is accessible. (1910.151)
( ) ( ) ( )
34. Personal protective equipment for eye, face, head, and extremities, protective clothing, respiratory devices, protective shields and barriers and provided, used, and maintained by employer wherever necessary. (1910.151)
( ) ( ) ( )
35. Fire extinguishers for necessary classes of fire are visibly mounted. (1910.157)
( ) ( ) ( )
36. Fire extinguishers have been inspected and are operable. (1910.157)
( ) ( ) ( )
37. Fire extinguishers for Class A fires are within 75 feet from any point in the work area and extinguishers for Class B fires are within 50 feet.
( ) ( ) ( )
38. Access to fire extinguishers is not hindered in any way. (1910.157)
( ) ( ) ( )
39. Air receivers have been checked for dents, cuts, or gouges, corrosion or pitting, hairline cracks. Storage safety: dry, clean area: drain pipe or valve at low point. (1910.166)
( ) ( ) ( )
40. All materials are piled, racked, or stored in a safe manner. (1910.176)
( ) ( ) ( )
41. High-lift trucks have overhead guards. (1910.178)
( ) ( ) ( )
42. Operators of lift trucks have been instructed and trained as to use. (1910.178)
( ) ( ) ( )
43. Machines, presses, grinders, saws, etc. are properly guarded to protect employees from hazards created by: A. Point of operation
( ) ( ) ( )

FIGURE 5–3 Continued

B. Nip points
( ) ( ) ( )
C. Rotating parts
( ) ( ) ( )
D. Flying chips and sparks (1910.212)
( ) ( ) ( )
44. All mechanically powered transmission apparatus 7 feet or less above floor or work platform is guarded. These include:
A. V-belts, shafts, pulley
B. Chain and sprocket drives
C. Flywheels (1910.219)
( ) ( ) ( )
45. Compressed-air cleaning equipment has been limited to less than 30 psi and is chipguarded. (1910.242)
( ) ( ) ( )
46. Portable power wood-working tools or equipment have deadman guards or switches and are so placed as to prevent accidental operation when not in use. (1910.243)
( ) ( ) ( )
47. Welding equipment is inspected for hazards. (1910.252)
( ) ( ) ( )
48. Circuit-breaker switches identify what circuits they control. (1910.303)
( ) ( ) ( )
49. Access to electrical panels is unobstructed. (1910.303)
( ) ( ) ( )
50. Employee safety meetings are scheduled and held at regular intervals. (safety check only)
( ) ( ) ( )
51. Adequate lighting in work areas for work or job being performed. (Safety check only)
( ) ( ) ( )
52. Elevator's load limit is so marked. (Safety check only)
( ) ( ) ( )
53. Self-service elevators are marked as to operation and floor. (Safety check only)
( ) ( ) ( )
54. Plant safety rules are posted and distributed. (Safety check only)
( ) ( ) ( )

Adapted from material prepared by California State Compensation Insurance fund, based on federal OSHA standards. Changes in standards are being made frequently, so that certain of the ones listed here may not be currently applicable. This is provided as an example checklist only.

FIGURE 5–3 Continued

OBJECTIONS TO CONSENSUS STANDARDS

Objections to consensus standards have been expressed often but never more forcibly than by Admiral Hyman G. Rickover. Among the comments he made before Congress were:[22]

> To forestall intrusion of Government, the industry concerned will usually propose voluntary safety requirements. These requirements represent the minimum all are willing to accept. This is not enough. There are more accidents. Only after the lapse of much time are laws finally enacted. Much harm will have been done in the interval—harm which could have been prevented.

> The typical industry-controlled code or standard is formulated by a committee elected or appointed by a technical society or similar group. Many of the committee members are drawn from the manufacturers to whom the code is to be applied. Others are drawn from engineering consulting firms and various Government organizations. However, since near unanimous agreement in the committee must generally be obtained to set the requirements or to change them, the code represents a minimum level of requirements that is acceptable to industry.
>
> In a subtle way, the use of industry codes or standards tends to create a false sense of security. Described by a code committee and by language of many of the codes themselves as safety rules, they tend to inhibit those legally responsible from taking action to safeguard health and well-being. Many States and municipalities have incorporated these codes into their laws, thus, in effect, delegating to the code committees their own responsibilities for protecting the public.

Admiral Rickover's comments were based on his disagreement with the standard of the ASME as they applied to nuclear equipment. Similar feeling existed regarding voluntary standards on some consumer products, such as shop tools, that were often used in small industrial plants. Pittle remarked in 1977:[23]

> In some cases an existing voluntary standard may be entirely adequate to prevent or reduce an unreasonable hazard associated with a product, but is ineffective in protecting the public because it is not widely accepted by industry or because the promulgating agency lacks authority to require adherence to its terms. In such cases, government's role may simply be to ensure that all segments of the industry adhere to the standard. Even widespread membership in a trade association which has adopted a voluntary safety standard may not be sufficient where the trade association has no mechanism for ensuring compliance with its standards .... Perhaps the greatest factor tending to differentiate government standards from voluntary industry standards is the greater willingness to push for requirements that demand new technology. One of the classic examples here is the Refrigerator Safety Act which compelled manufacturers to design doors which could be easily opened from the inside of refrigerators. Under this statutory prodding the industry has virtually 100 percent compliance. I am doubtful that left on their own, these manufacturers would have pressed forward as quickly as they did.

TEST STANDARDS

Standards are often set and used for testing and certification of products. The standard indicates not only the levels to which the product is to be subjected, but the number of tests to be accomplished, the number of successes to be achieved for approval, and other points to be determined. In many instances, the test results and the intent of the test standards themselves may be misinterpreted. As a result, claims that standards have been met regarding safety frequently are not justified, and accidents take place in use of the product. An outstanding example was the controversy regarding fire tests for some plastic building materials.

Test standards issued by the American Society for Testing (ASTM) and the British Standard Institute defined a "self-extinguishing" plastic as one that would fail to continue to burn after the ignition source was removed. A test piece of material was held in the horizontal position and ignited at one end. If the flame went out before it reached a specific, measured point, the plastic would be considered self-extinguishing. Many plastics were given that rating. However, if such a material was hung vertically, like a drape or vertical panel, the material would continue to burn after having first been ignited. Theoretically, the test standard had been established for quality-control purposes at producers' plants. Actually, the plastics were widely accepted, and often advertised, as self-extinguishing. Many fires took place because designers or users employed the standard as a guide to the flammability of building material and similar products.

In May 1973 the Federal Trade Commission instituted a class action complaint against the ASTM, the Society of the Plastics Industry (SPI), and 26 plastics manufacturing companies. The complaint charged that these organizations, either directly or indirectly:

- Knew of serious fire hazards relating to plastics (cellular or foamed polyurethane, and all forms of polystyrene and its copolymers) since 1967, but failed to disclose such facts to users.
- Misrepresented that these combustible plastics were "nonburning" (would ignite under no circumstances) and "self-extinguishing."
- Had knowledge that the test standards permissing these false descriptions were invalid for determining how the product would act in an actual fire.
- Failed in their duty to undertake adequate and effective precautionary, remedial, and corrective action to eliminate or minimize risks of injury or damage to life or property.
- Instituted invalid testing standards that enabled these products to be classified as "noncombustible" in numerous building codes throughout the country.

The last allegation is of special interest here. Any organization that observed building codes in which these materials were cited as "noncombustible" would wrongly assume they should be used to minimize fire hazards.

DIFFERENCES IN STANDARDS

Standards written by different standard-making organizations may differ, even when written for the same type of equipment, material, or use. The ASME standards for pressure vessels are written for fixed installations. Those for the Department of Transportation (DOT) are for mobile equipment that must be transported by common carrier over public highways. Portable pressurized gas cylinders (Chapter 20) produced under DOT standards are sometimes connected to systems with equipment produced under ASME standards. For similar usages, then, equipment with two levels of safety, both meeting required standards, may be operating together.

Another example of the existence of different criteria for the same or similar hazard is in the requirement for how long it should take to discharge an electrical capacitor or capacitive circuit after the power is shut off. A person who touches such a charged circuit or capacitor and acts as a ground can get a nasty shock. The requirement for eliminating this problem with different equipment should be the same, since the hazard, cause, and effects are the same.

Figure 5-4 indicates the different criteria and requirements. In 1979 the National Aeronautics and Space Administration (NASA) sent out a SAFE-ALERT to warn government and industry to be careful of a specific voltmeter on the market. The device required 45 to 60 seconds to bleed off power after it was shut off. It can be seen from the figure that the voltmeter would not have met the requirements of the military service or of IBM.

In a New York City case, Sicurauza v Union Crane and Shovel Company, there were found to be two separate standards setting forth the minimum requirements for wire rope to be used on boom cranes. The conflict between the standards set by Wire Rope Manufacturers and the Power Shovel and Crane Industry demonstrated the manner in which each industry seeks to establish standards which correspond to its self-interest. In this case, the Wire Rope Manufacturers had an interest in standards which provided good and safe products for many applications, while the Power Shovel and Crane Industry sought standards which corresponded to the interests of the manufacturers of such equipment; and its decision in this regard was clearly based on economics.[24]

STANDARD	REQUIREMENT
1. Military Standard 454, for Electrical and Electronic Equipment.	The capacitor or circuit must discharge to 30 volts in 2 seconds.
2. IBM Product Safety Criteria (Company manual).	Discharge to less than 60 volts within 10 seconds after the circuit is opened.
3. National Electric Code, Art. 460 (ANSI Std. C1).	Discharge to 50 volts or less than within 1 minute after disconnect for capacitors rated more than 600 volts.
4. Underwriters Laboratories Std. #478, Electronic Date Processing Units and Systems (ANSI Std. C33.107).	(a) Discharge to less than 50 volts within 1 minute. (b) Where necessary to remove a panel with tools, provide instruction on the panel that it is not to be removed for whatever time it takes for discharge (5 minutes maximum).
5. American National Standards Institute (ANSI) Std. Z136.1 (1973) for the Safe Use of Lasers.	If servicing of equipment requires entrance into an interlocked enclosure within 24 hours of the presence of high voltage within the unit, a solid metal grounding rod shall be utilized to assure discharge of high-voltage capacitors.

FIGURE 5–4 Criteria Regarding Electrical Capacitors

There are other examples of differences in technical standards. Figure 12–3 indicates a problem that exists with electric drills and some solutions. To minimize the problem, U.S. manufacturers of portable tools called for the use of a three-wire cord, by which the frame of the tool could be grounded. The British instead (and other European nations) had adopted a standard requiring "double insulation" and other safeguards that minimized or eliminated any chance of accidental shock. Thus there were two standards for protection of the same hazard. In the United States, it was found in many instances that the system and thus the tools were not being properly grounded, so the method did not eliminate the problem even while the prescribed standards were being met. The standard would have permitted an operator to be shocked fatally. Some manufacturers then began designing and producing, even before the standard was again revised, doubly insulated equipment, raising the question as to why in the first place the use of three-wire systems, such as for many plants, businesses, or homes, was needed.

(With the latest standard, the probability of a person being shocked has been made more remote. The hazard has not been eliminated entirely, because if an energized tool, such as an electric razor or hair dryer, is dropped into water, a person trying to retrieve it could be electrocuted. Where the presence of such a possibility exists, a battery-powered tool could be used. At the minimum, users should be made aware and warned of the hazard.)

An important difference in standards comes from the philosophical basis that determines the content of the standard. The philosophy behind a "specification" standard is to provide detailed design specifications that must be met. On the other hand, a "performance" standard is one that sets performance criteria that the design must meet.

CHANGING STANDARDS

Not only are there great variations between standards, but any change can itself bring repercussions, even when there is improvement. A change in the standard for wire ropes can have downstream effects on all types of equipment that use them. If a manufacturer, in good faith, observed the provisions of the old standard, the old equipment might be considered unsafe and if any accident occurred, subject to liability suits. Lawyers have maintained when changes in standards were made after accidents took place that they were admissions of the original inadequacy. Because of the very great number of standards in existence, no current method of listing all of their changes exists. Maintenance of currency of standards constitutes an undeniable problem.

INADEQUACIES OF STANDARDS

Some of the benefits of proper standards have been pointed out. It has also been said that any standard is better than no standard at all. In one sense this is true, but in another it is not. Many persons, designers, employers, and workers, rely on pronouncements that once a product has met a specific standard, it is safe, and they subsequently may not take adequate precautions. Actually the standard may have a hazard because one or more of its provisions resulted in unsafe design.

In 1970, the Final Report of the National Commission on Product Safety pointed out the following products that had met standards of the Underwriters Laboratories (UL) but were unsafe, resulted in accidents, and had to be changed:

- Hundreds of thousands of television sets that had caught fire.
- A vaporizer cited for scalding children.
- A lethal charcoal igniter.
- A toy oven with a temperature of 660° F inside and 300° F on top.
- Fifteen out of 29 portable electrical heaters that were rated as "not acceptable" by the Consumers Union.

In 1977, home smoke alarms which were listed as having met the UL standards were themselves causing fires.

STANDARDS AND ANALYSES

In 1978, the Consumer Product Safety Commission announced the recall of an automatic coffeemaker because of a potential fire hazard. The coffeemaker bore a metal plate on which appeared: "UL LISTED 429E, MADE IN SINGAPORE." An item bearing this plate was unsafe because it was either poorly manufactured or inadequately designed. If it was poorly manufactured, another lot could be a good one. If it was badly designed because of any provisions of a standard or failure to meet them, it would constitute a design failure.

A standard or code may be lacking in its requirements or scope or may be inadequate for actual conditions, so the designers may themselves create hazardous products. Each new design or proposed operation must be examined critically to determine whether it will safely meet the proposed aim. Too often designers or safety engineers blindly follow cited criteria without analyzing whether they are safe for a particular situation and their use is justifiable. Standards are only guides—partial indicators of hazards that might exist, with possible actions or safeguards.

PROLIFERATION OF STANDARDS

Standards are proliferating because standards organizations are, almost without exception, producing new standards regularly. The American National Standards Institute publishes more than 12,000 standards. The International Standards Organization publishes more than 15,000 standards. The International Electrotechnical Commission publishes more than 4,000 standards. These are only some of the organizations presently producing standards. The Occupational Safety and Health Administration also produces standards, as mentioned above.

STATUS OF OSHA STANDARDS

Many OSHA standards have not been updated since 1971. They were based upon existing national consensus standards and reflect the results of research conducted in the 1950s and 1960s. Many are in need of updating and correction, as well as translation into plain English. Part of the Clinton administration's "reinventing government" initiative has aimed to reduce the burden of regulations on industry and to put them in language which can be understood more easily. This made Occupational Safety and Health Regulation writing and rewriting a strong activity at the turn of the millennium.

In 1997, the following regulations in Occupational Safety and Health were in the Prerule, Proposed Rule, or Final Rule stage: "Standards Advisory Committee on Metalworking Fluids," "Control of Hazardous Energy Sources (Lockout/Tagout)," "Occupational Exposure to Ethylene Oxide," "Fire Brigades," "Grain Handling Facilities," "Cotton Dust," "Steel Erection (Part 1926) (Safety Protection for Ironworkers)," "Prevention of Work-Related Musculoskeletal Disorders," "Safety and Health Programs (for General Industry)," "Occupational Exposure to Tuberculosis," "Confined Spaces in Construction," "Fire Protection in Shipyard Employment," "Permissible Exposure Limits (PELs) for Air Contaminants," "Plain English Revision of Existing Standards," "Nationally Recognized Testing Labs Programs: Fees," "Flammable and Combustible Liquids," "Fall Protection in the Construction Industry," "Process Safety Management of Highly Hazardous Chemicals," "Revocation of Certification Records for Tests, Inspections, and Training," "Plain English Revision of Existing Standards (Phase II)," "Electric Power Transmission and Distribution; Electrical Protective Equipment," "Safety Standards for Scaffolds Used in the Construction Industry—Part II," "Safety and Health Programs for Construction," "Control of Hazardous Energy (Lockout) in Construction," "Respiratory Protection (Proper Use of Modern Respirators)," "Longshoring and Marine Terminals (Parts 1917 and 1918)—Reopening of the Record," "Access and Egress in Shipyards," Recording and Reporting Occupational Injuries and Illnesses," "Powered Industrial Truck Operator Training," "Permit Required Confined Spaces," and "Standards Improvement Project."

The "Standards Improvement Project" is part of a continuing effort to eliminate confusing, outdated, and duplicative regulations. In 1978 and again in 1984, the Occupational Safety and Health Administration conducted projects that resulted in the elimination of hundreds of unnecessary rules. In 1995, OSHA developed a list of standards it proposed to revoke or revise because they were out-of-date, duplicative, inconsistent with other OSHA standards, or preempted by other federal agencies. This work is ongoing. The writing of new standards and the revision and revocation of older standards has no end in sight. Some of the proposed rules listed above involve very significant issues and considerable impact on industry.

BIBLIOGRAPHY

[21] *Federal Register*, Vol. 37, No. 202, Oct. 18, 1972, Part II, is titled *Occupational Safety and Health Standards,* whereas vol. 30, no. 75, April 17, 1971, is titled *Safety and Health Regulations for Construction*, although the types of requirements are the same.

[22] Extracted from testimony before the Joint Committee on Atomic Energy, Congress of the United States, Washington, D.C., March 19–20, 1970.

[23] David Pittle, Commissioner, Consumer Product Safety Commission, remarks to 1977 Design Safety Conference, Chicago, Illinois, May 9–12, 1977.

[24] Ralph Manaker, "Standards—Effects on Liability," *Proceedings of the 1974 Product Liability Prevention Conference* (Syracuse, N.Y.: Law Offices of Irwin Birnbaum), p. 203.

EXERCISES

1. List some types of documents which might list safety criteria or requirements. Give examples.
2. Indicate the uses to which codes and standards can be put.
3. What are "consensus standards"? What are their principal advantages?
4. What is the difference between a performance and a specification standard? Which would be preferable to a designer?
5. Will the fact that a designer closely followed the provisions of a standard or other criteria absolve the employer or manufacturer of liability for any subsequent related accident?
6. What are the principal objections manufacturers or engineers have to standards? Do you feel standards are used for guidance or for control?
7. Can any company prepare its own standard? Is this a good or bad idea? What might happen legally if an accident occurred because there was a failure in observance of a company standard, even though there was no mandatory standard such as one required by OSHA?
8. How can a detailed checklist be prepared to assist in reviewing provisions of a standard? Who can prepare them and who can use them?
9. Why are changes in standards required?
10. An accident occurs, so a company changes and upgrades its standards. Why would a plaintiff's attorney object to this?

CHAPTER 6

Engineers and Safety

It has been said that when a physician makes an error, only one victim suffers, but when an engineer makes an error in design of a product, many persons may suffer. The question then becomes, which is more important for the well-being of the public: the physician or the engineer?

ACCOMPLISHMENTS OF ENGINEERS

The profession of engineering is a relatively new one that has increased tremendously with the growth of technology. Until the end of the Middle Ages with its comparatively slow technological developments, there were relatively few changes in manufactured devices or human activities. Industry was still almost entirely agricultural. Power was generated principally by the muscles of humans, horses, oxen, or other animals.

The invention of steam-powered equipment and its attendant enterprises was a tremendous boon to humankind. Savery invented the steam engine in 1698. The first primitive, inefficient steam engines were later improved by Thomas Newcomer (c. 1711), James Watt (1769), and many others. The burning of coal became a prime means of transferring heat energy to steam to drive engines. From that time forward, the number of inventions of all types grew rapidly, and with them their applications.

The time from the invention of steam engines until their use by Stephenson in locomotives was about 110 years. Operation of the laser was theorized in 1958, patented in 1960, and put into commercial production almost immediately. With the accelerated growth in technology, an engineer's training is said to be obsolete in five years.

Over the years, engineers designed and constructed tremendously larger, more energetic, hazardous, costly, and complex systems, and plants and new facilities started using or producing new materials, products, and processes. It was said that a new chemical was developed every twenty minutes. Many of them were hazardous.

ENGINEERING AND ACCIDENTS

Time and again new products, processes, methods, operations, and systems could not be used because of the dangers they presented.

Since time immemorial, mining for metal ores has been a hazardous occupation resulting in many accidents. The advent of steam engines and the new demand for coal increased the hazards and accidents even more, as coal mining grew into a tremendous industry. The principal hazard was the possibility of asphyxiation because of the lack of proper ventilation and the presence of carbon monoxide. Miners carried caged canaries whose death provided warning of the lack of adequate respirable air.

As coal mining went deeper, it grew more hazardous. A new danger was that of ignition of pockets of the flammable gas known as fire damp, swamp gas, or methane. This occurred often when the candles the miners had to carry for illumination ignited the methane. To lessen possibilities of ignition of the gas, fireflies and the phosphorescence of putrescent fish were tried as means of providing light, but without success.

The safety and well-being of horses and other beasts were more highly considered than those of the miners. The number of fires, explosions, and disasters grew, causing so much loss of life that in 1813 some English clergymen asked Sir Humphrey Davy to investigate whether any means of fire prevention was possible. After due study, Davy found wire gauze to be an effective barrier in inhibiting the spread of fires. In 1815 he developed a miners' safety lamp using metal wire gauze. The lamp proved effective although not completely so. Improvements by him and others, such as Stephenson, did much to lessen the problem of ignition of fires in mines.

STEAM EQUIPMENT AND ACCIDENTS

Stephenson also built locomotives to pull trains of cars and created the first railroad line. Railroads proved to be so highly beneficial for transporting people and freight that copies and improvements were rapidly adopted almost worldwide. With railroads came fatal accidents. The first occurred in England the day Stephenson's first railroad line was dedicated, when a prominent English legislator was killed. Only a year later the boiler of the first locomotive built in the United States blew up, killing one man and badly injuring other fuel servers. As a safeguard for the passengers on the train, the coaches were separated from the locomotive by bales of cotton. It was an early indication of the hazards involved with use of boiler-powered equipment.

Railroads and steamboats were created at about the same time and with similar problems, principally violent ruptures of boilers by steam pressure. As mentioned earlier, in 1833 President Andrew Jackson asked Congress for action to control the "criminal negligence" which caused accidents because of exploding boilers. In his Fifth Annual Message to Congress he called for the adoption of corrective legislation. Although mishaps and losses aboard steamboats on the Mississippi, Ohio, and Missouri rivers occasioned his immediate concern and comment, similar problems with accidents and explosions had arisen east of the Alleghenies with railroads. The number of fatalities and injuries because of transportation equipment accidents was called "horrendous" by a well-known magazine of the time. The occurrence of explosions continued to create a record of disasters.

On October 8, 1894, twenty-seven small boilers in a plant in Shamokin, Pennsylvania, exploded progressively. According to the American Society of Mechanical Engineers (ASME), by 1910 there were between 1,300 and 1,400 boiler explosions each year. This was almost four each and every day. There was little respect for safety and accident prevention.

TECHNICAL SOCIETIES AND SAFETY

Technical societies had been formed to provide interchanges of information between persons involved in engineering. The ASME held its first annual meeting in 1880 primarily to standardize dimensions of screws, pipes, and similar items. Engineers' desires to eliminate accidents were generally focused on avoiding damage and destruction of property and equipment, because it adversely affected profitable operations and increased insurance costs.

Other than with great disasters, for uncounted years most accidents have been mistakenly accepted as being expected and unavoidable occurrences. There was inadequate concern for the safety of the low-paid workers who were continually subjected to unhealthful working conditions, sickness, and mishaps that shortened their lives. To a great extent, many of the first safety measures were advocated and initiated by insurance companies to minimize monetary losses and costly litigation. Moreover, to gain economic advantages, manufacturers approved, released, and distributed newly invented or discovered advances before the potentials for injury, death, or damage had been evaluated properly. This tendency worsened as the rate of industrial research and development accelerated. As manufacturers become more competitive, engineers were pressured to finish new designs rapidly and at lower costs, with the result there were more errors, design deficiencies, and causes of accidents. Inadequate time was allowed for engineers to analyze the existence of hazards and/or to investigate potential safety problems. After the space shuttle *Challenger* exploded because of faulty design, the solid propellant boosters had to be redesigned. A newspaper said: ". . . rocket engineers, under pressure from NASA to redesign the rockets in time for a projected July 1987 launch, are complaining that the deadline may be creating an atmosphere in which concern for safety is tempered by schedule demands." Compounding this problem is the fact that little time is spent in teaching basic safety principles in engineering schools.

INADEQUACY OF ENGINEERING SCHOOLS

It is often said, erroneously, that information on safety is included in engineering courses. Because so many subjects and facts must be taught in engineering schools, instruction in safety and accident prevention is often omitted. Meanwhile, the need for knowledge of hazards analyses and accident causes and avoidance has grown extensively. The amount of safety knowledge required has become so great that instructors cannot convey it offhandedly and quickly. The worst aspect of the situation is that most faculty members who instruct engineering students in safety matters have themselves been inadequately educated.

Although Lederer's comments[25] were oriented toward aeronautical engineers, they apply to all technical disciplines: "Oversights in design lead to accidents for several reasons: the engineer's inexperience or ignorance, the engineer's attitude and the engineer's subjection to economic control . . . . In the past, many errors might have been avoided if engineers had been early imbued with a greater awareness of safety principles . . . . To permit repetition of oversights in design is hardly excusable, but they are repeated as each generation of engineers progresses to positions of administrative responsibility leaving in its wake a semi-vacuum of design 'know-how.'"

The comments of Hollister and Traut (Chapter 1) regarding the gradual increase in concern because of increased complexity and magnitudes of hazards, accidents, and costs are also highly relevant. As an indication of design inadequacy, about 1959 an Air Force safety engineer pointed out:

> The classic of all design deficiencies which have come to our attention was a combination safety shower and eyewash constructed at a northern missile site. In order to operate the eyewash, it was necessary for a man, who might already be blinded by acid, to put his head in the eyewash bowl and then to turn on the water valve with his right foot. The only problem was that the foot-operated valve was about four feet to his rear and higher than his waist. As an additional feature, if a man did happen to hit the valve, he got a full shower from overhead as well as getting his eyes washed out. However, the whole problem became academic in winter because the whole system froze up.

Sir Henry Royce, who set one of the highest standards of all times in sound, reliable design, epitomized the right approach: "If you give a man a power tool to work with and he comes to disaster—you don't blame the man, blame the designer."[26] The same British reference in *Engineering* cites the Robens Report, which led to many improvements in safety and health at work. The aim stated was to ensure, as far as practicable, that plant, machinery, equipment and materials are so designed and constructed as to be intrinsically safe in use.

ENGINEERS AS CAUSERS OF ACCIDENTS

Figure 6-1 indicates how engineering deficiencies can cause or contribute to accidents, as indicated by the shaded portion. Any engineering deficiency can affect management; create malfunctions and failures, maintenance problems, and failures to protect against adverse environments; and/or cause operator errors. Figure 6-2 presents a limited introduction to some aspects of malfunctions, failures, and reliability. All other factors indicated in Fig. 6-1 are treated in some of the following chapters. "Miscellaneous other" in the figure denotes that the exact accident cause would either be indeterminable or involve three or more factors.

Figure 6-3[4] lists the hazard factors that were accident causes in the chemical and allied industries. The largest factor, "equipment failures," includes "hazards built into the equipment." Other factors might also be attributable to design, such as "chemical process problems" or "structure not in conformity with use requirements."

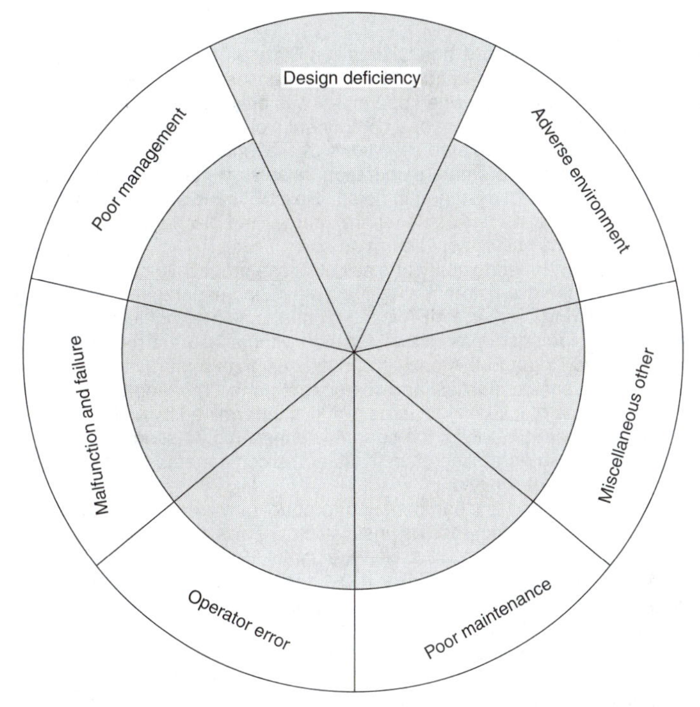

FIGURE 6–1 Design Deficiencies Also Partly Attribute to Other Causes of Accidents

REGISTRATION OF ENGINEERS

As a means of lessening accidents with their fatalities, injuries, and economic losses in damages and destruction, after the turn of the century registration of engineers was initiated. Like the workers' compensation laws, which aimed initially to induce greater regard for safety and accident prevention, registration of engineers has failed to achieve its major intent.

Registration of engineers was first enacted in 1907 in Wyoming, where it originated in a plan to control land surveying. Unqualified persons had been issuing land-plat plans that turned out to be erroneous, deceptive, false, or incomprehensible. Instead of a law simply to control land surveying, the public demanded a means of lessening accidents on the railroads. The public had become so incensed that it forced legislators to enact the first state law for registration of engineers. It was believed that better engineering would reduce the number of inadequate and deficient designs with their attendant failures in railroad operations, exploding locomotive boilers, and collapsing trestles. Soon other states began enacting similar legislation. The last state to require registration of engineers to protect public safety was Texas, motivated by a massive accident. In 1937 an explosion of leaking natural gas killed more than 400 students and teachers.

The first major problem regarding registration of engineers was that in most states, only engineers in the civil (including structural) discipline had to be registered

The great number of malfunctions and failures in military products during World War II called for better equipment that failed less frequently and required less maintenance. Often, failures occurred at inconvenient times and locations when operability was critical. The term "reliability" came to connote the probability that a piece of equipment or a component would perform its intended function satisfactorily for a prescribed time under stipulated conditions. A malfunction might include a complete failure, an unprogrammed operation or premature operation (false start), an erroneous output, or a cessation of operation at a time it was not designed to do so. Reliability and safety are closely related but are not synonymous, for example, an unreliable firearm that will not discharge may be completely safe, because it is inert and its bullets could harm no one.

The method developed to study and calculate the probability of success of a product, the cause of any malfunction, and the downstream effects a malfunction might generate is known as Failure Modes and Effects Analysis (FMEA). An FMEA is often used to help compute approximately the time a complex piece of equipment or product will function without failure. Programs of timely replacements can then be prepared. In an FMEA each proposed design of a product is progressively separated into its major assemblies, subassemblies, and component parts. The length of time each of the parts will last and the percentage of failures in a large group is determined by actual test. This is the estimated probability of success or reliability of the part. All subassembly, assembly, and product reliabilities are then computed in reverse order, based on those of the components, until an overall probability of successful operation is determined.

The first phase of use any component or product may result in "break-in" or early failures. The number of early failures will decrease as unsatisfactory parts are replaced, until there is a constant failure rate. During the second phase, called the useful-life period, there may be random failures in accordance with the estimated unreliability (1.0– reliability). Next the product is said to enter the wearout phase, where its probability of failure begins to increase. Although in some cases the product may continue in use, its reliability cannot then be computed.

To improve the calculated reliability of a complex assembly, various means can be used. The simplest method is to increase the reliability of the component parts by using improved materials and manufacture, better assembly, good quality control, or derating or redundancy.

Derating involves reducing the stress on a component or product, such as by using part capabilities greater than those actually calculated to be needed. Derating can often be done by reducing a part's load, or by lessening the temperature and any heat it might have to dissipate.

Redundancy can involve either series or parallel circuitry or both (see the simple examples which follow). Computations for components in series generally follow the Product Law of Reliability: Reliability of the product, RP = RA x RB x RC x RD.

Assume three reliabilities of 0.95, 0.80, and 0.90 in the circuits in the accompanying figures. In a nonredundant series design all components must be working for success to occur. The probability of success for the nonredundant series diagram is (.95 x .80 x .90) or .684. In the parallel redundant circuit, the probability of success is [1 – the probability that all three components fail]. The probability of a component's failing is [1 - reliability]. In the parallel design, success occurs if any component path is functioning. For the parallel redundant circuit, the overall reliability or probability of success would be: 1.0–[(1.0–.95) (1.0–.80) (1.0–.90)] or 0.999.

Parallel redundancy could be (and has been) used to lessen the probability of a complete failure in a critical system. In most products where redundancy is used, usually two but as many as four circuits or assemblies are incorporated. For optimal success different types of circuits, assemblies, or functions are used. For example, it is desirable to minimize any possibility of electrical failures during an operation in a hospital. A properly designed redundant electrical system might then use power from a central distributing system, a diesel-powered generator, and a bank of batteries.

To act as an effective safeguard, any redundancy must be true, i.e., independent of other failures in the redundant system. On March 22, 1975, a fire started in a nuclear power plant at Brown's Ferry in Alabama. The two similar electrical circuits to the reactor's safety system were supposedly redundant. But because they were located close together instead of being separated widely, the fire almost caused a failure to the safeguard system. The system had not been truly redundant, although the two circuits had been designed and installed to act so.

Launch of a space shuttle involved two boosters of solid propellant and one of liquid. Not only was there no redundancy, but failure of any of the three before the shuttle reached orbit would result in disaster.

FIGURE 6–2 Malfunctions, Failures, and Reliability

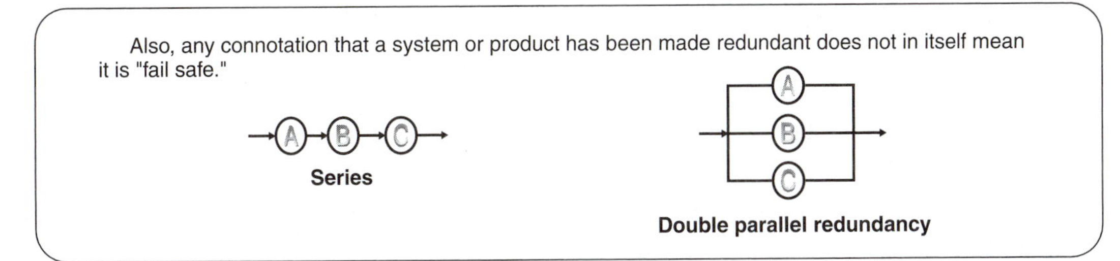

FIGURE 6–2 Continued

CAUSES OF CHEMICAL INDUSTRY FIRES AND EXPLOSIONS: 1960-1967*

HAZARD FACTOR	ACCIDENT DUE ENTIRELY OR IN PART TO THIS FACTOR	
	Number	Percent
1. *Plant site problems*–unusual exposure to natural calamities such as windstorms, floods, and earthquakes–poor location with respect to adequate water supply and other utilities–exposure to severe hazards of nearby plants–unreliability of public fire and emergency protection–traffic difficulties for emergency equipment–inadequate waste disposal facilities –climate problems requiring indoor facilities for hazardous processes.	2	1.64
2. *Inadequate plant layout and spacing*–congested process and storage areas–lack of isolation for extrahazardous operations–exposure of high values and difficult-to -replace equipment–lack of proper emergency exit facilities–insufficient space for maintenance or emergency operations–sources of ignition too close to hazards–critical plants areas exposed to hazards–in adequate hazard classification of plant area.	3	2.46
3. *Structure not in conformity with use requirements*–disregarding code requirements with regard to the buildings, electrical facilities, drainage, etc. –lack of fire resistive structural supports where required–failure to provide blast walls or cubicles to isolate extrahazardous operations–inadequate explosion venting and ventilation of building–insufficient exit facilities–electrical equipment not in conformance to codes–unprotected critical wiring.	3	2.46
4. *Inadequate material evaluation*–insufficient evaluation of the fire, health, and stability characteristics of all materials involved–lack of established controls for the quantities of material involved–inadequate assessment of effect of processing environment on hazard characteristics of materials–lack of information on dust explosion tendencies of materials–toxicological hazards of materials not properly evaluated–incomplete hazard–material inventory for the plant–improper packaging and labeling of chemicals.	26	21.31
5. *Chemical process problems*–lack of required information on process temperature of pressure variations–hazardous byproducts or side reactions–inadequate evaluation of the process reactions–inadequate evaluation of environment–requirement for extreme process conditions overlooked.	14	11.48
6. *Material movement problems*–hazards due to lack of control of chemicals during unit operations–inadequate controls for hazardous dusts –piping problems–improper identification of hazardous materials during transportation–loading and unloading problems in the plant–flammable gas and vapor problems–inadequate control of heat transfer operations–explosion problems in pneumatic conveyors–waste disposal and air pollution problems.	5	4.10

FIGURE 6-3 Causes of Chemical Industry Fires and Explosions: 1960–1967 Continued on page 94

7. *Operational failures*–lack of detailed descriptions and recommended procedures for operating all sections of the plant–poor training program–lack of supervision–inadequate startup and shutdown procedures–hazards due to poor inspection and housekeeping programs–inadequate control of hazards through permit systems–lack of emergency control plans–inadequate drills.	20	16.39
8. *Equipment failures*–hazards built into the design of equipment–corrosion or erosion failures–metal fatigue–defective fabrication–inadequate controls–process exceeded design limitation–poor maintenance program–inadequate repair and replacement program–lack of "fail-safe" instrumentation–poor check on construction criteria or material specifications.	44	36.06
9. *Ineffective loss prevention program*–inadequate support of top management–lack of assigned responsibility–poor accident prevention program–insufficient fire protection manpower, equipment, and organization–ineffective explosion prevention and control program–lack of emergency planning–poor check on boiler and machinery risks–lack of loss prevention coordination with other plant groups–ineffective investigation of accidents.	5	4.10

* *Hazard Survey of the Chemical and Allied Industries* (Technical Survey No. 3), American Insurance Association, New York 1968.

FIGURE 6-3 Continued

in order to practice. About the time of passage of Wyoming's law, about 50 percent of all engineers were in the civil discipline (they now constitute 12 to 14 percent of all practicing engineers). All other engineers, except in a very few states, were (and still are) exempted from registration. After a number of years, some states began to permit other types of engineers to register, although they were not required to do so. A number have done so because it permits them to use the title "consulting engineer."

Another major deficiency is that in most states there is no requirement that an applicant even be a graduate of an engineering school. (After World War II, the federal government initiated the requirement that only graduates from accredited schools could be designated and employed in any professional engineering capacity.)

Third, the state examinations all applicants must pass to practice are almost devoid of questions regarding safety and accident prevention.

Registered civil engineers approve design and construction drawings to ensure technical adequacy. No such state requirement exists or is mandatory for any other type of engineer, such as those in mechanical, electrical, or chemical disciplines. Thus, other than for civil engineering, there has never been any state requirement to certify or approve capability of activity of any engineer who designs the contents of any building. No engineer, even when state certified, is needed to review or approve any part of the safety of any nuclear facility; any part of a chemical processing or disposal plant; gas, petroleum, or chemical terminal, transportation or storage facility; public or private vehicle, such as an aircraft, locomotive, bus, or subway train; equipment or fire protection facilities in any high-rise building or industrial plant; or the equipment contained in any manufacturing facility.

POSSIBLE IMPROVEMENTS IN REGISTRATION

Registration of all engineers and reorientation of questions on safety and accident prevention could significantly reduce the hundreds of billions of dollars of accident-related costs throughout the United States. Because registration of engineers is a state function, current requirements differ. Although a few of these provisions are now in effect in one or two states, no state has all in effect. The following recommendations have been made for registration.

1. All engineers to be registered to practice.
2. All engineers should be graduates of approved engineering schools.
3. To register, an applicant must pass an examination oriented chiefly to and on safety and health. This will necessitate adequate instruction in accident prevention and safe design in engineering schools. Common causes of design defects that affect safety are listed in Fig. 6-4.
4. To maintain currency, all registrants to be required to annually complete engineering courses to maintain or upgrade capabilities. A suitable percentage of such courses should be oriented toward safety and health.
5. The title "consulting engineer" to be eliminated because all will be registered.
6. State boards of registration should include registered engineers more oriented toward safety and health. (In California and states with similar provisions, the Board of Registration should be reconstituted. At present, it is composed of seven public members, five registered engineers, principally of the civil discipline, and a land surveyor, who is covered by provisions of another registration law. It should be reoriented to contain five public members; three civil, mechanical, electrical, chemical, or similar engineers; and three engineers of safety, fire prevention, human factors, biochemical, or similar disciplines.)

Efforts of all engineers, registered or not, often are undertaken and controlled by the managers for whom they are employed. Sometimes there is a close relationship between managers and engineers in which each influences the other. Managers generally rely on the expertise of engineers. If a manager gets and follows the wrong technical information and advice, accidents and disaster may result.

Certification

Some professional organizations provide certification by examination. The certifying organization usually has a board made up of selected members of the profession. This board provides the applicant with a written examination of knowledge and skills and also reviews the applicant's education and experience. Certifications for the safety professional include the Certified Safety Professional (CSP), Certified Hazard Control Manager (CHCM), Certified Product Safety Manager (CPSM), and Certified Hazardous Materials Manager (CHMM). Closely related to safety are the Certified Industrial Hygienist (CIH) and the Certified Professional Ergonomist (CPE)/Certified Human Factors Professional (CHFP) certifications. All certifications are not equal, some being narrower in scope than others, and the applicant should carefully investi-

A defective design of a consumer product is one in which the product is more dangerous than an ordinary consumer would expect, when the product is used in an intended or reasonably foreseeable manner. A design is also defective if its benefits fail to outweigh the inherent risk. All defective designs are deficient. Some deficient designs may not be defective by this definition, because the deficiency does not effect the product's safety.

A design deficiency which affects safety may involve not only an error in an engineer's computations, but a failure to properly provide a safeguard when one is possible or advisable, or to protect against any adverse condition noted in Fig. 6-1. The deficiency may result because a designer or design:

1. creates an unsafe characteristic of a product.
2. is faulty so that it causes an accident.
3. is faulty so that the product or operation will not take place as envisioned.
4. does not envision, consider, or determine the consequences of an error, failure, action, or omission.
5. fails to foresee an unintended use of a product or its consequences.
6. fails to properly prescribe or evaluate an operational procedure where a hazard might be present.
7. is incomplete, in error, or confusing. The designer may not realize that the worker or user is probably not as technically qualified and cannot grasp the designer's intent.
8. violates normal tendencies or capabilities of a worker or user.
9. places an unreasonable stress on the operator.
10. fails to minimize or eliminate possibilities of human error or leads to errors.
11. creates an arrangement of operating controls and indicating meters that is conducive to errors or increases reaction time in an emergency.
12. leaves safety of a product up to its user to avoid an accident.
13. fails to warn of a hazard.
14. fails to provide adequate protection in a worker's personal protective equipment.
15. is faulty in that worker will not wear safety equipment because it is too heavy, cumbersome, restricts breathing or movement, interferes with work, or has any other adverse feature.
16. provides a warning, such as by a label, instead of providing a safe design to eliminate it.
17. fails to provide a suitable safety device where a hazard exists in that the:
    a. safety device is inadequate and does not provide the service intended when required.
    b. safety device is located where it is inaccessible in an emergency.
    c. safety device can easily be removed or bypassed by an operator or user.
18. fails to provide adequately against an adverse environment.
19. fails to avoid use of a toxic or other hazardous material without providing adequate safeguards.

Many of these deficiencies have resulted in accidents and litigation.

FIGURE 6–4 Design Defects which effect Safety

gate the organization and persons behind each one. There are few controls on this service; however, employers will sometimes require a particular certification. Generally, a certification not only should require examination of knowledge, skills, and experience, but also should set requirements for continuing professional maintenance and upgrading of the professional's knowledge and skills. Certification is not a substitute for a formal education or an academic degree in a given field, but it may be recognition of professional standing.

BIBLIOGRAPHY

[25] Jerome Lederer, Guggenheim School of Aeronautics, "Infusion of Safety into Aeronautical Engineering Curricula," *Third International Conference*, Royal Aeronautical Society, Brighton, England, Sept. 3–14, 1951.

[26] "Operator Safety," in *Engineering*, May 1974, pp. 358-363.

EXERCISES

6.1. Why has it been said that engineers can cause more fatalities and injuries than can physicians?

6.2. List six causes of accidents.

6.3. List five deficiencies designers might create in a system, product, or operation.

6.4. Why was the law for registration of engineers first enacted? Do all engineers have to be registered to practice?

6.5. In your state, what are the requirements for an engineer to be registered? If the engineer has been registered, does that mean he or she has been through any training, examination, or course on safety and accident prevention?

6.6. What is the definition of reliability? Why was it especially of concern during World War II? What is an FMEA?

6.7. What does derating a component mean?

6.8. What is series redundancy? For what is it used especially?

6.9. What is parallel redundancy? Why is it used in some hospitals?

6.10. If a system is redundant, does that always mean it is "fail-safe"?

6.11. What is a defective design?

C H A P T E R 7

Management and Its Responsibilities

Active participation by all managers of any industrial organization, especially those at the top, in carrying out accident prevention programs is so vital that the safety engineer who is not given effective support should (1) begin looking for a job elsewhere and (2) keep a "Pearl Harbor" file.

Without management support the conscientious safety engineer will be in a constant state of frustration; his or her efforts to accomplish the job successfully will be hampered from the start. Avoidance of accidents requires a sustained integrated effort by managers of all departments, supervisors, and other employees at all levels and in all activities. The safety engineer may indicate of what this effort should consist and may determine whether it is being carried out properly and adequately. Only the managers, however, can ensure that it is a properly integrated and coordinated activity.

SAFETY POLICIES

The influence of management must be apparent in the safety policies it sets, the degree to which those policies are observed, and the concern with which it treats any violation. Managers must leave no doubts in the minds of employees that they are concerned about accident prevention. This concern to prevent injury and damage must be sustained continually, rather than intermittently or only temporarily being presented with an accident report.

Unless management can provide this support, accidents will take place, perhaps repeatedly. A safety engineer working under such unfortunate conditions will do well to document all efforts to carry out a successful accident prevention program. Changing legal attitudes are leading more and more toward holding managers and safety engineers, especially managers, responsible for workers' safety. If the manager of the organization is not providing adequate support, the safety engineer must be able to demonstrate he or she was not at fault if an accident should occur.

Such a file can have beneficial results other than simply safeguarding the safety engineer against any managerial recriminations or claims of deficient job performance. It will tend to protect him or her in case there is an injury to an employee or costly damage, which might lead to criminal or personal charges or liability. Putting all affecting factors, conditions, and actions on paper in a logical sequence will make for an orderly presentation of any situation. Managers will respond more readily to presentations of all pertinent facts, in a foresighted, coherent, and well-thought-out manner, than they would to unsubstantiated, rambling dissertations.

Other than for normal routines, safety engineers generally have to request management approval for any safety actions or expenditures—modifying or replacing equipment or authorizing a new safety rule or procedure. In addition to having all pertinent facts regarding the situation that brought about the request, the safety manager should be aware of the possible answers to questions or directions by the manager. When the recommended corrective action would require expenditure of funds, the manager's comment, action, or question probably would be among those listed in Fig. 7-1. This list provides forewarnings so the safety engineer can make cogent presentations and be prepared for the manager's decision. The list was arranged in order of increasing costs. For example, if there is no violation of a mandatory standard or code, managers will usually attempt the least costly solution that will resolve the problem. Actions recommended or requested by the safety engineer could best be achieved with the same aim—lowest cost. Decisions by managers become matters to be acted on. Where actions are to be undertaken by others, the file becomes a means by which their progress can be monitored.

1. If the cost of correction is minor, approve with no quibbling.
2. If the cost of correction is high, the manager will ask the engineer to determine whether
   a. such a problem actually exists.
   b. it is mandatory in a standard, code, or regulation.
   c. the seller of the hazardous part of the equipment is contractually obligated to pay for the correction or for a safer item.
   d. more detailed studies can be made, including determination of the probability an accident might occur, with and without correction.
   e. a risk/cost/benefit analysis had been made.
   f. any liability resulting from the injury or damage caused by an accident would be covered.
   g. if a safeguard or correction must be made, what would it cost? Could the correction or safeguard be:
      - procedural: providing warnings and cautions about the hazards to operators.
      - an "add-on" device: low-cost item that would not be of great expense.
      - partial replacement: a redesign or add-on for only the portion of the equipment or operation that is legally required.
      - total replacement: replacing the entire equipment or operation with one which is safe.
      - abandonment: shutting down the equipment or operation because correction is too costly.

FIGURE 7–1 Managerial Decisions and Comments on Safety Measures

Managers will come to realize that if the safety engineer is concerned enough to make a problem a matter of record, it should be given greater consideration.

OSHAct AND MANAGEMENT

The OSHAct places the responsibility for employee safety principally on the employers. When a business is owned by one person, it is easy to determine who the responsible "employer" is. When the organization is a corporation, there may be a problem determining who can be held responsible for any violation of the OSHAct. It bears repeating here that under the California Cal-OSHA this may be any "employee having direction, management, control, or custody of any employment, place of employment, or other employee."

A monetary penalty, such as a fine, can be paid by a corporation; a penalty, such as imprisonment indicated in paragraph (e), can be imposed only on an officer of the company. It therefore becomes a matter of determining which corporate officer had the responsibility for ensuring that the OSHAct and OSHA standards were observed and not violated. This may require judicial review in a federal court in states that have not assumed responsibility for administration of the OSHAct or in the courts of states which have been granted authority to undertake this administration. In either case the legal proceedings may cost more, in terms of dollars, than the fine imposed.

Previously, managers were willing to invest in accident prevention programs when they were forced to, and to the degree to which they believed the savings realized would at least counterbalance any costs. The interest in accident prevention also depends to a large extent on the opinions of the individual managers regarding what they considered the risks their companies take. Generally the fines applied by OSHA or the states against firms have been far less than accident-generated costs to workers killed or injured, their families, the states, or the federal government.

Theoretically, if firms fail to react suitably and favorably by providing workplaces free from recognized hazards, and consequently accidents do occur, penalties may become extremely personal as managers are imprisoned. This would undoubtedly alert other managers that the government is adamant in its desire to make workplaces safe. Soon after the OSHAct and OSHA came to pass, it was believed that fines and penalties would be great enough to act as deterrents. Unfortunately, OSHA or state fines and penalties generally have been regarded more as threats to company managers to force adherence to standards than as actual economic losses to fear. Further, any managerial apprehension about personal losses was lessened by the protective corporate structure and insurance coverage.

ACTIONS AGAINST MANAGERS

In *United States v Park*, the Supreme Court pointed out that a top manager can delegate tasks to be accomplished but, in effect, cannot delegate the responsibility for ensuring that those tasks have been done. In that case, the Food and Drug Administration charged that Acme Markets and its president, Park, had violated the Federal Food, Drug, and Cosmetic Act by permitting food held in a warehouse to be contaminated by rodents. Acme pleaded guilty, but Park would not. He admitted responsibility for seeing that sanitary conditions were maintained but argued he had assigned the responsibility to "dependable subordinates." The case was first tried in a federal court, where Park was found guilty; the decision was reversed in a Court of Appeals and reversed again in the Supreme Court. In 1975 the Court indicated that companies may assign responsibilities but they must follow up on such assignments.

The Court's opinion made clear that all officers who have the power and authority to ensure compliance with the law (in this case, the Food and Drug Act) have a duty to do so. An executive with the responsibility and authority to prevent a violation but who fails to do so may be held criminally liable. Although this precedent was set in a case involving food, a violation of a safety law could result in similar decisions by other courts in similar situations. When a government agency searches for a target because of accidents or violations of a health or safety law, managers are sitting on the bullseye.

Managers are increasingly being held responsible for accidents and the existence of unsafe work places. Figure 1-3 points out an accident that took place before the advent of OSHA. Although the responsible manager was initially sentenced to 20 years and 6 months in jail after seventeen men were killed in an accident, after a retrial his sentence was reduced to probation for three years and a fine of $6,875.

After an accident in the nuclear power plant at Three Mile Island in Pennsylvania in 1979, there were demands for the removal and jailing of three top executives. Although findings laid the blame for the accident on the designers of the system, who failed to incorporate proper human engineering, and on the operators, one manager was relieved of his job.

After repeated failures to observe Cal-OSHA standards, two men were killed in an accident because of the presence of toxic gases after entering a manhole. The plant manager was convicted of involuntary manslaughter and initially sentenced to a year in the county jail. The sentence was reduced later to a fine of $500 and 300 hours of community service as conditions of probation.

In July 1985, three former executives of a silver recycling plant in Maywood, Illinois, were sentenced to 25 years in prison and each fined $10,000. The state prosecutor in the case claimed the death was neither involuntary manslaughter nor an accident, but murder. The conviction of the three men was the first in the country of such job-related deaths. In a plant used to recover silver on photographic film, a worker had died because he had not been directed, taught, or otherwise safeguarded so as to prevent his inhaling the cyanide fumes. It was claimed that the three company officers had clearly been aware of hazardous conditions in the plant but did nothing to warn the workers of the dangers of the cyanide. The court held that directors, officers, and high managerial personnel in a corporation act both for the benefit of the corporation and for themselves. The prosecutor in this case contended, and evidently the court accepted, that corporate responsibility is applicable in civil cases. In this instance, the charge had been murder and therefore one of criminality.

There had been other indictments of corporate officers for criminally negligent homicide when persons were killed in the workplace, but the case in Illinois was the first one in which corporate officials were charged with murder and convicted of killing an employee. It was part of the growing effort to fix individual responsibility for accidents and their prevention.

Numerous examples of managerial deficiencies can be cited. Fine[27] listed these as shown in Fig. 7-2. The results ensuing from accidents under both good and ineffective managerial control are described in Fig. 7-3. As Price stated in 1985, "Don't accidents in the work environment fall outside of management's control? And don't these accidents occur because of random events? How can management be held accountable? Yet, repeatedly the justice system is placing responsibility at management's door. . . . The courts have developed two important tests to determine management's responsibility: foreseeability and prudence." [7] These dual tests will be examined carefully in the chapter on safety analysis. The answers to Price's questions are: (1) management bears a responsibility, and (2) occurrence of accidents can be modified by systematic factors subject to good management. Accidents are not simply random events.

Safety cannot escape legal issues, and legal/regulatory knowledge and skill are required for the practice of safety.[28] The questions that companies and their management face in court are: (1) whether or not a reasonable effort was made to anticipate undesirable events (foreseeability), and (2) whether or not the action taken by management was reasonable and prudent. The emphasis is upon *reasonable.* A reasonable manager will analyze or have analysis performed. Good safety analysis is the job of the safety engineer. In many cases, the manager will delegate safety analysis to a properly educated safety engineer or safety engineering consultant. For one major professional certification in safety there are specialty areas of examination.

Thus, managers bear the ultimate responsibility for all activities under their control, although the faults may be those of subordinates. On the other hand, "A factor which is being recognized more and more by operations managers is that they, in the past, have borne the brunt of criticism due to inadequately designed systems. That is to say, when a badly designed system reaches the field and the inevitable operator error happens, the operator or the operational management supervisory system was blamed for carelessness or laxity . . . . It is becoming more and more important that the designer must design and defend her/his designs . . . [to a greater degree than has been necessary in the past]."[29] Thus in a company with a product that is deficiently designed so that injury, death, or monetary loss takes place, it is not the designer against whom any legal action is taken; it is the manager (although the designer may receive reprimand or be fired).

MANAGEMENT ATTITUDES TOWARD SAFETY

Fortune magazine stated[30]

> Many corporate managers continue to believe that careless workers are really to blame for accidents. But a 1967 survey of industrial injuries in Pennsylvania concluded that only 26 percent were the result of employees' carelessness. Even that figure does not tell the whole story. As a General Motors pamphlet on safety notes: "It is impossible to have an accident without the presence of a hazard." The pamphlet goes on to say that "carelessness" is not a good word to use in connection with analyzing the cause of an accident—it is human nature for people to make mistakes, to take short cuts. Even if "accidents will happen" in other words, it is management's responsibility to eliminate—as much as possible—the conditions that bring accidents about. Beyond that, workers have almost no control over the health hazards of dust or toxic gases whose effects they do not understand and whose presence they may not even be able to detect.

ACCIDENT CAUSES TRACED BACK TO MANAGEMENT RESPONSIBILITIES

IMMEDIATE CAUSE	EXAMPLE	POSSIBLE UNDERLYING CAUSES	POSSIBLE MANAGEMENT FAILURES
			INADEQUATE:
1. Poor housekeeping	An employee trips and falls over equipment left in an aisle Material poorly piled on a high shelf falls off	Hazards not recognized Facilities inadequate	Supervisory training Supervisory safety indoctrination Planning, layout
2. Improper use of tools, equipment, facilities	Using the side of a grinding wheel instead of the face and the wheel breaks Someone using forklift truck to elevate people–person falls of Someone using compressed air to clean dust off clothes–eye injury	Lack of skill, knowledge Lack of proper procedures Lack of Motivation	Employee training Established operational procedure Enforcement of proper procedure Supervisory safety indoctrination Employee training Employee safety
3. Unsafe or defective equipment, facilities	Portable electric drill without ground wire Axe or hammer with loose head Car with defective brakes, steering	Not recognized as unsafe Poor design or selection Poor maintenance	Supervisory safety indoctrination Employee training Employee safety consciousness Planning, layout, design Supervisory safety indoctrination Equipment, materials, tools Maintenance, repair system
4. Lack of proper procedures	No requirement to check for gas fumes before starting engine–explosion No definite instructions requiring power to be locked out before maintenance is done	Omissions Errors by designer Errors by supervisor	Operational procedures Planning, layout, design Supervisory proficiency
5. Improvising unsafe procedures	"Rube Goldberg" haphazard temporary expedients, without proper planning	Inadequate training Inadequate supervision	Established operational procedure Enforcement of proper procedure Supervisory safety indoctrination Employee training Employee safety consciousness Supervisory safety indoctrination Employee selection, placement
6. Failure to follow prescribed procedures	Short-cuts bypassing safety precautions Operation will only be done once; take a chance	Need not emphasized Procedures unclear	Enforcement of proper procedures Supervisory safety indoctrination Operational procedures

FIGURE 7–2 Accident Causes Traced Back to Management Responsibilities Continued on page 104

7. Job not understood	Employee uses wrong method, doesn't follow instructions	Instructions complex Inadequate comprehension	Operational procedures Planning, layout, design Employee selection, placement
8. Lack of awareness of hazards involved	Not realizing rotating shaft was dangerous Not realizing fumes were dangerous Not realizing that hydrogen from battery-charging operation could explode	Inadequate instructions Inadequate warnings	Supervisory safety indoctrination Employee training Employee safety consciousness Planning, layout, design Safety rules, measures, equipment Operational procedures
9. Lack of proper tools, equipment, facilities	Cart too small for hauling large items Auto maintenance done without proper wrenches–cut knuckles	Need not recognized Inadequate supply	Planning, layout, design Supervisory safety indoctrination Equipment, materials, tools
		Deliberate	Morale, discipline
10. Lack of guards, safety devices	Machine has exposed belt and gear–severe cut No warning horn on vehicle–pedestrian hit	Need not recognized Inadequate availability	Planning, layout, design Safety rules, measures, equipment Supervisory safety indoctrination Employee safety consciousness Equipment, materials, tools
	No guard rail on a scaffold 10 feet high	Deliberate	Operational procedures Morale, discipline, laziness

FIGURE 7–2 Continued

Prior to the passage of the Railway Safety Act in 1893 most managers blamed accidents on the unsafe acts of the personnel involved. (Railroad statistics still reflect this attitude. Bureau of Railroad Safety reports still list "negligence of employees" as the cause of more than half the railroad accidents that occur every year.) However, Fig. 1-1 indicates the reduction in deaths and injuries that resulted from the mandatory installation of automatic couplers on railroad cars. The number of trainmen involved in accidents increased, yet the number killed or injured in coupling accidents dropped substantially. Installation of the mechanical safety device was undoubtedly the principal factor.

Each manager endeavors to achieve the organizational mission most economically and effectively. Few will accept without objection any restraints or added requirements that could adversely affect that endeavor. A common attitude is that safety requirements will do just that. In most cases this is a misconception; generally, improvements in working conditions will increase productivity and improve employer-employee relations. Workers have to spend less time thinking of their own safety.

On May 23, 1939, during a trial dive in a series of acceptance tests, the American submarine *U.S.S. Squalus* sank to the bottom in 240 feet of water off the New Hampshire coast. A later investigation reported that according to the indicator lights in the submarine, the main air induction valves were closed as it submerged, but the valves were actually open. Twenty-six men of the 59 aboard were drowned as she sank.

The admiral in command had prescribed procedures for all test-dive operations and any emergencies that might arise. When, within the specified time, no surfacing report had been received that the test-dive had been successful, he became concerned. The admiral ordered another sub, about to start a long voyage, to proceed through the area of the missing boat, directing it to look for any signs of the *Squalus*. The *Squalus* was found and its predicament determined within hours.

Although the submarine was equipped with Momsen lungs for escape (which, like similar British and German designs, were found after the war to be and were declared ineffective), their use was considered inadvisable by the commander because of the water depth and temperature, except as a last resort. The admiral had rescue operations initiated immediately. Distant highly trained and specialized personnel, suitable rescue ships, and equipment were brought hurriedly to the site and in action within 24 hours; all of the trapped men who had not been drowned immediately were rescued 16 hours later.

Eight days after the *Squalus* accident, a similar acceptance test-dive of the British *H.M.S. Thetis* took place. The bow of the *Thetis* flooded when a torpedo tube was inadvertently opened and she sank to the bottom of the Irish Sea. There the water was only 130 feet deep, and because the craft sank at an angle, 18 feet of her 275-foot length remained above the surface. Unfortunately, although the regular crew normally consisted of 53 men, 103 persons were in the sub. The manager of the operation had permitted aboard other supervisors, observers, technicians, and seemingly almost anyone who had a desire to go, including even to waiters to feed the horde.

When a small escorting ship failed to find out what happened to the sub, it attempted to get a message to its headquarters. Receipt of the message was delayed when the messenger boy blew a tire, which he then had to repair. Divers on a tug sent to provide assistance proved to be ill-equipped to do the rescue work. The only other support ships available were a flotilla of destroyers under a commander who knew little about submarines. Each new rescue attempt was developed only after the previous one had failed.

The lack of managerial control cost 99 men their lives (seven men tried to use the escape apparatus, but only four reached the surface alive). All the others died of suffocation. The overcrowding had reduced the survival time available for a successful rescue. In addition, lack of planning for possible emergencies and the use of hurriedly improvised procedures (that failed) increased the time it took to reach the trapped men.

Thus, differences in management control can have different end effects.

FIGURE 7–3 The *Squalus* and the *Thetis*

The Survey of Working Conditions conducted by the University of Michigan[1] found that of 18 areas of concern investigated, 86.7 percent of all workers interviewed considered that "becoming ill or injured because of my job" was foremost on the list. In addition, 80.7 percent were concerned about "physical dangers or unhealthy conditions on my job." Minimization of such concerns will undoubtedly improve employee morale and productivity. In addition, an employee who has to pay less attention to protecting himself or herself against an imminent danger can work more effectively and efficiently. Experience has shown that an employee will work much faster in a safe, compatible environment than in one in which he or she feels endangered. It has also been found that any safeguard engineered into any piece of equipment is outstandingly more beneficial than any warning or caution to be careful.

MIDDLE MANAGERS

Top executive managers often create safety problems for middle managers. These lower-echelon managers are often pressed for production that is hazardous to their personnel. Top executives often make a decision and then tell middle managers to get it done "and I don't care how you do it." The result is often increased production at unsafe rates of work or supervision, or reduction in expenditures for necessary safety equipment.

In March 1987, newspapers published comments that great increases in producitvity requirements by managers had greatly increased stresses and accidents to workers. To reduce costs, managers were requiring workers to do the same work as three or four had previously done. Federal statistics for 1984 and 1985 show that injuries increased 12 percent over those in 1982. In one industry, from 1982 to 1985, workdays lost because of injuries increased from 77.7 to 97.7 per 100 workers while employment dropped from 394,300 to 304,900 workers. Management-generated stresses on workers are also indicated by manufacturing output that rose over 30 percent while worker employment increased only 0.5 percent.

Another contention is that, in their desire to reduce costs, managers are making unsafe cutbacks that increase accidents. They have reduced equipment maintenance and worker training programs. The deterioration of maintenance so that more accidents occurred, first noted with railroads, has spread to industrial plant activities. Training has deteriorated as workers laid off from one job are transferred to another where the work and required skills are different.

In the early 1990s, many companies trimmed management, staff, and work force in downsizing efforts to meet the challenge of a global economy. Often, middle management was the target, thus increasing stress on the management and workers remaining in the "streamlined" organization. Attention to safety can be easily lost when personnel see their jobs in jeopardy or the management personnel who cared for safety are eliminated. Some companies moved activities "offshore" to foreign countries where standards of safety and health are less developed. Workplace safety, at the beginning of a new millenium, is now a global concern, drastically in need of international voices.

FOREMEN/FOREWOMEN AND SAFETY

A prime requisite for any successful accident prevention program is to leave no doubt in the mind of employees that supervisors are concerned about their safety. Supervisors such as low-level managers and forewomen or foremen have close contact with workers and can provide the closest control of all activities, including safe operations. Similarly, the supervisor is often at the receiving end when there are recriminations because of accidents. Unfortunately, foremen/forewomen have so many duties and schedules to carry out that they are often overworked, and certain duties are not accomplished adequately. Neglect of accident prevention, which often occurs, can have the most serious consequences.

Although it is frequently and truly said that safety is everyone's responsibility, many people have a tendency to ignore or forget this. The supervisor, manager, or foreperson must ensure that such ignoring or forgetting does not happen. Where designed safeguards are lacking, failing to address hazards by the use of procedures and warnings can lead to disastrous results. Sometimes, when safety is everybody's business, it becomes nobody's business. The supervisor, manager, or foreperson is key to ensuring that this does not happen.

For the worker, a foreperson represents management. This person has to exert his personal authority and influence and has to see that the intentions and orders of the management are carried out. If the foreperson does not take safety seriously, those under her or him will not either. On the other hand, if the foreperson is convinced of the importance of safety, shows that safety has to be considered all the time, and personally does everything that reasonably can be done to prevent accidents, workers will follow his or her example.[31]

PROCEDURAL SAFEGUARDS

Use of procedural safeguards is a less satisfactory means to avoid accidents than is a well-designed safeguard in the equipment or operation. (As indicated previously, when tested by court decisions, procedural safeguards may not be acceptable.) Procedures or safety work rules issued by managers depend on workers to observe them and on foremen to ensure they are observed. Managers evaluating a procedural safeguard must also consider the effects of the added load on the front-line supervisor.

Prior to passage of the OSHAct, management decisions to provide safety devices or equipment were to some extent predicated on one or more of four factors: economic considerations (including added cost of insurance if adequate built-in safeguards were not provided), local codes, past or potential litigation, and employee (union) relations.

With the enactment of the OSHAct and the stricture that places of employment be free from "recognized hazards," the question of cost becomes largely academic. The issue is no longer whether the cost for safety can be justified economically. Many managers have said that standards and safety measures should be imposed only when they are economically justified as computed by risk/cost/benefit/effectivity analyses (see previous chapters). However, both because the costs of lives of persons can never be calculated definitely before an accidental fatality and because costs analyses differ so greatly from later actualities, managerial calculations are never accepted by OSHA or courts. In 1987 this was definitively ruled out, the indication being that when safety of personnel is at stake, cost cannot be considered.

MANAGEMENT AND SUPERVISION

Above all, managers must maintain an active, effective interest in the safety effort. Some means by which a manager responsible for any operation can have an effective program are to:

1. Establish in writing and disseminate specific and firm safety policies for the organization, and then ensure they are carried out.
2. Provide a coordinated effort, integrating the safety efforts of all organizations concerned. Many of these functions are described in the following paragraphs.
3. Direct the participation of all subordinate organization heads in the safety effort, with specific responsibilities assigned to each. Ensure that each manager passes on suitable guidance to personnel under his or her jurisdiction.
4. Establish a safety element that reports directly to the manager to be sure the safety program is carried out properly and effectively. This safety element should ensure that all organizations are familiar with OSHA standards pertinent to their operations; should be responsible for seeing that the standards are observed or action is taken to ensure compliance (such as making higher-level managers aware of any deficiencies); and should be the designated party to accompany any OSHA compliance officer or other safety inspector who may visit the plant.
5. Issue work rules or operating procedures to guide the safe conduct of all employees, including having:
   a. Detailed emergency procedures specifying when a dangerous condition exists or a critical failure or accident occurs. In addition to work rules, these procedures should include preestablished warning signals or alarms, safe areas and evacuation routes, communications to notify support activities (such as a fire-fighting unit), and availability and use of protective clothing and rescue equipment. These procedures should be familiar to all personnel participating in plant activities.
   b. Each potentially hazardous operation reviewed and analyzed to ensure that suitable procedures and safeguards are provided.
   c. Each dangerous operation conducted and controlled according to procedures developed and approved for that operation.
   d. A procedure requiring permission permits and safety surveillance for any hazardous operation.
   e. The buddy system to be used for tasks that involve considerable danger. These tasks include such work as operations on high-voltage equipment, where toxic gases or fumes might be present, cleaning or repairing the insides of tanks or cisterns, or similar situations. When the buddy system is used, constant communication must be maintained between the persons involved. A person in a hazardous situation must remain within sight of his or her buddy at all times or otherwise maintain evidence of well-being or a sign of need for assistance.
6. Carry out safety training on a continuing basis for all supervisors and workers, especially those newly employed or transferred. No supervisor should assume

that any new employee has received adequate safety training but should check each individual's knowledge and work habits. Managers should be sure that:

   a. Workers have had at least the minimum training necessary to be aware of any hazards before they are assigned to their jobs.
   b. Workers are taught the nature of possible hazards, how to avoid exposure, and the actions necessary if a mishap occurs.
   c. Periodic safety training is conducted for all workers.
   d. No worker is permitted to continue in any hazardous work if he or she is found unqualified, unsuitable, or incapable or performs unsafely.
   e. Training drills are held to ensure high proficiency of personnel during emergencies. Designated items of protective and emergency clothing normally used should be used during the drills.

7. Ensure:
   a. Good housekeeping practices are maintained at all times.
   b. Every piece of plant, vehicle, or other item of equipment is operated within the limitations for which it was designed.
   c. New equipment is inspected on receipt to ensure that all desirable safety features and devices have been incorporated or provided. The safety engineer should ensure that the equipment meets OSHA standards. Any deviations from the operating procedure considered unsafe, or any obvious safety deficiency, should be corrected before the equipment is accepted. Appropriate procedural safeguards should have been established for any hazard that could not be eliminated by design.
   d. Access to equipment components during operation, maintenance, repair, or adjustment should not expose personnel to hazards such as electrical charges, moving parts, radiation burns, extremes of heat and temperature, chemical burns, toxic gas, cutting edges, sharp points, or any other hazard unless a suitable safeguard has been provided.
   e. Persons involved in any hazardous operation should be instructed to report promptly any unusual condition or malfunction that would place them in or indicate that they are in jeopardy, such as an unusual odor or irritating substance.
   f. Personnel are directed to report any device, control, equipment, or protective clothing that does not work properly.
   g. Any piece of safety equipment that adversely affects performance is reported.

8. Check that every reported dangerous condition or accident is investigated.

9. Establish a program to monitor and audit operational activities for their safety aspects. A personal visit, even if short, is a highly effective use of time for all managers. Ensure that all subordinates or subordinate organizations are participating effectively in the safety program.

10. Establish a safety review board to evaluate, discuss, and take action on safety problems. The board should review mishap records, hazard and failure reports, and safety studies and analyses to establish that improvements are necessary or desirable. Whenever possible, a manager should act as chairman of the board or delegate the duty to a subordinate high-level manager.

11. Provide budgets adequate for achievement of all safety objectives.
12. Ensure that all OSHA record-keeping requirements are being observed.

SAFETY EFFORTS OF OTHER MANAGERS

Management structures and titles vary from company to company, but no matter the actual titles, some of the organizations involved in the safety program, and their contributions, may be:

- *Personnel:* ensures that workers are trained and physically capable of conducting their duties.
- *Medical:* ensures that prospective employees physically or mentally incapable of the work or questionable in those respects are denied employment. Conducts examinations for workers at time of hiring and then periodically after employment. Provides first aid and emergency treatments.
- *Production:* ensures that unsafe practices are not permitted, even at the expense of increased output.
- *Plant engineering:* ensures that no equipment that could affect health or safety is selected or installed which might adversely affect personnel unless potential hazards are adequately safeguarded. Ensures that safeguards are maintained properly and repaired where necessary.
- *Research and development:* ensures that when new products involve use or testing of materials, any hazards will be brought to the attention of managerial personnel.
- *Plant maintenance:* ensures that good housekeeping is maintained at all times.
- *Security:* ensures that emergency accesses are not blocked, unauthorized persons do not use company equipment, and all vehicles on company property are operated with due care and within safe speed limits. Figure 7-4 indicates checks involving safety that security personnel can make during conduct of their duties.
- *Purchasing:* ensures that safety equipment and materials are procured expeditiously. Materials, parts, and equipment that could be hazardous are obtained using specifications with applicable safety requirements.
- *Legal:* ensures that all managers are aware of the latest laws and judicial interpretations of laws that could affect the company. Takes necessary action if there is any citation for a violation or any litigation because of an accident.
- *Employee relations:* ensures that employee suggestions and complaints regarding safety are evaluated; that needs for use of safe practices are continually indicated in company newspapers, on bulletin boards, and through other media; and that employer-employee-union relationships are maintained at the best level possible. Assists workers in compensation insurance claims.
- *Records:* ensures that all data regarding safety are recorded, collated, and analyzed to spot adverse trends in accident occurrences.

Often there are several management systems within a business management system. These systems overlap and intersect so that they are not mutually exclusive.

SECURITY PERSONNEL SAFETY CHECKS

Plant security personnel can assist safety engineers by reporting hazards which they detect during duty periods. Following are some of the hazards that should be watched for:

- Obstructed lanes for emergency vehicles.
- Obstructed emergency exits and passages to the exits.
- Fire doors blocked or opened.
- Unlocked doors or gates to enclosed hazardous areas, such as electric transformer banks.
- Accesses blocked to fire fighting or other emergency equipment.
- Boxes, bales, cartons, or other items piled so high they destroy the effectiveness of sprinkler systems.
- Boxes, crates, bales, drums, cartons, or other packages piled so high that they may fall over and injure someone in the vicinity.
- Loose wires, piping, metal, tools, or other material, equipment, or supplies over which personnel can trip or fall.
- Wet, oily, or heavily waxed floors on which persons can slip and fall.
- Obstructions or loose objects on stairs or at the head of stairs; or poorly lighted stairs.
- Missing or inoperative egress or exit lights.
- Loose or broken stair treads, hand rails, or guard rails.
- Barriers, lights, and other protective devices missing at excavations, manholes, ditches, or other openings into which a person or vehicle could fall.
- The presence of oily rags, excelsior, or other highly flammable materials outside proper containers.
- Uncovered containers of solvent, fuels, or other flammable liquids.
- Cigarette butts, matches, or other evidence of smoking in nonsmoking areas.
- Failure of welders to have fire extinguishers close at hand.
- Missing fire extinguishers or evidence that they have been used, are not in their assigned locations, or due or past due for inspection.
- Steam, oil, water, fuel, or chemical leaks.
- Unusual odors or fumes.
- Improperly stored or inadequately secured gas cylinders.
- Broken electrical fittings, outlets, plugs, or other devices.
- Broken glass or other sharp, unprotected edges, points, or surfaces.

FIGURE 7–4 Security Personnel Safety Checks

Health and Safety Management is an integral part of each. The intersection of these systems forms the company's Core Management Systems. Figure 7-5 illustrates this concept. Each system in the figure intersects with the Health and Safety Management System, often in vital ways. Core management balances risks. If attention is paid only to safety and environmental issues, for example, then quality, productivity, and costs might suffer too much.

Two management systems are receiving a great deal of attention because they are addressed by International Standards Organization (ISO) standards. The first, Quality Management System, is supported by the ISO 9000 series; the second, Environmental Management System, by the ISO 14000 series. Because of their safety implications, the safety professional should become familiar with these standards; however, their contents are too extensive to be covered here. Many nations and regional jurisdictions have adopted these standards. They provide a basis for the core management of integrated business systems, including safety.

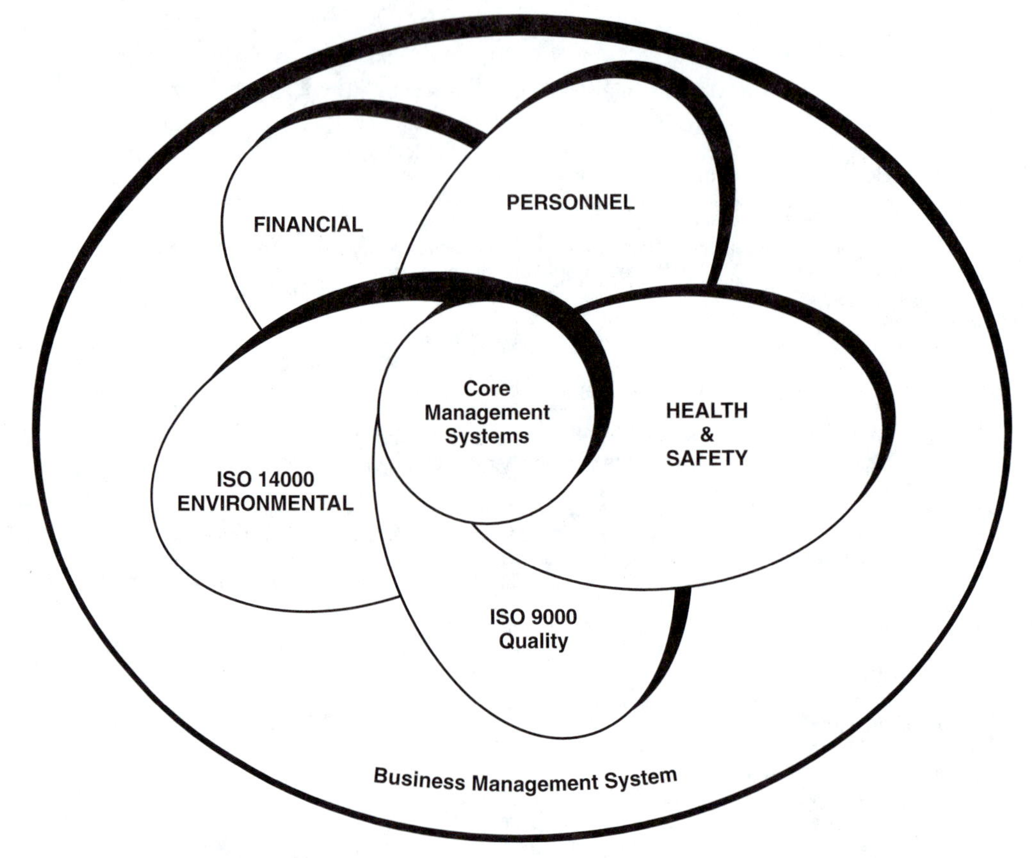

FIGURE 7–5 Interrelationship of Management Systems

Source: Modified from I. Fredericks and D. McCallum, "International Standards for Environmental Management Systems: ISO 14000," *Canadian Environmental Protection*, August, 1995.[32]

HAZARDOUS OPERATIONS

The safety person generally is immediately responsible for accident prevention in all hazardous operations. Each safety person should have the authority to stop any operation considered so unsafe that an imminent danger exists. In addition, safety engineers should be delegated by managers to ensure that:

1. Each operation is monitored to comply with OSHA standards and other established criteria.
2. Each operation considered hazardous is conducted strictly in accordance with procedures and checklists included in approved publications or approved by authorized agencies, using only designated equipment.
3. Each operation the safety person considers to be hazardous and to require surveillance is provided with suitable personnel to permit adequate surveillance.

4. Permit systems are instituted for operations requiring special care. Permits are required for operations that have to be controlled closely, such as for blasting or welding other than those industrial operations carried out automatically or semiautomatically.
5. Hazardous operations are conducted only in designated and approved areas.
6. Controls are instituted limiting access to areas where hazardous operations are to be conducted, with the number of permitted personnel posted conspicuously at each entrance.
7. Personnel permitted to remain at one time in a controlled area are limited to the number that could be evacuated safely in an emergency or supplied with necessary protective equipment to safeguard them.
8. Where advisable, guards are used to control entrances.
9. All personnel and organizations concerned or affected are notified before any hazardous operation is begun.
10. Exits, stairways, and escape routes are marked clearly, with provisions for emergency lighting. Evacuation procedures are posted conspicuously and tested periodically by drills to ensure designated equipment is in place, available, and operable.
11. Operations are forbidden while winds or other meteorological conditions could endanger personnel.
12. Open flames or unprotected electrical equipment are not permitted in areas where flammable or explosive materials are permitted. Welding or flame-cutting operations are permitted only in areas and at times approved by the responsible safety engineer. Persons smoking in locations other than designated areas are recommended for disciplinary action.

PERSONNEL

Managers and safety engineers must ensure that:

1. Undue exposure of personnel to physiological, psychological, or physical stresses is avoided where conditions are unsafe due to inadequate design, improper operating instructions, faulty equipment, or lack of suitable protective equipment.
2. No person is required to perform an operation that could result in injury to himself or herself or to any other person because of close proximity or incompatibility of their tasks.
3. Only qualified and certified personnel are permitted to undertake any hazardous duties or operations such as handling toxic, explosive, or highly flammable materials; to maintain, service, or repair any dangerous equipment; or to transport or operate any vehicle or mobile equipment or its component assemblies.
4. Programs are instituted to quality and certify workers for their duties. Qualified personnel are indicated as certified by suitable identification issued after proficiency examinations and demonstrations.

5. Certification programs include training and testing on safety subjects such as: hazards involved in the operation for which the worker is being certified; practices and procedures required to safeguard themselves and others; remedial actions to be taken in any contingency; safety devices; possible malfunctions that could cause an accident; color coding and other means of identification and markings of wiring, piping, and equipment; and meaning of warnings, sound alerts, or any other emergency signal; and any other information the safety manager considers advisable.

PERSONAL PROTECTIVE EQUIPMENT

Some operations require equipment to be worn by workers (1) to avoid injurious effects on the body and (2) to safeguard workers in the event of accidents. Here again, managers must ensure that certain rules are observed:

1. For normal operations, first choice will be given to eliminating the hazard in the environment rather than using personal protective equipment.
2. Approved protective equipment and devices must be made available and used to guard against specific hazards that cannot be eliminated but should be controlled when encountered during the operation
3. No supervisor will permit an operation to be conducted unless such equipment and devices are in proper working order and used as stipulated by the safety engineer.
4. Only protective and rescue equipment approved for the purpose by responsible agencies and in accordance with OSHA or other mandatory standards will be used. Managers will ensure that procedures are available for the supply, maintenance, and operation of such equipment, and that personnel are proficient in their operation and upkeep.
5. Locations of personal protective, emergency, and first aid equipment must be easily accessible and readily distinguishable. Equipment should be stored as close as practicable to the possible point of use. Operating procedures should identify the equipment stored and its location. Inspections are to be made periodically to ensure that stipulated items are present. Unauthorized persons removing, tampering with, using, or damaging the equipment are to be disciplined.
6. No person will enter a hazardous environment without the prescribed protective equipment, remove it while in the hazardous environment, or use it if it is faulty or damaged. Tests to demonstrate that the equipment is operating properly will be required before a worker enters a questionable environment.
7. All workers must be familiar with the capabilities, limitations, and proper method of fitting, testing, using, and caring for protective equipment. Managers will require and ensure that courses of instruction are provided to familiarize personnel with safety equipment, especially new types. Safety engineers and supervisors will schedule practice sessions or have training units conduct sessions to maintain user proficiency.
8. Devices are available to detect, warn, and protect against an impending or existing adverse environmental condition. Such equipment will be used to evaluate

atmospheres that might be toxic, flammable, or explosive or in which excessive levels of radiation, heat, pressure, noise, or other hazard might exist. Devices will be provided to apprise personnel of the status of such conditions that might be hazardous or of the loss of control of a hazard. Equipment provided should be adequate for detecting the presence of the hazard under conditions other than normal for the operating environment.

9. Detection and warning equipment will be maintained in a state in which operations and readings are dependable and accurate. To assure this, they will be tested and calibrated periodically.
10. Detection and warning equipment will be installed, maintained, adjusted, and repaired only by personnel trained and assigned for that purpose. Warnings will be posted against other persons tampering with, inactivating, or damaging this equipment.
11. Operating procedures specify the actions personnel should take when a warning signal indicates a time of danger or an emergency.

CHECKLIST FOR MANAGERS

Figure 7-6 presents a checklist managers might use to review a few of the aspects of plant safety and to determine whether they are taking appropriate action. Both managers and safety engineers can prepare additional items they believe should be covered in any periodic safety audit.

The disaster involving the space shuttle *Challenger* caused reassignment of many of the top managers of NASA. Managers had known for years about many of the deficiencies, most of them due to bad original designs. The accident was attributed to the managerial decision to launch in cold weather, despite an engineering deficiency in the booster. It illustrated the fact that management and engineering rely on each other. In the past, putting the blame for accidents on workers or users has been management's device for transferring responsibility for errors. Today any blame must be scrutinized more carefully, especially in light of the accident causers or contributors pointed out in Fig. 6-1.

No matter who is blamed for accidents to workers, management suffers a loss: increased premium costs and loss of skilled employees. A study was made in California of over 1,000 seriously injured workers who suffered a permanent impairment. One out of three had dropped entirely out of the job market, while a similar number indicated themselves as "unemployed and not looking for work." The loss to management because of workplace accidents has been substantial.

SAFETY INFORMATION SYSTEM

Management and safety professionals should institute and maintain a safety information system. The system must enable management (1) to determine if the safety program is adequate to meet stated goals and objectives, (2) to accept, eliminate, or control risks on a reasonable and prudent basis. In order to evaluate its risks and make

1. Has the chief executive issued a directive that indicates his or her policy toward worker safety?
2. Have rules of conduct to insure safe conduct of the facility been prepared?
3. Have the policy and the safe-conduct rules been widely posted in prominent locations where they can easily be read?
4. Does the policy directive indicate the functions and responsibilities of each organization as they relate to safety?
5. Is the chief executive kept informed of the progress of safety programs and any deficiencies?
6. Is the chief executive aware of the major restrictions on safety and health imposed by OSHA?
7. Does the executive's directive designate a top-level manager to be responsible for workplace safety?
8. Is the authority of this manager adequate to carry out safety functions effectively?
9. Is the person in charge of safety activities experienced and knowledgeable in those matters, or does he or she have someone to rely on who is?
10. Is the safety staff large and diverse enough to handle any problems that might arise, or will it be necessary to obtain consultant services to assist?
11. Is there a means by which progress of the workplace safety effort can be measured?
12. Is the list of accidents that have occurred in the workplace up-to-date so that it is ready and available for use by OSHA personnel?
13. Is there a method by which documents relating to workplace safety can be circulated to all concerned and then be stored and available?
14. Have budgets been prepared that include safety activities as normal functions, and have those activities been funded adequately?
15. Has a company safety committee been established, headed by someone at least at a management level and also including representatives of all workplace areas?
16. Does the committee meet regularly to hear and review presentations on plant hazards and accidents and to review employee complaints and suggestions regarding safety?
17. Is a procedure in effect to generate hazard reports, assign action to investigate the matter, and review the subsequent findings?
18. Have training courses been established for all personnel, especially new hires, to teach them of potential hazards, safeguards, warnings, and other safety matters?
19. Do contracts or purchase agreements require equipment sellers to inform the chief executive's staff of any potential hazard in the equipment?
20. Is new equipment reviewed by the plant safety engineer before it is purchased or before it is put into operation?
21. Do personnel training for operating newly purchased equipment have adequate information on safety and accident prevention?

FIGURE 7–6 Checklist for Managers

reasonable and prudent decisions to accept some risks and not others, management must have an adequate information system available. If management is to eliminate or control hazards, which otherwise present unacceptable risks, administrative, engineering, and operational factors must be in place to enable a sufficient and defensible safety program. These factors are discussed by the chapters in this text. They should be integrated into a safety information system.

Safety Technical Information System. A safety information system for management has several parts. The first part provides a detailed description of hazards under the responsibility of management. This prerequires a definition of what management's responsibility is and a systematic and iterative review of all things within the scope of that definition. The second part is a collection of codes, manuals, reference books on safety science and practice, research reports, lists of experts, accident case studies, accident history and trends, human factors reports, engineering reports, etc., which are relevant to the type of industry being managed. This part is an up-to-date and iterative collection of relevant information from *outside* the plant, grounds, or immediate jurisdiction of management. The third part consists of data and data analysis derived regularly from safety monitoring systems active *within* the jurisdiction of management. These systems produce specific, local data from sources such as the accident/incident/close-calls reporting system, safety inspections system, independent safety audits and appraisals, and reports of significant observations. These data are not useful unless coordinated with a clearly stated management safety policy and evaluated in accordance with pre-determined action triggers or action levels.

Safety Management Information System. The heart of this system is management's policies and plans for safety and health. The general company safety policy should be an up-to-date, clearly stated commitment to providing a safe and healthful place to work. It should be broad enough in scope to address the challenges to safety of workers, property, and the environment that are likely to occur.

Where there is a line-and-staff organization model, this information system should include a clear statement of line responsibility assignments for safety. These assignments are given to line individuals, from the duties of the highest line officer, through the supervisor, to the line employee. A similar statement of the responsibility for safety by staff functions and staff personnel should be included. Specific safety functions of staff departments (such as personnel, training, maintenance, purchasing, engineering, transportation, medical, etc.) should be stated.

It should also include a specification of the safety information that management needs and how it is to be obtained and delivered. It should include all management directives used to implement safety policy. These directives should state methodologies and functions to be used in policy implementation. It should include methods for providing, implementing, and tracking safety recommendations and improvements.

Individual plans for elements in a complete safety program are located in this system. Plans might include those for accident and illness prevention, safety audits and inspections, safety training, human factors and ergonomics reviews, Job Safety Analysis programs, Confined Space Entry programs, facility occupancy use and readiness, maintenance safety, safety data collection and analysis, employee motivation for safety, emergency response, medical services and rehabilitation, environmental safety, public relations, risk assessment and risk management, safety of design, safety and engineering change management, contractor and subcontractor safety, hazard recognition-identification-control, fire prevention, hazardous waste management, etc. It is not enough to have a well-written plan that is specific and *can* be implemented. A plan must *be* implemented. The safety professional can fall into the trap of spending too much time writing plans that are never implemented.

Safety Administrative Information System. This system includes not only budget and personnel information but also administrative interventions for accident, illness, and injury prevention, property and environmental protection. Records of budget requests (including those denied or cut back), past and present budgets, and supporting documentation should be maintained here. Personnel resumes of persons given professional safety duties, records of employees given safety training, and qualifications of personnel involved in the safety program should be maintained here.

Where there are administrative controls over hazards, these should be documented here. For example, if personnel assigned to an operation must be rotated out of that duty on a regular schedule for health and safety reasons, that requirement is recorded here. [A few examples of administrative controls are: (1) placing a time limit on a person's remaining in a noisy work area, (2) placing shelf-life deadlines on stored chemicals, (3) storing hazardous waste on-site for no longer than some specified period, and (4) limiting the manual filling of a tank to 80 percent of capacity.] Many such administrative controls are used for safety. This part of the system should have a means to track and alert management when deviations from administrative controls occur.

Regulatory and legal aspects of the safety program, such as lawsuit documentation, OSHA inspection, compliance, fines, and enforcement documents and correspondence can be placed here. Flags for deadlines for response or compliance should be maintained.

Safety Science and Engineering Information System. Records of engineering controls for safety are included in this system. This includes not only records and drawings of the safety features of physical facilities, stores, and equipment under management's jurisdiction, but also records of the safety reviews of engineering design and engineering change documents. It should include engineering reviews, analyses, and solutions for safety. These include reviews of energy sources that have danger or accident potential, of the adequacy of barriers that contain or redirect the flow of dangerous energy, and of objects or persons that might be affected by undesirable energy flow. Physical energy sources include chemical, thermal, mechanical, electrical, kinetic, and nuclear. Where science and engineering are involved in mitigating hazards, the records are maintained in this information system. Material Safety Data Sheets could be included.

These systems contain information about elements in the safety program that are not mutually exclusive or exhaustive. Therefore, similar information may appear in one or more of the systems, and elements of the information system may be added, subtracted, or otherwise customized as necessary. The purpose is to give management information that allows them to determine the adequacy of their safety program to meet specified goals and objectives and to determine the adequacy of safety policies. Also, it should enable management to decide whether a given risk of harm is acceptable to take or whether some other risk is controlled to an acceptable, reasonable, and prudent level.

To have value, information must be communicated to appropriate persons. The safety information system must implement a plan to provide its information to appropriate targets within and outside the organization.

BIBLIOGRAPHY

[27] W. T. Fine, *A Management Approach to Accident Prevention*, White Oak TR 75-104 (AD A 014562) (Silver Springs, MD: Naval Surface Weapons Center, July 1975).

[28] Fred A. Manuele, *On the Practice of Safety* (New York: Van Nostrand Reinhold, 1993), p. 132.

[29] Aerojet Nuclear Company, *Human Factors in Design*, Idaho Falls, Idaho, Feb. 1976.

[30] Dan Cordtz, "Safety on the Job Becomes a Major Job for Management," *Fortune*, Nov. 1972, p. 112.

[31] From *Accident Prevention—A Worker's Education Manual* (Geneva, Switzerland: International Labour Office, 1961).

[32] I. Fredericks and D. McCallum, "International Standards for Environmental Management Systems: ISO 14000," *Canadian Environmental Protection*, August 1995.

EXERCISES

7.1. Why is management support necessary for a safety engineer to accomplish his or her job?

7.2. Can a manager delegate all responsibility for safety to a lesser manager, a safety engineer, consultant, or other person?

7.3. Can a manager (a) be fined for a violation of an OSHA or state standard, (2) be cited in a civil lawsuit, or (3) be cited in a criminal lawsuit because of an accident? Explain what might happen to the manager.

7.4. How can managers show they are concerned with accident prevention?

7.5. Describe what would happen if managers or supervisors were lax in prescribing or enforcing (a) safety and work rules, (b) needs to wear protective clothing or equipment, and (c) a safe environment.

7.6. Why is the foreman or first-line supervisor key to any accident prevention program?

7.7. If you were a manager or supervisor, what corrections would you make in the area of safety?

7.8. Why are procedural safeguards less satisfactory than good design as safeguards? What have the OSHA and courts to say about this?

7.9. List five ways in which a manager can have an effective safety program.

7.10. Discuss how eight submanagers of a large company might participate in its safety program.

7.11. List five operations often considered extra hazardous.

7.12. List some causes of accidents that might be traced back to managers.

7.13. Managers often blame accidents on worker carelessness or other worker fault. Do you believe this assertion to be true often, occasionally, or rarely?

7.14. Do you believe there is good or bad management control of safety in the place where you work? Why? What evidence is there?

7.15. Do any of your supervisors press for improved production to the point where your safety or that of any of your co-workers is jeopardized? If, for that reason, you caused an accident, on whom would the blame fall?

CHAPTER 8

The Changing Roles of Safety Personnel

Traditionally, fire engines were red; now many are being painted yellow. Safety personnel's roles, capabilities, and training have also changed with time, as technologies advance and needs for accident prevention increase. Technical personnel who were concerned with preventing injury and damage in early safety efforts were interested in a comparatively narrow range of problems. Sir Humphrey Davy's miner's lamp can be considered the first technical advance in accident prevention that was deliberately sought because a hazard urgently needed controlling. Since then, other inventors have turned to accident prevention and thereby have become safety personnel.

One hundred years ago, when some of the major U.S. industrial companies became concerned, most of the causes of accidents (aside from fires) were mechanical. Pictures of machine shops of those days show them crowded with equipment operated by flatbelts from overhead lineshafts. Belts and machines were generally unguarded, constant hazards to the workers. Industrial plants were dangerous places for other reasons. Rupturing boilers and disintegrating flywheels on steam engines were violent, catastrophic failures which frequently generated devastating injuries and damage in the crowded facilities.

As science, invention, and industry entered new fields, new safety problems arose. Generally these involved new hazards, which first became known as scientists and inventors created and sometimes suffered from them. But even the scientists, and later the engineers, were concerned chiefly with the utility derived from the new inventions rather than with assurance of safety. Only in rare cases did consideration of safety outweigh other benefits. Thomas Edison rightly claimed that direct current was less hazardous than alternating current and preferred it for general use. (Alternating current was selected, however, because of the greater ease with which it could be transmitted over long distances.)

The next progression came as laboratory equipment and processes of scientists were transformed into industrial equipment, where the safety problems involved became concerns of the design, process, and plant engineers. The hazards resulting from high-temperature, high-pressure reactions or toxicities of the reactants and their products in the growing chemical industry were initially handled by chemists and chemical engineers (who were also the creators of the hazards). It became necessary to have other persons responsible for accident prevention. Similarly, safety in electrical plants was handled by electrical engineers after devices had been invented by others and accidents had taken place. As usage of formerly unique equipment and processes became more common and ceased to be the exclusive province of the engineer, efforts to provide safeguards and controls in industrial plants were assigned to less specialized persons. Industrial plant managers generally found it uneconomical to employ a technical specialist in each problem area. The end result was often that newly graduated engineers or lesser trained personnel were assigned to handle day-to-day accident prevention activities. Consultant specialists were called on only when a solution to a problem was beyond the plant safety person's capabilities.

SAFETY LAWS AND SAFETY ENGINEERS

Workers' compensation supposedly would spur increased safety, but it did even less toward that end than did insurance companies and litigation of the times. The growth of the labor laws and safety codes after Franklin Delano Roosevelt assumed office increased demands for safety personnel. This demand grew with the involvement of the United States in the Second World War. Accidents affected war efforts, as many untrained persons who had never before been employed moved into hazardous industrial jobs. When accidents took place, it became necessary to address the problem.

In 1943, Blake[33] wrote: "It is obvious that since widespread knowledge of good safety practice is necessary, at least the fundamentals of accident prevention should be taught widely. It seems reasonable to predict that the inclusion of safety in collegiate curricula generally will follow. Two methods were under discussion, namely, offering courses leading to the degree 'Safety engineers,' or including in each engineering subject taught the part of safety appropriate to the subject in question." The first organized courses in safety and accident prevention were taught in 1943 at a few engineering schools to undergraduates studying industrial engineering. Even fewer four-year schools now use Blake's concept of including information on safety. In many instances, teaching of accident causes and prevention is undertaken in two-year technical schools.

The next great advance took place with the passage of the OSHAct and of the OSHA standards. The standards required employment of safety personnel in both federal and state positions. For federal employment large numbers of engineers were necessary and had to be trained in safety matters. States had had some safety engineers to ensure observance of state codes, but the new laws resulted in needs for additional trained personnel.

Although the OSHA standards in general were a tremendous boost for safety personnel, in certain aspects they worked against the creation of new ideas on accident prevention. Because the OSHA standards indicated in detail most of the precautionary measures that had to be taken, the main function of plant safety personnel, consultants, and engineers who had to incorporate safeguards was to ensure that all OSHA provisions were met. The study of equipment, operations, or processes lagged. When a new type of activity began that was not covered by OSHA, any new hazards often were not covered.

At the start of the twenty-first century, new acts and new OSHA standards, OSHA guidelines, and OSHA publications have impacted the safety professional's role. Safety professionals are already greatly affected by the Americans with Disabilities Act, Guidelines for Ergonomics, Permit Required Confined Space Entry, and the Workplace Violence Awareness and Prevention publication referred to elsewhere in this book. At the time of this writing, OSHA has many regulations in the Proposed Rule Stage, including those for Safety and Health Programs (for General Industry), Permissible Exposure Limits for Air Contaminants, Plain English Revision of Existing Standards, Flammable and Combustible Liquids, and numerous regulations for the construction industry and shipyards. The safety professional should expect this situation to continue. It is part of the safety professional's role today to track what OSHAdministration and other governmental agencies are doing in their rulemaking and publications and to provide OSHA and others with feedback to influence these processes as a stakeholder. The internet makes this very feasible, and governmental agencies maintain on-line information systems which are available to anyone. The present-day safety professional must be engaged in governmental activities which ultimately impact the profession and industry in general.

SAFETY PERSONNEL

Whoever is designated in a plant with diversified activities must be a generalist, knowledgeable in a wide range of technical, legal, and administrative activities. To some, the breadth of effort involved and knowledge required provides variety that is found in few other industrial jobs. To others, this same scope engenders mental stress. Top-level managers too often assume a safety professional is knowledgeable in depth in all areas of accident prevention and capable of solving all problems that may arise. On the other hand, practically everyone has seen or been involved in an accident, or been alerted because of some hazard, and therefore has some knowledge or opinion regarding safety matters. Because of the need for correct and certain knowledge, properly educated safety personnel are increasingly necessary. Individual safety engineers must realize their capabilities and limitations, based on their education and experience, so they can understand exactly when a specialist should be consulted.

The difference in types of industry with their attendant hazards, organizational structures, management attitudes toward safety, and governmental emphasis on accident prevention have created a wide diversity of safety positions, duties, and responsibilities in industrial plants. When other employment of safety personnel, such as in government agencies, is considered, the number of positions, titles, duties, and responsibilities grows even larger.

To a certain extent, the following order of assignment of safety personnel indicates how their responsibilities have changed over the years. Since the days when there were few laws, codes, and standards for accident prevention, workers' compensation premiums were low, and there were far fewer and lesser awards for liabilities resulting from industrial plant accidents, costs have mounted. So have needs for their prevention and control.

"SAFETY MAN"

Years ago, the person called the "safety man" in most industrial plants, even large ones, was responsible for filling out accident reports and other papers required by insurance companies and state agencies. (The plant at which one of the authors first worked had been built around the turn of the century. For a long time its original safety deficiencies remained. In the 1930s, long after passage of workers' compensation and other labor laws, which were generally ineffectual, minor corrections finally began in our plant. One of our jobs, to be accomplished when there was little else to be done, was to design guards against rotating equipment.) In some cases the "safety man" may even have been someone partially disabled in an accident who was given the job to keep him or her on the payroll. This avoided any need for workers' compensation and the increase in premiums it might entail. The safety man's education and experience may have consisted almost entirely of the accident in which the injury had taken place. In today's plants, the safety professional is not represented by such a gender-specific form of address.

SAFETY ENGINEER

The term "safety engineer" is frequently applied to the person in charge of accident prevention, who may in fact be the so-called "safety man"—a person with some technical experience, or an engineering school graduate assigned to accident prevention duties. Government regulations requiring safe designs in industrial plants frequently resulted in reassignment of engineers of other disciplines into accident prevention activities. Because of deficiencies in safety education in engineering schools, these engineers lacked the knowledge of safety laws and standards and related information they needed in order to perform competently in their new jobs.

As discussed in the chapter on engineers and safety, some persons have passed examinations as Certified Safety Professionals. Many are graduate engineers, but others have achieved this rating by having shown their knowledge in safety and accident prevention. Other safety-related certifications include those of Certified Product Safety Professional, Certified Industrial Hygienist, Certified Professional Ergonomist, and Certified Hazard Control Manager.

In the past, engineering schools rarely taught safety courses to undergraduates. This is slowly changing. In many schools, safety and accident prevention may be taught to students taking industrial engineering courses. Some safety matters may be touched on because past massive accidents made them noteworthy. Some schools have initiated postgraduate engineering degrees in safety for insurance companies. The federal government has also had courses leading to postgraduate degrees in safety. These graduate courses have responded to the inadequacy of instruction in safety and accident prevention in undergraduate schools. This educational deficiency is slowly being reduced. In 1998 approximately 125 colleges and universities in the United States offered degrees in safety management, occupational safety, environmental protection, or a related field.

In the fall of 1987 a university in Pennsylvania commented: "While the demand for well-educated safety professionals is at an all-time high, the safety profession is still one that most people have never heard of. . . . The current demand for safety professionals is not likely to be satisfied by the present number of available graduates. They continue to be sought out all around the country." To help meet this need, the National Institute for Occupational Safety and Health provides, on a competitive basis, academic grants to institutions of higher education around the nation for the provision of graduate programs in safety. These grants usually include provisions for student financial aid. In 1998 there were NIOSH-funded programs in 18 colleges of engineering and other technical departments. The educational requirements of a safety engineer are changing with changes to the profession. The site review process for NIOSH-funded programs sets standards for these programs. In addition, the Academic Accreditation Council of the American Society of Safety Engineers and the Board of Certified Safety Professionals have jointly published curriculum standards for baccalaureate and master's degrees in safety.

A 1992 survey of the 27,000 members of the American Society of Safety Engineers indicated that 30 percent majored in safety or health in college, 20 percent trained as engineers, and 19 percent have business or management degrees. Three percent hold doctorates; 25 percent, master's degrees; and more than 50 percent, at least a baccalaureate degree.[34] In the future, more and more safety engineers will have baccalaureate and graduate degrees in safety as an engineering discipline.

In addition to plant safety engineers, other safety personnel can be described. One group of consultants may be limited to reviews of plants to determine compliance with OSHA and other standards. A second group may be knowledgeable in specific areas, such as flammable gases, explosives, or mines. Both of these groups may also be concerned with investigations after an accident has taken place. System safety engineering grew because of efforts to evaluate hazards that might be present and potential accidents that might take place with advanced new products and processes, usually with new military systems, even as they are being designed. Some methodologies derived from system safety engineering have led to their adoption by other types of safety engineers.

Four states now have registration of professional engineers in a safety engineering discipline. Registration gives the registrant the right to use the title "Safety Engineer," but the enabling law has no requirement that the services of such an engineer be used. The contracting office of one military service responsible for development of advanced high-tech systems does require that certain hazard analyses and documents be signed off and approved at specific points in the designs. Such approval will be valid only under the signature of a registered safety engineer or other engineer shown to have had extensive experience in safety programs. The principal problem is that in a new advanced design project, there may be 400 engineers with no training in accident avoidance who may make critical errors and only one or two safety engineers to find them.

PRODUCTION AND PROCESSING LOSSES

Early in the Industrial Revolution, accident prevention by owners was oriented toward preventing losses to equipment and facilities such as mines, and to chattel such as horses, mules, or oxen. Gradually, preventing accidents to workers regained the importance for safeguarding lives and preventing injuries it had formerly had even in ancient and biblical times.

Thus, over the years decisions have resulted in safety engineers having to revise their priorities regarding accident prevention. While protection of personnel safety still comes first, the second priority has come to be protection of the environment. This also means protection of animals, especially in endangered species and against man-generated environmental problems such as "acid rain." The new priority includes prevention of leakage or release of liquids, oil, chemicals, or noxious gases and other deleterious substances. Safety engineers must now be almost as concerned about accidental release or unprotected use of materials that affect the environment as they are with an injury that affects persons directly and quickly. Protection against damage to the environment comes right after protection of personnel and animals and before prevention of damage to equipment. Priority for rescues of equipment and plant is last, usually following stoppage of operations.

The other new concern of safety engineers is in the area of accidental in-process damage or loss. Parts and materials being processed can sometimes be damaged or rendered unusable, so that they are as worthless as if destroyed by a catastrophe. For example, timing equipment in a metal conditioning oven may become faulty, so that the metal is as damaged as if by fire. Avoidance of such damage usually has been the responsibility of the production manager and his or her staff. However, accident prevention principles and methodologies are being applied more and more to process control. Lack of a simple feature that should have been reviewed by a safety engineer might lead to an accident and to a disaster in a large, complex plant. Protective devices and similar principles are increasingly being incorporated. Damage to a component in a production process might cause a failure in the assembled product during operation, and consequently an accident. The expertise of safety engineers for plant safety may be beneficially applied to product safety. A knowledgeable safety engineer can help greatly in such matters.

Increased realization that accident prevention is productive and gainful in all types of industrial activities may increase the standing of all safety personnel.

GROWING AREAS WITHIN SAFETY

Computer-related areas such as computer integrated manufacturing (CIM) and software safety are expanding rapidly. Product safety, environmental protection, and system safety are presently experiencing growth.

There has been much discussion about whether a major shift in safety is in progress from engineering controls to an emphasis upon psychological and social factors. Certainly behavior-based safety interests have grown through the 1990s, complementing but not diminishing the more traditional approaches to safety engineering and management.

BIBLIOGRAPHY

[33] Ronald P. Blake, *Industrial Safety* (New York: Prentice-Hall, 1943).

[34] Roger L. Brauer, "Educational Standards for Safety Professionals," *Professional Safety*, Sept. 1992.

EXERCISES

8.1. How have the roles of safety personnel changed over the years?

8.2. List ten functions of safety personnel. Which would you consider most important? Why?

8.3. Discuss some safety aspects of building design with which a safety engineer should be concerned.

8.4. Why is it necessary that a safety engineer keep informed on the latest technical developments?

8.5. Discuss some means by which a safety engineer can keep informed.

8.6. If an accident were to take place, what priorities should be given to rescuers? How has this priority changed over the years?

8.7. Until the Railroad Safety Act was passed in 1893, was safety of equipment more highly regarded than that of personnel? Why?

CHAPTER 9

Personnel

Almost every mishap can be traced ultimately to a personnel error, either on the part of the person immediately involved in a mishap, a designer who made a mistake in calculation, a worker while incorrectly manufacturing a product, a maintenance man, or almost anyone involved. It was pointed out in Chapter 6 that inadequate design may be a leading cause of accidents by contributing to worker error. In addition, worker error alone is a major cause. A worker's error may cause well-designed equipment to fail. On the other hand, a safely designed product may have no adverse effect, even if an operator makes a mistake.

Operator errors have a multiplicity of causes—for example, lack of consideration of worker capabilities or their limitations. Errors and accidents frequently occur because equipment is designed for statistical persons rather than actual ones. Measurements and statistics may show the average person can reach so far, push so hard, and react so fast. However, there may be occasions when the worker's capabilities are less in some respect than the statistically synthesized person. Much man-sized equipment proved unsuitable for use by woman operators in World War II because of their smaller physique.

Automobiles are designed to go faster than most drivers are capable of operating safely. This was demonstrated by the decrease in accidents after speed limits were reduced because of the energy crisis. Industrial machines beautifully designed for high-speed operations could involve their operators in accidents because the operators could not work fast enough. Speeds may be excessive when required output increases or when capabilities are reduced because of fatigue or a hot environment. Machine operations involving workers must be undertaken with due consideration to all factors involved, or errors and accidents will result.

It has been pointed out that the latest modern safety practice is to provide and ensure (1) designs that will eliminate the possibility of errors and accidents, (2) procedures that will minimize the possibility of errors by operators, and (3) designs and safeguards that will prevent injury if an error is made that causes an accident. It has also been mentioned that properly designed features are far more satisfactory in preventing accidents than are procedures such as safety work rules, which are often ignored.

DISABLED PERSONNEL IN THE WORKPLACE

At one time, disabled workers were often denied employment in the workplace because it was believed they would require extensive and costly facilities and equipment, or they would be prone to cause or be involved in accidents because of their incapacities. Not only has this been found to be untrue in practice, but rational considerations also indicate otherwise.

For many reasons it has become economically beneficial to employ disabled workers within their capabilities. To a great degree, such employment not only brings savings to plant employers but also lessens welfare costs and, as pointed out in Fig. 9-1, lessens costs in workers' compensation premiums, creates better quality in product production, lowers absenteeism, and produces other benefits in the workplace.

In March 1987, the Supreme Court held that ". . . Congress had passed laws to ensure that handicapped individuals are not denied jobs or other benefits because of the prejudiced attitudes or the ignorance of others." Employers could still dismiss persons who are incapable of doing work or who pose a significant risk. Not only must a risk be shown, however, but it must also be shown that safe alternative work cannot reasonably be found for the employee.

On July 26, 1990, the Americans with Disabilities Act (ADA) was signed into law. It requires nondiscrimination in employment of the disabled by employers with 15 or more employees. When an individual's disability creates a barrier to employment opportunities, the ADA requires employers to consider whether reasonable accommodation could remove the barrier. There can be no discrimination in job application procedures, promotion, discharge, or other terms and conditions of employment, against a qualified individual with a disability. The ADA establishes a process that industry must follow to assess a disabled individual's ability to perform the "essential functions" of the specific job held or desired. It does not relieve a disabled employee or applicant from the obligation to perform the essential functions of the job. If the person's functional limitations impede job performance at levels that the employer expects of persons who are not disabled, the employer must take steps to reasonably accommodate the effects of the disability, to enable the person to work to the desirable standard of performance.

The employer is not required to make accommodation if it would impose an undue hardship. An employer may require that an individual not be a "direct threat" to the health or safety of others or of himself/herself. A health or safety threat can be invoked only if it is "a significant risk of substantial harm." A person with a disability, according to the ADA: has a physical or mental impairment that substantially limits one or more of his/her major life activities; has a record of such an impairment; or is regarded as having such an impairment. Persons who currently use drugs illegally are not protected by the ADA. That act states that homosexuality and bisexuality are not impairments and are not protected by ADA. ADA also excludes several behavioral disorders from the definition of an "individual with a disability."

Personal disabilities, physical and mental, generally are the result of birth defects, sickness (heart disease still constitutes the greatest cause of disabilities), accidents, combat injury, or old age. Once the problem is stabilized, most disabled persons can become productive members of the community, especially for occupations in light industries.

One of America's greatest presidents, Franklin Delano Roosevelt, who did the most to improve conditions and safety in the workplace, was disabled. His disability resulted from poliomyelitis in his middle years, but it did little to lessen his accomplishments.

Because of disabilities:

- a worker tends to be more careful in his or her actions. The result is not only safer operations, but often work is of better quality and with fewer rejects.
- these workers usually have increased incentives, not only to overcome limitations but to widen their abilities. Their motivation makes them more reliable and stable, having fewer absences to compensate for overactive weekends, and more safety consciousness to avoid any further injury or disablement.
- a great fund of mental capabilities is available to be used, beneficially enhancing and widening the population of productive personnel.
- designs of mechanical and electrical devices are given more consideration for easier and safer use. Tools, fixtures, and equipment that are made to enhance the ease of operations can also provide more convenience and faster production. Improvements provided for disabled persons can be beneficial for and incorporated into all work activities.
- equipment assisted operations make disabled workers less readily fatigued and less prone to errors.
- mentally disabled (or illiterate) workers provide inducements to create more correct, better, simpler, and easier to understand instructions, similar to those that now must be provided for use of computer.
- workers' compensation and rehabilitation costs will be less.

The many types and degrees of disability preclude an entire listing of what has been done to compensate for disabilities. However, the desire to overcome impairments has led to the generation of means and devices that lessen most of them. Listed below are some examples.

IMPAIRMENT ACCOMMODATION

Loss of vision – for partial impairment: glasses, large-sized letters; for total loss: white canes, seeing Eye dogs, barriers or stops, verbal communications or warnings, the sound of a ringing bell, sensation of vibrations and touch, such as the use of Braille, guides or controlled paths, shapes of control or activating devices.
Hearing – hearing aids, written communications, sign language, hand or body or electrically activated lighted signals, flashing lights, detectable vibrations.
Complete loss or loss of use of a limb or part of one – use of remainder of muscles and bones of the limb, including additions of prosthetic devices, increased use and abilities of the other limb, or transfer of control such as from hand operation to foot or vice versa.
Walking – canes, crutches, walkers, wheelchairs.
Climbing – canes, crutches, ramps, elevators.
Grasping, lifting, pulling or pushing – mechanical or electromechanical devices or tools.
Mental retardation – Limited and easy to understand instructions similar to those to or from computers, use of logos or other visual devices.

FIGURE 9–1 Disabled Workers and Safety in the Workplace

OLDER PERSONNEL IN THE WORKFORCE

The graying of the workforce requires that new attention be given to their occupational safety. There were 33.9 million persons 65 years or older in America in 1996. They represented 12.8 percent of the U.S. population—that is, one out of every eight Americans. This fast-growing segment increased by 8 percent from 1990 to 1996, compared with a 6 percent increase for the rest of the U.S. population. Since 1900 the percentage of older Americans (65+ years of age) has tripled. The number of older persons will more than double between 1996 and 2030. In 1997, 12 percent (or 3.8 million) of older Americans were working actively or seeking work.

Based upon 1993 data of the Bureau of Labor Statistics, the older workers are less likely to be hurt seriously enough to lose time from work. But when they are seriously injured, they typically require two weeks to recover before returning to work, as compared to one week for younger workers. The median absence due to work injury is 5 days for workers under age 35 and 10 days for workers over age 55. Falls represented 17 percent of the lost-work-time injuries for workers 55 years and over; they accounted for 8 percent for workers under 55. In 1995, U.S. occupational injuries resulting in fatalities per 100,000 employed in an age group was: 3, for ages 19–20; 4, for ages 20–34; 5, for ages 35–54; 7, for ages 55–64; and 14, for ages 65 and over.[35] Older workers apparently have a greater risk of being killed in the workplace in the United States. The 1996 report of the Japan Industrial Safety and Health Association indicates that, in that country, the highest rate of casualties per 1,000 workers (those killed or injured with absence of four days or more) was in the ages of 60–69 with 4.26 and ages 50–59 with 3.93. In Japan 45.2 percent of such casualties were to workers aged 50 and over. Older workers apparently run a greater risk of being killed or seriously injured in the workplace in Japan.[36] This may be an international phenomenon.

The American Association of Retired Persons (AARP) point out that older workers in the United States have fewer accidents from carelessness or poor judgment and more accidents from being less successful at escaping hazards. Overall, the AARP says, they have lower occupational accident rates; workers 55 and older, 13.6 percent of the work force, have only 9.7 percent of the on-the-job injuries.[37] In general, older Americans are safer drivers. More research is needed to clarify the status of occupational safety for older Americans.

HUMAN ERROR

Human error can be defined as any person's action that is inconsistent with established behavioral patterns considered to be normal or that differs from prescribed procedures. Errors can be divided into two categories: predictable and random.

Predictable errors are those which experience has shown will occur under similar conditions and can be foreseen because their occurrence has taken place more than once. It is known that a person will generally tend to follow procedures that involve minimal physical and mental effort, discomfort, or time. Any procedure contravening this basic principle is certain to be modified or ignored at some time by the persons supposed to carry it out. It is predictable that for those reasons an error may result unless the possibility has been designed out.

Random errors are nonpredictable and cannot be attributed to a specific cause because of their uniqueness. For example, a person may be highly competent as an operator but may be annoyed by a fly or mosquito. Swatting at it, the worker hits a critical control or piece of sensitive equipment. There are fewer types of random errors than of predictable errors, and they are being reduced as experience increases. If flies or mosquitoes become a common annoyance and swatting in the presence of a critical task a problem, it may become a predictable cause of error for which suitable precautionary measures can be provided. In any case, precautionary measures to minimize the effects of random errors are the same as for predictable errors. The one difference is that many random errors can be included under a general safeguard, whereas for a predictable error a specific safeguard may be provided.

An error is generally due to:

- Failure to perform a required function (omission). A step is left out of a prescribed procedure, intentionally or inadvertently, or a sequence of operations may not be completed. In some instances intentional omissions by workers may be due to procedures that are overlengthy, badly written, deviously expressed, in defiance of normal tendencies and actions, or not readily understood.
- Performing a function not required, including unnecessarily repeating a procedure or procedural step, adding uncalled-for steps to a sequence, or substituting an erroneous step (commission). (A frequent error of commission occurs when a worker tries to retrieve a mistakenly placed piece of work as the top of a machine die is descending. Measures required by OSHA have reduced the number of hands lost for that reason.)
- Failure to recognize an immediate hazardous situation requiring corrective action.
- Inadequate response to a critical contingency.
- Wrong decision to solve a problem that arises.
- Poor timing, resulting in a response that is too late or too soon for a specific situation.

As early as the fourteenth century, a court ruled: "When any man does an act, he is bound to do it in such a manner that by his act no prejudice or damage is done to others."

DESIGNING AND PLANNING ERRORS

The person who designs equipment or plans an operation may not only commit an error in calculations but be guilty of failing to remove or control a hazard, or of failing to incorporate desirable features as safeguards to prevent accidents or protect personnel. When a designer or planner cannot completely eliminate a hazard or the possibility of an accident, he or she must attempt to minimize the possibility that other personnel will commit errors leading to mishaps. In effect, the designer, through foreseeability, must attempt to make to system "idiot-proof," although knowing he or she will always be subject to the inevitability of Murphy's Law.

There are hundreds of whimsical versions of Murphy's Law, each containing a grain of truth that makes it sadly applicable. Four of the many versions are:

- Any task that can be done incorrectly, no matter how remote the possibility, will some day be done that way.
- No matter how difficult it is to damage equipment, a way will be found.
- Any item that can fail can be expected to fail at the most inopportune and damaging time.
- Instructions will be ignored when the most dangerous and complicated task is being accomplished.

The word "error" in designing or planning includes more than making a mistake in calculation. It also includes any design or plan that is technically practical but is improper, inadequate, or unsuitable for the intended operating conditions. For example, each control chosen for use on an operating panel may be excellent for its proposed individual function, but their overall arrangement may cause confusion and errors. Designers' errors may invite mistakes in reading dials or meters or may increase operator reaction times. With a multiplicity of widely spaced instruments, reviewing them properly and frequently enough might require an operator with four eyes or two heads. If two persons are used, there might be interference or failure to communicate when more than one control is required. Inability to act or respond properly may generate operator difficulties, errors, or accidents.

A design error can also be one that violates a normal tendency or expectancy. People expect that on a vertically numbered instrument, the higher-value numbers will be at the top; on a circular dial they expect values to increase clockwise. An improper design can unduly stress the operator. (Designs contrary to normal expectations can sometimes be used to make workers think about their actions, thereby acting as safety devices.) Instead of providing means to avoid an environmental problem such as noxious gas, poor design may require the user to wear burdensome protective respiratory equipment. The resulting fatigue may lead to errors. Among other fatigue-producing designs are those that entail glare, inadequate lighting, uncomfortable chair seats, vibration or noise, undue strength requirements for activation of controls, unusual positions in which to operate, or proximity to very hot surfaces. A person who may perspire so much that his or her efficiency is badly affected. If too close to the hot equipment, the worker must constantly be on guard to avoid being burned. The psychological stress imposed reduces efficiency.

Figure 9-2 indicates numerous causes of errors that can be committed by operators and the measures that designers or supervisors can use to eliminate or minimize them. Certain of these measures are mandatory in any design, in that they constitute good engineering practice. Other measures may be more costly and are provided only when required.

PRODUCTION ERRORS

A production error can occur in one of two ways: directly when a worker makes an error with an immediately known effect, or indirectly when the effect only becomes evident later. Manufacturing errors can ruin any design and make a product unsafe.

ERROR PREVENTION

Causes of Primary Errors	Preventive Measures to be Taken by Designer or Methods Engineer
1. Improvising procedures that are lacking in the field	1. Provide adequate instructions.
2. Following prescribed but incorrect procedures	2. Ensure that procedures are correct.
3. Failure to follow prescribed procedures	3. Ensure that procedures are not too lengthy, too fast, or too slow for good performance, and are not hazardous or awkward.
4. Lack of adequate planning for error or unusual conditions	4. Provide backout or emergency procedures in instructions.
5. Lack of understanding of procedures	5. Ensure that instructions are easy to understand.
6. Lack of awareness of hazards	6. Provide warnings, cautions, or explanations in instructions.
7. Untimely activation of equipment	7. Provide interlocks or timer lockouts. Provide warning or caution notes against activating equipment unless disconnected or disengaged from load, or other damaging conditions.
8. Errors of judgment, especially during periods of stress	8. Minimize requirements for making hurried judgments, especially at critical times, through programmed contingency measures.
9. Critical components installed incorrectly	9. Provide designs permitting such components to be installed only in the proper ways. Use asymmetric configurations on mechanical equipment or electrical connectors; use female or male threads or different-sized connections on critical valves, filters, or other components in which direction of flow is important.
10. Exceeding prescribed limitations on load, speed, or other parameter	10. Provide governors and other parameter limiters. Provide warnings on exceeding limitations, inadequate strength of stressed parts, use of excessive mechanical leverage.
11. Lack of suitable tools or equipment	11. Ensure that need for special tools or equipment is minimized; develop and provide those that are necessary; stress their need in instructions.
12. Interference with normal habits	12. Ensure that recognition and activation patterns are in accordance with usual practices and expectancies.
13. Lack of data on which to make correct or timely decisions	13. Ensure that response time is adequate for corrective action; if not, provide automatic corrective devices.
14. Hampered activities because of interference between personnel	14. Ensure that space is adequate to perform required activities simultaneously.
15. Inability to concentrate because of unsafe conditions or equipment	15. Ensure that personnel must not work close to unguarded moving parts, hot surfaces, sharp edges, or other dangers.
16. Error or delay in use of controls	16. Avoid proximity, interference, awkward location,, or similarity of critical controls. Locate control close to readout. Locate readout above control so hand or arm making adjustment does not block out readout instrument. Ensure that controls are labeled prominently for easy understanding.

FIGURE 9–2 Error Prevention Continued on page 134

17. Error or delay in reading instruments	17. Ensure that instruments are labeled and designed for easy understanding; do not require reader to turn head or move body; and that visibility problems due to glare or lack of light, legibility, viewing angle, contrast, or reflections are avoided. Provide direct readings of specific parameters so operator dies not have to interpret.
18. Inadvertent activation of controls	18. For critical functions provide controls that cannot be activated inadvertently; use torque types instead of pushbuttons. Provide guards over critical switches.
19. Controls activated in wrong order	19. Place functional controls in sequence in which they are to be used. Provide interlocks where sequences are critical.
20. Control settings by operator not precise enough	20. Provide controls that permit making settings or adjustments without need for extremely fine movements. Use click-type controls.
21. Controls broken by excessive force	21. Ensure that controls are adequate to withstand maximum stress an operator could apply. Provide warning and caution notes for those devices that could be overstressed.
22. Failure to take action at proper time because of faulty instruments	22. Provide procedures to calibrate instruments periodically, or provide the means to ensure during operation that they are working correctly.
23. Confusion in reading critical instruments because of instrument clutter	23. Make critical instruments most prominent or locate in easiest-to-read area.
24. Failure to note critical indication	24. Provide suitable auditory or visual warning device that will attract operator's attention to problem.
25. Involuntary reaction or inability to perform properly because of pain, due to burns electrical shock, puncture wound, or impact	25. Insulate or guard against hot surfaces, "live" electrical conductors, sharp objects, and hard surfaces.
26. Fatigue	26. Avoid placing on operator severe and tiring physical and mental requirements such as loads, concentration times, vibration, personal stress, awkward positions.
27. Vibration and noise cause irritation and inability to read meters and settings or to operate controls	27. Provide vibration isolators or noise-elimination devices.
28. Irritation and loss of effectiveness due to high temperature and humidity	28. Provide environmental control. Prevent entrance or generation of heat or moisture from external sources or from internal equipment or processes.
29. Loss of effectiveness due to lack of oxygen or to presence of toxic gas, airborne particulate matter, or odors	29. Prevent generation or entrance of contaminants into the occupied space. Provide suitable life support equipment. Avoid presence near occupied areas of lines or equipment containing hazardous gases of liquids.
30. Degradation of capabilities due to extremely low temperature	30. Ensure that design provides for adequate heating or insulation, protective shelter, equipment, or clothing.

FIGURE 9–2 Continued

This potential for accidents can sometimes be minimized by providing special care and attention to critical components during their production and their assembly into larger units. During assembly, personnel can make errors that could later cause failures and problems for operators. Undertorqing of connections so they loosen and leak is common. Over-torquing may cause them to crack and leak. Failure to keep electrical connectors clean and dry and free of loose wire strands can permit short circuits when the system is energized. Scratches, dents, corrosion, tool marks, and other rough finishes may lead to stress concentrations and structural failures under operational loads.

When installing new equipment that could be hazardous, the safety engineer should have a competent company maintenance person inspect it closely. This is to ensure not only that the equipment will operate correctly, but that there are no lacks or faults that could cause failures, accidents, injury, or damage. If the equipment manufacturer has stipulated that the equipment be installed in a certain way or under certain conditions, it should be done. The supervisor of the unit where the equipment is to be operated, and possibly an experienced operator, should also examine it to ensure that it operates properly, lacks undesirable features, and is not faulty in any way.

OPERATIONS ERRORS

Many procedural and control errors can occur during operation of a system. Procedural mistakes during normal operations can generate abnormal situations leading to accidents. An operator can activate a circuit inadvertently, fail to close a valve or to shut off a pump, or forget to set the brakes when parking a vehicle on an incline. Procedural errors are especially critical in emergencies. At such times, personnel are almost always in a state of shock and extremely susceptible to committing errors. If it is their error that caused the emergency, they may make more errors in trying to rectify the situation. If the emergency is due to a part or equipment failure, the operator will require time to determine, evaluate, and decide the cause and then to take corrective action. Again, a mistake in one of these steps may make the emergency worse.

No matter how calm the individual may appear, the ability to make decisions will be impaired. At such times there is an increased propensity for making errors. The extent of the operator's disruption or shock will depend on the complexity of the operation being performed, the individual's past training, reflexes, and temperament, the types and severity of the emergency, the reaction time available, and numerous other factors. The effects of errors, emergencies, and accidents can be mitigated by good training and practice so that workers make fewer mistakes, react more rapidly, or more quickly realize the corrective actions that must be taken.

Control errors can occur because a procedure is not carried out in the stipulated sequence or indicators are read incorrectly. Setting adjustments can be inaccurate; a wrong button can be pushed; control can be lost due to excessive speed. Many errors in critical operations can be mitigated if designers interpose an intervening safeguard operation. For example, if a device is not to be operated inadvertently, it may be desirable to interpose a release mechanism before a switch can be thrown or a button depressed. Requiring that the operation of a critical switch, button, or other controller involve two distinct steps, such as lifting a guard before a button can be pushed, may provide a safeguard against inadvertent activation. However, if extremely rapid action is required for safety, such as pushing a panic shut-down or a "destruct" button, an intervening step might be harmful.

There can be excessive delay in response to a control, or excessive sensitivity so that the operator can overcontrol a piece of equipment. In some instances, an operator can be so irritated by an inability to make fine adjustments, to get an adequate response, or simply by high environmental temperature and humidity, that he or she may "slam" the controls violently. Such intentional actions can produce accidents as damaging as inadvertent errors.

TWO-PERSON CONCEPT

To minimize the possibility of human error in any procedure involving a nuclear device, the Department of Defense has developed the "two-person concept." Two or more persons are assigned, each capable of undertaking the prescribed task and of detecting an incorrect or unauthorized step in a procedure. One person accomplishes a step and the other checks the action to make sure it has been accomplished correctly. The possibility of an omission or error is thereby reduced. It is not necessary that both persons have "equal" knowledge, only that each is able to detect and ensure that the actions of the other have been correct, authorized, and undeviating. Similar arrangements have been adopted in industrial plants where critical assemblies take place, such as setting up very high voltage equipment for test in a laboratory. The two-person concept differs from the buddy system (page 204).

It has been recommended that gas companies use a modified version of the two-man concept. In this method, because of the hazard of leaking natural gas, a repair is made by a qualified maintenance person immediately after a leak has been reported. At some time after the worker has left for another job, a supervisor or second qualified worker reinspects the repair to ensure it has been done correctly and completely.

HUMAN VS. MACHINE

Humans have often been compared to machines in terms that underrate the humans. People often make mistakes, whereas computers make none (or very few). On the other hand, humans are to some degree self-repairing, while machines are not, and are able to accomplish many functions machines cannot. Persons can reason inductively, mentally reaching far beyond the abilities of any existing computer. A worker can adjust to unusual situations, improving conditions where necessary to overcome unforeseen difficulties in the way of successful accomplishment. (During World War II, there was a report that a bomber's hydraulic system would not work because of a lack of fluid. The problem was overcome temporarily when a crewman urinated into the hydraulic reservoir, the entire event requiring a sequence of analysis, solution, response, and action impossible to any mechanism or computer.)

The capabilities of any normal person's senses can equal or exceed those of manufactured equipment. No simple machine can determine odor, taste, color, or their gradations with the versatility of a human. The ear can detect sound with an amplitude of vibration as small as one-thirtieth of the diameter of a single molecule. The eye is equally sensitive, having the ability to detect as few as six quanta or photons of light.

Unfortunately some persons do not employ this capability properly, or at all. Workers make errors by failing to recognize problems, deviating from normal operations or procedures, or sometimes failing to recognize or understand indications that special action or care is required. Persons unpracticed in the use of good judgment may fail to respond in adverse situations and must be especially safeguarded against.

Information on human errors as a causative factor in accidents can be determined from different areas of investigation in psychology, physiology, accident analyses, industrial operations, and other sources. Unfortunately, although the basic causes of accidents are alike, personnel have generated a myriad of variations that plague any safety effort.

Accidents do not involve willful violations or a desire to cause mishaps. (If they did, they would not be accidents.) Mishaps often occur when human capabilities are overwhelmed by a need to respond properly and rapidly to an adverse situation. These inadequacies can be permanent or transient. Physical limitations or inadequate training can be considered permanent inadequacies. Transient inadequacies can be due to long hours of activity with little rest; or to poor hearing, language differences, or other blocks to understanding between personnel; or to external physiological or psychological pressures.

A human being is subject to so many variables in many different situations that not all responses can be foreseen. Two similar machine-manufactured parts are generally the same in result and use, but the same cannot be said of humans. Personal actions will vary from perfectly correct to absolutely erroneous.

Because response characteristics, like fingerprints, differ for every individual, a designer would find it impossible to introduce a system applicable to every individual's case. Analysts have found it impossible to determine with any degree of certainty whether any specific person will or will not cause an accident. Therefore, with a new product or system, the designer must create provisions against as many foreseeable combinations as possible of workers' characteristics that would produce errors, assuming that all persons can and will at times make mistakes (Murphy's Law).

The determination of who will cause an error is analogous in life insurance to determining which persons of a certain age will die. Statistics can be used to determine death rates in a wide population. But it is impossible to pinpoint exactly which persons of a specific age will be the casualties, although some may have poorer health prognoses than others.

THE BIOCHEMICAL MACHINE

A person's body is a highly efficient engine which uses food as fuel and oxygen from the air as an oxidizer to produce power. As a mechanical structure, a human can go only so far, lift so much, and react so quickly and effectively. The body's sensory systems also have definite limitations. Information from the senses involves sensation, perception, analysis, decision, and, usually, response to a situation. Difficulties and accidents result from any demand for abnormal uses of a person's senses, quick decisions based on inadequate or overwhelming information, excessively rapid responses, or physical capabilities the person does not have. Also, common problems in industrial plants are the injuries, hernias, backaches, and muscle strains that people inflict on themselves because they try to exceed their psychophysical capabilities. This topic is discussed in greater detail in the chapter on work-related musculoskeletal disorders.

A vast amount of information in handbooks is based on people considered as machines. There are data on humans as structures, biological organisms, chemical processes, sensing devices, and power sources. There are data on precisely defined and measured physical capabilities with limiting conditions that must not be exceeded if accidents are to be avoided or prevented.

In many instances a person does not provide the same care for the superb mechanism that is his or her body as for a piece of equipment. This lack of consideration for human needs is still apparent, although the situation has improved because of new emphases on exercise and diet for body improvement and on improved medical care. It is said that in most cases a dog or cat is fed a more balanced diet than its master. An automobile's condition is generally checked more often that that of its owner.

The human body's structure, operations, and sensory system effectively are directly related to how well its biochemical medium is maintained. Violations cause inefficient operation of the whole system. The body needs adequate oxygen, without which life cannot be sustained for more than a few minutes; depriving the body of even small amounts can affect it adversely. Various contaminants, such as carbon monoxide, deprive body tissues of oxygen nourishment so that normal functions of the body are degraded. Ethyl alcohol, the primary constituent of intoxicating beverages, will upset the body's biochemical balance, acutely disrupting body functions or reducing their effectiveness. This reduced effectiveness makes the person susceptible to errors, even when the amount ingested is small.

Fatigue due to changes in the body's biochemical balances through accumulation of toxic wastes is another type of disruption. Long or irregular hours of strenuous effort are a cause of fatigue. The efficiency of a tired body is reduced; the probability of error is increased; the detection of any errors that may have been committed is less acute. (It has been said that automobiles assembled on Monday contain more errors than do assemblies on any other workday. Too many workers probably had strenuous vacations over the weekend.)

Bodies will adapt to some physiological stresses, but changes in cycles of work and rest are disruptive and contribute to fatigue and inefficiency. Both illness and medication reduce the body's efficiency, causing chemical imbalances, fatigue, and lessened ability to perform physical functions. It is essential therefore that all ailing persons, especially those who take medication and operate critical equipment, be under chemical control.

MOTIVATION

People are motivated by many desires that create a drive toward successful accomplishment of their work. We frequently exceed our own normal abilities and the abilities of others when motivated by esprit de corps, patriotism, love, hate, revenge, sex, competition, prestige, hunger, fear, pride of accomplishment, financial rewards, and numerous other reasons. When some of these aspects of motivation are excessive to the point where they adversely affect the person's mental or emotional balance, the equanimity and lack of motivation of machine appears to its advantage.

Aptitudes, desires, feelings, and motivations must be considered in ensuring efficient operations. An eager, well-motivated individual who is undistracted by personal problems or stresses can usually outperform one who is distracted or poorly motivated, other things being equal. Persons with emotional imbalances often remain potential sources of acts leading to inefficiency or accidents.

Related to human error is the broader aspect of human reliability. Human error has the connotation of a person's performing an unintentional act. Human reliability, in addition, involves intentional acts in accordance with or contrary to policy, instructions, or good practices. It was the reliability aspect in regard to military nuclear systems that led to development of the two-man concept in order to minimize the chances of errors or sabotage.

VIOLENCE IN THE WORKPLACE

The U.S. Department of Justice statistics on victims of crime, as found in the 1994 National Victimization Survey, indicates the extent of the problem of violence in the workplace. Not including homicides, the survey shows that during the year one million workers were victims of violence while working.[38] In fact, from 1987 to 1993, approximately one million persons annually were assaulted at work. One in six violent crimes in the United States occurs at work. Data of the Bureau of Labor Statistics show that there were 1,071 deaths due to workplace violence in 1994, a slight decrease from the year before. Forty percent of workplace fatalities of women were homicides, the leading cause of their being killed at work. However, 82 percent of homicide victims in the workplace were men.[39] In 1993, assaults and violent acts comprised 51 percent of all fatal work injuries in the New York/Northern New Jersey/Long Island area, 48 percent in Los Angeles, 41 percent in Houston, and from 34 to 36 percent in San Francisco, Detroit, Washington, D.C., Philadelphia, and Dallas.[40] However, this is not simply a large-city problem. Violence in the workplace happens anywhere.

Workplace Violence Prevention Program. Employers need an effective workplace violence prevention program. It starts with the commitment of management to the program and the involvement of employees.

A Written Policy on Violence in the Workplace. There should be a written workplace violence prevention policy statement, committing the employer to provide:

- authority and budgetary resources to the program
- encouragement to employee participation in the design and implementation of the company's workplace violence program

and stating that the employer:

- refuses to tolerate violence in the workplace
- is committed to the development and implementation of a program to reduce incidents of violence in the workplace

- will apply workplace violence policies consistently and fairly to all employees, including management and supervisors
- requires prompt and accurate reporting of violent incidents, whether or not physical injury has occurred
- will not discriminate against victims of workplace violence

The First Step. An initial step is to assess the threat of violence. In companies of sufficient size a team might be formed for this purpose, consisting of senior management, employee representation, safety engineering, human resources/employee assistance, finance, and legal personnel. The purpose of this effort (whether conducted by a team or by a program coordinator) is to assess periodically the potential for violence in a given workplace and recommend preventative actions. This **hazard assessment** will include a common-sense *workplace security analysis* and *review of relevant records*. The workplace security analysis will locate where and to whom the risk of violence exists by looking at processes and procedures, noting high-risk factors, and evaluating existing security measures. Locations that have high risk factors should be identified and analyzed with the goal of making appropriate recommendations. These factors include locations that have poor lighting, are isolated, or have easy access, as well as those where security is not strong, where money is exchanged or located especially during nighttime hours, where there is valuable property, or where there is a history of violence.

Any and all relevant records which can provide information leading to a successful program, without violating workers' rights to privacy, should be reviewed periodically and evaluated to help determine the potential for violence in the workplace. Some records to review include OSHA 200 logs, incident reports (both of assaults and near-assaults), police reports, workers' compensation records, medical reports, accident investigations, training records, and grievances.

The Second Step. A system should be put in place that will provide management and employees with situation awareness through the *tracking and analyzing of incidents* and *monitoring of trends*. Periodic surveys of employees that solicit their ideas on the potential for violent incidents and that identify or confirm the need of security measures are helpful. Independent reviewers such as safety professionals, law enforcement or security specialists, insurance safety auditors, and other qualified persons from outside of the organization can offer independent, unbiased viewpoints to improve a violence prevention program.

The Next Step. The next step is to provide hazard prevention and control through engineering and administrative intervention. *Engineering controls* should be provided in preference to all other options. These involve workplace design or redesign to remove the hazard from the workplace or create a barrier between the worker and the hazard. Physical barriers and security devices such as bulletproof enclosures, mechanisms to allow employees to view surroundings, surveillance equipment, lighting, access barriers and control mechanisms, speakers, alarms, call boxes, cellular phones, two-way radios, metal detectors, and drop safes are included in engineering controls. The design of facilities and work area, including furnishing to minimize violence potential, prevent employee entrapment, and facilitate emergency response, is part of engineering controls.

Administrative controls include, but are not limited to, varying employee routines and locations, providing identification cards for employees, and establishing sign-in and sign-out and escort/badging policies for nonemployees, providing written procedures for workers to follow when they enter threatening locations or feel threatened, banning weapons in the workplace, assisting employees who are victims of domestic strife and violence, and *training* employees in situation awareness/avoidance/response. All employees should be trained in:

- Techniques for recognizing the potential for violence
- Who is at risk, when, and where
- Procedures, policies and work environment arrangements developed to control workers' risk
- Conflict resolution
- How to react during incidents of violence, including emergency and hostage situations
- How to obtain assistance, including medical help
- Changes in potential for violence

Administrative controls also include *postincident response.* All effective violence prevention programs should provide means for the immediate and comprehensive treatment of victimized employees and those employees who witness the event. Emergency response procedures should be part of the violence prevention plan and should be designed to ensure that security, police, and medical emergency personnel are promptly summoned. Injured persons should receive prompt treatment. Prompt and appropriate postincident debriefings and counseling may reduce fears, unnecessary feelings of guilt, or other stress and psychological trauma. This process should be planned, if possible, before any incidents occur, to ensure prompt emergency response and the use of qualified professionals in trauma-crisis counseling. Also, the postincident response involves debriefing to determine exactly what happened and to make recommendations to prevent recurrence of similar events. Where feasible, these recommendations should be brought forth by a Violent-Incident Review Board specifically constituted for review of events of violence in the workplace.

JUDGMENT

Another outstanding capability of people is their ability to exercise independent judgment, especially on very complex, extremely infrequent, or normally unrelated facts. A person can assemble data from various sources, taking what is desired and rejecting what is not, correlate it with information drawn from experience, reject inconsistencies or incompatibilities, and quickly arrive at a logical conclusion, all without first being programmed to undertake these actions. The most complex computers can perform only those specific operations for which they have been programmed and for which they have been fed information coded in a particular form.

On May 7, 1966, a worker in a small fertilizer plant in northern Michigan descended into a cistern to finish work he had started the previous day. Soon afterward a second man, working on the surface, saw that the man in the cistern was in trouble and shouted for help. A third man ran out of the plant and jumped into the excavation to assist the worker. They both died. Another plant worker did the same thing, and he too was overcome. A passerby who had heard the alarms ran over to help. The plant manager attempted to restrain him, but he broke away and also leaped into the cistern. He too died. Firemen arrived soon afterward. The chief donned a gas mask, tied a rope around himself, and went into the cistern. He also died, because he didn't know that gas masks provided almost no protection against most gases. The coroner later reported the five men had been killed by accumulated hydrogen sulfide.

FIGURE 9–3 When a Man Needs a Buddy

Joe and Jack have often worked together in hazardous situations. Because he judges it to be so, Joe can sometimes tell that Jack is not acting at his usual level of competence. He may be a bit slower, make slightly more errors, look paler, or act more hesitant in his judgments. Joe may realize he may have to keep a closer eye on Jack than he normally does, help him a little more, or take over or double-check critical tasks. A case regarding a series of erroneous judgments is described in Fig. 9-3.

ACCIDENT-PRONE PERSONS

At one time much publicity was given to the concept that a small group of "accident-prone" persons, or accident repeaters, were responsible for most accidents. Study and observation over the years later showed this premise to be incorrect. The idea of persons being accident-prone was long maintained by many employers and others who found it easiest or advisable to lay blame for accidents on workers. There are persons who may have more accidents than do average workers, but generally this is because their jobs are more hazardous. Persons who are repeaters are involved in only 15 to 20 percent of all accidents. If a person has an accident, he or she takes care it will not recur (especially if it proves fatal). It is therefore self-limiting.

The desire to determine the causative factors of accidents involving the human element has led to much research on conditions that affect human behavior. One early (and erroneous) attempt in this direction discussed a means of determining a woman's abilities as a driver, which in part included the following ideas:[41]

> The fullness of the bony ridge in the eyebrow region is said to be an unmistakable sign of a woman's driving ability. Same with the ear in which the central section is larger. A firm mouth, with closed lips is credited with indicating carefulness, while full lips and a partially open mouth belongs to impulsive persons. If the crown of the head appears rounded, its owner is not very cautious, but if the back of the head at the crown is wider across, caution is well represented.

Swain (Fig. 9-4) lists representative factors that shape human performance. Schulzinger[42] stated: "It has proved possible to assemble no fewer than two hundred

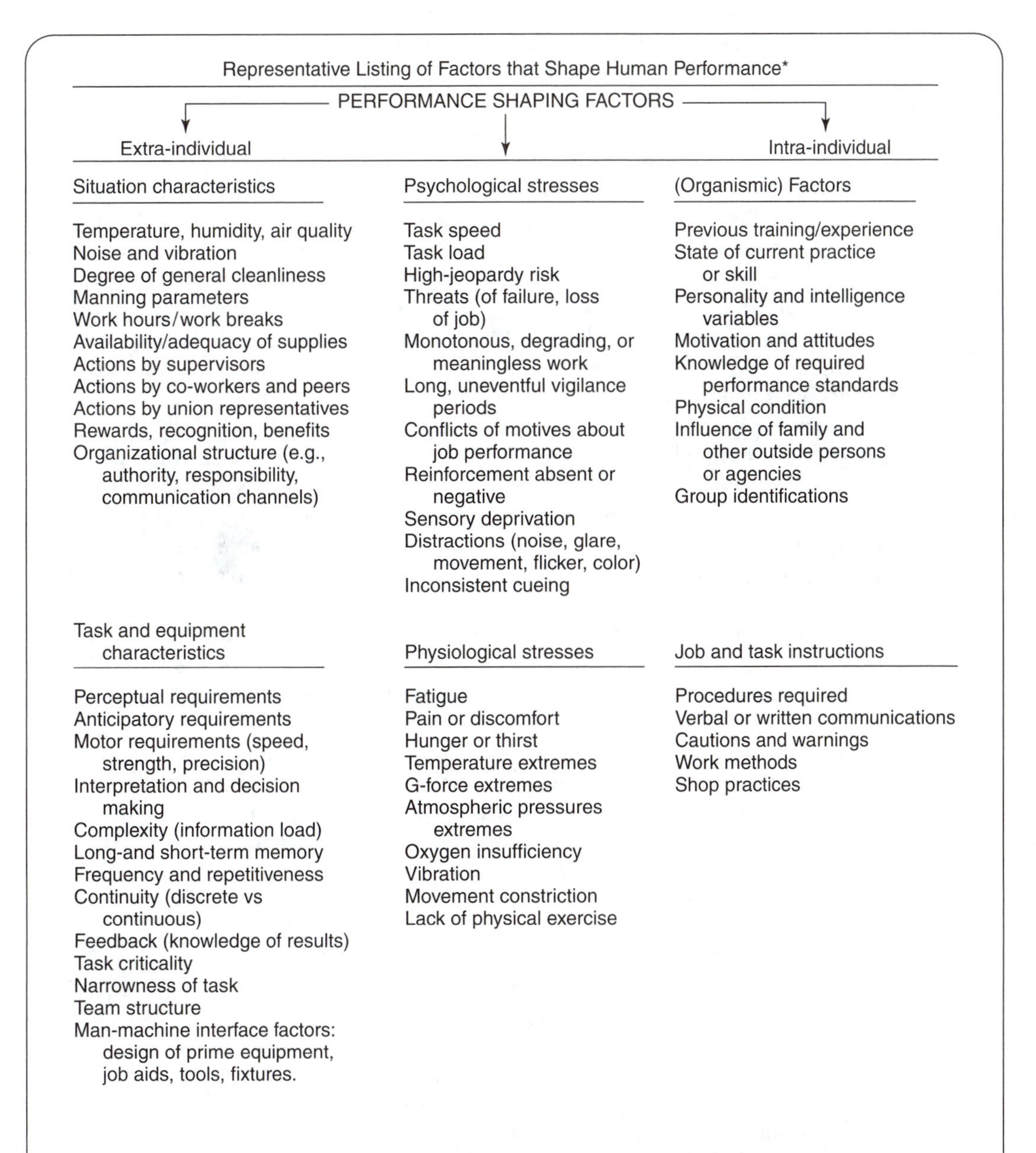

Representative Listing of Factors that Shape Human Performance*

PERFORMANCE SHAPING FACTORS

Extra-individual		Intra-individual
Situation characteristics	**Psychological stresses**	**(Organismic) Factors**
Temperature, humidity, air quality Noise and vibration Degree of general cleanliness Manning parameters Work hours/work breaks Availability/adequacy of supplies Actions by supervisors Actions by co-workers and peers Actions by union representatives Rewards, recognition, benefits Organizational structure (e.g., authority, responsibility, communication channels)	Task speed Task load High-jeopardy risk Threats (of failure, loss of job) Monotonous, degrading, or meaningless work Long, uneventful vigilance periods Conflicts of motives about job performance Reinforcement absent or negative Sensory deprivation Distractions (noise, glare, movement, flicker, color) Inconsistent cueing	Previous training/experience State of current practice or skill Personality and intelligence variables Motivation and attitudes Knowledge of required performance standards Physical condition Influence of family and other outside persons or agencies Group identifications
Task and equipment characteristics	**Physiological stresses**	**Job and task instructions**
Perceptual requirements Anticipatory requirements Motor requirements (speed, strength, precision) Interpretation and decision making Complexity (information load) Long-and short-term memory Frequency and repetitiveness Continuity (discrete vs continuous) Feedback (knowledge of results) Task criticality Narrowness of task Team structure Man-machine interface factors: design of prime equipment, job aids, tools, fixtures.	Fatigue Pain or discomfort Hunger or thirst Temperature extremes G-force extremes Atmospheric pressures extremes Oxygen insufficiency Vibration Movement constriction Lack of physical exercise	Procedures required Verbal or written communications Cautions and warnings Work methods Shop practices

*A.D.Swain, *Sandia Human Factors Program for Weapon Development*. SAND76-0326, Sandia Laboratories, Albuquerque, New Mexico.

FIGURE 9–4 Representative Listing of Factors that Shape Human Performance

and fifty factors contributing to maladjustment and accidents. The majority of these are purely mental, but numerous ones relate to mental states imposed by physical, environmental circumstances."

Alkov[43] has indicated that various events in people's lives can generate stresses to different degrees. His study (like those of others) was originally predicated on the relationship between life events and illness. However, "the knowledge that the emotionally stressed individual may be more prone to illness and accident is not new."

Davidson and others[44] investigated whether some unusual environmental factors could have influenced accidents experienced at the Sandia Laboratories in New Mexico. Data extending over almost 20 years were examined. The factors investigated were time of day, magnetic influence, lunar influence, solar influence, annual cycles, and barometric pressure. Preliminary data implied that relationships might exist in some instances, but more research was needed to establish their validity.

Another line of research to predict accident causes involved "biorhythmic" relationships.[45] Biorhythm was predicated on variations of behavior in natural cycles of 23, 28, and 33 days, based on a person's date of birth. Although acceptable explanations have never been established for these relationships, the concept has been tried in use. A transportation company in Japan used a computer to determine from biorhythmic data of 500 drivers those of them "scheduled" to have adverse days and accidents. On those days, the drivers were given cards urging them to use extra care. After the system had been in use for a year, it was claimed the drivers' accident rate had dropped 50 percent. Whether frequent issuance of the warning cards at any time calling for increased care would have achieved the same result has not been investigated and is not known.

Predictions of death, destruction, and accidents have been claimed by diviners using the entrails of fowl and sheep and by Nostradamus, Mother Shipton, astrologists, and other seers, but with no sustained success. Attempts have also been made to use accumulations of past data to predict future events by quantitative error evaluations.

QUANTITATIVE ERROR PREDICTION

A means has long been sought to measure the probability that persons would make errors, especially where they might be critical, and thus might generate accidents. The probability would also give a measure of how many errors might be expected. Corrective action could then be concentrated to reduce errors at a critical point or the total number of errors for the entire task sequence.

Probabilities give valid results only when based on extensive experience or trials. They cannot be used to indicate which individuals in a population will commit errors or cause accidents. A number of automobile accidents take place every year, leading to a statistic from which a probability will be projected for the following year to indicate possible future occurrences. However, there are no means of determining which driver or car will be involved. Any prediction can only be similar to that of actuarial tables used for life insurance purposes, which can predict approximately how many men out of 100,000 will die before a stated birthday but cannot predict which of the men it will be.

The American Institute of Research in Pittsburgh undertook tests of individuals performing various simple tasks and scored them on numbers of errors made. The data were then converted into probability (or human reliability) tables. From such tables, a complete operation may be broken down into individual tasks, each with a probability of successful accomplishment. All results are then synthesized into one for the complete operation.

Quantitative human error prediction is useful as a component of system reliability determinations. This is important in risk analysis for the evaluation of the safety of complex systems. Almost all systems, even so-called "unmanned systems," involve human operation or maintenance in some way. Risk assessment and risk management should account, if possible, for the human error potential. Later in this book there is a discussion of fault-tree analysis. Quantitative human error prediction, or human reliability analysis, provides essential information for the quantitative evaluation of fault trees that incorporate the human as a system component. A risk assessment that omits the human component can lead to risk management strategies that are ineffective. However, care must be exercised about the source of data used for human reliability quantification. Some data are very situation dependent and cannot be used generally; other data are little more than guesses. Sometimes the issue is whether soft data are better than no data.

HUMAN FACTORS ENGINEERING

Investigations of error causes are of interest to designers if they can lead to means of eliminating future errors and accidents. *Human factors engineering* is a technical discipline dedicated to this end. It involves the study of the relationship between persons and equipment design with the aim of producing optimal and safe worker performance. Besides improving the ease with which workers can undertake tasks such as reading dials, good human engineering will reduce fatigue, erroneous actions, and wrong decisions. In some circles the terms "human factors" and "ergonomics" are used interchangeably. Human factors engineering applies what is known about human physical, psychological, and physiological limitations and abilities to the design of equipment and systems to ensure their operability, safety, and comfort. *Human factors* is the science—i.e., the body of knowledge—used in human factors engineering. *Ergonomics* is a slightly narrower term (from the Greek: *ergo* = work; *nomos* = laws or rules) denoting the science of fitting the task to the person. Human engineering has resulted in the design of better tools, controls for equipment, seats and chairs, and many other items that affect comfort and safety. Investigation of the accident in the nuclear power plant at Three Mile Island cited two principal causes: human error and bad human engineering. Much of the operator error occurred in making decisions and responses required because of inadequate designs.

Human factors engineering attempts to obtain maximum effectiveness in any human-machine operation by integrating the best capabilities of both. It is concerned with the design of equipment so that it can be operated easily and rapidly with a minimum of undue effort or strain. A control panel designed for operation by a single person can overwhelm the worker's mental and physical capabilities, if an overabundance of data from instruments must be monitored or its controls operated at one time. Should the instruments on a panel be widely separated, an operator would either quickly be exhausted trying to perform an almost impossible task, make errors, or ignore some of the instruments, possibly leading to an accident. Much work has been done in the development of human engineering since it was originated after World War II. Numerous excellent texts are available on the subject. A human factors analysis of a very common operation is presented in Fig. 9-5.

PROCEDURAL MEANS OF ACCIDENT PREVENTION

Workers can be motivated to minimize actions that could lead to accidents. In an industrial plant, to avoid errors and accidents, workers can be alerted to follow safe procedures by warning signs and posters, lectures, and supervisory means. Firm supervisory control can be a definite motivating force in error prevention. Supervisors can not only indicate the error or warn workers of hazards and forbidden acts but also, if the worker continues to act unsafely, can sometimes impose a penalty such as time off or demotion. A plant in which it is known that disregard of safety rules will result in disciplinary action (a "tight ship") will generally be a safer workplace than one in which the rules are held lightly. It is for this reason that management support is a necessity.

Common examples of disregard for safety include smoking in areas not so designated; failure to wear hard hats, face shields, goggles, or respiratory equipment; blocking fire exits and routes to them; failures to secure ladders properly; and poor housekeeping, to mention a very few. No violation should be tolerated. Front-line supervisors should be held directly responsible by higher managers for ensuring that workers observe the safety rules and maintain safe working habits; supervisors in turn must hold workers responsible for safe conduct.

In some instances workers may unknowingly be conducting themselves in unsafe ways. This is often the case with newly employed or inexperienced workers. Some newly hired, previously employed workers may disregard established procedures to do things through habit as they had done them previously elsewhere. Inexperienced workers may never have been taught safe conduct. Both of these types of new employees must be checked frequently and carefully by supervisors until they are certain the workers have safe work habits. Then, as for all workers, supervisors should periodically check their operations, working conditions, tools, and safety equipment to ensure they have maintained those good habits.

CRITICAL OPERATIONS

Errors can be far more critical in certain operations than in others. Closer control must be maintained in those cases, in order to minimize both the possibilities of error and the injury or damage that could result if an error is made. Close collaboration among the safety engineer, supervisors, and workers is required. The frequency of plant visits

PRODUCT: Single-Section, Straight Ladder

Misuse Mode	Behavioral Factors	Design Consideration
1. User sets ladder at angle too near vertical so that it tips backward as he ascends or gets near top.	Lack of experience, user doesn't know proper erection-angle limits; doesn't know or realize his body needs to remain close to rungs; not familiar with c.g. factor.	Warning-use instruction label conspicuously located (consider use of alternate orientation of ladder) built-in safe-angle indicator. Design instruction with "pictures."
2. User sets ladder at shallow angle so that ladder bends or bounces when he is near center of span.	Same as above	Same as above; over design structurally, minimize bending characteristics.
3. Ladder is too short for situation so user stands on upper rungs with no hand support	Doesn't think ahead or recognize potential hazards; tries to "make do" – takes a chance.	Warning-use instruction.
4. Due to uneven surface, ladder is not set up properly (e.g., it lists right or left causing ladder to twist, shift c.g., or introduce structural stress). Feet of ladder slip or penetrate surface unevenly.	Not conscious of support-surface condition; poor judge of verticality; too lazy or to much in a hurry to prepare surface; willing to take chance due to inexperience ; doesn't consider which end of ladder should be down or up.	Self-leveling, broad-footprint foot design; conspicuous warning; built-in vertical level indication; provide combination device which will adapt ends of ladder either to ground or structure, in event user puts either end on ground.
5. Inadequate handholds (e.g, rungs too large, vertical elements too large and awkward in shape) for user to get in secure grip.	Overconfident, doesn't bother to try and get firm grasp; may have one hand full so body c.g. shift requires more grip force than he can apply.	Proper size and shape of potential handhold areas, nonslip treatment; warning instructions on use and hazard.
6. Insufficient bearing surface allows upper vertical elements to penetrate structure against which ladder is placed. Weight of user may cause this penetration as he approaches top although it didn't occur before.	Doesn't understand concentrated load effect; doesn't think about additional weight as he nears top; can't see top of ladder very well; doesn't know condition of support structure or material; awkward to place upper element where he intends due to manipulation/weight disadvantage.	Design wide-support footprint area, self-adjustable to support surface; special color to make support element conspicuous.
7. User doesn't pick ladder up at c.g. for carrying, causing him to drop it or dig one end into ground.	Lack of experience; doesn't know where e.g. is.	Warning; mark pickup point.
8. User injures himself picking up ladder which is too heavy to carry.	Doesn't know ladder too heavy; doesn't think about it before he tries to pick it up; doesn't pick up properly.	Warning—indicate weight.

FIGURE 9–5 Human Factors Analysis of a Straight Ladder Continued on page 148

9. User's foot slips off rung or misses and slips between rungs	Climbs without looking, at feet; puts them where he thinks rungs should be.	Use nonskid surface; use standard rung spacing and vertical separation.
10. In attempting to erect long ladder, user unbalances ladder, upsets himself, and may have ladder fall on him; feet of ladder may slip and result in injury to user.	Inexperience, takes wrong approach; doesn't recognize hazard.	Operation instructions in conspicuous location; design feet to dig into surface and/or nonskid type; limit ladder length according to weight.
11. Metal ladder comes in contact with high-tension electrical wire or component, may cause shock or burn as user touches ladder.	Unaware of hazard.	Use materials which will not carry electrical current.
12. In backing off roof or high place, misses rung with foot; may kick ladder and change its orientation, causing it to be unsteady.	Can't see where he is putting foot; unsteady; preoccupied with holding on to structure; in awkward position for determining orientation.	Maximize ladder width; point out hazard in operating instruction.

Man Factors, Inc., *The Implication of Product User Behavior: A Study Conducted for the National Commission on Product Safety,* San Diego, Dec. 30, 1969.

FIGURE 9–5 Continued

by OSHA Compliance Officers and other safety inspection personnel is usually based on the criticality of the operations conducted.

Very often a system is initiated under which critical operations will be accomplished only after a permit for them has been approved. The safety representative will then review existing conditions and preparations; check the workers to ensure they are physically qualified, understand what is to be accomplished, know the hazards involved and how to recognize any adverse situations; and ensure that the buddy system is used (see page 204) and that the "buddy understands what to do and is properly equipped." A few of these critical operations include, but are not limited to:

- *Tank entry:* this may be done to clean or repair the tank or other restricted space. See Chapter 26.
- *Welding operations:* welding can have disastrous results unless proper safeguards are employed (see Figs. 9-6 and 9-7).
- *Laboratory operations:* operations involving high-energy reactions.
- *Explosives:* operations involving the presence, use, handling or transportation of explosives (see Chapter 23).
- *Toxic substances:* operations involving the presence, use, handling, or transportation of highly toxic substances (see Chapter 24).

1. On Feb. 9, 1942, the sparks from a welder's torch accidentally ignited lifeboat material during conversion of the former French liner *Normandie*. By the time the fire had been put out, so much water had been added that the ship turned on its side and sank. Loss: $53,000,000.

2. In 1953 the General Motors plant at Livonia, Michigan, was destroyed from sparks from a welding operation undertaken to replace a steam line under a roof. The sparks ignited rustproofing oil in an open tank, from which the fire then spread to the oil-soaked wooden blocks that formed the plant floor. Almost the entire plant of 1,500,000 square feet was destroyed. Three persons died; damages were estimated at $45,000,000 (not including the financial impact on employees, merchants, and other businesses that resulted from loss of wages).

3. In 1960 the aircraft carrier *USS Constellation* was being fitted out at the Brooklyn Navy Yard when a lift truck broke a plug on a gasoline tank on the top deck. The gasoline flowed down an open shaft to a lower deck where the vapors were ignited by the flame of a welder. The fire spread so rapidly that 50 workers were killed and 150 injured. There was $40,000,000 in damages to the ship and its equipment.

4. On August 9, 1965, 53 civilian workers died as a result of a fire in a Titan II missile launch silo near Little Rock, Arkansas. The welder was adding steel to "harden" the silo when he ignited hydraulic fluid from an adjacent line. Only about 90 gallons of fluid burned, and the workers were killed principally by carbon monoxide and heat in the enclosed area.[46]

FIGURE 9–6 Welding and Four Notable Fires

- *High-voltage systems:* repair, modification, or maintenance of high-voltage systems (see Chapter 21).
- *Radioactive substances:* operations involving the presence, use, handling, or transportation of a radioactive material (see Chapter 27).

RESPONSIBILITIES OF THE INDIVIDUAL WORKER

Each worker must be individually responsible for:

1. Knowing that he or she is personally responsible for keeping himself or herself, fellow personnel, vehicles, and other equipment free from mishaps to the best of his or her ability. He or she should try to anticipate every way in which a person might be injured on the job, and conduct his or her work to avoid accidents.
2. Being constantly alert for any hazardous conditions or practice. He or she should report any unsafe condition or practice to the supervisor or safety office.
3. Acting as necessary to protect his or her own well-being and that of other personnel.
4. Applying at all times the principles of accident prevention in daily work. Practical jokes and "horseplay" should not be permitted on the job. Workers should advise inexperienced fellow employees of safe methods and procedures, warn them of hazards, and caution them when they are engaged in dangerous practices.

WELDING PRECAUTIONS

Welding is done safely every day. A good welder learns certain precautionary measures for safe welding operations and follows them until they become habits. Without these good habits, welding can be one of the most dangerous of all occupations. Precautions that a good welder should learn include the following:

- Welding should be performed by qualified welders only.
- Welding operations should be performed away from flammable materials. If the object to be welded cannot be moved to a safe location, all movable hazardous materials should be moved to a safe location. If this cannot be accomplished completely, and materials in nearby areas could be affected by welding arcs, flames, sparks, spatter, slag, or heat, the welding zone should be enclosed in fireproof blankets or other protective shields.
- Fire protection equipment should be kept immediately at hand and ready for use. In critical areas, the fire protection equipment should be manned while welding operations are being conducted.
- Care must be taken against allowing mixtures of fuel gas and air to accumulate. Leakage of hydrogen or acetylene into enclosed areas where it might create a flammable or explosive atmosphere must be guarded against. No device or attachment should be permitted which involves mixing the fuel gas and oxygen or air before it reaches the point of combustion, unless the device has been specifically approved for the purpose.
- Oxygen cylinders should be separated from fuel-gas cylinders or other combustible materials by at least 20 feet or by a fire-resistant barrier at least 5 feet high. Oxygen cylinders should not be stored near highly combustible materials or near materials likely to cause a fire or in any other hazardous location. Oxygen should be checked frequently to make certain it is not leaking from supply cylinders, especially in enclosed spaces, where it can cause ignition of materials that are not normally highly flammable. Grease and oil should be kept away from, and never used to lubricate, oxygen cylinder valves or regulators.
- Flammable and other deleterious materials should be cleaned from surfaces to be welded before welding is started. The very high temperature of the welding arc or flame can cause ignition of materials that are not normally flammable. No welding, cutting, or similar work should be undertaken on tanks, barrels, drums, or other containers which have been contaminated through use, unless the contamination is first removed so that there is no possibility of fire or emission of toxic vapors.
- The face, body, and hands should be covered to prevent burns from spatter, slag, sparks, or hot metal. Flameproof, heat-insulating gloves should be worn during welding operations. Wet or excessively worn gloves should not be used.
- Suitable precautions should be taken to avoid shock from electric welding equipment. The welder should not stand in water or any wet surface while doing electric welding. Hot electrode holders should not be dipped in water. Cables with damaged insulation or exposed conductors must not be used, and should be replaced before any such work is attempted. If lengths of cable must be joined, use only connectors designed specifically for the purpose; avoid makeshift methods. Cables should not be left where vehicles can run over them, where they could be cut or pinched, or where other persons could step on or trip over them.
- Adequate ventilation should be provided as protection against accumulations of toxic gases. Where such accumulations cannot be prevented, the worker should be equipped with respiratory protective equipment; however, use of such equipment should be secondary to preventing gas accumulations.
- Enclosed areas should be ventilated before a welder is permitted to enter, and continual checks run on condition of the environment. Respiratory equipment should be immediately available if there is a possibility a problem might arise.
- The eyes and skin should be protected against the glare and radiation from a welding arc or flame. Helmets or hand shields should be used except where the arc or flame is completely contained. Helmets and hand shields should be made of materials that are heat and electrical insulators. They should not be combustible and should be capable of being cleaned easily. Helpers and attendants should also be provided with eye protective equipment. Other personnel in the vicinity of a welding operation should be guarded against reflections by suitable shields and barriers. No person should be permitted inside an enclosure where welding is taking place unless he is suitably protected. Entrances to such enclosures should be posted with warnings to this effect.
- Gas cylinders must be handled carefully (breaking the neck from a full cylinder can turn the bottle into a lethal missile). The cylinders should be secured to keep them from falling. Acetylene cylinders must always be maintained in an upright position. Empty gas cylinders should be marked and have their valves closed tightly. Valve protection caps should always be in place on those cylinders designed with caps, except when the cylinder is in use or connected for use. Cylinders should be stored out of the direct rays of the sun and away from other sources of heat. Never strike an arc against a gas cylinder. (For additional safeguards regarding cylinders, see Fig. 20-4).
- Where hazardous operations are controlled by permits, no welding should be undertaken until a permit that approves the work is obtained.

FIGURE 9–7 Welding Precautions

5. Following prescribed procedures to the best of his or her abilities in any emergency. He or she must endeavor to remain calm, remember instructions, and take steps he or she has been instructed to take to safeguard himself or herself and fellow employees.
6. Promptly reporting to their supervisor or medical office any injury, regardless of the severity, and any illness that might adversely affect safe performance.
7. Ensuring that, after receiving instructions, he or she understands them completely before starting work.
8. Reviewing educational safety material posted on bulletin boards or distributed to each worker.
9. Knowing where and how medical assistance and other emergency help can be obtained.
10. Abstaining from removing, inactivating, damaging, or destroying any warning or safety device, or interfering with another person's use of it.
11. Observing all posted warning signs.

PROCEDURE ANALYSIS

A procedure analysis is a set of instructions for sequenced actions to accomplish a task such as a method for an operation, maintenance, or repair, assembly, test, calibration, transportation, handling, emplacement, or removal. Procedure analysis is a review of the actions that must be performed, generally in relation to a mission task, the equipment that must be operated, and the environment in which both must exist. Analyses must ensure that the procedures are not only effective and efficient but safe. Such analyses involve determination of the required tasks, exposure to hazards, criticality of each task and procedural step, equipment characteristics, and mental and physical demands.

A designer generally envisions procedures in light of the design he or she prepared. Procedures often themselves may be lacking, incorrect, poorly written, or difficult to understand. After they have been prepared and written, they should be subjected to analyses and evaluation by others and then, if possible, tested. The procedure analysis from the safety standpoint explores all related matters that could degrade performance or cause an accident and injury.

The possibility of error during an operation is one problem. The other is that although the worker may perform correctly, the designer may have stipulated a task incorrectly. The designer's vision and understanding must be passed on to the less well-informed operator in as simple an explanation as possible. Complexities in explanations and procedures are generally detrimental to understanding and tend to lead to errors and accidents.

The first part of any procedural analysis, therefore, is to review the task and procedure to be followed by the worker. Some of the basic precepts that must be considered are indicated in Fig. 9-2; however, the person who prepares and reviews them must realize that:

- Any equipment or procedure that can be used incorrectly will someday be employed the wrong way. (A corollary of Murphy's Law.) The effect of such actions must therefore be investigated and preventive measures taken to eliminate possibilities of deviation and error and to minimize the effects if an error does occur.
- No matter how simple a procedure appears, it should be examined critically for possibilities of error and danger.
- Personnel tend to take shortcuts to avoid arduous, lengthy, uncomfortable, or unintelligible procedures. These shortcuts frequently are open to error, which can generate accidents. A procedure calling for equipment that is difficult to maintain will suffer from lack of maintenance.
- Most human-machine relationships now involve procedural problems in the use of equipment rather than in failures of the equipment itself. On the average, equipment which has passed "burn-in" operations fails on the average of one in a thousand times; personnel, one in a hundred.
- Personnel believe they themselves are so knowledgeable, careful, and adept at the tasks involved that they will make no error, although other persons might. Personnel who are so assured of their own capabilities are the ones who especially must be protected.
- All procedures and each step within a procedure should be examined for necessity. All extraneous operations should be eliminated, because their inclusion might provide confusion as workers debate why they were necessary. Highly critical steps should be accented.
- Requirements for special training for personnel should be kept to a minimum. In addition, the procedures analysis should assume that at certain times training may be lacking or faulty.
- Procedures requiring person-to-person communications should also be kept to a minimum and be as simple as possible. Errors in communication are frequent sources of procedural failures that can cause accidents.
- Procedures involving interruptions will generate circumstances under which steps may be forgotten when a procedure is resumed. Such interruptions are especially prevalent in maintenance and repair where there are no instructions regarding the parts or tools required for carrying out the procedure. Later, the need for obtaining the parts and tools after the work has begun will cause interruptions and possibly omissions and mistakes.

The next step in making this type of analysis is to ensure the procedure contains the following:

- Preparatory assembly, installation, and servicing instructions.
- Instructions for the proper operation of the equipment, including, where applicable, tables, charts of operating data, adjustments, warnings, and precautions.
- A list of warnings and precautions for the proposed operation, including environmental requirements, assembly or test equipment setup, and preliminary checks or tests.

- Step-by-step instructions for performing the operation.
- Instructions for stopping the operation and disassembling or disconnecting equipment.
- Emergency shutdown and backout instructions.

The safety engineer may undertake to review any of the procedures developed by the designer and to identify any hazard for which a safeguard has not been provided. Here again it should be pointed out: wherever possible, safeguards should be designed into a system or product and the need eliminated for warnings, cautions, and other procedural means to avoid accidents. Figure 9-8 indicates a format that can be prepared by either a designer or a safety engineer. Most of the headings are self-explanatory.

The problem which then might arise is that an operator might deviate from the safe procedure prescribed by the designer and do something that might result in a mishap. The deviation might be deliberate or by mistake, but in either case it is necessary to evaluate what it might be and the consequences it might generate. If a deviation

PROCEDURES ANALYSIS
(INITIAL PHASE)

Task	Danger	Effect	Cause	Corrective or Preventive Measures
Charge nitrogen pressure vessel	1. A loose hose may whip	Personal could be injured or equipment damaged.	Hose failure; connection failure; failure to tighten connection adequately	Tie down, chain or sandbag hose at close intervals. Personnel wear hard hats and face shields. Establish torque values for tightening connections. Warning and caution notes in procedures.
	2. Vessel bursts	Fragments may injure personnel or damage nearby equipment.	Inadequate strength	Use high-safety-factor design. Provide warning against over pressurizing system. Do not expose pressure vessel to heat. Incorporate relief and safety valves. Test vessel o ensure that it will carry required pressure.
	3. High-velocity gas escapes	Gas may blow solid particles into eyes or against skin. Loss of gas may cause system to become inoperative due to lack of pressure.	Leak; hose failure; loosening fitting on pressurized system; crack	Procedures to provide warnings to depressurize system before attempting to disassemble connectors. Personnel to wear face shields.

FIGURE 9–8 Procedures Analysis (Initial Phase)

PROCEDURES ANALYSIS (DETAIL PHASE)

TASK PROCEDURE BEING REVIEWED: Charging nitrogen pressure vessel

Procedural step	Possible alternative	Potential hazard and effect	Probability of occurrence	Safeguard
Ensure vessel to be filled is secure	Failure to check	Vessel may fall over; may be damaged so it fails under pressure	Occasionally	Train personnel to do so at all times. Supervisor to ensure personnel follow this practice consistently.
Check that pressure vessel valve is closed	Failure to check	Pressurized stream may be emitted when protective cap is removed	Occasionally	Train personnel to do so consistently. Personnel to wear face shields and hard hats. Warning in instruction manual.
Remove protective cap from fitting of vessel to be filled	Cap not removed	No hazard. Cannot fill vessel.	Occasionally	
Wipe threads on connection fittings	Failure to clean	Threads damaged and eventually leak under pressure	Occasionally	Train workers to make this standard practice. Supervisor check to see it is done. Put caution in instruction manual.
Start screwing fitting from fill line to vessel fitting slowly	Fill-line fitting applied so rapidly it is crossthreaded/crossconnected	Threads damaged and eventually leak under pressure	Occasionally	Same
Tighten fitting to prescribed torque value	If inadequately tightened, fittings could leak or separate. If over-stressed, fittings could fail	Pressurized stream when valve is opened. Line might whip if fittings separate and line is not restrained. Either could cause injury.	Occasionally	Establish torque values and put in instruction manual with warning; ensure operator is trained to comply

FIGURE 9–9 Procedures Analysis (Detail Phase)

would generate no problems, no further action need be taken; if it might, then a measure should be incorporated that will either eliminate or minimize the adverse effects.

By going through any procedure step by step and evaluating it in this way, with suitable controls where advisable, the analyst can prevent accidents too often attributed to worker error. Figure 9-9 illustrates how Fig. 9-8 might be extended to analyze the possibilities of deviations.

OUTPUTS OF PROCEDURE ANALYSIS

A good procedure analyses can lead to:

- Corrective or preventive measures that should be taken to minimize the possibility that any error will result in either an emergency or an accident.
- Recommendations for changes or improvements in hardware or operations to improve efficiency and safety.
- Assurance that warning and caution notes are included where required, only when required, in the most effective locations in the procedures, and by the best suitable means.
- Review of requirements for special training of personnel who will carry out the operation.
- Recommendations for special equipment, such as protective clothing, which would be required for the conduct of the operation to be undertaken. Unnecessary clothing and equipment, often very costly, would be minimized.

Very successful procedure analyses have been used to ensure the correctness of each and every space flight. The best and most commonly used procedures have been with airliners, in which flight procedures include complete checkoffs before flights are attempted.

The OSHA Standard, Bloodborne Pathogens Final Standard, summarized in Fig. 9-10 is a special case requiring procedural analysis.

Probably the most publicized accidental event involving the lack of proper procedures took place at the nuclear power plant at Three Mile Island in 1973. Because of the lack of preparation of proper procedures, a contingency took place that left the operators at a loss as to what was happening and what to do about it. The contingency devolved into an accident in which luckily no one was hurt before it was contained, but which proved extremely costly.

CONTINGENCY ANALYSIS

Contingency analysis is predicated on the concept that if a hazard cannot be eliminated completely, the possibility, even if small, will always exist an accident will occur. The contingency analysis considers that something has gone wrong in an operation and emergency measures may be necessary. A contingency may occur if one circuit of a redundant arrangement fails, such as the loss of one electrical generator on the engine

In 1992 OSHA began enforcement of the Bloodborne Pathogen Standard. This standard requires employers to identify in writing those *tasks and procedures as well as job classifications* where occupational exposure to blood occurs, regardless of personal protective clothing and equipment. The purpose of the standard is to limit occupational exposure to blood and other potentially infectious materials, because of the potential to contract disease, even fatal disease, that this exposure carries. The standard covers all employees who could be reasonably anticipated to face contact with blood and other infectious materials as a result of performing their job duties. "Good Samaritan" acts such as assisting a co-worker who is bleeding are not considered, under this standard, to be an occupational exposure. At target is exposure to any body fluid visibly contaminated with blood. This is an effort to minimize the spread of human immunodeficiency virus (HIV) and hepatitis B (HBV) virus because of workplace exposure. The employer must have an exposure control plan that is accessible to employees and available to OSHA. It must be updated annually.

This standard mandates universal precautions, emphasizing engineering and work practices controls. It stresses handwashing and the provision of facilities and procedures for handwashing. It promulgates procedures to minimize needlesticks, minimize splashing and spraying of blood, ensure appropriate packaging of specimens and regulated wastes and decontamination of equipment or the labeling of contaminated equipment before shipping to servicing facilities. Employers must provide personal protective equipment (gloves, gowns, masks, mouthpieces, resuscitation bags) at no cost to the employee. The employer must have a written schedule for cleaning which identifies the method of decontamination to be used in addition to cleaning after contact with blood or other potentially infectious materials. Certain research laboratories and production facilities must follow standard microbiological practices plus additional practices to minimize exposures. Hepatitis B vaccinations must be made available to all employees who have occupational exposure to blood. Specific procedures for post-exposure evaluation and follow-up must be made available to all employees who have had an exposure incident. Hazard communication is specified. The standard requires warning labels which include the orange or orange-red biohazard symbol be affixed to containers of regulated waste, refrigerators and freezers, and other containers used to store or transport blood or other potentially infectious materials. Red bags or containers may be used instead of labeling. Bloodborne pathogen training of employees and confidential record keeping containing certain information for each employee exposed must be kept for the duration of employment plus 30 years. Training records must be kept three years and must include certain specified information. Medical records must be made available to the subject employee, to anyone with written consent of the employee, OSHA, and NIOSH. But medical records are not available to the employer.[47]

FIGURE 9–10

of a three-engined aircraft. No emergency would exist, but the failed part should be replaced. If a second engine failed, there would probably be an emergency in which immediate preparations should be made for a landing. If the third engine failed, an accident would probably result. All three of these conditions—contingency, emergency, and accident—have taken place with three-engined airliners.

If effective action is taken early enough, the contingency might be contained easily and an eventual mishap avoided. If a proper contingency analysis has been made, the analyst will have determined what could cause a degradation into an emergency or worse and made provisions for minimizing any adverse effects. Types of emergency equipment that would be needed could be selected and where they might be needed and stored. Escape routes could be developed and rescue procedures evaluated. The training to be given workers, operators, and other personnel to cope with any severe contingency could be developed. But most important, a procedure must be prepared to cope with each contingency. The methodology mentioned under the procedure analysis can be used effectively here.

Thus, both types of analyses go hand in hand: the procedure analysis must be done and done properly to minimize the possibility of accidents, and the contingency analysis must be done to indicate corrective measures that should take place in the event a failure, material or human, takes place. Unfortunately, designers and planners so often believe the products they design and the operations they plan are fault free that they neglect to provide for contingencies. When contingencies occur which could have been avoided or controlled by adequate preplanning, operators may be at a loss as to what to do, and disasters may result. The operator may then be blamed.

BIBLIOGRAPHY

[35] U.S. Department of Labor, Bureau of Labor Statistics, "Older Workers' Injuries Entail Lengthy Absences from Work," http://stats.bls.gov/osh, March 1998.

[36] Japan Industrial Safety and Health Association 1996 Report, www.jisha.or.jp.

[37] Donna Kotulak, "On The Job Older is Safer," *Safety and Health*, Oct. 1990, pp. 29–33.

[38] Ronet Bachman, *National Crime Victimization Survey: Violence and Theft in the Workplace.* U.S. Department of Justice, Washington, D.C., 1994.

[39] OSHA, "Guidelines for Workplace Violence Prevention Programs for Night Retail Establishments," DRAFT, U.S. Department of Labor, Occupational Safety and Health Administration, June 28, 1996.

[40] OSHA, *Workplace Violence Awareness and Prevention,* U.S. Department of Labor, Occupational Safety and Health Administration, Washington, D.C., www.osha-slc.gov, April, 1998.

[41] "How to Pick Women Who Can Drive Cars," *Literary Digest,* April 5, 1923, p. 58.

[42] M. S. Schulzinger, *The Accident Syndrome* (Springfield, IL: Charles C. Thomas, Publisher).

[43] Alkov, "The Life Change Unit and Accident Behavior," *Lifeline*, U.S. Naval Safety Center, Norfolk, VA, Sept./Oct. 1972.

[44] J. E. Davidson, et al., "Intriguing Accident Patterns Plotted Against a Background of Natural Environment Features," SC-M-70-398, Sandia Laboratories, Albuquerque, NM, August 1970.

[45] George Thommen, *Is This Your Day?* (New York: Crown Publishers, Inc.)

[46] Willie Hammer, "Missile Base Disaster," *Heating, Piping, and Air Conditioning,* Dec. 1968.

[47] OSHA, *Bloodborne Pathogens Final Standard: Summary of Key Provisions,* U.S. Department of Labor, Occupational Safety & Health Administration, Fact Sheet: 92-46, 1992.

EXERCISES

9.1. Discuss the types of errors that can place a worker in jeopardy through no fault of the worker.

9.2. What is Murphy's Law?

9.3. List ten actions designers can take to reduce personnel error. What causes them?

9.4. List ten actions supervisors can take to reduce personnel error. What causes them?

9.5. What is being done to predict the occurrence of operator error in task accomplishment?

9.6. Are older workers less safe than younger workers? Explain.

9.7. What is human factors engineering (sometimes called "ergonomics") and what does it attempt to accomplish? In your plant, do you believe there is some item of work—a tool, a seat, an arrangement—that could be improved by better design?

9.8. What are the elements of a "workplace violence program" for worker safety?

9.9. What are some natural factors that have been investigated for their possibilities of accidents? Which one do you believe most affects workers?

9.10. Why would a procedure analysis be beneficial? Describe some items that should be studied. Could any safety person perform such an analysis? Could you?

9.11. What is the purpose of the Bloodborne Pathogens Standard?

9.12. What are the benefits of a contingency analysis? What is the difference between a contingency, an emergency, and an accident? Describe how they relate to each other.

9.13. Do you believe the following should be blamed as much as they are for causing accidents, or more, or less:

(a) Workers and operators.

(b) Managers.

(c) Designers.

(d) All or none of these. On whom and to what degree should the blame be laid?

CHAPTER 10

Promoting Safe Practices

In former years, any effort to promote safe practices probably would have consisted almost entirely of a campaign to alert employees to the hazards in their work and a call for them to conduct themselves safely. The modern concept is that efforts to promote safe practices must start long before this. The fact that many design deficiencies cause accidents has been pointed out. There have been other causes of accidents, of which many have been blamed unfairly on workers.

Hazards that cannot be eliminated should be controlled, first by design and then by procedural means. Procedural means consist, in effect, in relying on employees to perform properly and safely. Use of procedures is a far less desirable means of accident prevention than is good design. The preceding chapter provided a rough indication of priorities that could be used for accident minimization, and use of good procedures comes last. Although many of the deficient designs may not be correctable by workers, they can be reported so people in authority can take other corrective action. And because there will be times when hazards cannot be eliminated by design, there will be times when the safe practices of workers will have to be relied on.

A very strong interest in behavior-based safety has emerged at the close of the 90s decade. This is shown clearly by a survey conducted by the *Industrial Safety and Hygiene News* for its 1997 Annual White Paper. This survey showed an increase in the percentage of environmental, safety, and health professionals who say behavior-based safety is a training priority, from 20 percent in 1994 to 56 percent in 1997.[48] Eighty-nine percent of the professionals surveyed for the report said the behavioral assessment and coaching skills are "very important" or "somewhat important" to their career growth.

Some zealots of the regulatory approach to safety, initiated by the Occupational Safety and Health Act of 1970, react to behavior-based safety (BBS) as a frontal attack upon the regulatory approach that has brought considerable progress and punch to the safety movement. Some BBS advocates have displayed a penchant for telling derogatory anecdotes about OSHA inspectors and have shown a lack of appreciation for the safety history that brought the safety movement forward. Pronouncements that the regulatory approach is a failed approach or that the BBS movement is "frightening" are extremes that do not serve safety. There is room for the engineering-for-safety, the regulatory, and the behavioral-based motivational approaches. A complete safety program will promote safe practices and comply with regulations, while eliminating and controlling hazards through engineering.

THE BEHAVIOR-BASED SAFETY APPROACH

This approach relies heavily on classical psychology of learning and motivation through stimulus and response, response conditioning, and response generalization. Human behavior is viewed as emanating from a complex of personal and situational factors. Personal factors include long-term personal traits, as well as short-term psychological states as antecedents to immediate behavior. Situational factors include the physical, task, organizational and cultural, and immediate psychosocial environments. It has often been said that every worker is a manager. Each worker can be expected to manage her/his situation to personal advantage. That advantage varies with how the worker perceives his/her situation. That situational awareness can result in well-trained employees who, nevertheless, remove machine guards, fail to use hearing protection, or perform similar unsafe acts.

BBS has been applied in the industrial setting with success to reduce unsafe acts. It focuses intervention on observable behavior, directing and motivating managers and workers through "activators" and "consequences." One approach uses what E. Scott Geller, Professor of Psychology at Virginia Tech, calls the activator-behavior-consequence (ABC) principle.[49] "Activators" are events or conditions which tell a person what to do or when to do it. Consequences are the rewards and/or penalties that result from, or are perceived to result from, the activated behavior(s). There are competing consequences for the worker. The timing, consistency, and significance of the consequences determine which will influence behavior. More competitive consequences are those that occur immediately and consistently after the activated behavior and are positive. If long-term improvement in safety behavior on the job is to occur, it must persist whether or not external controls such as rewards and penalties are there. The worker must have internal activators and internal rewards. Therefore, workers must have a sense of ownership of this process that brings them satisfaction and pride.

The next chapter discusses the validity of the statistics often used in safety. If a company's culture is simply to provide reward when there is a reduction of a statistic and to punish if the statistic goes up, that statistic may go down, but without any improvement in safety. Behaviors may be elicited such as bringing an injured worker to the workplace to sit or lie on a couch in the nurse's office to avoid the injury counting as a lost-time injury. Behavior-based safety seeks goals that will produce safe behaviors and reduce the potential for accidents.

In BBS, both workers and management must participate actively and "buy into" the process. This process is to produce a safety culture, which involves the uniqueness of a specific work site or facility.

1. A steering committee is formed, made up of both managers and workers. The committee identifies hazardous behaviors which workers encounter regularly.

These are critical at-risk behaviors that expose the workforce to injury and equipment and facilities to harm. This inventory emerges from a careful analysis of the past few years of safety data at a work site.

2. The committee investigates the number of injuries that are related to the hazardous or unsafe behaviors to analyze why both the behaviors related to the injuries or damage and the injuries or damage occur. The goal is to change the unsafe behavior.
3. The committee, or designated observer(s), observes a particular selected job task (preferably volunteered by the employees involved) a few times and then writes up the observations. The workers are encouraged to work as they normally would without fear of retribution or disciplinary action.
4. Then the committee notes the safe and unsafe (at-risk) behaviors
5. The committee presents its findings to the work-site workers and establishes a log of safe behaviors for each task.

This approach focuses on observable behaviors, provides positive activators to motivate workers, and applies continuous interest and evaluation. Positive activators are those which provide the workers with a sense of empowerment, freedom, and control. These are said to be longer lasting than negative activators, which generally are those events that tell the worker to avoid failure (i.e., an accident).

Measurement is a key to sustained interest. Geller emphasizes that employees and management should systematically track a variety of success indicators, such as numbers of behavioral observations, percentage of employees volunteering to be observed, the number of coaching sessions conducted per week, and the percentage of safe behaviors per critical behavior category or per work area.

Geller believes that behavior and person factors represent the human dynamic of occupational safety. A Total Safety Culture includes the persons (knowledge, skills, abilities, intelligence, motives personality), the environment (equipment, tools, machines, housekeeping, temperature, physical layout), and behavior (safe/unsafe work practices, "actively caring," complying, coaching, recognizing, communicating, demonstrating). The factors should be based upon ten principles:

1. The culture, not OSHA, should drive the safety process.
2. Behavior-based and person-based factors determine success.
3. Focus on process, not outcomes (outcomes are statistics such as OSHA recordables).
4. Behavior is directed by activators and motivated by consequences.
5. Focus on achieving success, not on avoiding failure.
6. Observation and feedback lead to safe behaviors.
7. Effective feedback occurs via behavior- and person-based coaching.
8. Observing and coaching are key actively caring processes.
9. Self-esteem, belonging and empowerment increase "actively caring" for safety.
10. Shift safety from a priority to a value.

THE REGULATORY APPROACH: SAFETY RULES

Having safety rules is not a new idea. Rushbrook[50] mentions:

> Even the pirates of the early eighteenth century had fire prevention regulations, which were rigidly enforced. The articles of Captain Shaw (1723)—which incidentally were "swore to" upon a hatchet for want of a Bible!!—contained the following fire prevention clause: Article 6—"That man shall snap his arms (flintlock guns) or smoke tobacco in the hold without a cap to his pipe, or carry a lighted candle without a lanthorn, shall receive Moses's Law (forty stripes lacking one) on the bare back."

Safety rules are codes of conduct to avoid injury and damage. Under common law the employer is obligated to provide rules by which all employees can perform safely and the means to ensure that those rules are observed. Unless the employer formulates such rules and sees they are observed, he or she may be considered willfully and grossly negligent if an employee is injured because there was a lack of such rules or they were not enforced adequately.

Employers have been cited and fined under the OSHAct where employees were injured or killed because of failures to enforce safe work rules. Safety rules may have been published which state horseplay is prohibited. Joe is a high-spirited young man who likes to indulge in horseplay. If his supervisor is aware of this and fails to restrain Joe from such conduct during working hours, so that another worker is injured, the company may be open to a personal injury suit. In such an event, the other worker would be entitled to workers' compensation; and Joe could be dismissed immediately.

It is important enough to repeat: *Safety rules must be provided to govern the conduct of employees, and they must be enforced.*

Rules to cover all employees in a plant, especially a large one, can only be general in nature. In addition to these may be specific rules and orders such as those regarding occurrence of a fire or use of protective equipment. Certain rules of conduct and procedure may be more critical than others, and it may be necessary to apply two different penalty levels for nonobservance. Needs for most rules are generally apparent, so it is unnecessary to go into long, detailed explanations of why they have been imposed.

Certain criteria should be observed in the preparation of safety rules:

- The number of general rules should be kept to a minimum.
- Rules for a participating operation and for workers involved should be included in the procedures for conduct of that operation.
- Each rule should be clear and unambiguous.
- Stipulate only those rules currently required. A prohibition against smoking may be unnecessary where no combustible material is present and where no one objects to smoking.
- Stipulate only those rules that will be strictly enforced.

EMPLOYEE PARTICIPATION

After the OSHA standards went into effect in 1971, an inspection was made of an oil company's plant after the union charged that numerous "imminent" dangers existed. The subsequent three-way controversy between the company, union, and the OSHA is of little interest. One thing which is of interest is the fact that the workers had known of previous problems which they then brought to the attention of the OSHA. Very frequently the workers in any operation are the ones most aware of the existence of hazards in their work. They are the most concerned with ensuring the elimination of hazards because their own lives and health are involved. Workers (and their unions) therefore constitute excellent sources from which information on hazards, potential and imminent, can be derived. There are numerous ways by which employees can participate in any safety effort.

It has been pointed out that design deficiencies frequently generate hazardous conditions under which employees have to work. It has also been pointed out that such deficiencies should be brought immediately to the attention of supervisors and their superiors for correction. Many workers fail to do so, often because they believe the presence of hazards is usual and normal, and they accept the potentially injurious condition. Some persons may not recognize the danger; others may recognize them but will do nothing unless they are rewarded in some way.

On the other hand, some workers frequently object to design changes that merely change the mode and rate of operation without affecting safety.

CRITICAL INCIDENT TECHNIQUE

The simplest way to find from employees if they are aware of any hazards in their work is to ask them. The Critical Incident Technique developed by Tarrants[51] is a means of doing this most effectively. The method is based on collecting information on hazards, near misses, and unsafe conditions and practices from operationally experienced personnel. It can be used beneficially to investigate human-machine-operational relations and apply the information learned to improve equipment, procedures, and operations.

The technique consists of a capable reviewer interviewing personnel regarding involvement in accidents or near accidents; difficulties, errors, and mistakes in performance; and conditions that might cause mishaps. The surveyer first explains to a group of experienced workers what is to be done. Then each worker is interviewed individually and asked to answer a series of questions individually on safety matters.

Also requested of the interviewee are any other pertinent comments regarding safety aspects of other occurrences they have observed, even though they themselves had not been involved. The worker is asked to describe all hazards, mishaps, or near misses he or she can recall.

In effect, the Critical Incident Technique accomplishes the same result as a review of a series of accident investigations: identification through personal involvement of hazards that could result in injury or damage. Tarrant states, "Studies have shown that people are more willing to talk about 'close calls' than about injurious accidents in which they were personally involved, the implication being that if no loss ensued, no blame for the accident would be forthcoming."

It has been estimated that for every mishap there are at least 400 near misses. When the witnesses who observed mishaps or near misses, but were not participants, are added to those who were involved, an extremely large population is available from which information on possible accident causes can be derived.

When more than two or three interviewees report similar difficulties, hazards, or near misses with similar types of equipment or operations, the area can be accepted as one that should be investigated. The results of the investigation will determine whether corrective action is necessary or advantageous.

OTHER METHODS

Attempts have also been made to obtain safety information through the use of questionnaires to be filled in by selected personnel. This method has generally proved to be unsatisfactory for a number of reasons. Many plant workers and numerous other persons hate to write anything lengthy. Much information on hazards, accidents, or near misses may be extensive, so that written worker reports are often incomplete. Where a question is answered by a true-or-false notation or multiple choice, much pertinent information may be omitted. If this method is used, one fundamental problem is the need for extreme care in selecting and phrasing the questions. Too often the person completing a questionnaire would give the questions interpretations neither considered nor intended by the person who prepared them. Any question should be avoided whose answer requires involved reasoning that is not immediately apparent to the reader.

Much information is also submitted to control and action agencies in the form of hazard or trouble reports. However, such reporting itself generates discrepancies that can be avoided through use of the Critical Incident Technique. Reports may require entries in narrative or checklist form or both. Personnel find it time consuming and difficult to prepare narratives. Even conscientious writers tend to select the easiest and most rapid means of accomplishing reports, which therefore usually lack detailed and precise expression that could indicate the source of the problem. Checkoff items can be completed more rapidly, but these, too, result in omissions of information that may be critical. In both types, entries may include information on the immediate or principal cause of an accident but neglect other contributory causes and factors.

Taped oral reports are often helpful, but they may omit useful information unless someone else is present to keep the reporting person on the subject in question and from wandering too much. Thus, the Critical Incident Technique appears to be the most eminently satisfactory.

SUGGESTION PROGRAMS

A prime concern of industrial workers is in avoiding accidents and ensuring that any potentially hazardous situation they may note is corrected. Suggestion programs are therefore to the employees' benefit. There are generally a number of ways an employee can participate. The employee can report orally a suggestion to his or her supervisor, safety office, union representative, an OSHA office, or the representative on a safety committee, or submit it in writing by dropping it into a collection box.

Submitting suggestions in writing is an excellent idea, especially if there is a reward for any idea accepted. The disadvantage is that many people heartily dislike writing. Suggestion programs have often been so beneficial, attempts have been made to make suggestions easier to submit. Often workers are given small awards even if none are accepted, large ones if they result in savings.

Some companies have appointed clerical workers to distribute widely the forms used and to collect periodically any submissions. The name of each suggestor is recorded and filed. The remainder of the suggestion is then routed for comment to any office that might have an interest. The comments are then collected and, if worthy, suitable action is taken. Suggestions can sometimes be stimulated by having contests to see who can submit the most within a stipulated time period.

For suggestion programs to be successful, certain facts must be recognized. The employee's effort must be acknowledged. Even if the suggestion is not adopted, it must be given careful consideration. If it is accepted, the employee should be rewarded, preferably in public, so that fellow employees are aware of the award. Such awards act as inducements for more suggestions and suggestors.

UNION PARTICIPATION

Management and unions frequently disagree on production goals, methods, rates, and work rules, work hours and conditions, and numerous other matters. However, they do cooperate when they have a common goal, and safety is generally one of these. A survey of almost 300 companies showed that, of 22 areas of mutual interest, accident prevention was the foremost area where management and unions cooperated.

Historically, a major reason for the birth of industrial unions has been the health and safety of its members. Some of the most common objections of workers to their representatives involve this concern. The safety engineer, therefore, will often find the union representative to be a strong supporter of his or her safety efforts. The safety engineer may, however, have two problems: (1) he or she may not consider some of the objections justifiable from a safety standpoint; and (2) he or she may agree, but supervisors and managers may not because of cost considerations. Conversely, a safety engineer who finds a worker performing in an unsafe manner can frequently have the situation corrected by pointing it out to the union representative. The representative may be able to correct the situation without creating any friction. Often much tact is required on the part of the safety engineer, the union representative, and management, although all three are interested in lessening the occurrence of accidents.

SAFETY TRAINING

Safety training should begin with the new employee and continue throughout the time he or she is with the company. The type of training, frequency, material presented, and by whom presented will vary with the employment. Safety training should be given to all new employees, regardless of previous experience. Items in this training should include information on and review of:

- Company safety rules and practices.
- Employees' duties and rights under the OSHAct or state safety codes.
- Necessity for strict observances of warning signs.

- Emergency signals and their meanings.
- Methods of use of emergency equipment.
- Company programs to aid employees in purchase of safety glasses, safety shoes, and hard hats.
- Means of summoning assistance in times of need.
- Location of medical office.
- Benefits under workers' compensation laws.

The new employee should be indoctrinated further by his or her immediate supervisor or by an experienced and knowledgeable fellow worker whom the supervisor may designate. This indoctrination should include safety matters such as:

- Hazards that might exist in the operations in which the new employee will participate.
- Safeguards that have been provided and precautionary measures that should be taken against those hazards.
- Locations of emergency exits, phones, fire extinguishers, first aid kits, and other emergency equipment workers might use.
- Procedures to follow in the event a specific type of emergency occurs or might occur.
- Means of reporting hazards or defective equipment.
- Need for good housekeeping.
- Recapitulation of pertinent items the new employee was told during initial training.

Even older and experienced employees can benefit from additional on-the-job safety training. One of the most effective means is by periodic safety meetings. However, such meetings should observe certain rules:

- Meetings should be only long enough to present the desired information; they should not be too lengthy.
- The attendees should be employees in approximately the same categories—that is, construction workers, drivers, office employees, or production workers—and the subject matter should be pertinent to their activities. Small meetings oriented to specific problems the attendees may encounter are effective but may be less effective than mass assemblies. Mass assemblies are often conducted because of the importance of the message or the speaker. The assemblage usually feels that if a top manager is willing to spend the funds and time required, the message must be important.
- In any case, the speaker must be capable of impressing his or her listeners with the importance of the material presented. (Any supervisor or group leader who considers the safety material unimportant and presents it only because he or she has been told to do so is wasting time. The supervisor unfamiliar with the material to be presented would do better to ask the safety engineer to conduct it.)

- Informal meetings at which attendees can contribute comments should be encouraged, since they are highly effective.
- A very effective action is to have a worker demonstrate how to use a piece of protective equipment or to describe or show the safety action to be taken in a simulated emergency.

Such meetings might include:

- Giving information on accidents with similar types of equipment or operations that have occurred previously elsewhere.
- Instructing employees on new types of equipment, their uses, capabilities, hazards, and safeguards.
- Indicating the precautionary measures that are required for any operation they will undertake.
- Pointing out (preferably without mentioning names) any unsafe practices by the workers that might have been noticed.
- Determining whether any other members of the group have any comments on unsafe equipment, conditions, or practices.
- Disseminating any pertinent safety information that may have been brought to the supervisor's attention.
- Ensuring that personal safety equipment is suitable for the activity in which the wearer will take part and is in good working order.
- Reviewing and demonstrating capabilities in first aid and emergency procedures to ensure everyone is knowledgeable and proficient in the use of equipment, CPR, and other responses to accidents.

IN-DEPTH TRAINING

Safety training can be given to a limited extent during informal meetings. Often, however, more formal and extensive training is required. For example, OSHA standards for construction require the presence near work sites of either medical personnel or persons who have valid certificates in first aid training from the U.S. Bureau of Mines or the American Red Cross. An employer may find it necessary to send employees to one of those organizations to obtain necessary certification.

The company may have volunteer fire crews to help the professional fire fighters in an emergency. Training of both the professionals and the volunteers may be undertaken by attendance at schools, by visits by members of outside fire departments, such as those of the city or county, or by consultants who may review existing capabilities and make recommendations if improvements are needed.

Training in depth may be required for personnel who must work in bulky or complicated protective equipment, such as during sandblasting or the transfer of highly toxic fluids. Or, personnel may have to be trained in the use of emergency equipment and procedures, such as resuscitation by paramedical personnel.

Representatives of companies that sell new mobile equipment will train operators in safe handling.

MAINTAINING AWARENESS

Intermittent safety efforts are generally ineffective. It is necessary to maintain an almost continual program of keeping personnel alert to safe practices. Hazardous areas must be posted with suitable warning signs; however, other means are available at other locations:

- Large billboard-type signs regarding the need to avoid and prevent accidents can be located at employee entrances to the plant or at points of entry into buildings.
- Posters for display at frequented locations are available at low cost from organizations such as the National Safety Council and National Fire Protection Association.
- Small folders or booklets from safety organizations, insurance companies, and the federal government can be given to each employee as he or she enters the plant or at other appropriate times.
- Slips with printed safety messages can be added to pay envelopes or attached to pay checks.
- Safety displays can be located at entrances to buildings, lunch rooms, or similar locations.
- Articles or photographs regarding safety matters or accidents can be posed on bulletin boards.
- Large advertisement-type displays can be included in company newspapers.
- Automatic projectors can be used for continuous showing of slides on safety subjects.
- Placemats and napkins with interesting messages on safety and accident prevention, available from the National Safety Council, can be used in lunch rooms, cafeterias, and at vending machines.
- Members of departments of other work units that have been accident free for a certain length of time or number of work-hours can be presented with green rosettes and ribbons, buttons, or pins; coffee or milk and a doughnut during a work break; or ballpoint pens inscribed with a safety message.
- Accident prevention competitions can be held between departments and plants based on low accident and injury rates.

GENERAL COMMENTS ON SAFETY COMMITTEES

Safety committees may be composed of both management and workers, of workers' representatives, of first-line supervisors, or of any mix of these. The workers' representative committee is probably the most knowledgeable of safety needs, but management backing will produce the greatest effect.

Too often, the presence of managerial personnel on such committees degrades their capabilities. Unless the managers are extremely careful, their presence alone tends to dominate the proceedings (intentionally or not); or they fail to attend, leave early because they are called away, or take part fretfully as though they had more important business elsewhere. On the other hand, a top-level manager interested in safety can be highly effective. Just the presence on a committee of such a manager indicates and raises interest, increases the committee's value, and often acts to raise the morale of all workers, because they believe managers are interested in their well-being and the roles of the workers on the committee.

Any committee, no matter how composed, should observe certain rules. The safety engineer should be a member to provide guidance and information on any matters brought up by the committee. The engineer can offer an opinion on whether a reported situation does constitute a problem, whether a proposal suggested by a worker or committee member is technically feasible, or what the standard or code requirements are regarding the matter raised. The safety engineer should not be a voting member of the committee.

A secretary should be present to take notes on matters considered and their outcomes and to prepare minutes of the meeting. Any worker who makes a suggestion, written or oral, or indicates the existence of a hazard should receive a notice regarding the outcome. If the matter requires further consideration or review, the worker should be so notified. An effective means is to forward to the worker a copy of the minutes of the meeting at which his or her suggestion or hazard report was discussed. The worker should be identified in the minutes by name, job, title, and organization.

Meetings of the safety committee should be scheduled on a periodically regular basis. Enough time should be provided to ensure that due consideration is given to each item; however, the chairman must ensure that the meetings are confined to safety matters, discussions are confined to items at hand, minimum time is wasted, and the meeting is adjourned as soon as its business is completed.

The committee should represent each area of the plant or company. If it makes the committee too large, subcommittees can be established for each major department or division. If the matter affects more than one committee, the problem under discussion can be brought to a higher-level committee. Similarly, the results at the higher level can be passed down as required. Committee members can either be elected or appointed for definite periods. Alternates should also be designated to take the places of members who cannot attend. Memberships should be on a rotating basis within specific areas; however, to maintain continuity of capability, only one or two members should be replaced on a committee at one time.

SAFETY COMMITTEE DUTIES

An effective committee can undertake numerous duties highly beneficial to the personnel of any company and the company itself. Consideration of suggestions made by employees has already been mentioned as a major function. Others include:

- Promoting accident prevention and safe performance within their own work areas.
- Assisting the safety engineer and immediate supervisor in investigating any reported safety deficiency in their own work area.
- Assisting the safety engineer, supervisor, and other investigators in determining the cause of an accident or near miss that may have occurred.
- Reviewing, as a committee, and discussing the findings of any investigating board to determine whether they appear to be reasonable, whether all pertinent facts

were considered in the investigation, the feasibility of recommended corrective action, and whether similar accidents are possible in other work areas (which should then be informed of the accident and the corrective action to be taken).

- Making inspection tours to determine whether safety rules are being observed, good housekeeping is being maintained, any dangerous conditions exist, protective devices and equipment are being used where required, and hazardous areas are posted.
- Making recommendations for additions or modifications to company safety rules.
- Assisting the safety engineer, training-office personnel, and supervisors in preparing, reviewing, or presenting safety training to fellow employees.
- Selecting those departments or employees who are to be presented with awards for safe performance.
- Assisting the safety engineer in selecting posters and other material to be used to stimulate interest in accident prevention.

BIBLIOGRAPHY

[48] *Industrial Safety & Hygiene News,* 14th Annual White Paper. Dec. 19, 1997, p. 14.

[49] E. Scott Geller, "How to Motivate Behavior for Lasting Results," www.safetyonline.net/ishn/9603/behavior.html, and in "Total Safety Culture," *Professional Safety,* Sept. 1994, pp. 18–24.

[50] Frank Rushmore, *Fire Aboard* (London: The Technical Press Ltd., 1961), p. 26.

[51] W. E. Tarrants, *Utilizing the Critical Incident Technique as a Method of Identifying Potential Accident Causes,* U.S. Department of Labor, Washington, D.C.

EXERCISES

10.1. Why are safety rules needed?

10.2. Are safety rules that are not enforced of any value?

10.3. List some of the criteria to be observed when preparing safety rules.

10.4. What is the Critical Incident Technique and how does it work?

10.5. What are some problems encountered when safety survey questions are to be answered in writing?

10.6. Describe some other types of employee participation in accident prevention programs.

10.7. Tell some of the stages of safety training that should be undertaken. Are such programs undertaken in your plant? Do you believe you have received adequate safety training?

10.8. What types of in-depth training are required? Have you ever had any? Does your company ever send personnel for such training?

10.9. What are the principal duties of safety committees? Have you ever been a member of such a committee? How were you selected? Did you find it beneficial or not? How would you improve it?

10.10. List ten ways in which the awareness of the need for safety can be maintained. Do you consider the safety engineer the principal person responsible for your safety? Do you do anything to help him or her?

10.11. What are the elements of a behavioral based safety program?

CHAPTER 11

Appraising Plant Safety

The great English physicist J. J. Thomson said quantifying a problem is the only way to understand it. At the present time, appraising plant safety or the changes of nonaccident operations by quantitative methods with any degree of accuracy involves high degrees of uncertainty. Safety engineers will produce more accident-free plants if they raise their awareness of, and follow, effective means of accident prevention by using the material and methods described in this book. However, the quantification safety aspect has its own benefits and should be known to all safety personnel.

Managers would like to know a plant's safety posture to direct whether or where corrective action should be taken. Many personnel would like to know whether a particular action will result in improvement or degradation. Judge Learned Hand queried whether comparisons of risks can be made by means other than economic factors. However, although many quantitative studies are based on cost considerations, many are not. State safety agencies, insurance companies, and OSHA all make appraisals of plant safety using numbers of accidents, or resulting fatalities or injuries. To all of these the economic factors may be added.

Thus, while both qualitative and quantitative methods may be used, the numerical data may be indicative of where the accident causes were. The safety engineer should be concerned with any condition where there is not a zero accident rate. That such an aim is achievable is indicated annually by the National Safety Council (NSC). The NSC publishes a list of the largest number of continuous man-hours worked without a disabling injury for each of 36 major industry groups. High-safety-quality plants achieved these accident-free periods by eliminating or minimizing the existence of unsafe conditions before accidents could occur. This is opposed to the idea of taking corrective action after accidents have taken place and have become numerical data.

Almost all quantitative methods use counts of past occurrences to indicate accident numbers, frequencies, and severities. These data are often used as measures of expectable future accident occurrences and risks. Originally, safety appraisals, both qualitative and quantitative, were made of existing plants to determine where improvements could be made; and these usages are still the most common, most significant, and most useful.

NEW PLANTS AND EQUIPMENT DESIGNS

Over the years it has been found that numerous problems can be avoided in plants being built or modified if plans are reviewed for safety aspects before any construction or change was initiated. Appraisals of plant safety should begin long before construction is started. This means the safety aspects should be considered as soon as a new plant or modification is conceived or design begun.

Some companies now require their safety personnel to review drawings for new facilities and equipment to ensure they meet legal requirements and good safety practices. Too often, appraisals are made only after an accident in a similar plant elsewhere has resulted in expensive litigation. Managers may then call for appraisals of their own plants to determine when similar accidents may occur there. Company insurers, or the state safety agency in which a plant is to be located, may have to review and approve plans and specifications before construction is initiated.

In spite of preconstruction approval, it is necessary that on-site inspections be made while construction is underway, after completion and prior to the start of operations, and continually after that. The safety posture before any plant has been built or any operation started can only be uncertain and often inaccurate appraisals. A prime, costly, and far-reaching example of what might result because of any such errors is the nuclear power industry and its accidents and costs for corrective action. Assessments of plant safety, nuclear or nonnuclear, by engineering designers are generally erroneous. This is because they believe too optimistically in their own designs, are untrained by schools in aspects of accident prevention so that they lack knowledge of hazards, and are generally unduly pressed by managers to finish any project.

At one time it was common practice to consider any safety aspects after plants had been built, equipment purchased, or operations begun. Too often, knowledge and existence of any hazards was derived from accidents that devastated plants, destroyed equipment, killed or injured operators, and halted operations. Actuaries found that the lack of accident experience data on newly designed plants caused them to add a substantial uncertainty factor that resulted in high insurance premiums. Theoretically, new plants and equipment would be built or created in accordance with regulations, codes, and standards for the intended localities and usages.

However, sometimes there are errors in their creation, the requirements change, or they are not observed properly. These deviations from legally stipulated safety measures might require expensive corrective actions. It therefore behooves employers to make safety appraisals of their plants, equipment, and operations to ensure that the possibility of accidents is minimized.

EXISTING PLANTS AND EQUIPMENT

Safety engineers should review plants, new or old, features of equipment, and procedures or operations to be undertaken. Often, safety engineers first are made aware of how safe a plant is when they are designated safety personnel after plants are put into operation. Whether they are old or new, the safety person should ensure:

- Plants include the egresses and exits and their markings required by OSHA standards and local codes and are properly maintained so.

- Electrical systems, especially for hazardous locations, comply with the provisions of the National Electrical Code and other designated codes.
- Pressure vessels are designed according to the provisions of the American Society of Mechanical Engineers (ASME) Code. The vessels should be stress-tested prior to use and at required designated intervals.
- Firefighting equipment is installed and maintained as required.
- Newly purchased equipment or equipment manufactured in-house has been installed and will meet prescribed OSHA and local standards.
- High-energy process vessels are separated by the greatest distance possible to prevent damage to other equipment in the event of a violent failure.
- Fire lanes and other routes to locations where other emergencies could occur are provided, marked, and maintained so that passages are not impeded or blocked.
- Emergency equipment and locations for their emplacements or storage are provided in readily accessible locations and checked periodically.
- Ventilating equipment is kept clean. Hoods, ducts, blowers, filters, and scrubbers are provided and adequate to remove air contaminants from the plant and to keep them from contaminating the environment. Ventilating equipment should be kept clean.
- Adequate work spaces are provided between pieces of equipment so that employees can have free passage and so there will be no physical interference to create errors and cause accidents.
- Hazardous operations are isolated so they do not constitute dangers to other personnel or activities. Welding shops should never be located close to paint spray activities, fuel locations, or where personnel can be affected by welding-arc radiation.

INDICATING PLANT HAZARDS

Certain areas are always more hazardous than others. Safety engineers should know where these places are and should devote time commensurate with the degree of hazard. A common practice to study the problem is to color-code a map of the plant showing areas as high-hazard, moderate-hazard, or low-hazard. Hazardous areas would be considered places where fuels, explosives, or highly toxic or reactive chemicals are handled or stored; wood-processing plants with large amounts of sawdust or other highly flammable materials; high-energy process reactors; and high-voltage and power-distribution stations. For example, paint spray or paint storage areas can be considered high-hazard locations. Certain of these high-hazard locations may be in the middle of less hazardous areas.

Inspections where safety engineers, supervisors, insurance personnel, and others can easily recognize highly hazardous areas can be highlighted. The colors can be used to indicate the periodicity of visits to be made. The most frequent visits would be, of course, to high-hazard locations. Moderately hazardous locations would include areas where there might possibly be high-noise-exposure levels; contaminated, but not highly toxic, atmospheres; locations where accesses to egresses and exits might be blocked by materials or equipment; areas where crates, boxes, pallets, or other containers might be stacked too high; or equipment where workers have a history of inactivating, bypassing, or removing safety devices, such as guards, that interfere with their work.

Even low-hazard locations sometimes have conditions that should be corrected. A very common one involves the use of coffee-making or other electrical cooking equipment in offices. Some of these devices, especially when more than one is in use, draw large amounts of current for which the circuitry was not designed. Frayed cords, "Christmas-tree" arrangements on plugs, and extensions resting on floors where rollered equipment passes over them constitute a few items in otherwise low-hazard areas.

All conditions that are hazardous to any degree should be appraised periodically.

SAFETY INSPECTIONS

Another form of appraisal is by safety inspections, which can be either informal or formal. An informal inspection can be conducted by a supervisor who ensures every morning that the facilities and equipment are in proper and safe condition and working order prior to the start of operations. Supervisors can observe and appraise workers to determine they are in condition to conduct themselves suitably and safely. Similar inspections can also be conducted by higher-level supervisors and managers, safety personnel, safety committee members, and even members of the security force as they make their rounds. Any discrepancies noted at such times should be brought to the attention of the supervisor. A written report is not necessary for an informal report. In some instances security personnel have taken photographs of unsafe conditions with Polaroid cameras and then left the photographs with the supervisor.

More formal inspections can be conducted by plant safety personnel; safety committee members; plant or department managers; fire prevention personnel; insurance company engineers; elevator, boiler, and pressure-vessel inspectors; or municipal, state, or federal agency representatives. Personnel in the last three categories are highly experienced and trained professionals and use proved methodologies to appraise the safety of the areas they cover. Procedures they follow are beneficial to anyone making a formal inspection:

- An inspection checklist and record of findings is used to indicate the conditions of equipment and operations they inspected.
- High-risk operations and activities noted on the checklists are given special attention, but others are not ignored.
- Findings during previous inspections are reviewed to determine whether any discrepancies have been found, and if so, whether they were corrected.
- Personnel making inspections note whether workers are suitably equipped with hard hats, safety shoes, safety glasses, or other protective equipment prescribed for the areas visited.
- A report of inspection is prepared and presented to the responsible supervisors and managers. A preliminary report may be given verbally, followed later by a written report.

- Findings in the report should be specific and not vague generalities. Any report should name locations, equipment, and operations and cite pertinent discrepancies. (*Note:* Stating that housekeeping was generally poor in a described location is a specific finding.)
- The report, oral or written, may present recommendations for corrective action that should be taken.
- If a discrepancy violates any governmental standard, code, or regulation, the specific document, paragraph, and requirement should be cited.

CHECKLISTS

Checklists are often used to evaluate many of the safety features (or their lack) in industrial plants. Checklists are provided at the ends of many of these chapters to assist personnel in evaluations. Some checklist questions may not be applicable to the specific plant being appraised, and their answers can be omitted. In other cases, the person making the survey may add other items to be reviewed, especially if such reviews are made periodically and it is desirable to compare answers on different occasions.

Checklists can be prepared from numerous sources. Even persons new to safety matters can prepare them simply by asking themselves what they would like to know. One item will usually lead to another.

To determine whether or not a plant complies with the OSHA standards, the specific requirements in the standards can be presented as questions. Other sources might be items in the text of this book, articles from magazines, educational brochures used by the National Safety Council or government agencies, or newspaper notices of accidents or litigation.

The questions asked in checklists are usually very broad in scope, requiring more details. For example, a question on a checklist might be whether a pressure vessel had been proof-pressure tested. The plant might have 20 or 30 pressure vessels; therefore a simple yes or no answer would probably not suffice. In this case, asking whether or not each test had been made, whether each had been satisfactory, and by whom the test had been would require multiple answers. (A common checklist of this type might be one for fire extinguishers for which there are generally a great number in a large plant. If a checklist item questions whether fire extinguishers have been checked or refilled, a complete answer would require an extensive survey.)

The National Institute of Occupational Safety and Health (NIOSH)[52] has issued an excellent document which permits a manager or safety engineer to appraise the effectiveness of workplace safety programs. In one chapter are listed many of the activities that should be reviewed and evaluated. Standards are given by which ratings should be assigned as poor, fair, good, or excellent. The chapter also presents a rating form by which a numerical relative risk assessment can be developed if it is desirable to make comparisons.

For the safety engineer, the use of accident statistics is less productive in appraising plant safety than is knowledge of hazards, controls, checklists, and similar information. Statistical data are usually indicative of accidents after they have occurred. Zero accident- and injury-frequency rates are the only ones with which the safety engineer can be satisfied. Any accident is a profound disappointment to a safety engineer.

QUANTITATIVE APPRAISALS

To determine the safety level of any plant or industry by quantitative means by use of after-the-fact data, accident statistics and frequency and severity rates are employed. The data may be indicative only after there has been a sizable accumulation of past numerical accident information. These data generally provide limited answers about relationships between causes and effects, so only broad accident preventive measures can be taken. Checklists and analyses are more suitable, detailed, and effective for safety accomplishments.

PROBLEMS WITH VALIDITY OF STATISTICS

Probabilities are based on past performance; the longer the record the better will be the estimate of future performance if there have been no changes. To be statistically valid, quantitative data must have been collected either over long periods or from a large number of similar activities. When they must be collected over a long period, by the time statistical validity has been established, many persons may have been killed or injured before corrective action is taken. If any action has been taken, the past numerical data no longer apply until after long and exactly similar performance. For example, a number of accidents occurred in a plant within a specific period of time, so it appeared that because of unguarded rotating equipment they were at a rate of 3 mishaps per 1,000,000 man-hours. As a result, guards were emplaced. The previous accident rate was no longer valid for estimating or predicting future accident-occurrence rates in that shop. (A good safety engineer would have had guards installed before or immediately after the first accident and not waited until such mishap data accumulated.)

Accident statistics do provide valuable information to regulatory agencies and insurance companies which identify causative factors and whether additional safety measures are needed to lessen future accidents. Insurance company actuaries can use accident data to establish costs of future premiums for employers. These actuarial determinations constitute a good deal of educated guesswork for new operations and equipment where there is no or little accident experience, such as for new and untried plants or aircraft. Because of the uncertainty, insurers will charge as much as the traffic will bear to cover all financial risk losses. After there has been experience, rates can be adjusted to reflect the experience: down if there have been few or no accidents, up if there have been losses or there is knowledge that the risks of losses will increase. The very threat of increased premiums has forced some companies out of business. This has occurred because of accidents, or the threat of accidents, with zeppelins, nuclear power plants, machinery manufacturers, plant operators, and surgeons. The means of calculating accident and injury frequency and severity rates is indicated in Fig. 11-1.

Some safety engineers use accident and injury frequency and severity rates for comparative purposes to determine:

FREQUENCY AND SEVERITY RATES

Frequency rates: rates can be computed in various ways to determine the frequency of accidents or injuries, or the severity of injuries. The methodologies by which rates are computed are similar in all cases: however, the bases used may be different. If *A* is the event for which the frequency rate is to be computed, *B* the numerical base, and *C* the exposure, then

$$\text{frequency rate} = \frac{A \times B}{C}$$

If 1,000,000 man-hours is used as the base, an accident frequency rate can be computed by:

$$\text{accident frequency rate} = \frac{\text{number of accidents} \times 1{,}000{,}000}{\text{man-hours of employee exposure}}$$

Thus for a plant that had 18 accidents in a year during which employees worked a total of 1,200,000 man-hours,

$$\text{accident frequency rate} = \frac{18 \times 1{,}000{,}000}{1{,}200{,}000} = 15.0 \text{ per million man-hours}$$

If during the same time there were 6 disabling injuries at that plant, the injury frequency rate would be

$$\text{injury frequency rate} = \frac{6 \times 1{,}000{,}000}{1{,}200{,}000} = 5.0 \text{ per million man-hours}$$

The Bureau of Labor Statistics uses a base of 100 full-time employees as opposed to the 1 million man-hours used by the American National Standards Institute (ANSI Standard Z16.1). It is assumed that 100 full-time employees would work 200,000 hours per year (40 hours per week per worker, 50 weeks per year). Computed on this basis, the injury frequency rate for the plant mentioned would be:

$$\text{injury frequency rate (BLS)} = \frac{6 \times 200{,}000}{1{,}200{,}000} = 1.0 \text{ per 200,000 man-hours}$$

It is therefore evident that when rates are cited, it is necessary to know the bases on which they were calculated.

Injury severity rate: certain industries may show high injury frequency rates, but the injuries may be minor. Other industries may have few injuries and an extremely low injury frequency rate, but when injuries do occur they are severe. The American National Standards Institute has therefore established a means of measuring severities through use of time charges. With this method fatalities and injuries are assigned time charges to be used in determining the rates. These charges are based on average experience. For example, each fatality or permanent total disability is assigned a time charge of 6,000 days. This was based on the life expectancy of the average worker times the number of working days per year. (The Bureau of Labor Statistics does not include a fixed charge for a fatality.) Time charges for permanent partial disabilities are tabulated in ANSI Standard Z16.1: loss of an arm above the elbow, 4,500 days; loss of an eye (or sight), 1,800 days; loss of both eyes (or sight) in one accident, 6,000 days; complete loss of hearing in one ear, 600 hours; and in both ears (one accident), 3,000 days. Injuries resulting in temporary disabilities are charged the number of calendar days lost for computations by ANSI methods. With the Bureau of Labor Statistics method, only actual work days lost are changed. The Bureau of Labor Statistics method requires time charges be included even if an employee is assigned another job; any change in occupation resulting from a work accident or illness is recordable. Therefore, by ANSI Z16.1:

$$\text{disabling injury severity rate (BLS)} = \frac{\text{total days charged} \times 1{,}000{,}000}{\text{employees hours of exposure}}$$

FIGURE 11–1 Frequency and Severity Rates Continued on page 178

If the six disabling injuries indicated above resulted in 240 days lost, the disabling severity rate would be:

$$\text{disabling injury severity rate (BLS)} = \frac{240 \times 1{,}000{,}000}{1{,}200{,}000}$$

$$= 200 \text{ days per million man-hours}$$

The average severity per injury can also be determined. This can be done in either of two ways:

$$\text{average days charged} = \frac{\text{total days lost or charged}}{\text{total number of disabling injuries}} = \frac{240}{6} = 40$$

OR

$$\text{average days charged} = \frac{\text{injury severity rate}}{\text{injury frequency rate}} = \frac{200}{5} = 40$$

FIGURE 11–1 Continued

- How rates for his or her plant or company compare with the averages for the industry. Rates for various industries are computed by the National Safety Council[53] and the Bureau of Labor Statistics. Companies that are members of the National Safety Council are generally more safety oriented than are nonmember companies.
- How accidents, injuries, and severity compare from period to period and whether they are improving or deteriorating.
- How different types of hazardous operations compare.
- How well different departments are doing regarding safety. An increase in rates may indicate a lack of supervisory emphasis. The data can also be used to determine which department or plant had the best or worst performance if an accident prevention competition is held. In such cases the safety professional must ensure that all the organizations being compared observe the same rules for reporting. The rates may also have to be adjusted to compensate for the difference in magnitude of hazards between different types of operations. Average industry rates might be used to provide such adjustments for the units being compared. Comparisons can also be made between specific plants and for the entire industry.

PROBLEMS WITH QUANTITATIVE RATES

Unfortunately, even where accident and injury statistics can be useful, they are often so incomplete that they can lead to inaccuracies. For example, to reduce reportable injuries and lost work days, some companies reassign injured workers to temporary jobs less strenuous than the regular ones. Workers often receive pay which is not reported as lost because of injuries. Because no time was listed as lost, it appeared that no injury had taken place. OSHA reporting requirements have attempted to correct and eliminate such schemes by trying to ensure all injuries are reported. Failures to do so make quantitative safety data unreliable.

In 1986, the Bureau of Labor Statistics expressed concern over the validity of accident record keeping on injuries. Lack of proper reporting by industrial companies was distorting and undermining the use of statistics. A worker's union stated that the exemptions that OSHA had been forced into "made injury rate figures a fairly useless measure of plant safety."

There have also been differences between recordable and reportable injuries as defined in the American National Standards Institute (ANSI) Standard Z16.1, according to which reports are made to the National Safety Council, and the BLS/OSHA definitions and criteria. The result is that NSC injury rates differ and are less than those of the BLS and OSHA.

In 1941 Heinrich[19] stated that 88 percent of all accidents were caused by "unsafe acts" of persons involved. *Fortune*[30] indicated that a 1967 survey in Pennsylvania had concluded that this figure should be only 26 percent. Even in 1910, Eastman[11] pointed out that managers had contended that 95 percent of the fatal accidents in the Pittsburgh area were due to "carelessness," while her own studies indicated that only 22.5 percent were caused wholly or in part by employee's negligence. The latter is more in line with a study she cited regarding workers in Germany. That study indicated that 29 percent of all work accidents were attributable to the workers themselves.

The data used by Heinrich (and the opinions expressed by the managers in the Pittsburgh area) had evidently been biased by employers themselves, so any appraisal was faulty. Heinrich had made his analysis from reports submitted to insurers by the managers of the companies at which the accidents had happened. However, no manager would want to indicate to the insurer, or to any safety agency having jurisdiction over the locale where the accident had take place, that the company had been at fault, or that hazards existed or had not been suitably controlled in the workplace under his or her own authority. It is also doubtful whether any safety person would want to indicate that any accident was due to a hazard he or she had not detected or corrected. Further, any manager would be reluctant to make such an admission, especially in writing and to an insurer or safety agency.

VALIDITY OF STATISTICAL COMPARISONS

A common problem in the use of statistics is that the person preparing them may present only one slanted aspect of the data, so readers receive erroneous impressions. In some instances this is unintentional, in others it is done to prove a point the analyst wants to prove, to impress the readers, or simply to avoid running contrary to accepted opinion. An example involving safety statistics can be cited regarding automobile accidents. A speaker quoted in the newspapers pointing out that "in the United States in 1898 one American was killed in a car accident. In 1972, more than 4 million Americans were involved in traffic accidents and 56,300 of them died."

In 1912, the first year in which the National Safety Council published statistics on automobile accidents, there were 33 fatalities for every 10,000 registered vehicles in the United States. The actual rate in 1898 was probably much higher, although the total number of fatalities was lower. The current rate is far less than 5 deaths for every 10,000 vehicles. It is evident that the two statistical statements provide far different views on whether automobiles are becoming more or less safe. The value of the statistical comparison is that often the sheer magnitudes of accidents, fatalities, and injuries will induce corrective actions.

Many statistics are quoted which leave out any measure of exposure. For example, suppose in a hypothetical case a company reports that 15 percent of persons aged 20 to 35 suffered lost-time injuries, and only 1 percent of persons aged 50–65 suffered lost-time injuries. This leaves the impression that those aged 20–35 are unsafe workers as compared with those 50–65. However, a review of the job assignments indicates that the dangerous jobs are handled by the younger group. The older group is not exposed to the same risks of injury; therefore, such simple comparisons are not valid. Statistics should be adjusted for exposure to accident potential, before they are used as a basis for comparisons.

Consideration must also be given to type of exposure. Risks are not comparable across types of industry. This is illustrated by Fig. 11-2. Neither are risks comparable within a type of industry. This is illustrated by Fig. 11-3.

RISK ASSESSMENTS

When funds are requested for correction of safety deficiencies not required by law, managers would like to know the risks involved so they can make suitable decisions. (See Fig. 4-6.) The theory is that the computed economic risk of an accident should justify any expenditure for the proposed safeguard. Managers would also like to know how greatly this improvement would increase the safety of their plant or activity. Even before plants are built, such estimates are often made by cost/risk/benefit/effectivity analyses (Fig. 4-6).

Two numerical methods of risk indication are frequently used: those involving relative methods and those using probabilities of occurrence.

The relative method of indicating risks because of hazards is simpler, more widely used, and has many different forms. A task, the toxicity of gas, or the accident potential of a piece of equipment is rated by a group of knowledgeable personnel according to a numerical scale. The scale adopted may range from 1 to 10, 1 to 6, and so on to indicate degrees of hazard. A gas whose toxicity or flammability is rated at 5 may be more toxic than one rated at 3. A liquid with a flash point of 225°F may be more fire-safe than one that has a flash point of 150°F (see Fig. 22-5 and Chapter 22). With the relative method, doubling the number in the scale would not necessarily double the hazard involved.

Probabilities of future occurrences can frequently be estimated from past experience. However, to acheive any accuracy, the experience has to have been over a long period and comparatively large populations. The expected events to be assessed must occur under conditions similar to those under which the data were derived. The National Safety Council may predict that, based on experience during a specific holiday, a certain number of persons will die in traffic accidents. If the weather or the desirability of watching a ball game creates a situation different from that of the NSC past

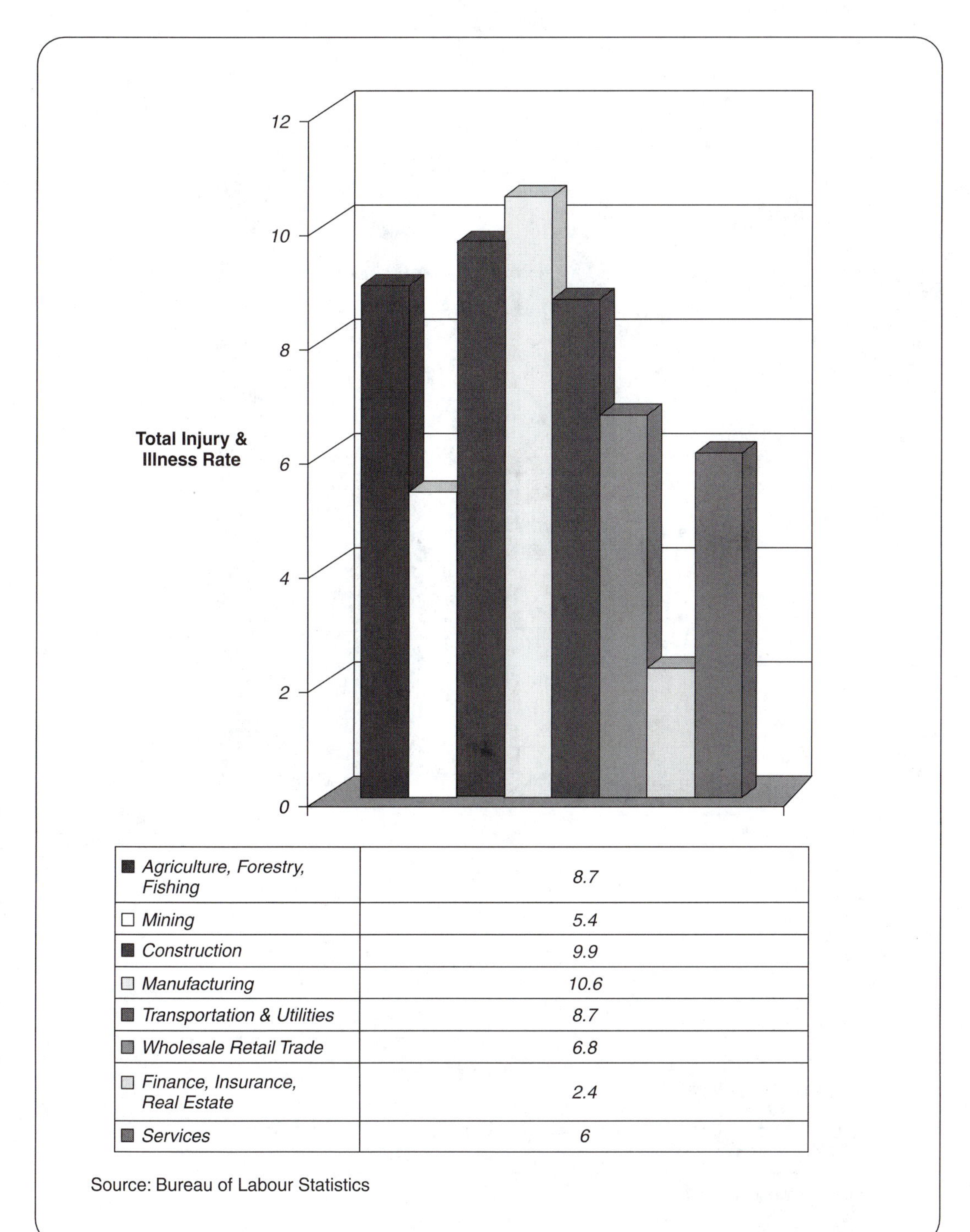

Agriculture, Forestry, Fishing	8.7
Mining	5.4
Construction	9.9
Manufacturing	10.6
Transportation & Utilities	8.7
Wholesale Retail Trade	6.8
Finance, Insurance, Real Estate	2.4
Services	6

FIGURE 11–2 Total Injury and Illness Rate by Industry, 1996
Source: Bureau of Labor Statistics

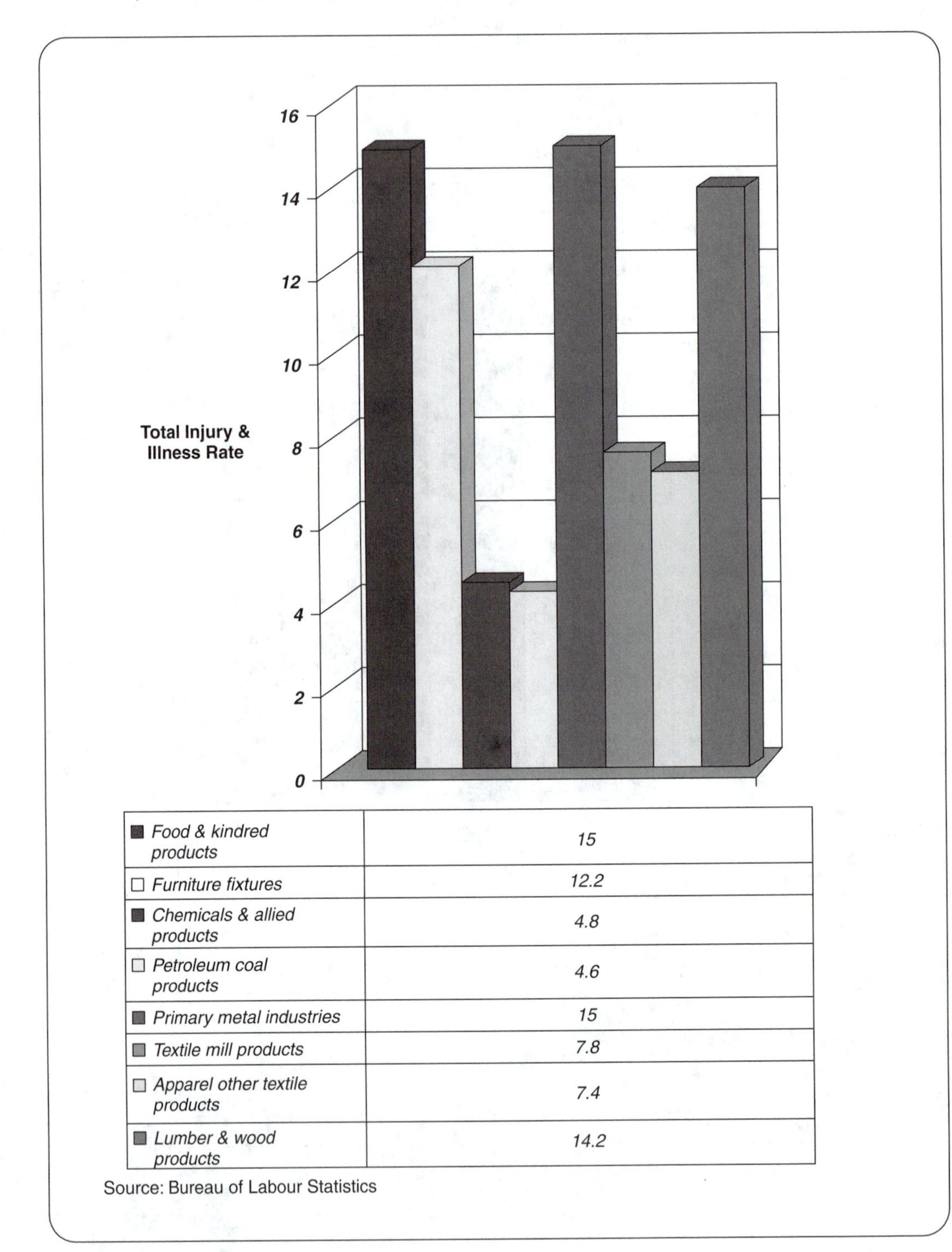

Food & kindred products	15
Furniture fixtures	12.2
Chemicals & allied products	4.8
Petroleum coal products	4.6
Primary metal industries	15
Textile mill products	7.8
Apparel other textile products	7.4
Lumber & wood products	14.2

Source: Bureau of Labour Statistics

FIGURE 11–3 Total Injury and Illness Rate by Manufacturing, 1996
Source: Bureau of Labor Statistics

data, the actual number of accidents, fatalities, and injuries will differ from the prediction.

Probabilities of accidents can sometimes be developed for new operations by suitably combining the probabilities of occurrence of subordinate events for which data are available. Unfortunately, because too often there is a lack of dependable input data, overall probabilities derived by these methods are rarely accurate to a highly dependable degree. Further, there is no way to prove by test the correctness of any of the computed estimates. Thus, although the theory of the methodology is acceptable, practice leaves much to be desired.

Another problem arising with risk-cost-benefit analyses involving probabilities to compute economic justifications for accident prevention measures is that estimated probabilities of accidents and losses are generally too low. The result is that employers will rarely show before-the-fact calculations of economic justifications for expenditures for safeguards. In 1987 the Supreme Court ruled that safety superseded other considerations such as any risk analysis.

A highly theoretical example would be if the operators of the *Titanic* had had to determine the economic justification for more lifeboat spaces before its initial, final, and sole voyage. The ship had been considered "unsinkable," that is, it had a zero probability of going under. If the probability of a catastrophic accident during any ship voyage was taken at the same value as that of an individual being struck by lightning (5×10^{-7}), and the ship made 50 voyages a year, such an accident would occur on the average of once every million years. The cost of additional lifeboats could never have been justified, based on the original estimate of the Titanic's announced safety features.

The risks an organization takes in any enterprise may be written as:

$$\text{risk} = P_A \times L_A$$

where P_A = probability of an accident;
L_A = probable loss resulting from the accident.

Figure 11-4 lists the probabilities of an individual having a fatal accident during a year because of various hazardous activities. The values shown were derived by dividing the total number of fatalities in the United States in 1969 from that specific cause by the total population of that year. On that basis, since 160 persons were killed in hang-gliding accidents that year, the risk of death from that cause would be as small as that of being hit by a lightning bolt, a relatively safe operation. However, because there were only about 20,000 hang-gliding enthusiasts who exposed themselves to the risk of being killed, the actual probability of one being killed is 8×10^{-3}.

Therefore, risk calculations must include factors for exposure of the persons or activities being considered. The risk equation then becomes:

$$\text{risk} = P_A \times E_A \times L_A$$

where E_A = exposure of a person or object to an accident.

The probability loss also requires careful consideration and inclusion in any appraisal of risk. An accident loss could vary from a negligible amount to a complete wipeout. It may therefore be necessary not only to estimate the probability of an accident in increments of different amounts over the entire range of foreseeable possible

INDIVIDUAL RISK OF ACUTE FATALITY BY VARIOUS CAUSES*

(U.S Population Average 1969)

Accident type	Total number for 1969	Approximate individual risk acute fatality probability/yr[1]
Motor vehicle	55,791	3×10^{-4}
Falls	17,827	9×10^{-5}
Fires and hot substances	7,451	4×10^{-5}
Drowning	6,181	3×10^{-5}
Poison	4,516	2×10^{-5}
Firearms	2,309	1×10^{-5}
Machinery (1968)	2,054	1×10^{-5}
Water transport	1,748	9×10^{-6}
Air travel	1,778	9×10^{-6}
Falling objects	1,271	6×10^{-6}
Electrocution	1,143	6×10^{-6}
Railway	884	4×10^{-6}
Lightning	160	5×10^{-7}
Tornadoes	91	4×10^{-7}
Hurricanes	93	4×10^{-7}
All others	8,695	4×10^{-5}
All accidents		6×10^{-4}

[1]Based on total U.S. population.

*WASH-1400-D, *Reactor Safety Study-An Assessment of Accident Risks in U.S. Commercial Nuclear Power Plants*, Atomic Energy Commission, August 1974.

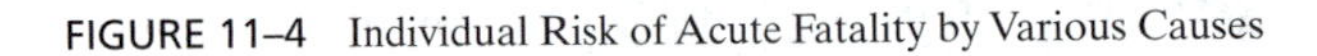

FIGURE 11–4 Individual Risk of Acute Fatality by Various Causes

losses, but to put against each average probability increment the possible severity or estimated loss. A total of all computed averages would, in theory, give the potential loss.

Rather than use this complex type of appraisal, most companies simply use their listed assets, apply an estimated loss, and add whatever insurers might agree to permit. Here again, accident predictions are often underestimated, and actual claims are generally overstated. Either the items listed in Fig. 1-2 are not all duly considered, or they are undervalued.

ACCEPTANCE OF RISK

The risks that persons will accept vary with the benefits expected. Starr[54] commented on the risk-benefit relationship; first he categorized risks as being "voluntary" and "involuntary." A voluntary risk is one freely accepted by an individual on the basis of

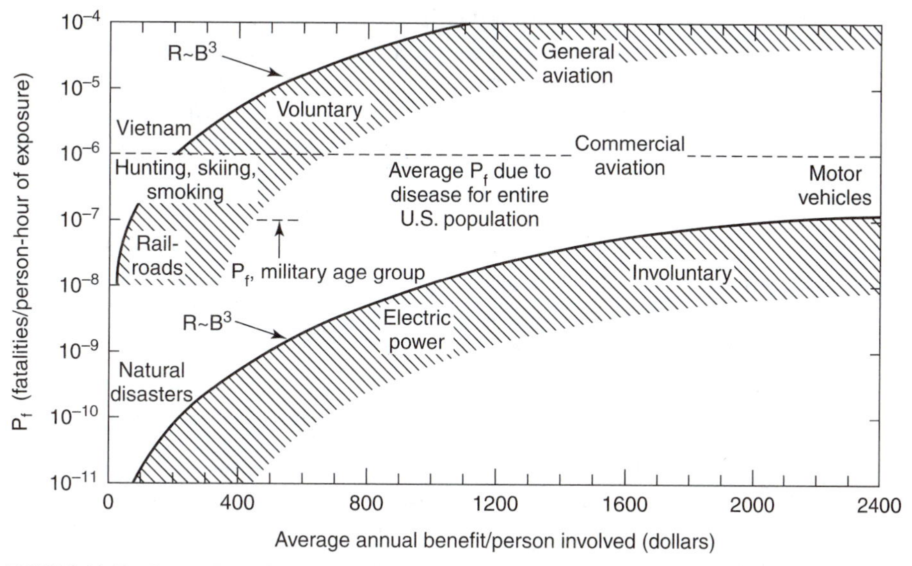

FIGURE 11–5 Starr's Benefits vs. Technological Risk

the individual's own values and experience. An involuntary risk is one by which an individual subjects himself or herself to another person's (or other persons') judgment. For example, a passenger on a bus, train, or airliner relies on the judgment of the operator and accepts risks over which he or she has control (other than the decision to take a bus, train, or airliner). Starr prepared the chart shown in Fig. 11-5 (its derivation is presented cited in the article) and also presented the following comments:

- The indications are that the public is willing to accept "voluntary" risks roughly 1,000 times greater than "involuntary" risks.
- The statistical risk of death from disease appears to be a physiological yardstick for establishing the level of acceptance of other risks.
- The acceptability of risk appears to be crudely proportional to the third power of the benefits (real or imagined).
- The social acceptance of risk is directly influenced by public awareness of the benefits of an activity, as determined by advertising, usefulness, and the number of people participating.
- In a sample application of these criteria to atomic power plant safety, it appears that an engineering design objective determined by economic criteria would result in a design target risk level very much lower than the socially accepted risk for electric power plants. (*Note:* The accidents at Three Mile Island and Chornobyl have substantially affected that comment.)

The study that produced the table in Fig. 11-4 also produced the following comments:

- Types of accidents with a death risk in the range of 10^{-3} (one in a thousand) persons per year to the general public are difficult to find. Such high risks are not uncommon in some sports and in some industrial activities, when measured for limited groups at risk (that is exposed to the hazards involved). Evidently this level of risk is generally unacceptable, and when it occurs, immediate action is taken to reduce it.
- At an accident risk level of 10^{-4} (one in 10,000) deaths per person per year, people are less inclined to concerted action but are willing to spend money to reduce the hazard.
- Risks of accidental death at a level of 10^{-5} (one in 100,000) per person per year are recognized in an active sense.
- Accidents with a probability of death of 10^{-6} (one in a million) or less per person per year are evidently not of great concern to the average person.

The last statement is a disputable generalization. Figure11-4 indicates that fatality probabilities in the tornado and hurricane zones are less than 10^{-6} per person per year. Yet the average person in those zones is very concerned with the effects of these phenomena. Also it appears that even though the total number of fatalities in either case might be the same, an accident that causes a great number of deaths at one time, even at rare intervals, is of more concern than are accidents involving a few persons on each of many occasions over an extended period. Thus, while there is much concern over the possibility, with very low probability, of a nuclear accident which might cause thousands of deaths, there is far less concern that 50,000 might be killed in traffic.

Quantitative appraisals of plant and operation safety, the probabilities of accidents, and risk estimates have always been considered questionable. There was much outcry against the authors of the study whose values are shown in Fig. 11-4. Appraisals of safety and risks in existing have to be readjusted with the advent of accidents as they were at industrial plants, Three Mile Island, and then Chornobyl.

RISK COMMUNICATION

It is not enough to simply appraise plant safety and risks and to "manage" plant risks. Risks must be communicated to the workers in the plant and to persons in the community. Regulations such as OSHA's Hazard Communication Standard of 1988 and the Environmental Protection Agency's The Emergency Planning and Community Right-to-Know Act of 1986 place requirements on the employer to know plant risks, appraise them, and communicate them to employees and employers. More than 30 million U.S. workers are potentially exposed to one or more chemical hazards. There are an estimated 650,000 existing hazardous chemical products, and new ones being introduced almost daily. These acts cover both physical hazards, such as flammability and explosivity, and health hazards. The acts are discussed briefly in the next chapter.

It is important to recognize that risk communication is part of the process of appraising plant safety. In 1989 a National Academy of Sciences Committee on Risk Perception and Communication defined risk communication as "an interactive process of exchange of information and opinion among individuals, groups, and institutions. It involves multiple messages about the nature of risk and other messages, not strictly about risk, that express concerns, opinions, or reactions to risk messages or to legal and institutional arrangements for risk management."[55] Behavior based safety (BBS) is a format that can be used to accomplish this interactive exchange of information and opinion.

It is possible for risks to be misunderstood and exaggerated by workers. Attention can be inappropriately invested by employees and employers on some risks, because of misunderstanding and false rumors in the community or among the workers. This phenomenon has been called by some "The Social Amplification of Risk." Similarly, insufficient attention to some risks can be given for the same reason. Risk appraisal should provide correct information about hazards within the plant and community. This information then can be provided in interactive exchange to all concerned parties. The Academy of Sciences report states, "Preparing risk messages can involve choosing between a message that is so extensive and complex that only experts can understand it and a message that is more easily understood by nonexperts but that is selective and thus subject to challenge as being inaccurate or manipulative." The interactive exchange format is important to having a culture of trust among all concerned that contributes to clarification where needed. Clarification occurs best when all persons concerned are able to listen to one another.

BIBLIOGRAPHY

[52] National Institute of Occupational Safety and Health, "Self Evaluation of Safety and Health Programs," U.S. Government Printing Office, Washington, D.C.

[53] "Accident Facts," National Safety Council, Chicago, Ill.

[54] Chauncey Starr, "Social Benefit versus Technological Risk," *Science*, Vol. 164, Sept. 19, 1969, pp. 1232–1238.

[55] National Research Council, *Improving Risk Communication* (Washington, D.C: National Academy Press, 1989).

EXERCISES

11.1. What organizations make plant safety inspections and appraisals?

11.2. How soon should considerations of plant safety and appraisals begin?

11.3. Cite some features of an individual plant that a safety engineer should review during the design stage.

11.4. What are the benefits of quantitative appraisals? How are they compared to qualitative appraisals? Can either be made without the other?

11.5. Describe how a safety inspection should be made to appraise the status of a plant's safety.

11.6. How can safety be measured?

11.7. Why would various organizations want different statistical information on accidents? What types of information would each organization want?

11.8. In what type of statistical information should an industrial plant safety engineer be interested?

11.9. Describe how statistical information on accidents and losses can be distorted unintentionally.

11.10. Why do the BLS/OSHA data differ from those of the National Safety Council?

11.11. How are accident frequency rates determined?

11.12. Do you feel that any specific probability of being injured should be set at a reasonable level of safety and that a person should not be able to sue any employer, operator, or product manufacturer who had met that level, based on prior analyses of possible risks?

11.13. What level of risk would you be willing to accept if you were told you might be injured or killed?

11.14. How does degree of risk affect insurance premiums?

11.15. Do you believe a risk-cost-benefit analysis is a good way to determine whether to undertake a proposed safety measure not required by law?

11.16. What is social amplification of risk?

11.17. What is the relationship between plant appraisals and risk communication?

CHAPTER 12

Hazards and Their Control

By the time they are given machinery to operate, workers can do little to change the adverse features that designers have imposed on them. However, this chapter may get workers to recognize hazards, to determine whether improvements can and should be made, and to decide whether there has been compliance with OSHA or state standards.

Most of the types of hazards that might be present have been touched on in general terms. Later in this book they will be described in more detail, together with causes and effects of accidents and the precautionary measures that might be taken for their prevention or avoidance. This chapter will be oriented toward hazard and accident control. To clarify some of the terms often used, and their relationships, Fig. 12-1 has been included.[55]

Determining exactly which hazard might be responsible for an accident is not as simple as it sometimes seems. Often, exactly what happened is a complex series of events. Consider the violent rupture of a high-pressure tank made of ordinary, unprotected carbon steel. Moisture can cause corrosion, which reduces the strength of the metal, which ruptures and fragments under pressure (see Fig. 12-2). The fragments hit and injure personnel and damage nearby equipment. Which hazard—moisture, corrosion, reduced strength, or pressure—caused the failure?

Figure 12-2 also illustrates how safeguards can be provided to prevent the mishap and to contain any possible injury or other damage. In this series of events, the moisture started the degradation process. If the tank had been made of stainless steel, there would have been no corrosion; moisture would not have been a problem; and there would have been no damage.

Rupture of the tank, which caused the injury and damage, can be considered the primary hazard. The moisture started the series of events and can be called the cause or the initiating hazard; the corrosion, the loss of strength, and the pressure are contributory hazards. The primary hazard is often indicated by other names: catastrophe, catastrophic event, critical event, or single failure. It can be seen that a primary hazard is one that can directly and immediately cause (1) injury or death; (2) damage to equipment, vehicles, structures, or facilities; (3) degradation of functional capabilities (disruption of plant operations); (4) loss of material (accidental release of large amounts of oil or chemicals).

EXPLANATION OF TERMS*

The following explanations are the authors' attempt to define more precisely terms that are widely used but often in diverse ways.

Hazard: condition with the potential of causing injury to personnel, damage to equipment or structures, loss of material, or lessening of the ability to perform a prescribed function. When a hazard is present, the possibility exists of these adverse effects occurring.

Danger: expresses a relative exposure to a hazard. A hazard may be present, but there may be little danger because of the precautions taken. A high-voltage transformer bank, such as those in power transmission systems, has an inherent hazard of electrocuting someone as long as it is energized. A high degree of danger exists if the bank is unprotected in the middle of a busy, inhabited area. The same hazard is present even when the transformers are completely enclosed in a locked underground vault. However, there is almost no danger to personnel. An above ground installation with a high fence and locked gate has a danger level between these two.

Numerous other examples can be cited showing how danger levels differ even though the hazard is the same. A person working on a very high structure is subject to the hazard that he could fall to his death. When he wears an anchored safety harness, the danger is reduced but is still present, since the harness might break.

Damage: severity of injury or the physical, functional, or monetary loss that could result if control of a hazard is lost. An unprotected man falling from a steel beam 10 feet above a concrete pavement might suffer a sprained ankle or broken leg. He would be killed in a similar fall from 300 feet. The hazard (possibility) and danger (exposure) of falling are the same. The difference is in the severity of damage that would result if a fall occurred.

Safety: frequently defined as "freedom from hazards". However, it is practically impossible to completely eliminate all hazards. Safety is therefore a matter of relative protection from exposure to hazards: the antonym of danger.

Risk: expression of possible loss over a specific period of time or number of operational cycles. It maybe indicated by the probability of an accident times the damage in dollars, lives, or operating units.

*W. Hammer, *Handbook of System and Product Safety,* Prentice-Hall, Inc., (Englewood Cliffs, N.J.) 1972.

FIGURE 12–1 Explanation of Terms

DETERMINING EXISTENCE OF HAZARDS

Each product or operation may have certain inherent hazards, although the probability of accidents with some may be remote. Each will have only a limited number of primary hazards and a large number of initiating and contributory hazards. A list of primary, initiating, and contributory hazards can be developed in two ways. Experience is the principal one, but it may not include all possibilities of what might occur. The database of experience can be extended with theoretical possibilities. Or the reverse process may be used: the theoretical aspects can be examined and then confirmed by actual experience. The hazards and safeguards for existing equipment and operations, or similar ones, may already be known. Proposed products and operations can be used

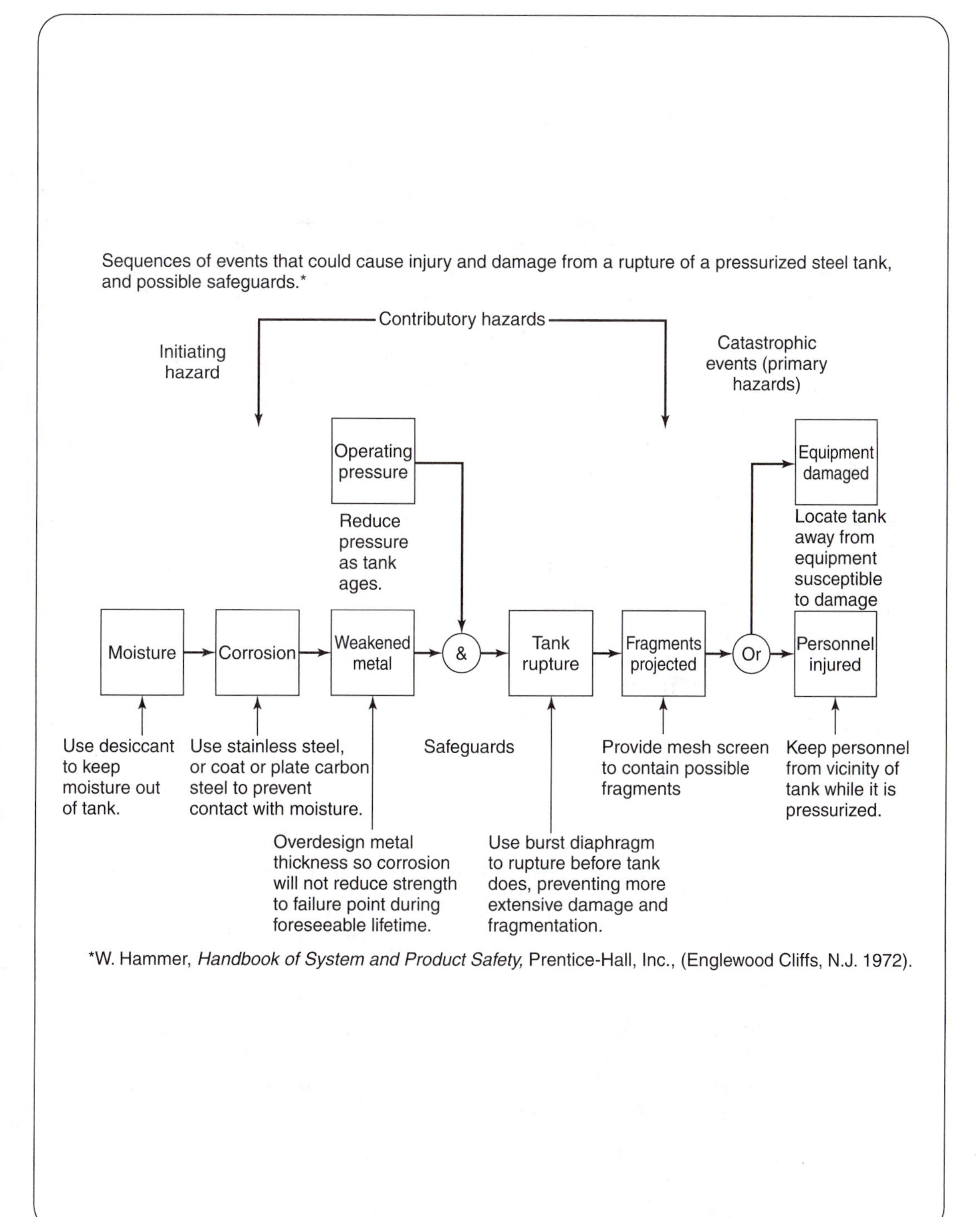

FIGURE 12–2 Pressure Rupture Accident

to synthesize a new machine tool or piece of electrical equipment. Potential hazards can be determined from experience and theory and evaluated by analyses.

ELIMINATING AND CONTROLLING HAZARDS

It has already been pointed out that hazards can be eliminated or controlled by good design or procedures for accident avoidance. Because of designer or managerial aims and capabilities, the optimum solution for any hazard may depend on the circumstances involved. For example, as pointed out in Fig. 12-3, drills can have different uses, power sources, advantages, and adverse features. All types of hazards can be categorized according to safety. The methods are presented in a rough order of priority of means to eliminate or reduce conditions that could cause accidents, injury, or damage. In general, safety of the system will be improved more, the higher the method is in the order. These indicated preferences may be precluded because of desirable operating aims or the practicability of certain modes of accomplishment. A less safe method may then have to be adopted. Each situation must be evaluated for affecting factors so that a desirable and satisfactory solution can be achieved.

In determining an order of preference, the following general features should be observed:

1. Designs to eliminate hazards are most preferred over any other method.
2. Where safeguards by design are not feasible, protective safety devices should be employed.
3. Where neither design nor safety devices are practical, automatic warning devices should be incorporated.
4. Where none of the above is feasible, adequate procedures and personnel training should be used.

Intrinsic Safety

The most effective method of avoiding accidents is with designs that are "intrinsically safe." Intrinsic safety can be achieved by either of two methods: (1) eliminating the hazard entirely, or (2) limiting the hazard to a level below which it can do no harm. Under either condition no possible accident can result from the hazard in question.

Elimination

A very common example of hazard elimination and accident avoidance is by good housekeeping. Tripping over misplaced objects, slipping on wet or oily surfaces, and spontaneous ignition of trash or oily rags can be eliminated simply by keeping facilities clean and orderly. Numerous examples can be cited, such as:

- Using noncombustible instead of combustible materials. This method has been observed with paints, fabrics, hydraulic fluids, solvents, and electrical insulation.
- Using pneumatic or hydraulic, instead of electric, systems where there is a possibility of fire or excessive heating. Fluid control systems are often applied for this reason.

ACCIDENT PREVENTION	MINIMIZING AND CONTROLLING DAMAGE
Intrinsically safe { 1. Eliminating hazards	1. Isolation and barriers
2. Limiting hazard levels	2. Minor loss acceptance
3. Isolation, barriers, and interlocks	3. Personal protective equipment
4. Fail-safe designs	4. Escape and survival equipment and procedures
5. Minimizing failures	5. Rescue equipment and procedures
6. Safe procedures	
7. Backout and recovery	

Although using the lowest numbered option for accident prevention is highly desirable, other considerations may make another option more practical. For example, personnel have been killed when using metal-cased electric drills. Tool manufacturers contend that this was due to improper repairs which resulted in a live conductor touching the metal case when the drill was used. Other potentially fatal conditions could exist when metal-cased tool drilled into a live conductor, or when an energized tool using 110-volt power was dropped into a container of water or other conductive liquid and someone tried to retrieve it while still energized. The following table indicates why and how trade-offs must sometimes be made.

TYPE OF DRILL		ACCIDENT PREVENTION PRIORITY NUMBER AND DESCRIPTION OF SAFETY FEATURES	OTHER ADVANTAGES	ADVERSE FEATURES
MECHANICAL HAND DRILL	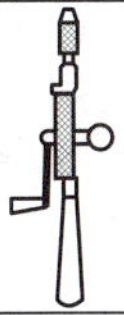	1. Elimination of use of electricity eliminates possibility of shock hazard.	Cost of drill is low.	Low mission effectiveness. Tiring to use. Must ensure gears are guarded.
CORDLESS BATTERY DRILL	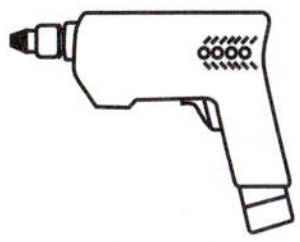	2. Uses electricity but power level is too low to cause injurious shock.	Can be used safety when operator is in water. Highly portable an convenient. No cord to be caught on projections. Can readily be taken into places with doors or others closed-off places.	Limited power which limits size drill which can be used and type of material which can be drilled. Needs periodic recharging.
THREE-WIRE METAL-CASED DRILL	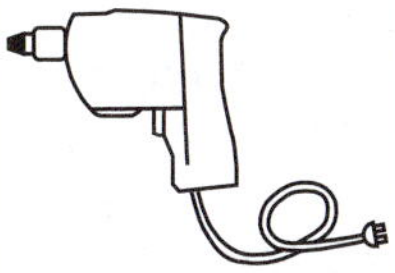	4. Fail-safe. Third wire provides path to ground for current if there is a short.	Cheap for manufacturers to change from 2-wire tool; only connection on interior of metal case needed for third wire. No need to redesign and provide plastic case.	Path to ground may not be complete so will not be fail-safe. Trying to retrieve a live tool which has been dropped in water may result in fatal shock.
TWO-WIRE METAL-CASED DRILL	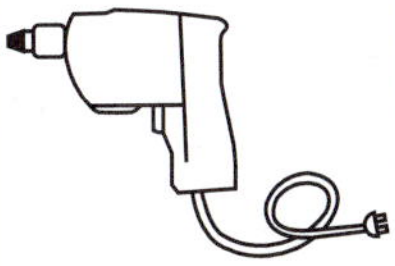	5. Increase reliability so there will be fewer failures that users will attempt to fix themselves.	Redesign not needed.	Manufacturers contention that problem due to incorrect repairs may not be valid. Higher reliability means higher cost. Failures may still occur but at reduced rate. Cord flexing where it enters drill may expose live conductor. Dangerous in water.
TWO-WIRE DOUBLE-INSULATED DRILL	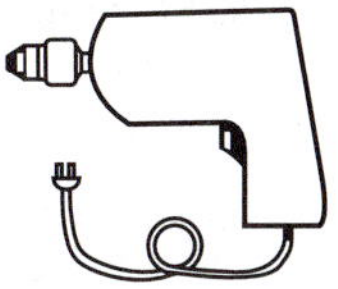	4. Plastic protects user against shock if an internal short causes live conductor to contact case.	Two-wire cord slightly cheaper than three-wire. Plastic case may be cheaper than metal.	Plastic not as abuse-resistant as metal. Cord flexing where it enters the case may expose live conductor. Dangerous in water.
COMPRESSED-AIR DRILL	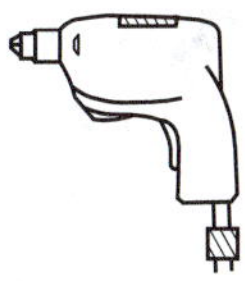	1. Use of compressed air eliminates electricity and possibility of injurious shock.	More power and higher reliability than electric drills.	Very few homes and not all shops have compressed air. Hazards of compressed air system. More expensive. Hose may make use inconvenient.

FIGURE 12–3 Drills—Accident Precautions

- Rounding edges and corners on equipment so that personnel will not injure themselves.
- Eliminating leaks by using continuous lines with as few connectors as possible.
- Eliminating vibration, shock, rail separation, and derailment by using welded and ground joints on railroad lines.
- Avoiding automobile accidents at highway intersections and railroad crossings through use of grade separations and limited-access highways.
- Eliminating protuberances, such as handles, ornaments, and similar devices in vehicles, which could cause injury after a sudden stop.

Hazard-Level Limitation

In certain instances the type of hazard cannot itself be eliminated. However, the level of the hazard might be limited so that no injury or damage will result. Although electricity under some circumstances can prove fatal, it may be possible to eliminate any adverse effects by using low-voltage, low-amperage power, such as 24-volt or battery power. Restriction could be used to keep pressurized systems below dangerous levels. The energy available under any condition, even if there is a failure of any kind, should not permit the hazard to be so great as to cause injury.

Examples of methods by which hazard levels may be limited include:

1. Providing overflow arrangements that will prevent liquid levels from getting too high.
2. Using solid state electronic devices where flammable or explosive gases may be present, so any power requirement will be far less than that required for ignition of a flammable mixture.
3. Ensuring that the concentration of a flammable or toxic gas is far below a dangerous limit. If the limit is exceeded, a blower could be started automatically or an inert gas introduced.
4. Adding dilutents to air where flammable dusts are present to minimize the possibility of an explosion.
5. Incorporating automatic relief provisions to keep pressure within a safe limit.
6. Using grounds on capacitor or capacitive circuits to reduce charge accumulations to acceptable levels after the power is shut off. This will lessen the tendency for a jolting shock.
7. Using sprays or other conductive coatings on materials to limit the level of static electricity that can accumulate.

ISOLATION, LOCKOUTS, LOCKINS, AND INTERLOCKS

These are some of the most commonly used safety measures. Interlock is effective principally because it requires positive actions by operators for procedures to continue, giving personnel opportunities to ensure they are correct. They are predicated on three basic principles, or combinations of the first two: (1) isolating a hazard once it has been recognized; (2) preventing incompatible events from occurring, from occurring at the wrong time, or from occurring in the wrong sequence; and (3) providing a release after suitable and correct action has been taken. Some examples of the major means of providing safeguards follow.

Isolation

Isolation as discussed here employs separation as an accident prevention measure. Later descriptions indicate isolation as a means of preventing injury or damage after an accident has taken place. In each case, there is a variety of means by which isolation can provide safety benefits.

Isolation is used to separate incompatible conditions or materials that together would constitute a hazard. Fire requires the presence of a fuel, oxidizer, and ignition source. Isolating any one of these from the other two will eliminate any possibility of fire. Some grades of bituminous coal are often stored under water, isolating the coal from the oxygen and ignition source needed for fires to start spontaneously. Other common examples of isolation as protective measures include:

1. Isolating workers inside protective garments or equipment to prevent environmental injuries.
2. Using thermal insulation to prevent persons from contacting hot surfaces which can burn them.
3. Using isolators to keep noise inside closed spaces.
4. Potting electrical connectors and other equipment to prevent entrance of moisture or other deleterious materials that could degrade the system.
5. Using "explosion-proof" or encapsulated electrical equipment in flammable atmospheres.
6. Keeping corrosive gases and liquids from incompatible metals or other materials that might be affected adversely.
7. Using lead, water, or carbon to isolate nuclear radiation.

Isolation is also used to limit the effects of controlled energy release where it is not due to an accident. A small amount of explosive to be tested can be placed in a suitable box or vault to absorb or contain the energy. Effects of the explosion can be measured. Or an explosion in the open air can be measured to determine the energy required to activate it, the distance and type of effect that might result, and the safety zones to be used provide isolation in the event of an accident.

Certain materials and processes are harmful unless isolated by suitable protection. Examples of protection include shielding persons from radiation from welding arcs to nuclear emissions, having workers in paint spray booths or sandblasting rooms wear protective equipment, and using gas masks, air packs, or oxygen generators in toxic atmospheres.

An operational activity that generates a problem can be isolated from personnel who can be injured or equipment that can be damaged, or vice versa. An engine producing a great amount of vibration, noise, and heat can be isolated by putting it in a separate room. Or it can be isolated through the use of suitable vibration mounts, noise suppressors, and thermal shields and sinks. Such isolation in a small area should be used only when the hazard, such as an adverse environment, cannot be eliminated by design.

Machine guards are widely used to isolate hazards in industrial plants. These guards are fixed over rotating parts, sharp edges, hot surfaces, and electrical devices to prevent personnel from coming in contact with the dangerous object. Security fences around electrical substations are similar. In such instances, the isolator is fixed. In other cases, such as railroad-crossing guard barriers, the isolator devices are removable when conditions are safe.

Lockouts, Lockins, and Interlocks

Lockouts, lockins, and interlocks include some of the most common means of providing isolation of personnel, equipment, and operations. Lockouts and lockins range from extremely simple devices such as bars on doors to those that are more complicated. Interlocks may be far more complex, since they generally involve stop and release mechanisms.

Lockouts and Lockins

Figure 12-4 lists a few of the measures that can be employed as lockouts and lockins. The difference between the two is relative. A lockout prevents an event from occurring or prevents a person, object, force, or other factor from entering an undesired zone.

A small sampling of the devices that can be used. They may be used in conjunction with each or any others and with variations that might be desirable.

1. Guards to protect personnel from moving machinery, parts, or cutting devices.
2. Fenced enclosures or vaults for high-voltage transformers to keep out and protect personnel from electrocution.
3. Shielding on reactors to keep personnel from nuclear radioactivity.
4. Storing subbituminous and similar grade coal, which is subject to spontaneous ignition, under water.
5. Storing oily rags away from air in covered metal containers until they can be disposed of.
6. Lockouts to keep personnel from open elevator shafts.
7. Lockouts to prevent opening doors while a hazard exists inside.
8. Locks on automobile engine systems and steering columns to prevent ignition or turning of wheels.
9. Safe-and-arm devices in explosive equipment to lock out movement that might lead to an explosion.
10. Safety wiring and other locking devices on nuts and bolts against part movement.
11. Locks securing switch levers to prevent activation of electrical circuits on which work is being done.
12. Lockouts to prevent pumping of highly flammable liquids into or from a tank or tank car unless the system is grounded adequately.
13. Park positions in vehicle transmissions to lock out vehicle movement.
14. Interlocks to deactivate hazardous electrical equipment when panels or drawers are opened.
15. Interlocks on equipment that must be operated in specific sequences.

FIGURE 12–4 Isolation, Lockout, Lockin, and Interlock Devices

A lockin, on the other hand, keeps a person, object, force, process, or other factor from leaving a restricted zone. Locking a switch on a circuit to prevent it from being energized is a lockout in this case; a similar arrangement on a live circuit to prevent current from being shut off is a lockin.

People have been killed when equipment they had deenergized was inadvertently activated by other personnel. Workers repairing electrical circuits they had opened were fatally shocked when the system was energized by someone else who did not realize work was being done on the system. A repairman working inside a vat that contained rotary mixers was killed when someone started the equipment. These accidents could have been avoided by using lockouts, if the switches opening the circuits had been secured so that only persons conducting the repairs, or their supervisors, had the locks or combinations to them. Provisions on circuit breakers permit lockouts and lockins to be made in this way.

Interlocks

Figure 12-5 indicates many of the different types of interlocks and their modes of operation. Interlocks are provided to ensure that event A does not occur under the following conditions:

1. *Inadvertently.* For event A to occur in such cases a preliminary, intentional action, B, is needed—for example, lifting the cover that prevents a critical switch from being activated accidentally.
2. *While condition C exists.* An interlock may be placed on an access door or panel to equipment where high voltage exists. If adjustment must be made and the door or panel is opened, the circuit is inactivated so that the unsafe condition no longer exists. Guard gates at railroad crossings are a combination of isolator, lockout, and interlock. They isolate the tracks and passing train, keep other vehicles and pedestrians from the tracks when a train is imminent, and open to permit traffic to pass when the danger has been removed.
3. *Before event D.* Such interlocks are desirable where the sequence of operations is important or necessary and a wrong sequence could cause a mishap. Manufacturers provide numerous pushbutton switching arrangements with many variations of interlocks. Analyses of desired operating procedures, controls, and the consequences of an error or failure will determine the type of interlock that should be used.

Fail-Safe Designs

Equipment failures produce a high percentage of accidents. And since failures will occur, fail-safe arrangements are often made to prevent personal injury, major catastrophes, damage to equipment, or degraded operations. Fail-safe designs ensure that failures will leave the system unaffected or convert them to states in which no injury or damage will result. In most situations, but not all, this safeguard will result in inactivation of the system. Fail-safe designs can be categorized into three types:

1. In the most common and prevalent, a fail-passive arrangement reduces the system to its lowest energy level. The system will not operate again until corrective

INTERLOCKS

Interlocks are one of the most commonly used safety devices, especially with electrically operated equipment. They take many forms. The following list indicates the principles of the types more frequently used. In some cases, two or more of these principles are involved in the design of one interlocking device. Some interlocks themselves prevent action or motion; other send signals to other devices which prevent initiating the source of the action or motion.

TYPE	*MODE OF OPERATION*
Limit switches, including: • Snap-acting switches • Positive-drive switches • Proximity switches	A wide variety of limit switches can be used for interlock purposes. They are generally operated by moving an external part of the switch in, out, or sideways to open or close the switch, depending on circuit design. In some cases the limit switch itself will open or close the circuit of which it forms a part; in others, a signal or lack of one from the limit switch will open or close a relay which, in turn, will open or close a power circuit.
Tripping devices	Action releases a mechanical block or triggering device which either permits or stops motion.
Key interlock	Inserting and turning a key in a mechanical lock permits action.
Signal coding	Specifically coded sequences of pulses emitted by a transmitter must match the sequence in a suitable receiver. When the sequences match, the receiver initiates or permits action.
Motion interlock	Motion of the mechanism being guarded against prevents a guard or other access from being opened.
Parameter sensing	Presence, absence, excess, or inadequacy of pressure, temperature, flow or other parameters permits or stops action.
Position interlock	Nonalignment of two or more parts prevents further action.
Two-hand controls	Two simultaneous physical actions by a person are required, sometimes within a specific length of time.
Sequential controls	Actions must be performed in the proper sequence or operation is inhibited.
Timers and time delays	Operation of the equipment can take place only after a specific length of time has passed.
Path separation	Removal of a piece of the circuit or of the mechanical path physically prevents operation.
Photoelectric devices	Interruption or presence of light on a photoelectrical cell generates a signal which can stop or initiate action.
Magnetic or electromagnetic sensing	Presence of a magnetic material stops or initiates operation of the equipment.
Radio-frequency inductive	Sensing of any conductive material, especially steel or aluminum, causes it to operate.
Ultrasonics	Senses the presence of nonporous materials.
Mercury switches	Mercury provides the path between two metal contacts through which current passes. The path can be broken by tilting the switch in which the mercury and contacts are sealed so that mercury flows away from one contact and breaks the path for the current.

FIGURE 12–5 Interlocks

action is taken, but no further damage will result because of the hazard causing the activation. Circuit breakers and fuses for protection of electrical devices are fail-safe devices. Each will open when the system is overloaded or a short-circuit occurs, deenergizing and making the system safe. For renewed operation, the circuit breaker must be reclosed or the fuse replaced.

2. Fail-active design maintains an energized condition that keeps the system in a safe operating mode until corrective and overriding action occurs or that activates an alternative system to eliminate the possibility of an accident. A fail-active design might include a monitor system that activates a visual indicator if a failure or adverse condition occurs in a critical operation. Or it might be a feature by which a malfunction in the warning system itself is indicated by a continuously blinking, differently colored, or auxiliary light or a loud noise. This should provide a very high degree of certainty that the fail-active system is still operative. A battery-operated smoke detector provides one example of a fail-active design. A fail-active design might also be incorporated in a street-crossing signal light that continues to blink when there is a failure.
3. A fail-operational arrangement allows system functions to continue safely until corrective action is possible. This type of design is most preferable, since there is no loss of function. The American Society of Mechanical Engineers requires a fail-safe operational orientation of feedwater valves for boilers. Incoming water must first flow under, rather than over, the valve disk. Detachment of the valve's disk from its stem permits the pressure of the water to continue to raise the disk and for the boiler to flow and keep operating normally. If initial flow had been over the disk, its detachment would have closed the valve, causing stoppage of flow and depletion of the water in the boiler. Before such arrangements were required, such occurrences resulted in lacks of water, increased steam pressures, and violent boiler ruptures.

Much railway equipment is based on fail-safe principles, predicated on the idea that gravity is the only force that can be depended on. As a result, semaphores, signal switches, and the lights to which they are connected are weighted devices. In event of a failure, a heavy arm drops, causing the fail-safe warning signal to be activated.

Other examples of fail-safe devices include:

- Air brakes on railroad trains and large trucks.
- Deadman throttles on locomotives.
- Automatic blocks on railroads.
- Control rods on nuclear reactors, which automatically drop into place to reduce the reaction rate if it exceeds a preset limit.
- Self-sealing breakaway fuel-line connections.
- Automobile headlight covers that open and expose the lights in event of a malfunction.

As pointed out in Fig. 6-2, the term "fail-safe" is sometimes incorrectly employed for redundant arrangements. A redundant arrangement reduces the likelihood of a complete operational failure. With a proper fail-safe system, when any operation fault or failure does take place, no accident will follow except in extreme cases.

It should also be recognized that for any fail-safe device to be a safeguard it must be of proper design. A deficient arrangement may fail or not operate rapidly enough. A fuse in an electrical circuit may not blow fast enough to prevent damage to the system. Or snow or ice could jam the weighted devices used by railroads so that the fail-safe system might not work.

FAILURE MINIMIZATION

Hazards sometimes are such that fail-safe design arrangements cannot be provided. On the other hand, the process may be so critical that even a fail-safe arrangement is less preferable than a system that will fail only rarely. Nevertheless, to ensure that possibilities of failures which could cause accidents are minimized, four principal methods are employed:

1. *Safety factors and margins:* this is probably the most ancient means to minimize accidents by design. Under this concept, components and structures are designed with strengths far greater than those normally required. This is to allow for calculation errors, variations in material strengths and stresses, unforeseen transient loads, material degradation, and other random factors.
2. *Failure-rate reduction:* this is the principle on which reliability engineering is predicated (see Fig. 6-2). It endeavors to use components and design arrangements to produce expected lifetimes far beyond the proposed periods of use, thereby reducing probabilities of failure during operation.
3. *Parameter monitoring:* a specific parameter, such as temperature, noise, toxic gas concentration, vibration, pressure, or radiation is kept under surveillance to ensure it is within specified limits and to determine when or if it exhibits an abnormal characteristic. If it does, preventive or corrective action can then be taken.

SAFETY FACTORS AND MARGINS

Theoretically, if an item is to withstand a prescribed stress, making it strong enough to withstand three, four, or five times that stress would reduce the number of failures and accidents; that is, a structure or container that had a safety factor of 4 would fail half as frequently as one that had a safety factor of 2. In practice, the inadequacies and uncertainties in use of safety factors previously pointed out have led to a refinement known as margin of safety. The difference between the two should be understood.

A *safety factor* is expressed as the ratio of strength to stress. Initially, strength (S) was nominal ultimate strength of the part: the average or prescribed value at which it would fail completely. (A similar ratio has also been employed based on yield strength to give a yield factor of safety. A material stretched passed its yield point would never return completely to its original length.) However, it was found that the strength of a specific material was not constant. For example, a lot of steel rods might be required in which each would withstand a stress (L) of 10,000 pounds per square inch (psi). However, differences will occur in the composition of the material involved, manufacturing and assembly, handling, environment, or in usage. As a result, some of the rods will fail at less than 10,000 psi; others at more; the remainder at the stipulated value. Failure will occur when stress becomes greater than strength. Using nominal or prescribed strengths and stresses, normally S/L was made so large that there was no failure. Now, S_{min}/L_{max}, or minimum strength/maximum stress, may be stipulated. By this means, which is not unusual in aerospace systems, it is possible to have a safety factor of 1.0 or 1.25. Such a safety factor as low as 1.25 would indicate a *safety margin* of 0.25 or S_{min}/L_{max}.

Failure-Rate Reduction

Operating components will not last forever. Actions therefore should be taken to limit failures while the system is operating, the rates of failures, and shutdowns of the system, especially any failures that can result in accidents. Methods by which these can be accomplished include:

- Increasing life expectancies of components to longer durations, especially to times longer than is normally required (derating).
- Screening.
- Timed replacements.
- Redundant arrangements.

Derating

Manufacturers are constantly working to produce components with longer lives. They hate their parts under specific conditions and stresses. Reducing the stresses under which components will operate will reduce their failure rates and increase their reliability. One of the stresses most affecting electronic equipment is heat. Failures increase with temperature; reducing temperature increases their lives. In one form of derating, therefore, cooling is provided even when the components operate at their normal capacities.

The principal cause of operating failures is the load factor: the ratio of actual load to rated load. This can be measured by voltage, current, or other indications of internal stress. Closely packed components, such as in some electronic equipment, generate heat, which causes mutual damage to the different parts. Formerly, cooling was necessary or component lives were shortened quickly. With the development of transistorized and embedded circuitry, little heat was required, generated, or transmitted. As the load factor decreases, so does the failure rate, and the longer it is before components will fail.

Derating can also be accomplished through the use of components whose capacity is greater than actually required. (This is the electronic equivalent of using a safety factor.) If components of greater capacity are unavailable, too costly, or too bulky, it may be possible to achieve the same functional result through load-sharing redundant arrangements.

Screening

One means of reducing failures is through close control of component quality. Quality control in the United States has improved greatly because of better quality of foreign manufacturers. Quality control at one time was meant to eliminate those components which failed to meet operational test requirements. Screening for reliability purposes also means eliminating those components that pass the operating tests for specific parameters but indicate they will fail within unacceptable times. Reduction in failure rates of all types, such as for electronics, automobiles, and plant equipment, has also lessened the occurrence of accidents.

Screening can be accomplished in a number of ways. The simplest is by visual inspection and measurement. Other types of tests include: application of normal pressure or voltage tests for short periods to time; accelerated tests during which a much greater stress is imposed over a short time; burn-in tests and tests for the entire periods during which they should operate or at least a warranted time; or step or progressive testing (which will not be explained here). It should be remembered that screening requirements will be less if quality is initially higher and errors fewer.

Timed Replacements

To maintain constant failure rates before wearout failures begin, it is necessary that components be replaced in a timely fashion. Some failures may be so critical because of the accidents they might lead to that they cannot be permitted. Others must be kept to a minimum, while still others merely generate inconvenience. In any event, to maintain operational capability it is necessary that components be replaced before they fail.

Replacements must be timed, because to do so too early is wasteful. It imposes an unnecessary maintenance and supply workload by excessive use of new units. This in turn can increase failures during burn-ins.

There are two means by which timed replacements can be made effectively and efficiently. One is by using data obtained by controlled laboratory testing of similar components or assemblies. The tests indicate the times after which wear out failures can be expected to take place. Replacement times can then be programmed at intervals shorter than these.

The second means is by noting component degradation or drift in operational systems. The output of components in circuits is measured, such as voltage or pressure. Once the output drops off past a stipulated point, which may be long past the prediction of failure, it is replaced. In many cases of large and critical equipment, tests are plotted to show the rate of degradation so that replacements can be predetermined and made at optimal times.

The SOAP (Spectrometric Oil Analysis Program) was developed with this concept in mind to determine when aircraft engines should be overhauled. (The parameter used was determining minute amounts of aluminum, iron, copper, and other metals in engine oil by which wear abrasion can be determined.) Some automobile brake linings now have indicators which indicate wear. In each of these, it is possible to estimate the remaining useful life and when replacement should be made.

Redundancy

Failure rates of complex equipment can often be greatly reduced through redundant arrangements. Some aspects of redundant design have been indicated in Fig. 6-2. In addition to parallel and series redundancy, others can be pointed out, all having the aim of lessening the probability of operational failures:

1. *Decision redundancy.* There are a number of types all based on three or five (odd-numbered) circuits having individual outputs and a monitoring unit which decides the action to be taken. With one type (Majority Vote) at least two of the three inputs will decide the action; in another (Median Select), the middle value is chosen. Other, more complex ones have been designed.
2. *Standby redundancy.* An operative unit operates until a failure is indicated; after that another unit is turned on, either automatically, semiautomatically, or manually.

MONITORING

Monitoring devices can also be used to keep any selected parameter, such as temperature or pressure, under surveillance to ensure it remains at proper levels, does not reach dangerous levels, and no contingency or emergency is imminent. Greater benefits can be derived if contingencies are prevented or corrected immediately, rather than in response to an emergency.

Monitors can be employed to indicate whether:

- A specific condition does or does not exist.
- The system is ready for operation and is operating satisfactorily as programmed.
- The measured parameter is normal or abnormal.
- A required input has been provided.
- A desired or undesired output is being created.
- A specified limit is being met or exceeded.

In addition, a monitoring system must lead to suitable corrective action when necessary. This may consist simply of conveying information to an operator, considered to constitute part of the system, who then accomplishes any required tasks. The actions in the overall process involve four principal steps, whether the system consists wholly of hardware, wholly of personnel, or of a mix of hardware and personnel.

Detection

A monitor must be capable of sensing the specific parameter which has been selected in spite of all other stresses that could be expected to exist during an operation or in an environment. (Much of the information on environmental aspects will be presented in Chapter 25.) A detector may be capable of measuring extremely small concentrations of toxicants in a laboratory. But in an operating environment, vibrations, temperature variations, moisture, electrical interference, or other stress may degrade performance or cause complete failure. The sensing function may be accomplished continuously, continually but intermittently, or intermittently at the desire of an operator. It must be able to sense and provide readings for only those parameters for which it has been selected, without being affected by extraneous undesirable conditions. It must be capable of detecting a hazard at a level low enough to permit awareness of its presence, generally soon enough to permit corrective action along before an emergency arises. The input element of the monitoring device should be located where it can sense the hazard for the selected parameter. Frequently, the value of a monitor is negated because input to sensing device is in a poor location for proper sampling. Fire detectors in homes are sometimes placed in living rooms, but fires generally originate in kitchens or closets where furnaces or hot water heaters are located.

Measurement

The parameter a monitoring device senses may be one in which only one or two bistable conditions can exist; that is, a device is either on or off. The monitor may also be able to determine additional information, such as the existing level of a parameter being monitored continuously or the exceeding of a predetermined level. The second type of monitor permits comparison of existing and predetermined levels. Methods of measurement vary from the very simple to the highly complex. A simple method is to mark a display, such as a dial, with predetermined limits; an indicator then points out the existing level. An operator observes and compares the existing level and the limit to determine whether there is an abnormality. One style of automobile gauge to monitor radiator water temperature is of this type. In the second type of automobile gauge, a light goes on to warn the operator only when the water temperature exceeds a preset level.

Interpretation

An operator must understand clearly the meanings of any readings provided by the monitoring devices. The operator must know whether a normal situation exists, an unusual condition is impending, or corrective action must be taken. Displays and signals should employ means by which personnel are provided information that is easily read and understandable, having the least ambiguity, the minimum possibility of misinterpretation, and the minimal necessity for additional information. Monitors should provide timely and easily recognizable displays and signals. Indicators and signals within a specific system are frequently standardized for this reason. Personnel must be trained and be knowledgeable of the exact meanings of any output of a monitor or warning device. The combination of information from the monitor plus previous training is required to produce a decision on a subsequent course of action. If either one is lacking or inadequate, a suitable course of action may not be possible within the time available, or there may be a delay until the deficiency is eliminated. The most notable case of deficiency in monitoring, interpretation, and response, one that almost resulted in a major accident, was that at the nuclear power plant at Three Mile Island in 1979.

Response

When a monitor indicates a normal situation, no response other than continuation of the action or program is necessary. When corrective action is required, the more information and time available to interpret and analyze it, reach a decision, and respond, the more likely will be the proper and effective decision and response. For this reason, whenever possible, data from the monitor should indicate as early as possible the approach of an adverse condition. In some instances, the level at which the monitor will indicate the existence of a problem should be set far below the actual danger level. For example, air contains approximately 21 percent oxygen; the danger level for respiration is 16 percent. A monitor could indicate when the level of oxygen in an enclosed space drops below 20 percent. The air is still breathable at that point, but the deficiency indicates the existence of a situation that should be investigated for a suitable response. Situations are numerous where early warning could permit proper responses. Threshold limit values (TLVs) (see Chapter 24) are examples; carbon monoxide or flammable gas detectors are others.

When the system is such that the response must be made by a person, analysis of the procedure should have been made to ensure time would be adequate to take corrective action under the circumstances. Occasionally, the attention of a person who should be aware of an unusual situation is focused elsewhere than on a visual indicator. In those instances in which a failure to take timely action could prove disastrous, auxiliary aural alerting or warning devices may be used. Where a serious, critical, or catastrophic condition would result if corrective action were not taken very rapidly, a monitor could be interlocked to automatically activate hazard-suppression or damage-containment devices.

Applications of Monitors

A few of the many applications of monitors are:

- Gas monitors to determine the presence of toxic or flammable substances.
- Infrared detectors to indicate the presence of hot spots or of flames.
- Detectors to determine emissions of pollutants from stacks.
- Liquid level indicators to warn when the fluid reaches a preset high or low limit.
- Governors that activate warning signals or lights or take corrective action automatically when a predetermined speed is exceeded.
- Odorants to indicate leakage of gases or high temperatures of metals, insulations, or other materials.

To ensure they perform properly, monitors (and warning devices) should:

1. **Be** accurate, quick acting, and easy to maintain, calibrate, and check. Procedures to test and calibrate must be available and used as prescribed.
2. **Be** selected for performance at a high level of reliability. In extremely critical applications, it may be desirable to permit testing by the operators to determine or indicate any failure of monitor circuitry.
3. **Have** independent and reliable sources of power and circuitry for monitoring and warning about critical functions.
4. **Require** an energy level lower than that which would constitute, contribute to, or activate a hazard in the event of a failure.
5. **Not** provide a path that could cause degradation of the entire system, change of a safe condition, inadvertent action, or other adverse effect because of a failure.

Buddy System

The buddy system may also be employed as a means of monitoring and safeguarding persons who undertake hazardous operations. The buddy system has been employed for many years by Boy Scouts, Girl Scouts, and similar organizations for hazardous activities like swimming or in industrial plants.

Two methods have been employed. In the first, two persons, who constitute a buddy pair, are subjected to the same hazard at the same time and under the same conditions. Each must ensure the well-being of the other, monitoring the other's activities, or providing assistance when required. Power company personnel who must work on live high-voltage electrical systems use this type of mutual aid and surveillance.

In the second type of buddy system, only one person of the pair is exposed to the hazard. The other acts as a lifeguard whose sole duty is to protect and assist the person in danger, should the need arise. A common example in industrial work is the task during which a worker must enter a tank to accomplish cleaning or repair. A buddy is stationed outside to monitor the well-being of the person inside. The buddy may provide warnings on any adverse condition that he or she may note, assist the worker if aid is needed, or call for outside assistance when it is required.

The outside buddy in this system should have no duty except that of monitoring the worker in danger. He or she might be authorized to perform such minor activities as passing tools when required. However, under no circumstance should the buddy have to leave the assigned station to obtain those tools. This prohibition against leaving includes performing any errand for the worker being monitored, even at the worker's own request. A means for communicating with the other personnel for other supplies or assistance should formerly have been devised and instituted.

Procedures for operations involving the buddy system must indicate what each person is responsible for and must do, and the hazards that must be monitored. The procedure should include a list of tools, supplies, and devices that might be needed to avoid any omission or call for them. Monitoring by devices should also be done often by the worker in danger. Supervisors must ensure that participating persons are aware of their duties and know the procedure, that required equipment, materials, and safety devices are available, and that means of communication have been arranged. Everyone should know what action to take in the event of a contingency, how to use emergency equipment, and how to ensure it is available and adequate.

Although they may not be designated as parts of the buddy system, other personnel may be employed to monitor hazardous conditions A worker can monitor road conditions and use red flags to tell vehicle drivers and equipment operators whether passage is safe. Pilot cars can be monitors for wide or dangerous loads on highways. They precede the dangerous load and monitor the route for obstructions and other hazards. When there appears to be inadequate clearance, danger to other personnel, or possibility of collision with another vehicle, the pilot car warns the vehicle that follows.

WARNING MEANS AND DEVICES

Warning means and devices are means of avoiding accidents by attracting or focusing attention of the operator or other person on an item that constitutes a hazard. Warnings are required by law to inform workers, users, and the public about any dangers that might not be obvious. Failure to warn, is in itself, considered a defect by the courts. Thus, engineers are legally responsible when, having failed to eliminate a hazard, they instead use warnings. In this case they use a procedural instead of a design means. Unless automatic equipment is used to control the results of monitoring, any alerts to potentially unsafe conditions or situations are by means of human senses.

Every method of identifying and notifying personnel that a hazard exists requires communication. All human senses have been and are used for this purpose. Figure 12-6 lists how the senses can be used as monitoring and warning devices. They are presented in a rough order of frequency in which these senses are used. With them are cited examples of their uses.

Vision is the principal sense, and signs and labels are the prime means, by which information of the existence of hazards is transmitted. Figure 12-7 points out some of the most important aspects and requirements doing this properly.

It has been reported 27 million Americans are functionally illiterate, so that they cannot read a label on a bottle. A stockyard worker destroyed a head of cattle because he couldn't read the word "POISON" on the bag he thought was feed. A Navy recruit caused $250,000 damage to equipment trying to follow the pictures in the text of a repair manual to hide his illiteracy. Compounding the problem is the fact that many personnel are literate only in foreign languages. It is estimated that damages in billions of dollars occur annually because workers can't read instructions.

The American National Standards Institute recently issued ANSI Z535.2-1998 Standard for Environmental and Facility Safety Signs. This revision attempts to establish specifications for signs to alert persons to: (1) the type of the hazard, (2) the degree of the seriousness of the hazard, (3) the consequences that can result from that hazard, and (4) what to do to avoid the hazard. It specifies the placement of text and symbols in relation to the headers. Text is flush left with mixed case. It recommends the use of multiple pictographs on a single sign to help overcome language and educational barriers.

SAFE PROCEDURES

The need to follow prescribed procedures has been mentioned before. Safe procedures should include any warnings established by the analysis. Unfortunately, since many people do not read operating procedures until they have run into difficulty ("when all else fails, read the instructions"), and ignore warnings, this method has a lower priority rating in means of preventing accidents.

BACKOUT AND RECOVERY

A failure, error, or other adverse condition may eventually develop into a mishap. By this time a contingency or emergency may exist. By suitable action an accident can be avoided from this abnormal situation, which may be an extremely dangerous one. Failure to act correctly or adequately can permit the situation to deteriorate into a mishap. This interim period extends from the time the abnormality appears until normality is recovered or accident develops. If recovery takes place, the incident can be considered a near miss.

THE SENSES AS WARNING DEVICES

Sense	Means	Description	Example
1. Visual	a. Illumination	a. A hazardous area is more brightly illuminated than nonhazardous surrounding areas.	a. Having well-lit highway intersections, obstacles, stairs, and transformer substations.
	b. Discrimination	b. Paint a physical hazard in a bright color or in alternating light and dark colors.	b. A structure (such as a pole), piece of equipment, or fixed object which could be hit by a moving vehicle is painted yellow or orange. OSHA standards require that the inside of the door to an electrical switch box be painted orange so that the fact that it is open can be recognized easily.
	c. Notes in instructions	c. Warning and caution notes are inserted in operations and maintenance instructions and manuals to alert personnel to hazards.	c. A warning in a car owner's manual to block the wheels before jacking the car to change a tire.
	d. Labeling	d. Warnings are painted on or attached to equipment.	d. "NO STEP" markings on hydraulic or pneumatic lines; high voltage; jacking points.
	e. Signs	e. Placards warn of hazards.	e. Road signs indicating hazards; "NO SMOKING" signs; EXPLOSIVE, FLAMMABLE, or CORROSIVE signs on trucks carrying such material.
	f. Signal lights	f. Colored or flashing lights (or reflectors) attract attention to a hazard or indicate urgency.	f. Red flashers on construction barricades at night; swinging red lights at railroad crossings; yellow caution lights at intersections; traffic lights.
	g. Flags and streamers	g. Tags or pieces of cloth are used to warn of danger.	g. A tag on a switch to indicate circuit is being worked on; a red streamer at the end of a long load protruding over the rear of the vehicle; colored strips of cloth on wires, ropes, and cables to make them more easily visible; flags used by flagmen at road construction sites to warn motorists when it is safe to proceed.
	h. Hand signals	h. A set of hand motions is used to pass instructions, warnings, and other information from one person to another.	h. Signals to the crane operator from a man guiding a load being lifted into place.

FIGURE 12–6 The Senses as Warning Devices

2. Auditory	a. Alarms	a. A siren, whistle, or similar sound device provides warning of existing or impending danger.	a. A siren indicates that there is a fire in a plant; a siren or whistle warns personnel to clear an area where blasting is to take place.
	b. Buzzer	b. Alerts a person that a specified time has passed or that time has arrived to take the next step in a sequence of actions.	b. Some compressed-air packs contain buzzers that sound when the pressure in the tank has decreased to a predetermined level, or after a preset time has passed.
	c. Shout	c. Voice action to warn of a danger.	c. One person warns another of an obstruction.
3. Smell	a. Odor detection	a. Presence of an odorous gas can indicate the presence of a hazard.	a. An odorant is added to refined natural gas (which has no odor) so that leaks can be readily detected.
		b. Burning materials give off characteristic odors	b. The presence of an unseen fire can sometimes be detected by characteristic odors of products of combustion.
		c. Overheating equipment can be recognized by the odor generated.	c. Vaporization of oil can permit detection of a hot bearing; odor of hot, steaming water can warn a car driver of a broken radiator hose.
4. Feel	a. Vibration	a. Rough running of equipment can indicate the presence of a problem and impending failure.	a. Vibration of a rotating shaft can signal a loss of lubrication, wear, and damage.
		b. Corrugations or vibration inducers in a road can warn a driver of a hazard.	b. Lane and shoulder markers in a road can warn a sleepy driver when he is going off the road or out of his lane.
	c. Temperature	c. Excessively high temperature can warn of a problem.	c. A maintenance man may be able to detect by its temperature a bearing that is acting abnormally; a temperature increase in an air-conditioned space may warn of equipment problems; excessively high temperature of a cooling fluid may indicate a possible problem in the equipment being cooled.
5. Taste	a. Ingestion	a. May indicate that material taken into the mouth is dangerous.	a. Acid, bitter, or excessively salty taste may indicate that material is not proper for consumption.

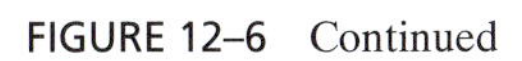
FIGURE 12–6 Continued

All employers and manufacturers have a duty to warn workers of hazards in the workplace. A failure to do so which results in an injury has long been considered a negligent act which may also be considered a design deficiency.

Labels are a visual type of warning device, a means of alerting personnel to the existence of a hazard. They are probably the commonest means now used to warn of hazards. In spite of this, they often neither are adequate as a tool of accident prevention nor satisfy the legal duty to warn of hazards. As a result there is much litigation because of accidents in which there were improper failures to warn. Below are some of the principal aspects that should be known regarding label warnings.

A warning label, to be considered adequate, must contain at least certain items of information, Others are advisable and recommended. There must be

1. A key word (see below) to attract the attention of the worker, user, or other person who potentially might be in danger, such as WARNING, POISON, FLAMMABLE, or EXPLOSIVE. Warning labels should be of legible size and located where they are easily apparent.
2. Information on the nature of the hazard to be guarded against. Labels should be simple, easy to understand, and not open to misinterpretation. Statements should be directed toward the educational level of persons to be warned.
3. Action to be taken to avoid injury or damage. The word WILL should be used instead of MAY if there is a reasonable possibility the hazard might produce injury.
4. A brief statement of the consequences that might result if the indicated action is not taken.
5. A brief instruction on appropriate emergency action to minimize harm if the hazard happens.
6. Accessible wording. The labeling must be in a language knowledgeable to the worker or user. Where workers or users may be from different countries, the labels should be multilingual in the languages expected.
7. Preferably, a logo of an exclamation mark within an equilateral triangle or other symbol to alert personnel to the hazard. Logos and symbols should be as universal as possible. Words, signs, logos, or symbols should be consistent.

Key word: ANSI Z535.2-1998 defines three choices for signal words to be used in headers on signs or labels:

1. CAUTION indicates a potentially hazardous situation which, if not avoided, may result in minor or moderate injury. It may also be used to alert against unsafe practices.
2. WARNING indicates a potentially hazardous situation which, if not avoided, could result in serious injury or death.
3. DANGER indicates an immediately hazardous situation which, if not avoided, will result in death or serious injury.

Header background colors should be yellow with "caution," orange with "warning," and red with "danger." Lettering for DANGER should be white.[56]

Further: Use of a warning label or sign is an indication of awareness a hazard exists. Manufacturers or designers should do their reasonable best to eliminate or properly control any hazards if accidents or litigation are to be avoided.

FIGURE 12–7 Label Warnings

Actions that can be taken include:

1. Normal sequence restoration, which may be possible, during which the situation can be corrected without damage. A change may simply eliminate an error, being made directly to the correct step in a sequence, or action can revert to a desired point. Another means is by stopping the entire procedure, such as by hitting a stop button. After that the procedure can be restarted as desired.

2. Inactivating only malfunctioning equipment that (a) is nonessential to the entire operation, (b) can be spared because of redundancy, (c) has already fulfilled its function, and (d) may be replaced by a temporary substitute.
3. Suppressing the hazard by immediately reducing it to a level where immediate danger no longer exists. After spillage of a small amount of gasoline (see Chapter 22), the possibility of an accident might be eliminated by flushing it away with water and creating a normal atmosphere.

If the emergency is not suitably countered, the result will be an accident. Following are means of lessening any adverse effects by damage minimization and containment.

DAMAGE MINIMIZATION AND CONTAINMENT

As long as a hazard exists, there is the possibility, however remote, that an accident will result, without our knowing when it will occur. Functional requirements and cost considerations make it impossible to eliminate all hazards or to incorporate safeguards for complete protection. Some hazards must be accepted, and accidents will happen. Some of the protective means of minimizing and containing effects of accidents include physical isolation, personal protective equipment, energy-absorbing mechanisms, "weak links," and escape and rescue.

PHYSICAL ISOLATION

Isolation has already been indicated as a means of accident prevention. It is also frequently used to minimize the results of violent release of energy, such as by use of:

1. Distance, by citing possible points of accidents far from persons, equipment, or vulnerable structures. Quantity-distance criteria for explosive safety are predicated on this principle. Standards are set for the amounts of explosives that can be located and at what distances from other critical areas or items so they will not be harmed by an accidental explosion.
2. Deflectors to lessen damage by deflecting or absorbing energy. The remainder should then constitute less than the amount that would be damaging. Energy may be deflected by such means as heat reflectors from fires, noise shields, or sloped barricades between explosive storage buildings.
3. Containment to prevent the spread of fire, such as sprinkler systems or water sprays.
4. Barriers of metal, concrete blocks, or other impenetrable or nonconductive material.

Personal Protective Equipment

This connotes the equipment persons might wear for protection against an accident. Although some types might be used for protection against both adverse environments and accidents, their potential use in accidents is pointed out here. Needs for such equipment vary from protection against environments that are hazardous, against questionable conditions where an adverse condition may or may not have taken place, or against accidents. The need for personal protective equipment can be divided into three categories:

1. For scheduled hazardous operation. Spray painting would require protective clothing even during scheduled operations, but also if it took place inadvertently.
2. For investigative and corrective purposes. It may be necessary to determine whether the environment is dangerous because of a leak, contamination, or other condition. The type of material might be toxic, or corrosive, or unknown. The leak might be simply suspected or its concentration uncertain. The protective equipment must be capable of providing protection against a wide range of hazards, which might be unknown.
3. Against accidents. This may constitute the severest requirement, because the first few minutes after an accident takes place may be the most critical. Reaction time to suppress or control any injury or damage is extremely important. Because of this, protective equipment must be simple and easy to don and operate, especially because it is often required at a time of stress.

The equipment must not degrade performance unduly, and it must be reliable and suitable for the hazard that might be involved. It must work as intended, or the worker might be exposed to an unsuspected and fatal hazard. Therefore it should have been more stringently designed and tested than equipment for normal purposes.

Many employees resist the use of personal protection equipment. Therefore, management must be have a proactive program and must shoulder responsibility to encourage and enforce this hazard control. Their responsibility in this regard is discussed in Chapter 7.

WEAK LINKS

A "weak link" is one designed to fail at a level of stress that will minimize and control any possibility of a more serious failure or accident. "Weak links" may be such simple items as perforations that permit tearing of paper products along desired lines. For safety purposes, weak links have been used in electrical, mechanical, and structural systems.

The most common use of a weak link has been with electrical fuses, which have been designed to fail before more valuable parts are damaged. The heat generated by passage of current through a low-melting-point metal causes the circuit to open before the current load becomes dangerous during a short-circuit. Other means of limiting extensive damage include: boilers with mechanical fuses that melt when water levels drop excessively, so that steam can escape and there is no rupture; sprinkler systems that open to release water for fire extinguishing; shear pins that fail at designed stresses to prevent damage to equipment being driven; or panels that will fail along designed fault lines to provide openings so personnel can escape.

Weak links all have an inherent problem. Although the damage that might result is minor, there will be damage, namely failure of the weak link. Thus, a circuit breaker limits damage, but it can be closed without adverse effects, which is not the situation with a fuse. A fuse or other weak link makes the system inoperable and must be replaced before it can be used again. Because of these, in some extremely critical designs, weak links are used as secondary equipment. A safeguard, such as a pressure relief valve, might fail to operate properly. At a higher, but still safe, pressure, the weak link would open, reducing the excess of an explosion.

ESCAPE, SURVIVAL, AND RESCUE

An emergency may continue to deteriorate until it is necessary to abandon or sacrifice structures, vehicles, or equipment to avoid injury to personnel. Following unsuccessful efforts to recover from an emergency, it may be necessary to leave the danger area, abandon ship, or bail out. This is the point of no return. And for such situations, escape, survival, and rescue procedures and equipment are vital: lives depend on them.

Escape and survival refer to efforts by personnel to save themselves using their own resources; rescue refers to efforts by other personnel to save those endangered. Although such actions might never be necessary, what is needed are adequate designs, procedures, suitable equipment, and knowledge of their use. The failure of equipment may be worse than if no equipment had been provided at all, putting the victim in an even worse situation. In addition, there is the stress or traumatic shock of the accident. Some actions might be possible up to a specific time, after which they are not. A pilot who is flying in an aircraft may have the option of either parachuting or attempting a forced landing. If the pilot jumps to escape, he or she is committed to use the parachute and must survive in the air. If the parachute fails, the probability of survival is low.

Escape and Survival

In some cases, escape and survival in the event of an accident may be a fairly easy process; under slightly different conditions it might be extremely difficult or impossible. After a fire in a single-storied plant, workers might escape by simply walking out a door and survive in the open air. Fire in a high-rise building might permit no escape.

A suitable analysis should have determined the hazards and accidents that could occur and how to combat their effects. Escape routes should be prescribed adequate for the number of personnel who would use them. Routes and exits should be marked conspicuously (OSHA standards require them) so they can be followed easily. Emergency lighting may be necessary. Safety zones should be established to which workers could withdraw.

Rescue Procedures and Equipment

There is the possibility that persons involved in an accident might not be able to escape under their own resources, so rescues may be attempted by other persons. The abilities of rescuers vary. Such persons might be: (1) fellow workers familiar with the plant, hazards, and equipment, and who may have been advised of what to do in any emergency, (2) untrained persons unfamiliar with the equipment, (3) personnel familiar with the hazards in general but not with specific equipment or materials involved, or (4) persons knowledgeable and capable of handling the need. Thus, a rescuer may be anyone from a passerby or a plant helper to a trained firefighter, chemical explosives expert, or mine cave-in expert.

For anyone to be able to provide assistance, especially for volunteers, certain facilities should have been provided. Prominently marked latches on the outside of aircraft are advised that inform rescuers how to open hatches so injured crewmen can be removed in rescue attempts.

BIBLIOGRAPHY

[55]. Willie Hammer, *Product Safety Engineering and Management* (Englewood Cliffs, NJ: Prentice-Hall, Inc., 1980), p. 113.

[56]. American National Standards Institute, *ANSI Z535.2-1998 Standard for Environmental and Facility Safety Signs,* New York.

EXERCISES

12.1. Discuss the different categories of hazards.

12.2. What is the best means of accident prevention? List five examples.

12.3. How can the magnitudes of hazards be limited? Describe how a design can be intrinsically safe.

12.4. How can isolation be used to keep personnel from accidents?

12.5. What is meant by making equipment fail-safe? How can it be done?

12.6. What are lockouts, lockins, and interlocks? Give examples of each. Describe some types of interlock devices and how they work.

12.7. How are monitors used to prevent accidents? Give five applications. List the characteristics a good monitor should have.

12.8. What is the buddy system and how is it used? What are the differences between the buddy system and the two-man concept?

12.9. Tell how the human senses can be used as monitoring and warning devices and give some examples of each.

12.10. List some features that are required or incorporated in any proper warning label or sign according to the common law.

12.11. What are back-out and recovery as they apply to accident prevention? How are they related to contingencies and emergencies?

12.12. List some means by which injury or damage can be minimized in the event of an accident.

12.13. What is a "weak link"? Describe some common types.

12.14. Describe the relationship between escape, survival, and rescue. Tell how equipment designs and procedures can be developed for them.

CHAPTER 13

Planning for Emergencies

If everything proceeded according to a designer's plans and intentions, there would be few emergencies. But emergencies do occur, often with injuries, fatalities, and great economic losses. (See Fig. 13-4.) As long as any hazard exists in any industrial plant there always will be the possibility—no matter how improbable, no matter how good the safety program—that an accident will occur. Planning to minimize the effects of accidents is only slightly less important than minimizing the hazards that cause them. It is unjustifiable and ineffective to begin studying means of controlling or responding to an emergency after one has begun. For proper control of both, preparedness procedures, capabilities in men and equipment, and training should have been instituted previously. Figure 13-1 is a diagram indicating occurrences subsequent to emergencies, some of which may have been declared because of accidents.[57]

MEDICAL RESPONSES IN EMERGENCIES

Accidents may happen where people are injured, or a worker may suffer from a sudden incapacitating illness. The well-being of workers should be the prime consideration in any such emergency. Fellow workers and supervisors should be aware what to do to help. Preplanning and training for such emergencies will be invaluable for assisting any disabled person (who may turn out to be the supervisor or any other worker). The outcome of any serious injury may depend on the speed of the response in aiding the victim. There may be no time to consult a manual or book on first aid. Persons who can provide assistance must therefore already have knowledge to do so when called upon. The response must be almost automatic. Because of the need for haste at such times, most states have "good Samaritan" laws which absolve from liability persons who try to help in accidents. Knowledge and training in first aid will not only minimize indecisions that could be costly in time and suffering, but also give those who would like to help the capabilities and confidence they need.

Each supervisor and worker should know how to call for and obtain medical assistance rapidly and what to do until it arrives. First aid measures may be taken until more qualified assistance (by a physician, paramedic, or registered nurse) is available. Generally, in industrial or urban plants, it takes only a matter of minutes for such

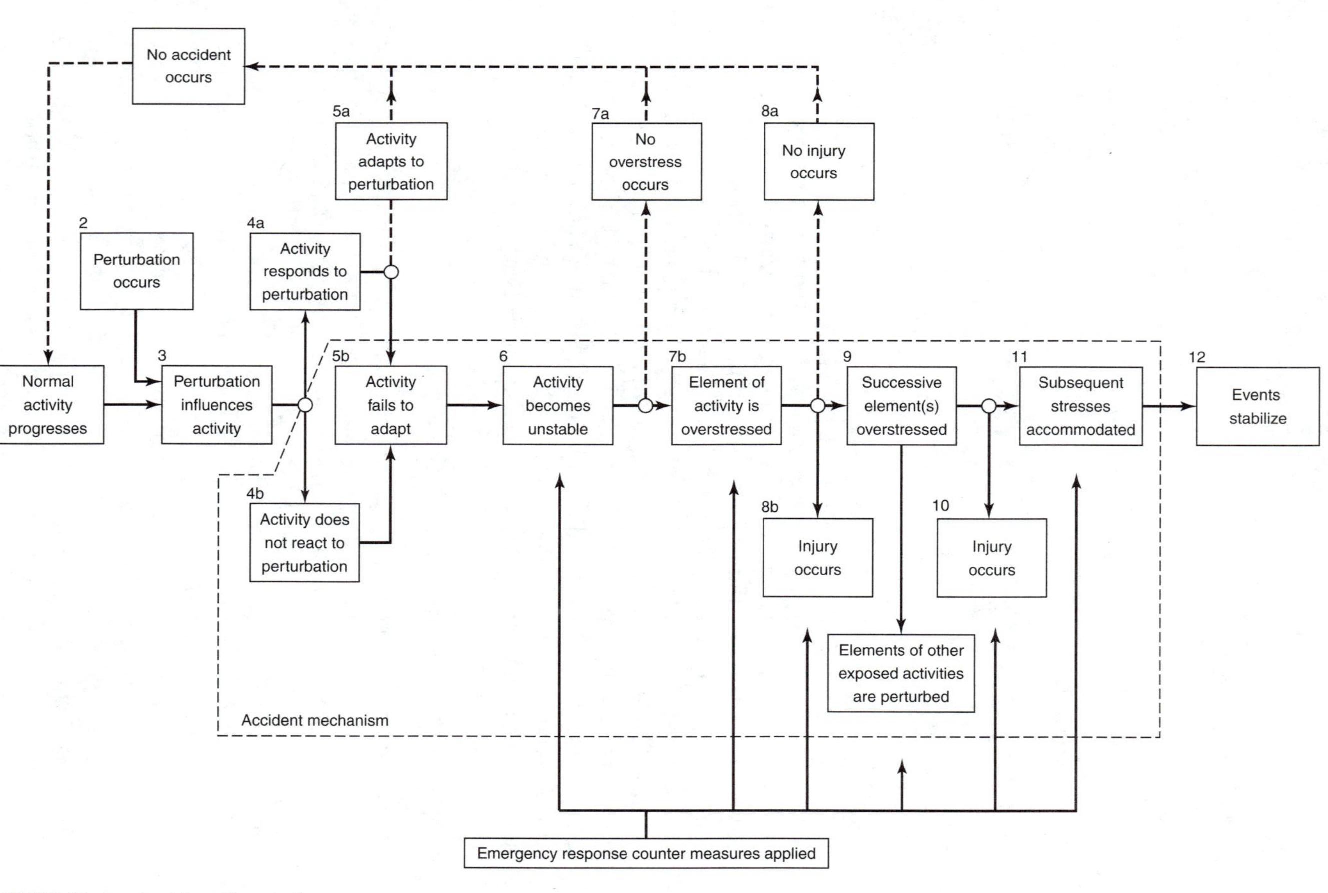

FIGURE 13–1 Accident Events Sequence

medical assistance to arrive. Industrial plants may have their own medical staffs, industrial physicians within call may be located nearby, and hospitals have ambulances that can provide rapid transportation of the injured if needed. OSHA standards require that where medical personnel are not close to a worksite, which often occurs during construction in remote areas, at least one person trained in first aid will be available.

Emergency medical assistance should be requested as soon as possible, but first aid measures should be initiated immediately in the event of a life-threatening injury. The person providing assistance should be careful not to become a casualty from the same hazard that generated the victim's injury. (See Fig. 9-2.) If the injured worker must be moved from a dangerous spot, it should be done as carefully as possible to avoid aggravating any injury. If time is available, a quick check should be made for injuries to determine where special care is required.

Moving an Injured Worker

One person can pull a victim to safety on a blanket, tarpaulin, or even a sheet of heavy plastic. (The victim should be pulled in the direction of the body's axis, not sideways.) If enough other persons are available to help, the victim should be lifted so that no strain is imposed on the injured part of the body. The head, back, legs, or arms should be supported so they have no tendency to bend. Here again it must be stressed that an injured person should not be moved unless there is an imminent danger.

Aiding the Victim

Accidents can be categorized as life-threatening or not. Some of the life-threatening injuries and a few of the lesser injuries and their care are described in Fig. 13-2. Supervisors and workers should turn over the care of the victim to paramedics, physicians, or nurses when they arrive but continue to assist when they are requested to do so.

Emergency Medical Facilities

The safety engineer, medical personnel, and supervisors should agree upon where first aid equipment should be located. The supervisors should then ensure that all workers and other persons are familiar with the locations of the kits, their contents, and how to use them. Safety engineers or medical personnel should check periodically to ensure emergency equipment has not been removed or used without resupply.

Emergencies may vary from a single person who slips, falls, or twists an ankle to a catastrophe in which an entire industrial plant is destroyed, or even one that initiates a disaster over a great area, such as Bhopal or Chornobyl. It sometimes happens that there is only a narrow interval between a minor emergency and a catastrophe or disaster.

Organizational Cooperation for Emergency Control

Considerations of the procedures, equipment, facilities, and personnel which would be needed in an emergency must be made part of the plant design and operations, not an afterthought following completion of the plant. The entire program for preplanning for

Injuries can be categorized as life threatening or not. A life-threatening injury generally requires a rapid response and examination by a paramedic or physician. But even before either might arrive, much immediate, beneficial first aid could be provided. This should be known by all workers and supervisors.

Life-threatening emergencies common in industrial plants may include

Unconsciousness In the event of unconsciousness, such as from a heart attack, fainting, or shock, the victim should immediately be checked for breathing. If the heart and breathing have stopped, immediately call 911 or delegate someone to make the call. Perform CPR (cardiopulmonary resuscitation) The ABCs of CPR are **A**irway, **B**reathing, and **C**irculation. The airway can usually be opened by tilting the head back and moving the chin up and forward, with the victim lying flat on his or her back on a straight, firm surface. Manual heart compressions are coordinated with mouth-to-mouth resuscitation. Depress the person's chest 1 and 1/2 to 2 inches in a slow, even rhythm, by pushing with the heel of one hand at the base of the breastbone where the ribs form a V. After 15 compressions, breathe twice for the person mouth-to-mouth. Repeat as necessary until help arrives. CPR should be learned in a class given by a competent teacher.

Bleeding Uncontrolled bleeding should be stopped by use of compresses. The wound should not be touched or permitted to become dirty in any way. The compress may be a dressing from a first aid kit or a clean handkerchief, towel, or shirt. It should be applied to the wound with firm pressure. If the compress becomes soaked with blood so it starts to exude blood too copiously, another compress should be added over the original (not in its place). Flow from very bad wounds can often be stopped or slowed through the application of pressure at suitable points. A tourniquet should be used only where a compress or pressure point does not work. The tourniquet should be placed between the wound and the heart and tightened only enough to stop the blood from gushing from the wound.

Electrical Shock Power should be shut off (preferably) or the victim removed from a live circuit (rescuer must use care to avoid also becoming a victim) and CPR applied if breathing has stopped.

Corrosive or Cryogenic Burn Wash chemical off skin with flood of water from an emergency shower or hose. Clothing saturated with the harmful liquid should be removed and the skin flushed with water.

Noncorrosive Internal Chemical or Poison If the affected person can swallow, induce vomiting with a spoon or finger at the back of the throat, or a glass of water with two tablespoons of salt. Have victim lie with face turned to permit free discharge from the stomach.

Third Degree Burns Apply cold water or ice (no grease or oil) to burn areas and cover with sterile gauze or clean cloth. Add cold water until victim can be taken to a hospital.

Trauma Any person who has been involved in an accident, especially the person injured, suffers from trauma (shock) to some degree. Any person in a serious accident should be considered in shock and examined and treated by a physician. (It has frequently happened that a person in an accident has appeared to be uninjured, walked away, and then collapsed and died.) A person who has been a victim of any of the above injuries in an accident should be kept quiet, warm, inactive, and as comfortable as possible. Clothing should be loosened and any bulky or heavy items removed. Protection should be provided against chill with a coat, blanket, or other covering. If there is no head wound, or broken leg, the head should be lowered slightly to help increase the flow of blood to the brain. Where there is a head wound, the victim should be kept lying flat with the head raised or at least level with the body.

Non-Life-Threatening Injuries Non-life threatening injuries might include severe injuries, but other than for the effects of trauma, time for assistance is not as critical. Some of the injuries and possible responses in this category are given below

Fractured Bones Treat patient carefully to avoid making injury worse and increasing the shock. The part should be treated as though fractured if one is even suspected especially if:

- The part has an abnormal shape.
- There is an inability to move the part.
- There is pain on movement with extreme tenderness over the injured part.
- There is swelling and change in the color of the skin.

FIGURE 13–2 First Aid in Medical Emergencies

A splint can be applied by a worker who has been instructed on how to perform one, a paramedic, or a physician. A sling can be used to support a splinted fractured arm or hand.
Severe Back Pain Have a victim lie on a hard, flat surface; keep warm; allow release only by a physician.
Minor Bleeding Apply a compress or, if bleeding is light, a Band-Aid.
Eye Injury Do not rub eye. Pull eyelid away from eyeball where it can be doused with clean warm water. Do not try otherwise to remove any dirt particle in eye; let it be done by a paramedic or physician. For more than removal of a small particle, the eye should be examined in a hospital.
Heat Cramps or Prostration Move to cool area. Give drink of one-half teaspoon of salt in a glass of water every 15 minutes until the patient has recovered. Best treatment is examination by a physician, especially in a hospital.

FIGURE 13–2 (continued)

emergency control must be a coordinated effort. In a small company, one person may be immediately responsible for safety, security, fire fighting, and medical service and how each is to perform in the event of an emergency. In other and larger organizations, one person may be responsible for only one or two of these functions.

Any emergency plan should indicate who will have the overall primary responsibility for directing actions, once an emergency has occurred. The person to be in immediate direct charge should be a professional who is experienced with the particular problem which generated the emergency. A production supervisor may be responsible for all activities in his or her department, but if an accident occurs, the supervisor must defer authority to the medical personnel, fire fighters, or safety personnel who respond. And among these, one should be designated to be in charge.

One aspect of safety can be used to illustrate the cooperative effort that can exist between organizations: providing egress from an area where an emergency could occur. A safety analysis may show, or jurisdictional safety codes, such as the OSHA standards, may require, that fire exits be provided to permit persons to leave a structure that might burn. On November 1, 1970, 146 persons were killed in Saint-Laurent-du-Pont, France, trapped in a dance hall when a fire broke out. Fire exits had been provided as required by government regulations. However, the doors had been secured shut to prevent them from being used by unauthorized persons to enter the hall.

Securing fire exit doors so they cannot be opened is no worse than blocking the doors or passages to fire exits in an industrial plant with pallets of heavy equipment or with portable or mobile bins of parts to be processed. This is very common. Here security personnel can do much to assist those responsible for safety (Fig. 7-4). A usual function of security personnel is to check all doors, and because they must use the accesses to them, they can also determine whether any of the passages or doors are blocked. Cooperation is required between the plant manager, production supervisors, and security personnel before an emergency takes place, and with fire fighters and medical personnel afterwards.

OSHA Standards

Many of the OSHA standards have been stipulated to minimize accidents and injuries and therefore emergencies. Good planning is necessary to limit the effects of emergencies. Provisions for egresses and exits and for their marking are probably the prime examples. Others include:

- Limitations on weights of explosives that may be stored in one location and minimum distances permitted to inhabited buildings, railroads, or highways. Limitations on sizes of above-ground tanks for storage of flammable fuel, distances to property lines and public ways, and requirements for dikes to contain a tank's content if there should be a rupture.
- Limitations on sizes of containers and portable tanks holding flammable liquids.

Hazard Communication Standard

OSHA's Hazard Communication Standard (HCS) includes all workers exposed to hazardous chemicals in all industrial sectors. It is based upon the idea that employees have both a need and a right to know the hazards and the identities of the chemicals they are exposed to at work. Chemical manufacturers and importers must convey the hazard information of their product to employers by container labeling and by Material Safety Data Sheets (MSDSs). The employer must have a hazard communication program to deliver this information to their employees through container labeling, MSDSs, and training. An MSDS should specify emergency procedures as well as the material's hazards and physical characteristics. The Chemical Manufacturers Association sponsors ANSI industry guidelines for the preparation of Material Safety Data Sheets. The safety engineer must be able to access the sheets at any time, without delay. They must be available to local emergency responders, as discussed later in this chapter.

The employer must include training of workers in emergency response in the event of a spill or leak of hazardous chemicals. Under the Hazard Communication Standard, all workplaces where employees are exposed to hazardous chemicals must have a written plan which describes how the standard will be implemented at the work site.

Adapting Plans to the Plant

Although most of the factors indicated next relate to major emergencies, lesser problems are more common, such as a single person being injured or suffering a heart attack. Planning for contingencies in different plants will entail many of the same considerations, but plans will differ because of the different hazards that might be present, the types of facilities and equipment, the organizational structures and responsibilities, and personnel available. Observance of provisions of pertinent OSHA standards and of state and local codes must also be ensured. In spite of numerous commonalities in some respects, each preemergency plan must be custom made.

Planning Procedures

A checklist for evaluating potential needs before an emergency is presented in Fig. 13-3. More detailed information for planning ahead for emergencies must include actions such as, but not limited to, the following.

Hazards Leading to Emergencies

1. Besides listing the hazards (the conditions or actions that could lead to possible injury or damage, not the hazard or damage itself), the types and magnitudes of personnel, equipment, facilities, and other economic losses that might result should be estimated. The expenditures for safety equipment and other safeguards and the concentration of safety effort may depend on this analysis.
2. Possible initiating and contributing hazards should then be evaluated. The safeguards that have already been provided to prevent emergencies because of loss of control of each hazard should be determined. These may include such items as pressure relief valves, inerting systems on reactive chemicals or processes, blowout plugs and panels, or governors on equipment.
3. Safeguards that have already been provided to minimize injury and damage if control of the hazard is lost and an emergency begins must be evaluated for adequacy. Such safeguards include monitoring and warning systems, emergency shutdown equipment, and steel doors and sprinkler systems for fires.

Procedure for Emergency Actions

4. A procedure must be developed for rapid reaction to any emergency by designated organizations and their personnel. The lack of such a procedure may prove disastrous (see Fig. 9-2). Available resources in personnel and equipment that could be used should be reviewed and deficiencies and omissions corrected. The procedure must consider the problems that might be involved, their locations, which unit is to respond if there is more than one, and the optimal means by which the problem should be combated. The availability of outside resources and when they should be used should be considered. It may be necessary or advantageous to make agreements with the outside organizations for mutual support assistance. Procedures must be established for intercommunications so that assistance, in plant and out, can be provided when required. The point at which the outside agency should be informed and at which assistance will be provided should be agreed upon.
5. Actions to be taken by designated personnel or organizations at the onset of an emergency should be established. The identity of the person(s) who is (are) to respond must be predicated on the type of emergency. If a person collapses, the organization and those who need to respond will undoubtedly differ from those which must respond to a fire or explosion. This necessitates that the person(s) to receive the first communication on the emergency be properly trained to obtain

1. The safety engineer must be familiar with and evaluate
    a. the types of hazards that could generate emergencies.
    b. those hazards with the greatest risks and potential losses.
    c. how each emergency could be aborted if one does occur.
    d. the extent of possible fatalities and injuries.
    e. the monetary loss of
        - an entire plant.
        - one department or section of the plant.
        - a critical piece of equipment or system.
        - one or more workers.

Other items to be reviewed or undertaken include

2. Program for combating any emergency and plans for the response to each of the various emergencies.
3. Evacuation in the event of a disaster. Municipality or state evacuation plans. Knowledge of plant personnel's participation in any disaster.
4. Statutory requirements to prevent, or that must be observed in the event of, an accident in your plant, municipality, and state.
5. Procedures that are published and displayed to be followed in the event of
    a. an injury to one or more persons.
    b. a fire, explosion, toxic spill, massive environmental contamination, or dispersion of any nuclear substance.
    c. other accident.
    d. multiple emergencies (such as earthquake and fire).
6. A list of available resources of personnel and equipment.
7. Training of supervisors, fire fighters, other medical personnel, and workers to ensure familiarity with the types and locations of emergency equipment in a plant. Ensure that all have been trained and are familiar with their means of operation. Supervisors and workers should have been designated to respond to specific emergencies.
8. Plainly marked and well-lighted emergency exits and routes to them. Unobstructed evacuation passages and routes.
9. Facilities to broadcast any alarm or calls for assistance, either in plant or to nearby affiliated organization. Familiarity with the locations of each facility and location and the routes that might be used.
10. A readily available list of organizations, personnel, equipment and utility companies to help provide assistance, with names and telephone numbers.
11. A list of fire extinguishers and their locations. A program for marking their sites and refillment and replacement, with name and phone number of person(s) designated to check on refill dates and maintain and refill them.
12. Holding emergency drills and notifying the plant manager as to the results.
13. A published list of company personnel who will be in direct charge for
    a. combating any emergency.
    b. security and protection of workers and equipment involved.
    c. safeguarding equipment and other assets remaining after a disaster.
    d. preparing a list of fatalities and extent of injuries.
    e. transportation needs.
    f. liaison with city, state, and federal officials, insurance companies, and news media.

FIGURE 13–3 Checklist—Evaluating Potential Emergencies

all pertinent information rapidly and accurately. Means must be provided for informing all personnel that an emergency exists and emergency procedures are to be instituted.

6. All personnel within a plant should receive instructions, commensurate with their work and the hazards to which they may be exposed, on what they should do in an emergency. These instructions should be supplemented at times by reminders such as posters. Supervisors should ensure that they and the workers they supervise are familiar with the instructions. Each person can provide valuable assistance if he or she is on the spot when an emergency arises and can take or assist in immediate corrective action. But they should also know when to obtain more active assistance first rather than spending time combating the problem. The person should know what to do to help the professionals when they arrive.
7. A supervisor, foreman, or other person should be assigned to ensure that everyone leaves an area when it is designated to be vacated. This person should check washrooms and other places where occupants might not be able to hear evacuation signals and should check for persons, especially handicapped workers, who might have trouble leaving expeditiously.
8. Consideration should also be given to the use of volunteer emergency personnel. Organized and trained volunteers can provided tremendous benefits during any emergency. They constitute a cadre of persons in a plant who can react quickly and effectively. (Each volunteer worker may be spurred by the fact it is his or her plant or job that might be destroyed or friends injured.) Because they are trained, they are far less prone to panic and will reduce the tendency of others to panic. The volunteers may be able to eliminate the emergency entirely, limit it until the professionals arrive, and provide assistance after the professionals take over. Here again it is fundamental that these volunteers get continued proper training and periodic exercise.
9. Training must be undertaken to ensure that operators and rescue personnel are proficient in carrying out emergency procedures (see Fig. 13-4), even though the plans may never be put into actual practice. Simulated emergencies will help increase proficiency. Investigations of many serious accidents reveal that frequently personnel died because of lack of proficiency in the use of emergency equipment and devices or because of failure to follow established procedures.

Communications

10. Means of communication must be established by which to alert the emergency organizations' personnel that their services are required. The most effective means must be established for personnel, vehicles, equipment, and stations. It should be investigated whether a special emergency number should be provided or the operator should be called. Increasingly, many cities have begun using the number 911 on dial or pushbutton telephones. The number to be used should be posted prominently on or near all phones.

Outside of a nuclear disaster, the worst accidents have been in chemical plants. First responses to fires or explosions are generally by fire fighters. The need for continued training for fire fighters to minimize possible accidents resulting from the original previous emergency is exemplified in the following extract from an article by J. Ranill "Cooperative Fire Fighting," *Chemical Engineering*, May 31, 1971 [58]:

. . . refinery people do everything they can to train city and county firemen in handling petrochemical fires.

One of the films that is shown to illustrate the need for such special training is a documentary of the Kansas City oil fire disaster in 1959. Here several firemen lost their lives simply because they did not know how to deal with petroleum fires.

In this case, it was a fire in an old bulk plant with inadequate venting on horizontal storage tanks. A fire broke out on a loading rack and spread quickly to one of four horizontal tanks. A group of firemen standing over 90 ft. from the end washed the burning gasoline back under the tank. Unexpectedly, the end of the over-pressurized tank blew off as a ball of fire as it exploded outward to envelop and kill six men and seriously burn 13 others. With proper training, they would have flushed the burning fuel away from the tank and cooled the tank from the side, not the end.

The article has an excellent description of mutual aid arrangements between a group of chemical plants for fighting fires.

FIGURE 13–4 The Kansas City Disaster

11. Consideration should also be given to installation or purchase of a secondary communications system for emergency use. Communication by telephone is generally dependable from the standpoint of mechanical and electronic reliability. Unfortunately, personnel often jam the lines as they attempt to determine more about the emergency, whether they can assist, and to obtain instructions. In addition, the condition that created the emergency might damage the telephone system. It would therefore probably be advisable to have backup service through a paramedic, firefighting, ham or similar radio network. If a plant is very remote, ham radio or similar radio services can be, and have been, used.

12. Alarm systems must be provided. Alarms must be distinctive, different from other sounds, and last long enough so persons for whom they are intended will be alerted properly. They must be instructive, informing personnel what and where the problem is.

Equipment and Transportation

13. The types of equipment for which personnel should be provided to meet specific types of emergencies must be determined. The equipment must permit quick response. It must be easy to don and simple to use, especially by personnel under the stress of an emergency. It must be highly reliable and effective. It must not unduly degrade the mobility or performance of the user, cause the user to fatigue rapidly, or itself constitute a hazard.

14. The most effective locations for the emergency equipment must be established. Analyses should determine whether it should be carried by personnel in vehicles in which they respond or whether it should be stored near the sites of possible emergencies. Equipment storage sites should be located as close as possible to where the equipment might be required. Sites must not be so located that the condition that creates the emergency prevents reaching it for use or renders it ineffective. Storage units should be easily accessible and marked for quick identification. Plant directives should require that access to storage units will not be blocked, and that equipment be inspected periodically. A log to indicate the dates and by whom the inspection was made, and to note any discrepancies, should be provided.

15. Plans must be made for transportation services which might be required. Means must be provided to move equipment, supplies, and personnel needed to combat the emergency, rescue anyone trapped or in dire need, and take injured personnel from the scene if it is a disaster. (See Chapter 26.) Transportation equipment may be needed to remove files and valuable equipment or hazardous materials to safe locations, and communications vehicles may be needed to act as control points.

16. Routes to safety should be determined and analyzed for adequacy for the number of personnel by whom they might be used. Routes and exits should be marked conspicuously (OSHA standards require marking of routes, egresses, and exits) so persons can follow them to safety without hesitation. The number of exits and egresses must be adequate for the time that would be available. They also must be adequate if one is blocked. Emergency lighting may be necessary if the loss of the normal lighting system could throw the route into darkness. In large buildings with no windows, independently lighted illumination and directional arrows should point to the direction to be taken to the nearest exit.

17. A map should be provided to each mobile unit that must respond to an emergency, indicating the best means of getting there and alternative routes in case the expected route is blocked. Emergency routes should be examined to determine if any obstructions do or could exist. Where obstacles are possible, the locations should be marked to warn personnel against parking vehicles, equipment, or material there.

Safety Zones

18. Safety zones and evacuation routes should be established. An analysis must also be made to determine locations in which personnel will be safe or to which they can withdraw during or as a result of an emergency. Protective structures can be provided beforehand to house personnel or to allow them to pass through to safety if there is any likelihood damage could be so extensive and so swift there would be difficulty evacuating the area. Or the number of personnel who could be affected might be extensive. (In earthquake regions, a common instruction is to take refuge under a sturdy table, desk, or similar piece of equipment if it appears the building might collapse. Similar refuges can be used if a building is damaged by an explosion and in danger of collapse.)

Utilities

19. Personnel must know how to control utility services so they cause no damage but are available when required. It may be necessary to cut off power through high-voltage lines so that personnel are not shocked, shocked persons can be released and treated, or fires are not ignited. A circuit breaker may have to be closed to provide for lighting or emergency equipment. A ruptured water or gas line may have to be shut off.

Abandonment and Rescue

20. In some cases, measures taken to combat emergencies may be inadequate and the situations may worsen. A state may be reached where attempts to correct the situation should be abandoned and effort redirected to safeguarding personnel. In certain emergencies, efforts must be diverted from saving facilities, equipment, or material which appears to be beyond saving and redirected to saving other facilities, equipment, or material. The times at which these transfers of effort should be made should be considered carefully before the emergency might occur. Persons involved in an emergency who arrive at such critical points generally have no time to evaluate carefully all conditions or to review possible actions to determine which is the optimum. They may wait too long in heroic attempts to save equipment or material and unduly endanger their lives. To ensure that personnel realize exactly when they are in danger and when they should abandon the equipment or material, instructions should be provided as to where the point-of-no-return lies. Procedures must indicate the point at which to consider control of a hazard lost and to abandon ship.

21. In an emergency the persons involved may not be able to escape using their own resources. Provisions must therefore be made for rescue by other personnel, should the need arise. Rescue devices, like those for escape, must be foolproof, require a minimum of effort to operate, and be operable when only a few words of instruction are provided. The instructions should be marked so they are easy to recognize and understand by persons under stress, especially if the attempted rescue is the spontaneous action of untrained personnel. The presence of effective devices with suitable markings can mean the difference between successful attempts and failures. Periodic tests should be made to ensure that escape and rescue devices work properly with the instructions provided.

Checklists

Each sentence of the preceding numbered paragraphs could be made into a checklist item. This will permit determination, especially by safety personnel, of how well plans for emergencies have been prepared. Figure 13-5 provides an additional checklist to ensure nothing necessary has been omitted.

The Emergency Planning and Community Right-to-Know Act

The Emergency Planning and Community Right-to-Know Act (EPCRA) was passed in 1986 to improve emergency response to accidental releases of toxic and/or hazardous chemicals into the environment. It establishes the Environmental Protection Agency (EPA) oversight role. By authority of this act, the EPA oversees State Emergency Response Commissions (SERCs), which, in turn, oversee Local Emergency Planning Committees (LEPCs). The act required the creation of these commis-

1. Do these plans provide for procedures for extinguishing different types of fires which might occur in the plant?
2. Do these plans have adequate evacuation and recovery procedures for each type of emergency?
3. Do the plans assign responsibilities to specific personnel to direct operations to counter emergencies? Are these persons aware of their responsibilities? Are they qualified to lead in the necessary actions which might be required?
4. Are emergency crews qualified, designated, and on station?
5. Are different communications channels assigned to support emergency operations? Is the emergency phone number widely and prominently displayed?
6. Are there plans to evacuate personnel from each work area in the event of emergencies?
7. Is information on evacuation routes and warning signals posted in each work area? Are the evacuation routes and exits marked?
8. Are the emergency plans and procedures posted in prominent areas?
9. Have personnel received training in emergency procedures:
   a. Workers?
   b. Supervisory personnel?
   c. Firefighters?
   d. Medical personnel?
   e. Communications personnel?
10. Are drills on simulated emergencies being conducted periodically for personnel?
11. Is there a procedure to ensure that all personnel have been alerted to the emergency and that those who will not combat it have been evacuated?
12. Are the egress provisions (i.e., doors, stairways, elevators) adequate for the evacuation in the event of an emergency?
13. Do all doors open in the proper direction to facilitate egress of personnel in emergencies?
14. Are there procedures to preclude obstructions to personnel or equipment in critical evacuation or emergency equipment access routes or areas?
15. Can egress routes from work areas be followed by personnel in the dark or in smoke?
16. Is the emergency equipment called out in the emergency procedures available at the facility, and is it operational? Can the equipment be reached easily if an emergency occurs?
17. Are warning systems (sirens, loudspeakers, etc.) installed and are they tested periodically? Are all personnel familiar with the meanings of warning signals and the required action to be taken?
18. Is there a fire detection system at each facility? Are fire extinguishers sized, located, and of the types required by OSHA standards, and are they suitable for the types of fires which might occur?
19. Is fire fighting equipment located near flammables or hazardous areas?

FIGURE 13–5 Checklist—Planning for Emergencies

sions and committees. Facility owners/operators that have on their premises chemicals designated under EPCRA as "extremely hazardous substances" must notify the SERC and LEPC if such a substance is in excess of EPCRA's threshold planning quantity. The owner/operator that has these chemicals on premises must appoint an emergency response coordinator who cooperates with the SERC and LEPC in preparing comprehensive emergency response plans.

EPCRA requires the facility to notify the SERC and the LEPC in the event of an accidental release of "extremely hazardous substances" or of a quantity greater than the threshold reportable quantity of a "hazardous substance" according to the Comprehensive Environmental Response Compensation and Liability Act (commonly called Superfund).

EPCRA requires a facility where a hazardous chemical, as defined by the Occupational Safety and Health Act, is present in an amount greater than a specified threshold to submit Material Safety Data Sheets (MSDS) to the SERC, LEPC, and local fire department. Facility owners/operators must make MSDSs available to the SERC, LEPC, and local fire department. They must provide to these groups inventories of chemicals and their location on their premises for which MSDSs exist.

BIBLIOGRAPHY

[57] W. G. Johnson, "The Management Oversight and Risk Tree," *Accident/Incident Investigation Manual,* prepared for U.S. Energy Research and Development, ERDA-76-20, Aug. 1, 1975.

[58] J. Ranill, "Cooperative Fire Fighting," *Chemical Engineering,* May 31, 1971.

EXERCISES

13.1. Why should personnel be trained to respond quickly when an emergency occurs where someone is injured? What types of persons in an industrial plant might respond? Could you? Do you know others in your plant who could provide adequate responses?

13.2. What basic things must an employer whose workers are exposed to hazardous chemicals do to comply with the Hazard Communication standard?

13.3. What types of injuries are considered "life threatening"? What types of injury might require immediate action but still not be life threatening? Would you know how to apply first aid?

13.4. Under what circumstances may it be necessary to move an injured victim? Would you know how to do it?

13.5. List as many items as you can to be considered during planning for emergencies. What types of emergencies might there be in your plant?

13.6. Representatives of what organizations should participate in planning against possible emergencies?

13.7. Are you aware of what the alerting signals are for emergencies in your plant? Have there ever been any drills for fire or other disaster?

13.8. Would you know the best route by which to evacuate the building or other structure in which you are currently employed? If you couldn't get out of a plant in a disaster, how might you safeguard yourself? Use one example of the type of disaster situation that might be envisioned.

13.9. Do you know the location of the nearest piece of emergency equipment? Who is designated to use it? Would you be able to do so?

13.10. Newspapers make frequent mention of emergencies, accidents, disasters, and catastrophes in industrial plants. List three that you have read about in newspapers or seen on television. What was the response and outcome of each? How might it have been avoided?

13.11. What are the priorities for what to save in a disaster?

13.12. What is a SERC? LEPC? When and how are they involved with industry?

CHAPTER 14

Accident Investigations

The findings of accident investigations have long been used to identify factors that affected safe operations of industrial activities. Processes, equipment, and safeguards have become more complex, and so have the causes and effects of failures and other factors (see Chapter 6) that lead to accidents. Thus accident investigations and the methods involved in determining causes have also become more complex.

For example, guards at one time were simple devices. Now even many of the mechanical ones are highly sophisticated. Electronic safety devices often consist of networks and components designed in accordance with Boolean logic as applied to switching circuits. A failure of such a safeguard would probably have to be investigated by someone knowledgeable in circuit analysis. A failure of this type occurred in 1973 with a train of the Bay Area Rapid Transit system in the San Francisco area. The accident that resulted was less severe than many that have occurred to trains since railroads were first built, but it is apparent that accidents will take place in spite of designs based on the latest technology. In January 1987 a railroad accident resulted in the deaths of 15 persons and injury to 176, although the rail line had had complex electronic systems installed. The capabilities of investigating personnel to determine the causes of these accidents had to be far greater than those required only a few years ago.

The Reason for Investigations

Many accidents are still of a nature which appear to have simple, easily determinable causes; but even here, changes in legal aspects may require investigative abilities never before required. For example, an accident occurs in which a chain breaks under a load: the load drops and hits a worker, who is killed. At one time the report of investigation may have described it in fairly simple terms and indicated that a contributory cause of the worker's death was an unsafe act on his part: he put himself in a dangerous position. Reports of a later period might also include as a contributory cause supervisory error: the supervisor should not have permitted the worker to put himself in danger. An attorney for the worker's dependents might have shown the supervisor knew the worker was in the dangerous position or knew the chain was bad and permitted the worker to continue with the operation and so was guilty of gross negligence. The attorney might also bring suit for the dependents against the manufacturer of the chain, claiming negligence in its manufacture. For this, analysis of the chain and the part where it broke might be required. The chain manufacturer might claim as a defense that the chain had been damaged through long use (if it was an old one), had been used improperly in that it was too small for the intended load or there had been a warning the load should not exceed a specific weight. It is apparent that in such cases there may be investigations by representatives of organizations with different interests, each attempting to ensure that whomever it represented is not held to be at fault. Groups with such interests might include, but not be limited to:

- The union that represents the worker, to ensure that blame is not attributed to the worker unless it is very evident he or she was the cause. The Airline Pilot's Association frequently objects to findings of the Federal Aviation Agency (FAA) that the blame for an aircraft accident was pilot error.
- The employer, who would be against any finding which indicated he or she had not maintained a safe place to work, or had been grossly negligent in any respect. It has sometimes been said that the person in charge of an investigation should be a supervisor of the operation where the accident occurred. However, because charges of supervisory error could be levied, a supervisor might be somewhat biased. The supervisor would be the suspect, the judge, and the jury. Also suspect because of possible bias would be the safety engineer, who, it might be contended, had properly failed to safeguard the plant. It therefore appears the best investigator in any company is a top-level manager from elsewhere in the organization. (*Note:* The head and most of the members of the team that investigated the accidental loss of the booster and space vehicle *Challenger* were mostly nonastronauts, although the astronauts were represented.)
- The employer's workers' compensation insurer, who might not be obligated to pay full benefits if it could be shown that the worker's negligence had contributed to the accident. The insurer might have to pay nothing if a third party could be shown to be at fault and liable for damages.
- The insurer of the third party (the chain manufacturer), who would endeavor to show that the accident could not be attributed to negligence on the manufacturer's part in producing the product which failed.
- State agencies, whose personnel would endeavor to determine whether any state regulations or codes had been violated.

OSHA, which would also try to determine whether any of its standards had been violated. Figure 14-1 shows an accident report form utilized in the past by the Department of Labor. Other federal agencies might also have investigations for matters under their jurisdiction. The National Transportation Safety Board might have someone to investigate accidents, especially major ones, which occurred to interstate carriers.

U.S. DEPARTMENT OF LABOR
Occupational Safety and Health Administration

1. CSMO NO.	2. OSMA-1 NO.
3. AREA	4. REGION
5. DATE	

ACCIDENT INVESTIGATION REPORT

SECTION I

EMPLOYER INFORMATION

6. EMPLOYER'S NAME		7. EMPLOYER CONTACT	8. TELEPHONE
3. ADDRESS	10. CITY	11. STATE	12. ZIP
13. TYPE OF BUSINESS	14. SIC NO.	15. PROCESS INTERRUPTION COST *(Est.)*	16. PROPERTY DAMAGE COST *(Est.)*

SECTION II

ACCIDENT FACTS

17. NAME OF INJURED ☐ M ☐ F	18. BADGE NO.	19. AGE	20. TIME OF ACCIDENT	21. DATE OF ACCIDENT	22. EXACT LOCATION OF ACCIDENT SITE

23. WORK ASSIGNMENT WHEN ACCIDENT OCCURRED	24. OPERATION INVOLVED	25. EXACT LOCATION OF ACCIDENT AT SITE
26. REGULAR OCCUPATION OF INJURED	27. EQUIPMENT INVOLVED	28. NAME OF SUPERVISOR AT TIME OF ACCIDENT

29. FATAL	30. TOTAL DAYS LOST	31. NO LOST TIME	32. NO INJURY	30. OTHER FACTS	34. FIRST NOTIFICATION OF ACCIDENT *(Phone, Radio, Paper, Other)*

35. BRIEFLY DESCRIBE ACCIDENT

SECTION III

CORRECTIVE ACTIONS

36. 1. What has been done to correct condition causing accident? If nothing, explain.

37. 2. What remains to be done? *(Investigator's comment)*

SECTION IV

STANDARDS INVOLVED

38. 1. Did a violation of a standard cause or contribute to the accident?

Check one ☐ Yes ☐ No If "yes," cite standard.

39. 2. Does the standard adequately cover the cause? ☐ Yes ☐ No

If "no," complete and submit OSHA-9.

SECTION V

40. WITNESSES

NAMES	ADDRESS	CITY	STATE	TELEPHONE

CSHO's Signature ______________________________

From OSHA-4
Aug. 1971

FIGURE 14–1 Department of Labor Accident Investigation Form

INVESTIGATING BOARD CHAIRMAN'S RESPONSIBILITIES

Figure 14-2 (from [84]) lists the duties and responsibilities of the chairman of the board named to conduct the investigation if the accident has been considered great enough. In some companies there are standing boards of members that can be called immediately should an accident happen. Rosters should be kept up to date, with a number more than might be required, including personnel who possess capabilities that might be needed. This is because members leave, change positions, retire, or die. Thought as to possible memberships should be given before an accident takes place, because as little time as possible should be lost before the investigation is begun.

CONTRIBUTING PERSONNEL

In any accident investigation chaired by a board, there will be other plant personnel who probably can contribute knowledge of what took place. Figure 14-3 (from [57]) indicates briefly how an investigation is to be conducted.

The safety professional must investigate each accident or near accident, but may be asked to provide information pertinent to an accident. Insurers, manufacturers, and government agencies may undertake their own investigations, participate in any being conducted by plant personnel, or review findings (with which they may not agree).

(1) Direct and manage the investigation.
(2) Assign tasks to members; establish deadlines.
(3) Use the abilities of a trained investigator to outline and expedite the work.
(4) Establish a "command post." Do not use your office or even your building. If feasible, separate the investigation office from regular work.
(5) Assure that the scene is safe and that investigation does not compound the event or interfere with emergency operations.
(6) Assure that the scene is secured until all evidence has been recorded or collected.
(7) Release the scene to management for rehabilitation and operation when possible.
(8) Handle requests for information, witnesses, technical specialists, laboratory tests, or administrative support with a liaison member of management.
(9) Handle all communications with the field organization and public officials. Remember, the field organization is responsible for public news releases.
(10) Keep the appointing official informed.
(11) Assure that the investigation does not function in ways which relieve line management of operational responsibility.
(12) Call and preside over meetings.
(13) Assure that all potential causal factors are studied.
(14) When the board has determined its findings, conclusions, and recommendations, supervise preparation of the report.
(15) Do not release the board until the report has been completed.
(16) Unless otherwise instructed in the appointment, before leaving the site, brief management on the facts determined (not conclusions and recommendations). Receive additional factual evidence, if offered by management, but revise the factual section of the report only as the evidence warrants.

FIGURE 14–2 Chairman's Responsibilities for Accident Investigations

(1) Initial actions. Getting started properly is very important; otherwise, evidence can be lost while the board is trying to organize itself. Following these sequential steps should assist in beginning an orderly investigation:

(a) Assemble the board for field organization briefing on synopsis of the occurrence and scope of investigation.

(b) When possible, assign tasks to board members then or while enroute to the accident scene. (If board members travel by different means or from different locations, do this as soon as possible after arrival.)

(c) Get a short briefing from the individual who has been controlling the accident prior to your arrival. Get local organization charts.

(d) Establish formal liaison with management.

(e) Go to the accident scene.

(f) Perform a general survey of the accident scene to get a "feel" for the accident.

(g) Prevent unnecessary handling or moving of evidence. Review security provisions.

(h) If personnel are readily available, find out what each witness might be able to contribute. Alert him or her to a possible follow-up interview.

(i) Photograph evidence and the scene.

(j) When needed, give the board a briefing on investigation methods.

(k) Establish command post and arrange for other needed resources.

(l) Finalize board organization and plan.

(m) Assign additional initial tasks or revise previous instructions based on the briefings you have received.

(2) Continuing tasks.

(a) Collect and preserve evidence.

(b) Interview witnesses.

(c) Prepare diagrams.

(d) Secure as-built drawings; copies of procedures, manuals, and instructions; maintenance records; inspection and monitoring records; alteration or change records; design data; material records; and personal histories.

(e) Conduct reenactment, where necessary or useful.

(f) Arrange for laboratory tests, where necessary or useful.

(3) The board should meet at least once daily to exchange information and coordinate results.

(4) An analysis of the accident goes forward simultaneously, both in the minds of board members and in analytics prepared by the trained investigator or consultants. The analytics help determine what additional information to seek and later help determine causes and recommendations.

(5) Prepare an outline, and start writing as soon as feasible.

FIGURE 14–3 Conduct of an Investigation

Each accident which involves any injury or in which there is damage greater than a specific amount (to be set by the plant manager) should be investigated. Besides members of the investigating board, there should be:

- The safety professional.
- The supervisor of the unit or area where the accident occurred.
- Any workers who saw or were involved in the accident.
- An employees' representative, such as a union committee member or the safety committee member from that unit or area.
- A representative from the plant engineering staff, if plant equipment or facilities were involved or damaged.

- And, if advisable, a specialist in investigating accidents of the type that caused the injury or damage. The services of the specialist or technical expert should be employed if the injury was serious, the damage level was high, or there might be litigation because of the accident. The exact number and type of personnel involved should depend on the severity of the injury or damage. In case of a near miss, investigations by the safety engineer, supervisor, and worker's representative, all of whom may be biased to some degree, would probably be adequate. If a fatality or very extensive damage has occurred, the inquiry should be conducted by the investigating board.

In no case should a plant manager attempt to put the blame for the accident on the workers, until and unless the causes have been determined through an investigation done adequately and properly.

CONDUCTING THE INVESTIGATION

Investigations require that certain equipment be available. Each member should have a copy of an accident report form to use as a checklist. Since the form must be completed for submission to the state and to the insurer, using it will also ensure that all minimum necessary information is obtained. Not only will insurers supply those forms on request, but other forms such as Fig. 14-1 are also available. This form indicates only that an accident has taken place, and it must be submitted rapidly. A proper investigation of a serious accident will take longer.

Investigators should also have on hand a notebook or paper on which to record any notes, comments, or information that might be pertinent. A measuring tape or rule is usually needed. A tape recorder is useful to record the comments of any person involved in the accident, witnesses, or any other person knowledgeable of factors pertaining to the accident. A video recorder would be helpful to record and show any pertinent objects, equipment, facilities, or other items. If such a record had been made before the accident, the postaccident film could be used to point out losses.

The investigation should be initiated as soon as possible after notification of occurrence. The investigator should attempt to avoid prejudging what may have happened. Prejudgments may lead to investigator failures to make proper use of witnesses' statements. Even where a statement contains highly significant information, it may be ignored if it does not meet an investigator's preconceived ideas.

Similarly, the investigator should not place total reliance on statements of only one witness or even of a person involved in an accident unless the statements are supported by physical facts or by other persons. Statements can be valuable. However, experience has shown that they can often be unreliable, increasingly so as time passes. This is due to a number of facts: persons have a tendency to see what they expect to see and to believe what they theorized had happened. In addition, accidents generally occur so rapidly there are often gaps in the witnesses' accounts, if they did not fill them in with events they believe took place.

There is a fair probability that when independent statements of fact from different witnesses are consistent, the fact is true. Here again, if there was much discussion between witnesses of the accident before their statements were taken, there was probably an unconscious influence of one person on another. Such discussions may have occurred with a person (or persons) involved in the accident. Fellow workers or witnesses with only a partial, incomplete knowledge may modify their recollections, since they would probably believe the person (or persons) involved knew more about what was going on than they. Yet, the workers (or persons) involved may have been in such a state of shock, events may have occurred so rapidly, or both, that they may not have actually known what happened.

Methods of Accident Investigation

A great deal of information, sometimes conflicting, may be gathered which must be sorted out in an orderly fashion. There is always at least one established, incontrovertible fact: an accident occurred. When, where, to whom it happened, and the results are known. The how and why it occurred are often more obscure and difficult to determine. A logical method of approaching the problem is to put down on paper a short phrase which indicates the final outcome: for example, "John Jones killed by falling load." Then, working backward, below this can be listed information that contributed to the cause of the accident. The information can then be separated into pertinent statements or facts which relate to each contributing cause: "Chain broke" and "Jones in dangerous position." The investigators may then want to know whether the chain broke from any readily apparent causes: was the load jerked or was any other unusual stress put on the chain; was Jones in the dangerous position for any reason? The depth to which the investigation should proceed may depend on the severity of the accident. Thus, if it had had serious effects, and the reason the chain broke is of interest, a laboratory analysis would be required. The investigators can indicate in their report that the chain's failure was unknown, and, if the matter was to be pursued further, that a laboratory analysis was to be undertaken.

Figure 14-4 (taken from [46]) is a diagram assembled from all the available facts about an accident investigated by the National Traffic Safety Board. The diagram indicates the relationship of events and causal factors after a tank truck partially filled with liquid oxygen (LOX) exploded. The same report also says: "Strict adherence to the rules for developing the diagrams is not necessary. Common sense works as well."

Management Oversight Risk Tree (MORT), another method for investigating accidents, is described in Chapter 15. The general technique, fault-tree analysis, is also given in Chapter 15.

ACCIDENT REPORTS

In a very high percentage of plants, accident reports are prepared by the supervisor concerned. In many instances the pertinent facts are then transcribed to forms which must be submitted to government safety agencies and to the insurer. As mentioned earlier, Heinrich estimated that 88 percent of all accidents were due to unsafe acts of the workers involved. The figures were probably due to a built-in bias. A supervisor or manager probably prepared the report and would be reluctant to indicate it was his or

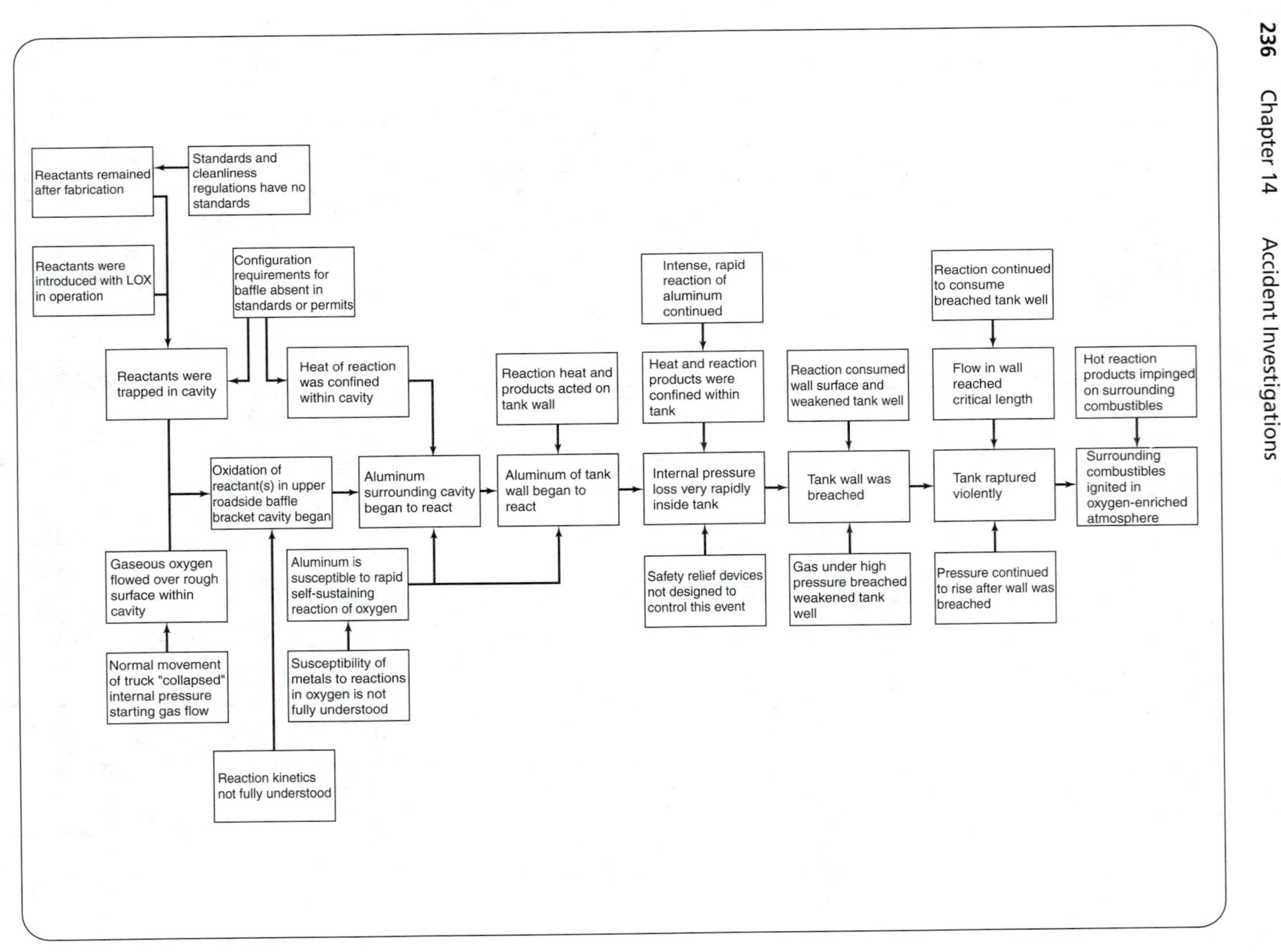

FIGURE 14–4 Relationship of Events and Casual Factors

her deficient actions that had caused the accidents. The accident investigation would probably also be oriented that way. A safety engineer might be less biased and therefore a more suitable person to evaluate and report the facts. Best, of course, would be a report prepared by a committee whose members would impartially represent all areas of interest.

How lengthy and detailed a final accident report should be has been a problem for a long time. (The initial report to the government agency or insurer is usually short and is used only to indicate that an accident has taken place.) Although the report should be kept as short as possible, it should be long enough to provide all the facts which contributed to the accident, so that corrective action can be taken. If any person involved in the accident was injured and is available, he or she should review the report to determine whether there is any objection to its contents. With the increase in claims and litigation because of accidents, attorneys may call for sworn depositions. The attorneys, often with technical experts to assist, will then proceed to pick apart in minute detail any claims or facts that will help further their cases.

CORRECTIVE ACTIONS

Here also concepts are changing. When supervisors fill out reports of accidents or accident investigations, they are often required to indicate the actions they have taken to prevent recurrences or occurrences of a similar nature. Generally, the corrective action stated is that the employee had been given a warning or additional instructions on how to carry out the work.

Two major aspects of accident prevention are usually omitted. The first is eliminating or controlling the hazard. Each accident (or near miss) must be evaluated to determine whether it would be possible to incorporate a suitable safeguard. Such a safeguard should either prevent a worker from performing incorrectly or else prevent an injury even if the person did act wrongly. Supervisors may not be qualified to assess such possibilities adequately and therefore will require assistance from a person who is technically qualified to do so: the safety engineer or a plant engineer. The safety engineer should be required to indicate whether a plant engineer can provide the safeguard, whether the company that supplied the deficient equipment should do so, or whether a procedural safeguard is the only solution. The second aspect of accident prevention usually omitted is "corrective action" by changing the procedures, including behaviors that led to the accident.

INSURANCE CLAIMS

The report submitted to the government agency or insurer may relate to either an injury (or fatality) to personnel, property damage, or both. The company may have suffered not only damage to equipment and facilities but also costly material and operational losses. In some instances there might have been no injury because of the accident, but only damage and other economic losses.

In either case it is necessary that all costs be certified and evaluated so suitable claims can be submitted to the company's insurer. The need for prompt and effective action is especially critical when a large loss has occurred to a moderate-sized company, especially one which has only one plant. Insurers may be especially critical if the company's financial posture is not the best and there is even a slight suspicion of possible arson. The regular company managers would probably be fully occupied with the problem of keeping the plant and company operational. It may therefore devolve on the investigating board, safety engineer, or fire investigators to determine what the losses had been. This job cannot be left to the insurance company's adjusters; their jobs are to minimize the size of claims against the insurance company.

Any person who attempts this job of preparing a loss claim will require assistance. The effort must be an integrated one, because no one person in the plant will be completely familiar with all the details or information which must be collected and recorded.

Equipment and facilities that were damaged or destroyed must be identified and evaluated. Supervisors and plant engineering personnel may have to determine whether equipment has been totally destroyed or is repairable, and if so, the extent of the repairs that will be required to restore it to its former capabilities. The plant engineering department may have to provide exact data on the type, size, and manufacture of the equipment; when it was purchased and at what cost; whether replacement parts are available; and the labor cost to make repairs.

Losses of raw materials, materials in process, and completed products may have to be determined from records or from estimates. It may be necessary to obtain information from suppliers, plant storekeepers, supervisors, or other personnel in many departments. Material that could be salvaged must be separated from that damaged beyond use and the degree of damage determined. The cost of the salvage operation itself must be computed.

Erection or use of temporary facilities to keep the company operating must be undertaken to replace damaged or destroyed facilities, equipment, or material before restoration work is begun.

If the investigation showed that the accident was due to faulty equipment or to a safeguard that failed that had been provided by a third party, the remains should be isolated and conserved. This would permit a closer search, more detailed examination, and possible identification and determination of the cause.

The labor costs in combating the emergency and in preventing any subsequent losses must be determined. Records of expenditures of material, such as medical and fire extinguishing supplies, must be collected. Each cost must be made available or prepared by the company accountant, or one hired by the company, to compute overhead charges for activities to which these apply. Knowledge of cost data is so necessary that companies often rent storage facilities in safe places, such as former salt mines, to maintain duplicate data in case of major catastrophes.

If an insurance adjuster is immediately available, he or she may be able to make a joint survey with plant personnel. Items on which there is agreement or disagreement can be identified immediately. However, under no circumstances should the matter be left entirely in the hands of the adjuster.

Insurance company officials have commented that, because of rising costs of labor and materials, losses due to accidents may be far higher than the original costs of the facilities, equipment, and materials. Current restoration costs must be known even before an accident takes place to ensure that coverage is adequate for potential losses, and they must be known after an accident to ensure that suitable and substantiable claims are made.

OTHER ASPECTS OF ACCIDENT INVESTIGATIONS

Any accident is an indication that the accident prevention program may not have been as good as it should have been. Safety engineers would do well to realize that this could reflect on their capabilities and efforts. It has sometimes been said that accident investigations are not undertaken to fix blame but to determine corrective actions to be taken. However, to determine corrective actions to be taken requires that blame for an accident be fixed. A proper accident investigation, though, will unearth its root causes. Rarely will blame lie at one place or person. For example, the root causes might lie in a cultural milieu into which many actors have bought. Whether disciplinary action should be taken if blame is fixed on an individual is beyond the duties of safety engineers, but those involved in an accident investigation should be careful not to engage in finding an easy target, a scapegoat. They are obligated to find out who was or may have been responsible for the accident—whether it was the worker(s), supervisor, manager, other third party(ies), or any or all of these. Finding the responsible party or parties is only the first step. The next step is to find out why she/he/they are responsible—then, to find out what corrective measures should be taken.

Although accident investigation does have its uses, it must be considered that the hardest way to learn about hazards and accident prevention is through accidents.

EXERCISES

14.1. What are the uses and needs for accident investigations? Who might call for an investigation? Should that person participate in the investigation?

14.2. Describe the various groups that might have an interest in an accident investigation. Why would some of these groups be interested? Which would you exclude because of their partiality?

14.3. The most publicized accident investigation took place after the explosion of the space shuttle *Challenger.* Was the membership of the board well constituted, or would you have eliminated or added anyone else?

14.4. Describe some facts to be considered in each investigation to be conducted. Give some methods of investigation analyses that have come into use.

14.5. Could an accident or any findings of an investigation be subject to litigation? What might happen if the findings were inadequate, attempted to hide facts, or were otherwise incomplete?

14.6. Why is it necessary that investigators have no preconceived ideas regarding the cause of the accident?

14.7. Why must witnesses' accounts be checked carefully? Why might they be erroneous?

14.8. What is a major reason a supervisor should not be the one to fill out the part of the accident report that indicates the possible cause?

14.9. Who is responsible for taking action to correct the condition that caused the accident?

14.10. Why is it not the best policy to simply accept any figures of cost of losses indicated by a claims adjuster? Cite the objections to such a practice.

CHAPTER 15

Safety Analysis

GENERAL

Safety analysis is a necessity in today's litigious society. Companies that do not analyze may not prevent accidents which are preventable and may find their ability to defend themselves in a postaccident lawsuit compromised. The standard to which the courts hold management is prudence, foreseeability, and reasonableness. Prudence requires that: (1) production systems, products, and facilities be analyzed for safety, and (2) proper preventive measures follow the analysis. A prudent manager analyzes. An attempt must be made to foresee the foreseeable. This attempt must be reasonable. The key is reason. Analysis is simply the application of reason. There are two types of reason: inductive and deductive. If a company can demonstrate that they made a reasonable attempt to foresee the foreseeable by applying system safety analysis inductive and deductive techniques and responding to the analyses with proper preventive actions, the company has a strong legal position. They can show they were reasonable and prudent in trying to foresee and avoid accidents, provided that they have carefully documented their efforts.

The philosophy developed by system safety engineers is that accident prevention can and must begin as soon as the idea for a new system or operation is conceived. Methods of analyses have been developed to this end which can be undertaken early. The analyses can indicate tentatively any hazards that might be present in any proposed operation, the types and degrees of accidents that might result from hazards, and the measures which can be taken to avoid or minimize accidents or their consequences. As more and more details become available after the initial analysis on the characteristics of the proposed system or operation, they can be expanded further and further to determine more intimately where there might be potential causes of accidents, their effects, and safeguards needed. Almost 90 techniques have been developed, used, or proposed.[59] This chapter includes a few important ones to indicate some of the methodologies now in wide use. This may make occupational safety engineers aware of some types of detailed analyses they might want to undertake.

Some of these techniques start at the component level and reason upward to find what "undesirable end events" might occur from that components fault or failure. These "bottom-up" techniques are based on inductive reasoning. Some techniques start with the undesired event which is to be avoided and reason down to identify which components of the system could contribute to the undesired event through their fault or failure. In systems where there is a critical potential for catastrophe, the prudent manager must make every effort to foresee and avoid the catastrophe. Everything is not foreseeable, but the manager must make a thorough effort to foresee when there is a potential for a catastrophe. This means the manager will use both inductive and deductive system safety analysis, if feasible. Preferably, the inductive analysis will be conducted blind to the deductive analysis, and vice versa. The results of the two analyses will then be compared. Hopefully, the undesirable end events identified in the inductive analysis will match those the deductive analysis started with; the component faults and failures identified in the deductive analysis will match those components the inductive analysis started with. Where the inductive and deductive techniques do not match, the deficient analysis is corrected. This double-blind cross-check is a strong indicator of a reasonable effort to foresee the foreseeable, and when accompanied by prudent corrective actions, places management in a very defensible position.

PRELIMINARY HAZARDS ANALYSIS

The first method generally used in the systematic process of determining hazard causes, effects, and deterrents is the preliminary hazards analysis. As its name implies, this is an initial study from which analysis efforts can be expanded further. It is fairly broad in scope, investigates what hazards might be present, whether they can be eliminated entirely, or if not, the best way to control them. If the hazard cannot be eliminated, the analyst determines whether there are standards or methods by which the hazard could, should, or must be controlled. A review is made of the functions to be performed and whether the environments in which they must be performed will have any adverse effects on personnel, equipment, facilities, or operations.

According to Greek mythology, Daedalus wanted to escape from Crete where he and his son Icarus were being held captive by the king because of his skill as an artificer. Daedalus made wings of feathers, flax, and beeswax with which he and Icarus could fly to Greece. Before they flew off, Daedalus told Icarus:

My boy, take care
To wing your course
Along the middle air;
If low, the surges
Wet your flagging plumes;
If high, the sun
The melting wax consumes.

Daedalus's instruction to his son can be taken to be an elementary preliminary hazard analysis. However, Daedalus, not being an experienced safety engineer, did not go far enough in his analysis. His analysis and the preliminary hazards analysis a modern safety analyst might prepare are shown in Fig. 15-1.

[It can easily be seen from this example that system safety engineers are smarter than designers. We safety engineers can figure out potential ways of getting killed which Daedalus didn't even consider, and we can even indicate safety measures by which he could have prevented Icarus's accident (he flew too close to the sun, the wax melted off his wings, and he fell to his death). Modern designers, on the other hand, are not nearly as smart as Daedalus; they have never been able to figure out how to make flyable wings out of feathers, flax, and beeswax.]

Each product or operation will have a limited number of hazards, which can be determined as soon as a few facts are known. For example, the proposed product is to use electrical power. The hazards, which can potentially be present when electrical power is in use, could include any or all of the following: electrical shock; burns from hot equipment; fire due to arcing, sparking, or very hot surfaces; inadvertent starts of equipment; failure of equipment to operate at a critical time so that an accident will occur; radiation effects; and electrical explosions. Numerous conditions can lead to each of these, but these are the basic hazards. If the electrical product uses small batteries, such as for transistorized equipment, most of the hazards mentioned will not present. The analysis can therefore be limited to those which could possibly exist, no matter how improbable. Progress of the design can be monitored thereafter to determine whether the hazard can be eliminated; and if it cannot be, how best it could be controlled; and if it is controlled, whether or not the control appears to be adequate and how the adequacy of the control can be verified. At this point, very rough estimates might be made of the probability of an accident due to that hazard, and of the severity of its probable effects.

A practice has been initiated in which the preliminary hazards analysis list, which is prepared in tabular form, is broken down into individual items. Each item is recorded (Fig. 15-2) and then tracked through development or modification of a system, product, or operation to ensure that adequate consideration is given to its hazard elimination/control. (To assist in the tracking, visualizations, such as in the fault-tree method described later, can be used.) When all necessary actions are completed to eliminate or acceptably control the hazard, the item is signed off, since no further action is required. (An item should never be signed off with a comment that an action will be taken, is scheduled to take place, or a similar entry. Signoff is considered accomplished only after the proposed action has been completed and shown to be adequate.)

As design or planning progresses, studies are made of the hardware and facilities, through reviews of assemblies of their major components, their proposed interrelationships and interfaces with each other and with personnel, environments that could affect them, and the effects they could generate on personnel, other components and assemblies, and the environment.

Some information does not lend itself well to an analysis presented solely in a tabular form. A narrative format can then be added to include addition data. For example, an analysis presented in columnar form might indicate that a hazard was an accidental fire. To have a fire requires a fuel, oxidizer, ignitable mixture, and source of ignition. Detailed information on each item could then be listed separately, with details on the many fuels, oxidizers, and sources of ignition.

PRELIMINARY HAZARD ANALYSIS

IDENTIFICATION Mark I Flight System

SUBSYSTEM Wings DESIGNER Daedalus

HAZARD	CAUSE	EFFECT	PROBABILITY OF ACCIDENT DUE TO HAZARD	CORRECTIVE OR PREVENTIVE MEASURE
Thermal radiation from sun	Flying too high in presence of strong solar radiation	Heat may melt beeswax holding feathers together. Separation and loss of feathers will cause loss of aerodynamic lift. Aeronaut may then plunge to his death in the sea.	Reasonably probable	Make flight at night or at time of day when sun is not very high and hot. Provide warning against flying too high and too close to sun. Maintain close supervision over aeronauts. Use buddy system. Provide leash of flax between the two aeronauts to prevent young, impetuous one from flying too high. Restrict area of aerodynamic surface to prevent flying too high.
Moisture	Flying too close to water surface or from rain.	Feathers may absorb moisture, causing them to increase in weight and to flag. Limited propulsive power may not be adequate to compensate for increased weight so that the aeronaut will gradually sink into the sea. Result: loss of function and flight system. Possible drowning of aeronaut if survival gear is not provided.	Reasonably probable	Caution aeronaut to fly through middle air where sun will keep wings dry or where accumulation rate of moisture is acceptable for time of mission.
Inflight encounter	a. Collision with bird	Injury to aeronaut	Remote probability	a. Select flight time when bird activity is low. Give birds right-of-way.
	b. Attack by vicious bird	Injury to aeronaut	Remote probability	b. Avoid areas inhabited by vicious birds. Carry weapon for defense.
Hit by lightning bolt	Bolt thrown by Zeus angered by hubris displayed by aeronaut who can fly.	Death of aeronaut	Happens occasionally	Aeronaut should not show excessive pride in being able to perform godlike activity (keep a low profile).

FIGURE 15–1 Preliminary Hazard Analysis

HAZARD REPORT

IDENTIFICATION/TITLE ____________________

EQUIPMENT/SYSTEM/SYSTEM ____________________

____

PERSON INITIATING REPORT: ____________________

REPORT NO. ____________________
DATE INITIATED: ____________________
DATE THIS REPORT: ____________________
SIGNATURE: ____________________

CLOSEOUT DATE: ____________________

DESCRIPTION OF HAZARD AND ACCIDENT WHICH MIGHT RESULT:

EVENTS AND CONDITIONS WHICH MIGHT CONTRIBUTE TO THE HAZARD OR ACCIDENT:

POSSIBLE MEANS TO ELIMINATE OR CONTROL HAZARD OR ACCIDENT EFFECTS:

ESTIMATED PROBABILITY OF ACCIDENT OCCURRENCE:

	CURRENT CONDITION	WITH CONTROL		CURRENT CONDITION	WITH CONTROL
FREQUENT	______	______	REMOTE	______	______
REASONABLY PROBABLE	______	______	EXTREMELY IMPROBABLE	______	______
OCCASIONAL	______	______		______	______

MEANS OF VERIFYING ADEQUACY OF CONTROL/APPLICABLE SAFETY REQUIREMENTS:

ORGANIZATION/PERSON TO TAKEN ACTION:

STATUS OF ACTION TO BE OR HAVE BEEN TAKEN:

FIGURE 15–2 Hazard Report

FAILURE MODES AND EFFECTS ANALYSIS

One of the methods used to accomplish safety analyses has been derived from reliability engineering: Failure Modes and Effects Analysis (FMEA). FMEA is an inductive technique. It is used to determine how long a piece of complex equipment will operate satisfactorily and what the effects of any failure of individual components might be. The method is intended for analyses of proposed equipment and systems. An occasion might arise in which it would be highly desirable for an occupational safety engineer to determine whether or not the manufacturer of industrial equipment and systems to be purchased and installed had had such an analysis made, and to understand how failures might occur, modes and frequencies of failures, and the necessity for proper and timely maintenance and replacements.

In this method of analysis, the constituent major assemblies of the product to be analyzed are listed. Each assembly is then broken down into subassemblies and their components. Each component is studied to determine how it could malfunction and cause downstream effects. Effects might result on other components, and then on higher-level subassemblies, assemblies, and the entire product or system. Failure rates for each item are determined and listed. The calculations are used to determine how long a piece of hardware is expected to operate between failures, and the overall probability that it will operate for a specific length of time. It is the best and principal means of determining where components and designs must be improved to increase the operational life of a product. It is best used to analyze how often and when parts must be replaced if a failure, possibly affecting safety, must be avoided.

Until the use of Boolean mathematics, described in the next paragraph, FMEA calculations were often erroneous. Also, because many component failures would have no effect on safety, that aspect of an FMEA does not involve accident possibilities. Also, failure modes and effects analysis is limited to determination of all causes and effects, hazardous or not. Further, the FMEA does very little to analyze problems which could arise from operator errors (unless the system analysis includes the human as a component), or hazardous characteristics of equipment created by bad design or adverse environments (unless the scope of the analysis includes these elements). The FMEA is excellent for determining optimum points for improving and controlling product quality. Another method, fault-tree analysis, has been found more effective for safety purposes.

FAULT-TREE ANALYSIS (FTA)

In 1959 the Air Force became concerned with the potentially catastrophic events that could occur with the Minuteman missile then being developed by Boeing. The Air Force contracted with Bell Laboratories to develop a method of analysis by which probabilities of occurrence of events with which they were concerned could be computed. The Air Force wanted to know the possibilities and probabilities that a missile could be launched or a warhead activated inadvertently, and the chance of either of these being done intentionally by an unauthorized person in an act of sabotage.

Two years later Bell Laboratories completed the project. The new method, fault-tree analysis, involved Boolean logic in ways similar to those being used increasingly in electronics industries. It is a deductive technique. Safety engineers of the Boeing Company adopted the proposed method and became its foremost proponents. They added to the symbols that Bell Laboratories had proposed. They organized the method so it could be computerized to permit calculating the probabilities of the problem with which the analyst was concerned. Although still used to determine probabilities of mishaps in complex systems or operations, fault-tree analysis is being used far more frequently to logically analyze the possibilities of potential accidents due to hazards, and the quantitative safety level.

At first the fact that Boolean logic (and arithmetic), with which few engineers were familiar at the time, was its basis gave many people the impression the method could be used only by mathematicians. (Some of the first Boeing employees involved in making those early safety analyses were mathematicians.) It turned out that Boolean logic was fairly simple to understand. One of the fundamental principles is that any statement, condition, act, situation, or process could be described as being in either one of two states. Something could be true or false, on or off, up or down—but not both; it would go or not go; or it could be fully open or not fully open (closed or partially open). There were no middle or intermediate positions.

FAULT-TREE SYMBOLS

Primary Event Symbols

- Basic Event, such as a component failure—*Circle.*
- Conditioning Event, a condition which must be satisfied before the event above the gate can occur—*Ellipse.*
- Undeveloped Event, an event not fully developed because of a lack of information or significance—*Diamond.*
- Normal Event, expected to occur during system operation—*House-shape.*

Intermediate and Top Event Symbol

- Output Event, those that should be developed or analyzed further to determine how they can occur (the top output event is called the end event; the others are called intermediate events)—*Rectangle.*

Basic Gate Symbols

- AND condition (or gate): All events leading into it from underneath must occur before the event leading out of it and at the top can occur. In a fault-tree Boolean equation, the AND condition is generally expressed by AB, A × B, or (A) (B) = C.
- OR condition (or gate): Any event leading into it from underneath will cause the event leading out of it at the top to occur. In a fault-tree Boolean equation, the OR condition is generally expressed by +, such as A + B = C
- A transfer symbol: the events which will occur where this is shown will be the same as where the inverted triangle with the same identification is shown—*Triangle.*

The either and-or, on-off situation applies to any condition, event, or other occurrence, but there can be any number of conditions, events, or occurrences. For example, a fire requires: (1) a fuel, (2) oxygen, and (3) sufficient heat to ignite. The fuel, oxygen, and ignition heat each either will be present or will not be present. But they must all be present to have a fire; if one is not, there will be no unacceptable risk of fire. The three occurrences are bound together in a so-called "CONDITIONAL AND" gate. The condition is that the ambient fuel and oxygen is within flammable range (see Fig. 15-3, G5). The CONDITIONAL AND is shown in Fig. 15-3 as an AND gate with a triangle inside (below G1).

In contrast to the "AND" gate is the "OR" gate. The OR gate may be simply a restatement of the gate's output, or it may have two or more events as inputs. When two or more events or conditions are possible as alternatives, any one will produce the specific result envisioned. All of the occurrences need not be present at the same time, as they would with an "AND" condition. The use of some of these symbols is shown in Fig. 15-3, a simplified Fault-Tree Analysis (FTA) of a top event of an unacceptable risk of fire with a system consisting of a fuel tank and fuel line. A fuel could be any one of hundreds. Figure 15-3 arbitrarily limits the analysis to fuel from a gasoline fuel tank. Tank leaks and line leaks are shown as primary component faults, indicated by the circle symbol at G6 and G7. Since air is normally present in the ambient for which the example fault tree is drawn, "Oxygen in the air" is shown in a "house" symbol at G3. A source of ignition could be a lit match, an open pilot light, a hot surface, a mechanical spark, an electrical arc, and so on, but it must be a source of heat sufficient to heat the fuel until self-sustained combustion occurs (for example, to greater than 536°F for 60 octane gasoline). Any one ignition source could contribute to a fire, but there must be one. The fault tree in Figure 15-3, for sake of brevity, shows that the analyst did not wish to pursue the identification of sources of ignition; therefore, there is a diamond below "Ignition Source >500 F Present" at G4. Ordinarily the identification of ignition sources would occur and be shown on the tree as inputs to a gate below the G4 event. "Fuel spill during operations" is elected not to be pursued further and is shown with a diamond at G8. Gates are numbered in this figure, and the tree is drawn using CAFTA for Windows.[60]

A fault-tree analysis works like this: the end effect (called the top event), such as a specific type of injury, accident, damage, or occurrence whose possibility is to be analyzed, is selected. The top event may be chosen from a known potentially hazardous condition, determined by a previous analysis (for example by a preliminary hazards analysis), an undesirable occurrence, or a specification requirement. (A few examples are shown in Fig. 15-4.) A specification requirement might state: "Reactor temperature will not exceed 600°C." (The designer prepared the specification on the basis that temperatures exceeding this level are undesirable and may lead to a potentially hazardous condition.) The top event for the fault tree might then be: "Reactor Temperature Exceeds 600°C." The analysis effort is then oriented to establishing those possible conditions, failures, errors, and other events which could contribute to or permit a temperature greater than 600°C to occur.

The analyst determines whether or not these contributory conditions and events must all occur (AND condition) to cause the top event or whether or not any of them alone can do it (OR condition). Then each contributory condition or event is analyzed further in the same way.

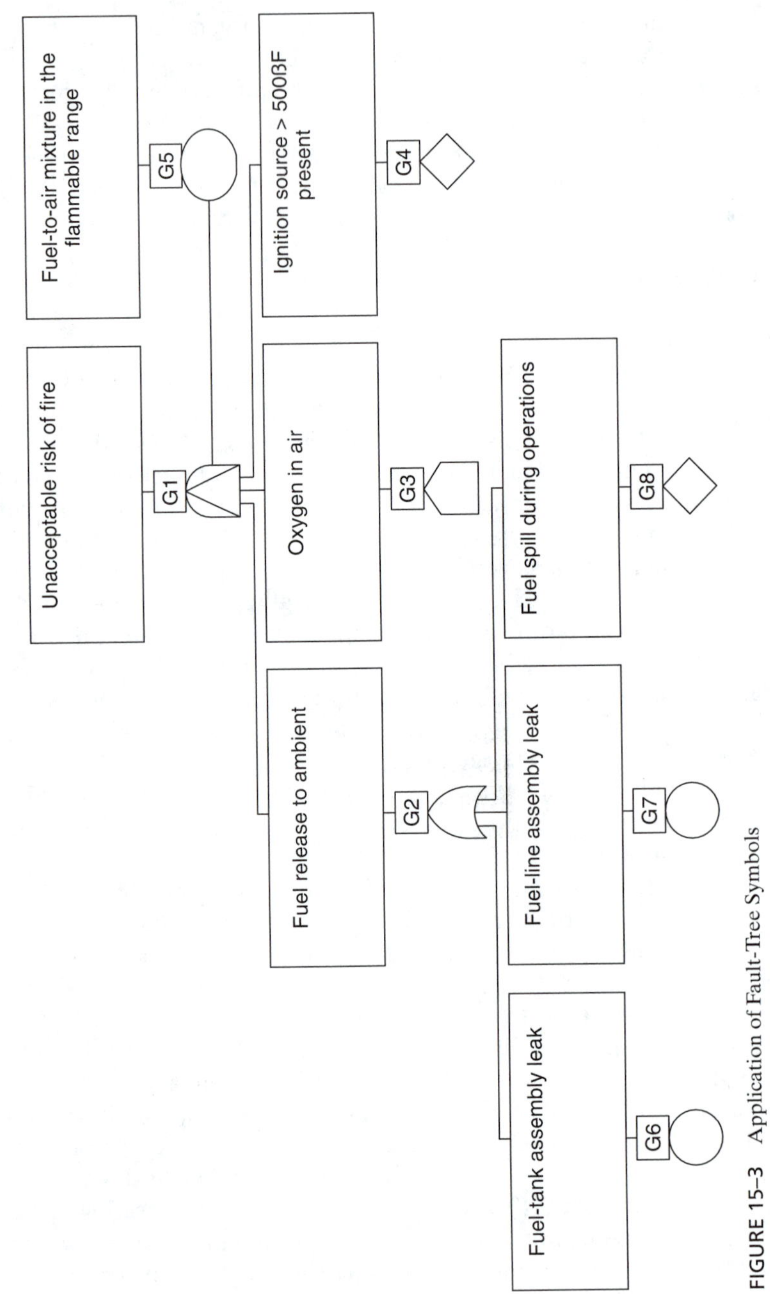

FIGURE 15–3 Application of Fault-Tree Symbols

SAMPLE TOP EVENTS FOR FAULT TREES

Injury to .
Radiation injury .
Inadvertent start of .
(Equipment to be named) activated inadvertently.
Accidental explosion of .
Loss of control of .
Rupture of .
Damage to .
Damage to from .
Thermal damage to .
Failure of to operate (stop) (close) (open).
Radiation damage to .
Loss of pressure in .
Overpressurization of .
Unscheduled release of .
Premature (delayed) release of .
Collapse of .
Overheating of .
Uncontrolled venting of (toxic, flammable, or high-pressure gas).
(Operation to be named) inhibited by damage

FIGURE 15–4 Sample Top Events for Fault Trees

If this procedure is continued and laid out in a diagrammatic form, it will develop into a "tree" (which looks more like tree roots). As the tree proceeds down from the top event, "branches" are created. Proceeding down the tree and branches, an analyst can see what the contributing causes might be. Conversely, going up a branch indicates the effects resulting from any notation or combination below.

The symbols used by some fault-tree analysts are more extensive than shown or discussed in this chapter. For example, there are several gates in addition to the AND gate and the OR gate, and gates and events are numerically coded. In another usage, each box is numbered; a separate table is prepared in which the numbers listed contain more detailed information on that event, action, or condition. The safeguard action recommended can be described.

Fault-Tree Construction Rules. There are several ground rules and procedural rules for fault-tree construction. One basic ground rule is to write a fault as a fault. This is a fault tree, and the events within the tree, except those normal events shown as a house, are fault events and should be written as faults. Procedural rules are: (1) no miracles, (2) complete the gate, and (3) no gate-to-gate lash-ups. "No miracles" simply means that what normally happens will happen and the analyst cannot invent some way out. "Complete the gate" means that the analyst completes all inputs to a gate before proceeding to develop the tree elsewhere. "No gate-to-gate lash-ups" means that there is always an intermediate event (placed in a rectangle) between gates.

Figure 15-5 shows the top levels of a fault-tree analysis of a powder-actuated fastener tool (similar to a nail gun) for the top undesired event, "PERSONNEL INJURY UPON FIRING OF TOOL." Three personal injury events are foreseen in the use of the tool on a prescribed underwater construction task of interest.[61] Personnel could be struck by the fastener, by fragment(s) of the workpiece, or by the tool itself. Each is developed using the FTA construction rules mentioned above. The first event, "PERSONNEL STRUCK BY FASTENER . . . ," has a path below it, through an AND gate and below it an OR gate. The OR gate output is FASTENER IN FREE FLIGHT; it has three inputs—INADVERTENT FIRING, FASTENER PASSES THROUGH WORKPIECE, FASTENER RICOCHET. These inputs are developed further elsewhere, as indicated by the transfer symbol. For example, the top event of a tree for INADVERTENT FIRING would transfer in to the same event shown at the triangle marked "1". Space does not permit showing the rest of the tree referred to by the transfer symbols. These transfers have trees at least as extensive as the top part of the tree shown in the figure. Thus, the reader can imagine why these are called "trees," because the logic causes the formation to spread wider at the base, giving the appearance of a Christmas tree triangular shape.

When all the branches and their interrelationships are considered, the analyst can arrive at all the possibilities of the top event occurring. Sometimes it may be desirable to analyze only one branch or sequence at a time. For example, in Fig. 15-5, the sequence on the tree beginning: "PERSONNEL IMPACTED BY TOOL UPON FIRING" has an OR gate with one input that is a restatement of the output. Below the restatement is an AND gate with two events as its input, "OPERATOR IN RECOIL PATH" and "SEVERE TOOL RECOIL." If both occur together, there is some probability of injury. That path leads directly to the top undesired event. Such a combination

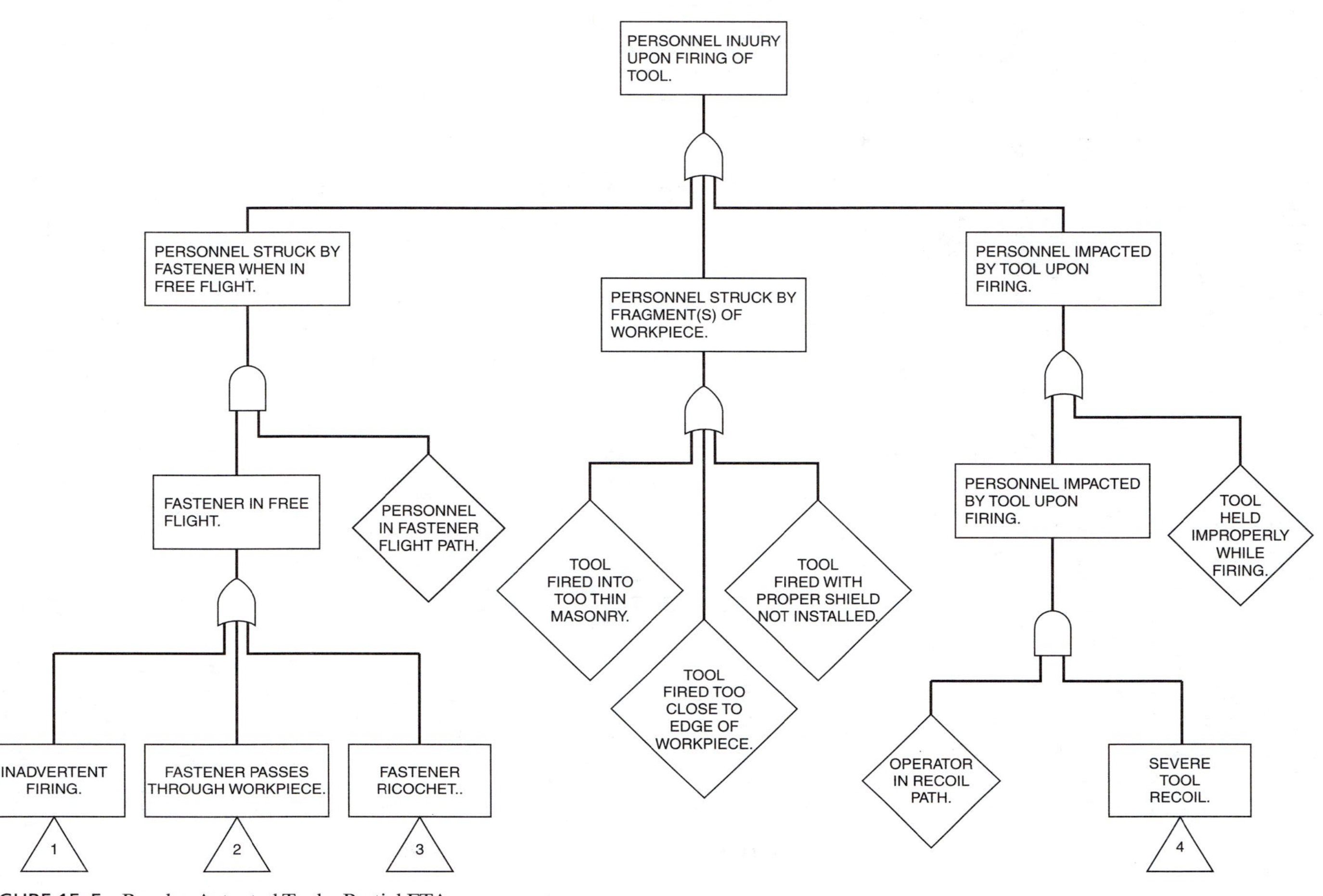

FIGURE 15–5 Powder-Actuated Tool—Partial FTA

of events that lead to the top is called a *cut set.* A minimal cut set is made up of the fewest events in a path to the top event.

Some times only one thing has to happen to reach the top undesired event. This is called a "single-point failure" and is a very undesirable system feature. There are four single-point failures in Fig. 15-5. Can you find them? (*Hint:* A single-point failure will have only OR gates between it and the top event). A tree can be evaluated qualitatively by finding all cut sets and ordering them by single, double, triple, . . . , *n*-point failures. In general, safer systems will have few, if any, low *n*-point failures. Single-point failure events must be examined most critically. Corrective action by a design change is often advisable, optimally by interposing an AND condition (such as a simultaneous action to be taken by an operator).

All of the possible sequences which could result in a top event, such as injury upon firing the tool, can be found. If a probability of each of the initiating bottom events is applied, it becomes possible to calculate the top events. Which sequence is most likely to occur can be determined. Or if there are two or more cut sets (sequences), their probabilities can be compared. If a probability of any calculated event is high or excessive, corrective steps could be taken to lessen it.

Quantitative Fault Trees

With a quantitative analysis, it is not correct to add the probabilities of the cut sets to obtain the overall probability of the top event. Too many duplications (redundancies) would distort the result. Therefore it is necessary to eliminate the redundancies by using Boolean mathematics (see Fig. 15-6). Sometimes the probability of an accident might be lessened adequately for acceptance by increasing the reliability of the parts and equipment involved. The probability of occurrence must be assessed and a determination made whether the risk probability of the event occurring is acceptable.

A redundancy exists when the same event is shown more than once (which may happen in two or more branches), or one event may be the subevent of another. Boolean mathematics permits elimination of redundancies. If it is not done, the ultimate probability value will be in error (Fig. 15-6).

When a quantitative analysis is to be made, each entry on a tree must be expressed in language or a symbol which permits a probability to be applied. Boolean mathematics is then used in writing and simplifying equations by which the probability of the top event can be expressed in terms of probability of the initiating events and their relationships.

FTA methodology can be applied to almost any system or operation. Very detailed analyses of very complex systems can extend down to the level of the individual component, human action, environmental effect, or hazardous characteristic to show their cause-and-effect relationships as they contribute to the potential problem which has been selected as the top event.

Fault Tree for Investigations

A fault-tree analysis can be used in accident investigation by selecting as the top event the accident which occurred. A tree could be prepared listing all possible contributing factors. These factors are then examined one by one to determine whether or not they

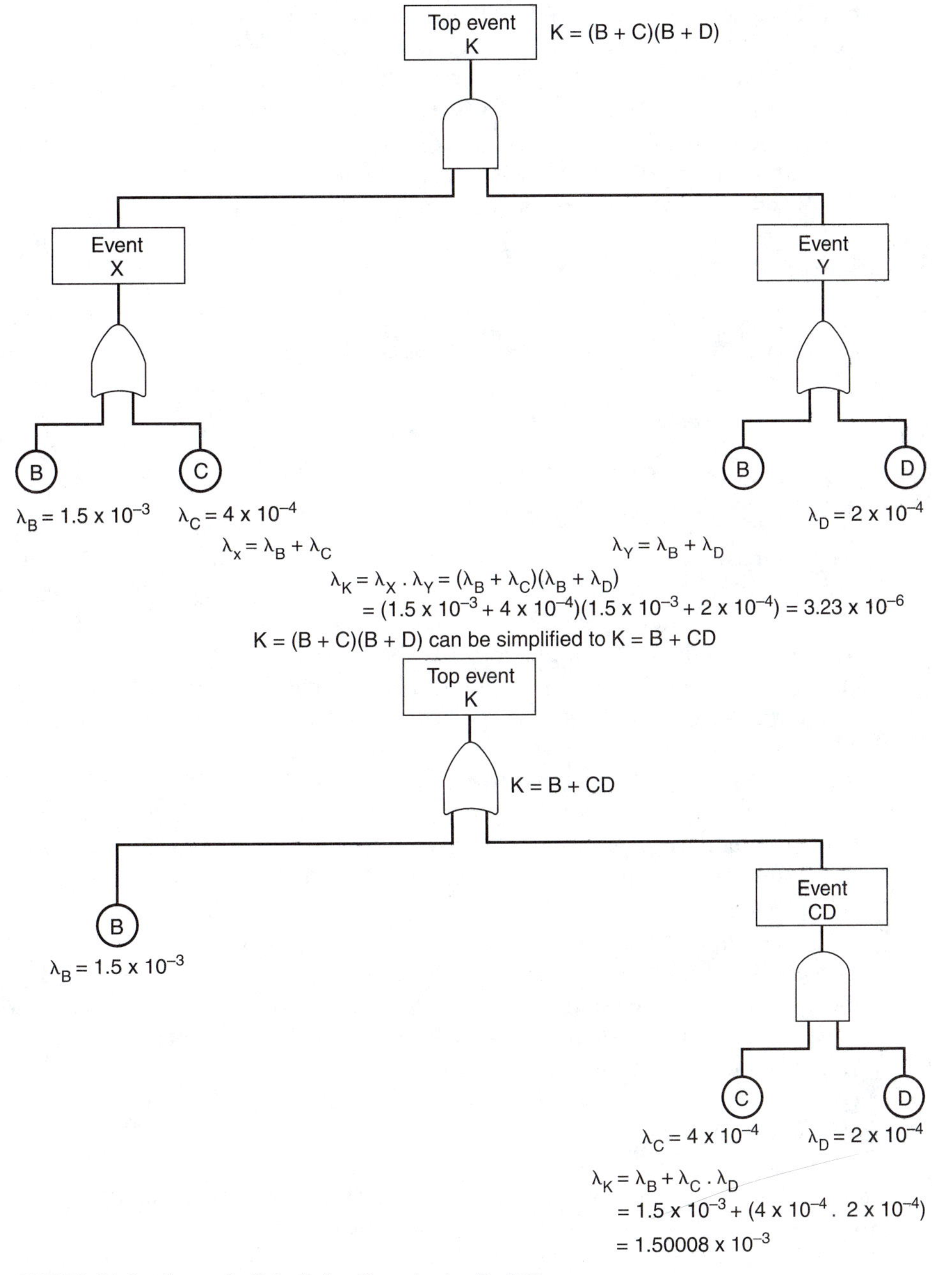

FIGURE 15–6 Correctly Calculating Quantitative Fault Trees

could possibly have occurred. If they had not been possible in the accident situation being analyzed, they are eliminated. The remaining events of the tree indicate what the initiating cause might have been. Probabilities assigned to each possibility would permit preparation of a list of priorities by which investigations for previously undetermined facts should be made. A generic, "universal" logic tree for accident investigations and safety program evaluations is Management Oversight Risk Tree (MORT). MORT is based on fault-tree analysis concepts.[62]

Limitations of Fault-Tree Analyses

One of the very few limitations of fault-tree analysis is that the tree indicates only the events included which will contribute to the occurrence of the top event. Should the analyst fail to include any possible event, the analysis is inadequate.

SAFETY ANALYSIS METHODS MANDATED FOR PROCESS SAFETY MANAGEMENT

Safety analysis techniques are now mandated by regulations for some industries. This program requirement is examined below, and the techniques it suggests that are not already covered in this chapter are described and explained.

In the mid-1990s, industries with processes that involved "highly hazardous chemicals" (HHCs) were placed under requirements by OSHA (29 CFR 1910.119) to conduct process hazard analysis and have a process safety management program.[63] The program is triggered by above-threshold quantities of any of 136 different, listed chemicals. The purpose of this standard is to prevent or minimize the consequences of a catastrophic release of toxic, reactive, flammable, or explosive highly hazardous chemicals from their use, storage, manufacturing, handling or on-site movement by industry.

The elements of a Process Safety Management Program (PSM) are shown in Table 15-1.

The importance of this standard is that it requires safety analysis and names certain analytic techniques to use or their equivalent. "The employer shall use one or more of the following methodologies that are appropriate to determine and evaluate the hazards of the process being analyzed. . . . What-if, Checklist, What-if/Checklist, Hazard and Operability Study (HAZOP), Failure Mode and Effects Analysis (FMEA), Fault Tree Analysis, or an appropriate equivalent methodology" [29 CFR 1925.64(e)(2)].

What-if. In "What-if?" analysis, an analysis team reviews the covered process from beginning to end. At each step in the process they think of questions that begin with "What-if. . . ." They address procedural, hardware and software errors and produce a tabular listing of these questions and their answers. This constitutes a set of potential accident scenarios, their consequences, and remedial recommendations. Report form can be narrative, but use of a matrix table often provides clarity.

Checklist. The "Checklist" method applies a list of specific items, usually taken from industry standards, consensus codes, industry guidelines, etc. The list is prepared by an experienced safety engineer, familiar with the covered process and the

TABLE 15–1 Elements of a Process Safety Management Program

Elements	Description
Process Safety Information	Requires compilation of written process safety information, including hazard information on HHCs, and information on technology and equipment of the covered process.
Employee Involvement	Requires a written plan of action for involving employees inthe conduct and development of process hazard analyses and process safety management, and for providing employees access to the information required by the standard.
Process Hazard Analysis	Specifies that process hazard analysis be conducted and updated and revalidated at least every five years. Must be retained for the life of the process.
Operating Procedures	Requires these be in writing and include clear instructions for safely conducting activities of the covered process. Steps for each operating phase, operating limits, safety and health considerations, and safety systems and their functions must be readily accessible to involved workers. They must reflect current operating practice, and must implement safe work practices for special circumstances such as lockout/tagout and confined-space entry.
Training	Operators of the covered process must be trained in the overview of the process and in the operating procedures. Training must emphasize specific safety and health hazards, emergency operations, and safe work practices. Refresher training is required at least every three years.
Contractors	Contract employers are required to train their employees to safely perform their jobs at the process site. Must document that employees received and understood training and assure that contract employees know about the potential process hazards and the worksite employer's emergency action plan. Must assure that contract employees follow safety rules of the facility. Must advise the worksite employer of hazards that contract work itself poses or hazards identified by contract employees.
Prestartup Safety Review	A safety review is required for new facilities and significantly modified work sites to confirm that the construction and equipment of a process are to design specs. Must assure that adequate safety, operating, maintenance, and emergency procedures are in place and that process operator training has been completed. For new facilities process hazard analysis must be performed and its outstanding action items resolved and safety implemented before startup. Modified facilities must meet management-of-change requirements.
Mechanical Integrity	Written procedures must be established and maintained for the ongoing integrity of process equipment, particularly those components which are part of a covered process.
Hot Work	Hot work permits must be issued for hot work operations conducted on or near a covered process.
Management of Change	Written procedures must be established and implemented to manage changes to facilities that affect a covered process. Employees must be informed of changes.
Incident Investigation	Requires employers to investigate as soon as possible, but no later than 48 hours, after incidents which did result or could reasonably have resulted in catastrophic releases of covered chemicals. An investigation team which includes at least one person knowledgeable in the process involved and others with knowledge and experience in incident investigation and analysis must develop a written report on the incident. Reports must be retained for five years.
Emergency Planning and Response	An emergency action plan must be developed and implemented. It must include procedures for handling small releases.
Compliance Audits	Employers must certify that they have evaluated compliance with process safety requirements at least every three years. Prompt response to audit findings and documentation that deficiencies are corrected are required. The two most recent audit reports must be retained.
Trade Secrets	Requires that certain information be available to employees. Employers may enter into confidentiality agreement with employees to prevent disclosure of trade secrets.

company's policies and procedures. The items are usually in form of a questionnaire to be applied to the covered process or system. Those parts of the design which are not adequate should be revealed when the checklist is applied. Then corrective measures are identified.

What-if/Checklist. The What-if/Checklist Analysis uses the list of What-if questions and supplements it with a checklist. Therefore, it is a combination of the above two methods, as its name suggests.

Hazard and Operability Study (HAZOP). HAZOP addresses each element in a process.[64] The elements must first be identified. The expected way each element operates is determined. Deviations from the intended or expected operations and parameters are identified using guide words. The consequences from the deviating element are estimated. Causes of the deviations, currently planned (or used) hazard controls, and inadequate controls are identified. The study is conducted by a team made up of five or six persons usually including a team leader familiar with HAZOP, a safety and health expert, a manager/supervisor, a technical person, a senior process operator, a maintenance person, a chemist, etc. The quality of the results will depend upon the quality of the team selected.

The team will start by collecting process information from piping and instrumentation diagrams (P&ID), written operating instructions, material safety data sheets, plant models, Safety Analysis Reports (SARs), etc. They then review each element step by step through each line or process. At each element they apply key words or guide words to tease out the potential deviations from normal expectations of a design parameter. The guidewords are shown in Table 15-2.

The hazard and operability consequences, if any, for each deviation are determined and documented. The actions or changes in plant methods which will prevent the deviation or reduce the consequences are identified. If the cost of the changes can

TABLE 15–2 HAZOP Guide Words

Key Word	Meaning	Comment
No/Not	Complete negation of design intention	No part of the design intention happens, e.g., "No flow," "No pressure," etc.
More, less	Quantitative increase or decrease	Refers to quantities, properties, and duration of design parameters, such as flow rate, temperature, etc.
As well as	Something else occurs in addition to design intention	Design intentions are achieved, plus additional things occur.
Part of	Only part of design is achieved	Partial addition or removal of material occurs. Activities are incomplete.
Reverse	The opposite of the design intention	Reverse flow of material, electrical current, voltage polarity. Backward installation (e.g., check valve), opposite chemical reaction (e.g., decomposition).
Other than	Complete substitution	Original design intention is not accomplished; a different activity occurs—e.g., leak, rupture.

be justified, the team then must agree on the change and who will be responsible for the action. The team then must follow up to make sure that the action has been taken. If the cost is not considered to be justifiable, the team must agree to accept the risk. One or more reports of the evaluation are made, usually including tabular representations of the results. Team decision making is at the heart of this methodology.

This technique as well as the others in this chapter are supported by software. The reader is advised to "surf the net" for the latest. Arthur D. Little Company is one source.[65]

Job Safety Analysis (JSA). The purpose of this technique is to uncover and correct hazards which or intrinsic to or inherent in the workplace. It should be done with a team that includes the worker, supervisor, safety engineer, and management. Its success depends upon the rigor this team exercises during five steps: (1) Select a job, (2) Break the job down into steps, (3) Identify the hazards and determine the necessary controls of the hazards, (4) Apply the controls to the hazards, and (5) Evaluate the controls.

Eliminating, controlling, and minimizing hazards are becoming more and more necessary for work plants to comply with the standards of regulatory agencies. Formerly, chemical plants were built in which hazardous environments existed and where operators were safeguarded, theoretically, by use of protective equipment, such as gas masks, whose need was established after operations had begun. Some government agencies, such as OSHA, have adopted policies of "deterrence" and will not countenance the existence of hazardous conditions. Intrinsically safe environments are being required in which operators can work without the need for protective equipment, which is often inadequate, and which workers find burdensome and inconvenient and will not wear. To ensure that regulatory agencies do not require extensive and expensive retrofits of corrective equipment, it is highly advisable that analyses be made early to recognize potential problems and their solutions. The billions required to revise and correct designs and procedures at Three Mile Island and other nuclear power plants are a prime example of need for proper and adequate analysis and planning for safety purposes.

BIBLIOGRAPHY

[59] System Safety Society, *System Safety Analysis Handbook,* July 1993.

[60] Electric Power Research Institute and Science Applicatons International Corporation, *CAFTA for Windows Fault Tree Analysis System User's Manual,* July 1995.

[61] W. H. Muto, B. A. Caines, and D. L. Price, *System Safety Analysis of an Underwater Marine Tool,* VPISPO/NAVSWC-78-1, 91 pp., 1978

[62] D. Conger and K. Elsea, *MORT User's Manual,* 1998, Conger & Elsea, Inc., 9870 Highway 92, Ste. 300, Woodstock, GA 30188.

[63] "29 Code of Federal Regulations 1910.119, Process Safety Management of Highly Hazardous Chemicals," *Federal Register,* vol. 57, no. 36, Feb. 24, 1991.

[64] C. Bullock, F. Mitchell, and B. Skelton, "Developments in the Use of the Hazard and Operability Study Technique," *Professional Safety,* Aug.1991, pp. 33–40.

[65] Arthur D. Little, "HAZOPtimizer: Documentation and Reporting Software," 1991.

EXERCISES

15.1. For what purpose is a preliminary hazards analysis used? When should it be prepared? Describe the information that might be included.

15.2. For analysis of what type of problems is failure modes and effects analysis beneficial? What aspects does such an analysis not include?

15.3. Describe the steps in preparing a fault-tree analysis.

15.4. Give an example of a situation involving an AND condition for an event to occur. An OR condition.

15.5. How can the top events for fault trees be selected?

15.6. Following are hypothetical specification requirements for equipment. Indicate what the top event might be if a fault tree were prepared for each:

NO WORKER INJURY BECAUSE CRANE FAILS.

ELEVATOR IS NOT TO MOVE WHILE DOOR IS OPEN.

EMERGENCY BUTTON TO STOP OPERATION OF MACHINE.

15.7. What is a single-point failure? Give an example.

15.8. What is a cut set? What would be the relationship between a cut set and a single-point failure?

15.9. How can a fault-tree analysis be used for accident investigations?

15.10. What are the advantages of performing safety analyses before equipment is designed or purchased or an operation begun?

15.11. What are the elements of a Process Safety Management program? (*Note:* This program mandates that system safety analysis techniques be used.)

15.12. What is HAZOP? How is it performed?

CHAPTER 16

Acceleration, Falls, Falling Objects, and Other Impacts

Injuries due to acceleration (or in its negative sense, deceleration) have been common since the first human fell out of a tree or off a cliff. The injuries which resulted from falls and other impacts are the principal effects of acceleration (deceleration).

Data for types of "nonfatal occupational injuries and illnesses involving days away from work" list five sources of injury which can probably be classified as resulting from falls or impacts (Bureau of Labor Statistics, 1996):

Source	Percent
Contact with objects and equipment	26.2 percent of all cases
Struck by object	12.7
Struck against object	6.8
Fall to lower level	4.2
Fall to same level	11.7
Total	61.6 percent of all cases

The great number of impact accidents and the results of tests have been major factors in the demand that corrective measures be used. This has led to the passage of federal and state laws requiring suitable safeguards, which resulted in substantial reductions in such injuries. Notable actions (among many others) are the use of safety nets and protective helmets (hard hats) for workers or other persons who might be hurt by falling objects and of seat restraint devices for vehicle drivers and passengers. Subsequent data therefore presents information on what might occur if protective measures were not taken.

FALLS

In 1996 there were 14,100 fall-related deaths, the second leading cause of unintentional-injury deaths in the United States. Falls do not have to be great to be fatal. Persons have been killed when they struck their heads in falls from upright positions on slippery floors; yet the *Guiness Book of World Records* states that Lt. I. M. Chissov of Russia fell 22,000 feet from a damaged Ilyushin 4 without a parachute and survived. He struck the edge of a snow-covered ravine and then slid to its bottom. He suffered a fractured pelvis and severe spinal damage, but lived.

The most serious damage from a fall is generally broken bones. Of these, injury to the head is the most serious and frequent. Roth and others[66] have indicated that "within the conditions studied, the total energy required for skull fracture varies from 400 to 900 inch-pounds, with an average often assumed to be 600 inch-pounds." White and Bowen[67] cite data in Fig. 16-1, which are taken from other investigators.

White and Bowen also indicate that 50 percent of all persons impacting against a hard surface with a velocity of 18 miles per hour (27 feet per second) will be killed. This is equivalent to a free fall of 11 feet. These last values were computed, being extrapolations from results obtained with animals. In addition, these values apply to impacting the entire body, whereas the data in the table related to specific bony structures.

It has been found that the ability of a human body to sustain an impact, such as a fall, depends on three major factors: (1) velocity of an initial impact, (2) magnitude of the deceleration (negative acceleration), and (3) orientation of the body on impact.

Figure 16-2 presents data derived from occurrences for which deceleration information was known or could be derived. The lower left corner of the chart states "Estimated head deceleration in fall from standing position" at a point where the deceleration distance is slightly less than 0.5 inch. A concrete floor would probably be unyielding, therefore the deceleration distance would be small, and the magnitude of deceleration much greater.

APPROXIMATE IMPACT VELOCITIES AND EQUIVALENT HEIGHTS OF DROP FOR FRACTURE OF HUMAN SPINE, SKULL, FEET AND ANKLES

Effects on Man	Impact Velocity		Equivalent Height of Drop	Comment
	ft/sec	mph	In.	
Experimental skull fracture	13.5–22.9	9.5–15.0	37–91	Range of 1–99 per cent fracture of cadaver heads dropped on flat metal surface.
Fracture – feet and ankles	12–13	8–9	25–30	Impact table data using cadavers with knees locked.
Fracture – lumbar spine	8	6	12	Estimated for impact on hard surface in sitting position.

FIGURE 16–1 Approximate Impact Velocities and Equivalent Heights of Drop

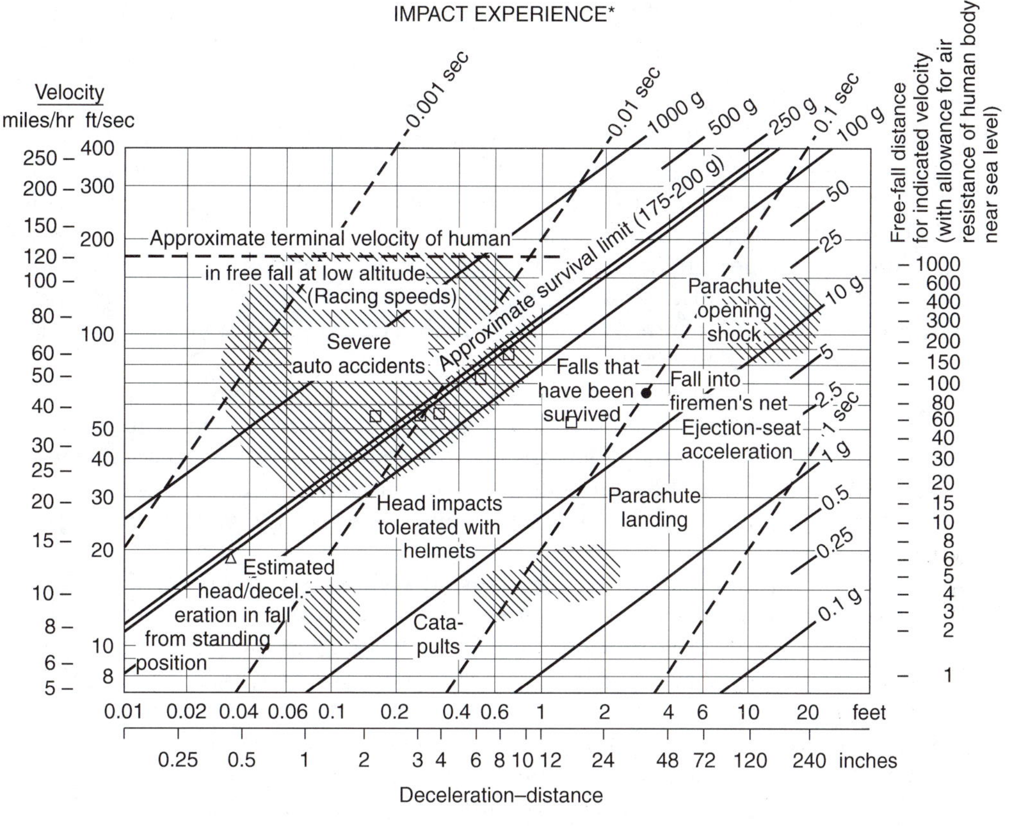

*Paul Webb, ed., *Bioastronautics Data Book*, NASA SP-3006, National Aeronautics and Space Administration, Washington, D.C., 1964.

FIGURE 16–2 Impact Experience

The kinetic energy of a body impacting on a hard surface can be derived from the equation $E_{\text{kin}} = wv^2/2g$, where w is weight in pounds, v is feet per second, and g is the gravity constant in feet per second per second. The human head weighs approximately 12 pounds. A skull fracture would occur, on average, at 600 inch-pounds or 50 foot-pounds. Thus, the velocity at which the skull would fracture would be

$$v^2 = \frac{E_{\text{kin}} \times 2g}{w} = \frac{50 \times 2 \times 32.3}{12}$$

$$v = 16.3 \text{ feet per second}$$

which falls within the range shown. The deceleration can be computed by $A = v^2/2s$, where s is the stopping distance. If A is divided by 32.2 feet per second, the answer is expressed in g's, as shown by the slant lines in Fig. 16-2. It can be seen that A varies inversely with the stopping distance. If s can be increased, the deceleration decreases.

Experiments have shown that human ability to withstand decelerations depends on the orientation of the body, the part of the body hit, and numerous other factors. If the deceleration is given, the distance in which a body must be stopped at any given velocity can be computed. This relationship is employed in the design of padding, football and crash helmets and hard hats, bumpers, collapsible steering wheels, and the locations of engines and trunk compartments for automobiles.

Lt. Chissov impacted with a much higher velocity than would a person who slipped and fell on concrete floor. However, he decelerated far more slowly because he hit and was stopped by the snow.

Similar escapes from death have occurred in other instances because something slowed and decelerated the falling body. In each and every instance there was some element of luck, so that the falling person struck an object or substance which slowed the body before it hit the ground. Such persons have hit and survived falling on wooden rails or fences, automobile tops, awnings, or water after falling from high levels (Fig. 16-3).

Numerous falls of industrial workers into bodies of water occurred during construction and maintenance of bridges and elevated structures. This led to the OSHA requirement for the use of safety nets in such locations. Studies have shown that approximately one-third of the persons who fell feet first into deep water suffered no injury. Injuries to feet and ankles, common with feet-first falls to the solid ground, were nonexistent in similar falls into water. In such cases, with little resistance, the water is quite yielding. The remaining two-thirds who survived feet-first impacts into water suffered bruises to the legs and buttocks, spinal injuries, bleeding of a lung, and shock.

High-fall impacts into bodies of water in other positions almost always resulted in body injury or death. The exceptions occurred when head-first entries were made, especially by professional divers, such as those from the cliffs at Acapulco. Some semi-professional divers survived when they dove off high bridges, where suicides had been killed, on bets. The exact numbers and techniques of the successful bettors is unknown: they simply picked up the money they had won and left.

PREVENTIVE MEASURES AGAINST FALLS

Parachutists are taught techniques to avoid landing injuries from what is equivalent to moderately high falls. Except for such controlled falls, which are done on purpose, it is evidently unwise to depend solely on luck to avoid the effects of accidental falls and impacts. It is a much wiser policy to prevent falls and the subsequent impacts in the first place; and, second, to provide safeguards to avoid hard impacts where the possibility of falls cannot be eliminated entirely. A person can slip and fall to the floor on the same level on which he or she is walking or working, fall from one floor level to

SURVIVALS AFTER FALLS*

Person Who Fell	Fall Distance (ft)	Details	Factor That Lessened Effects of Impact	Injuries
Woman, age 42, 5 ft 2 in., 125 lb	55	Jumped from sixth floor of a building; landed on fairly well-packed earth on left side and back.	Ground gave way for 4 inches.	None.
Woman, age 27, 5 ft 3 in., 120 lb	66	Jumped from a seventh-floor window, landed on a wooden roof head first, then on shoulder and back.	Broke through roof of ¼ inch pine boards supported on 6 by 2 inch beams and landed on ceiling below. Broke three of the beams.	Scalp lacerations; abrasions along the spine; fracture of vertebrae.
Woman, age 36, 5 ft 4 in., 115 lb	72	Jumped from eighth floor and landed on a fence made of wood and wire.	Fence broke down part way and woman tumbled to ground.	Minor bruises.
Woman, age 30, 5 ft 6 in., 122 lb	74	Jumped from ninth floor, fell onto an iron bar, metal screens, a sky-light of wired glass, and on a metal lath ceiling, face down, prone.	T-bar was bent 13 inches; screens, wired glass, and metal lath stopped her progressively.	Three ribs fractured, contusions on head and chest, scratches on head from wire, and other minor injuries.
Woman, aga 21, 5 ft 7 in., 114 lb	93	Jumped from a 10-story window and fell almost horizontally (on right side and back) on freshly turned earth.	Soft earth gave way 6 inches.	Fractured a rib on right side and right wrist.
Man, age 42, height and weight unknown	108	Fell from a tenth-story window and landed on hood and fender of a car, face down, then bounced to pavement.	Portions of car were depressed from 6 to 12 inches, where different parts of the body impacted.	Fractured skull, but survived.
Man, age 27, 5 ft 7 in., 140 lb	146	Jumped from roof of 14-story building; landed on top and rear deck of coupe in a semi-horizontal position.	Car dented in some places to 8 inches, 5 inches where head and shoulders hit.	Suffered numerous fractures, but recovered within two months.
Woman, no details	144	Jumped from a seventeenth floor, landed on a metal ventilator.	Ventilator crushed 12 to 18 inches.	Both bones of both forearms broken, left leg and foot injured.

*Tabulated from data from an article by H. De Haven, "Mechanical Analysis of Survival in Falls from Heights of Fifty to One Hundred and Fifty Feet," in *War Medicine* American Medical Association, 1942.

FIGURE 16–3 Survivals after Falls

TABLE 16–1 Precautionary Measures for Use of Ladders

1. Purchase and use only those ladders which meet OSHA standards.
2. Makeshift ladders should not be used.
3. Examine all ladders for defects before use, especially for such defects as:
   (a) Loose or broken steps or rungs.
   (b) Broken, split, or cracked rails.
   (c) Loose nails, screws, or bolts.
   (d) Missing, worn, broken, or damaged safety shoes.
   (e) Bent legs, steps, or rails on metal ladders.
   (f) Defective locks on extension or step ladders.
   (g) Condition of hinges and spreaders on step ladders.
   (h) Wobble.
   (i) Worn or loose metal parts.
4. Keep ladders clean and free of oil, dirt, and grease. Before attempting to climb a ladder, ensure that footwear is not oily, dirty, or greasy.
5. Avoid using metal ladders near electrified equipment or lines. Shut off power first if ladders must be used in such locations.
6. Be sure that the ladder is long enough for the job. Make no attempt to use a ladder when standing higher than the third rung from the top on a straight or extension ladder, or the second from the top on a stepladder.
7. Ladders should extend at least $3\frac{1}{2}$ feet above the point at which a person may want to step onto a roof, platform, or scaffold.
8. Be sure that the ladder has secure footing and good support at the top before starting to climb. The foot of a ladder should be placed one-fourth of its working length horizontally from the vertical line under its support.

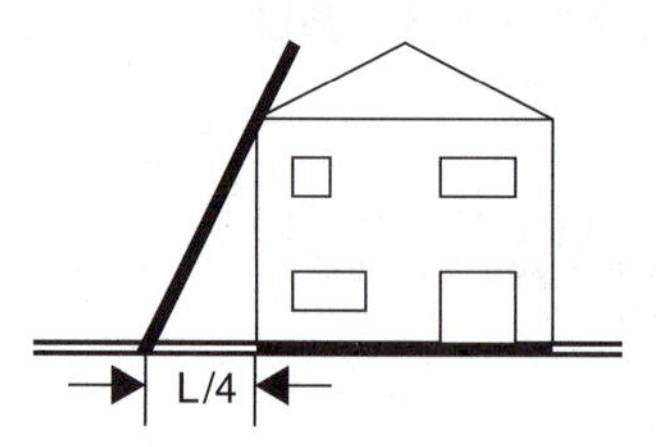

9. Place portable ladders so that each side has secure, solid footing.
10. In locations where the surface on which ladder rests might be slippery, tie the base securely with a rope. If this cannot be done, have another person hold the base of the ladder; then, if possible, secure the top of the ladder.
11. Portable ladders are designed for one 200-pound man. No more than one man should use a portable ladder at the same time.
12. Never lean a ladder against unsecure or unsafe objects or surfaces, or against piping that could be damaged. Both rails of the ladder must be supported at its upper portion unless the ladder is equipped with a suitable single support attachment.
13. Do not place a ladder against a window or sash; securely fasten a board (do not nail it on) across the rails so that the board rests on both sides of the window.
14. Do not tie or fasten ladders together to make one longer ladder unless hardware provided or endorsed by the ladder manufacturer for such extensions is used.
15. No ladder should be placed in front of a door that opens toward the ladder unless the door is locked, blocked, or guarded. A sign inside the door should route personnel to another exit.
16. No other person should be permitted to work or pass under a ladder on which a person is working. If there is other work activity in the area which could endanger a person on a ladder, rope off the area around the ladder and assign a man as a guard.
17. Stepladders should not be used as substitutes for scaffolds or work stands.
18. Stepladders should be locked in an open position with an automatic spreader or locking device prior to use.
19. No attempt should be made to use a ladder in a horizontal position, such as for a runway or scaffold.
20. When climbing or descending ladders, face the ladder and hold onto each side rail.

21. Tools, lines, or other objects should not be carried in the hand while climbing.
22. Never attempt to use a ladder in a strong wind.
23. No attempt should be made to reach beyond a normal arm's length while standing on the ladder, especially to the side.
24. When a portable ladder is being carried, the front end should be raised to avoid hitting other persons. Care should be exercised when going through doorways.

---

another, or fall to the ground from a natural elevation, a structure, or a piece of equipment. In addition, a person may fall because of the collapse of a piece of equipment, a ladder might slip with a person aboard, or there might be a failure of structural support, walkway, or hoist. Table 16-1 indicates accident prevention measures which can be observed when ladders are used.

Some persons are psychologically or physiologically unsuited for working at heights. They are incapable of working on open structures, as many structural steel workers do on skyscrapers. Some persons have problems on six-foot ladders. Others have no problem when they work in high locations where they have a sense of security by being close to a solid wall, on railed scaffolding, or enclosed in a cage around a high ladder.

Persons who must work at heights (other than within buildings or other enclosed, secure structures) should be permitted to do so only after they have indicated and demonstrated they will have no adverse reaction and after a safety net has been provided. A new worker on a high structure should be accompanied by experienced workers until the new worker's reactions have been observed. Any worker who appears nervous or "freezes" should not be permitted to continue such work. Persons who have or are recovering from colds or flu may have an affected sense of balance. For that reason, workers should be discouraged from working at heights while their illnesses last.

Another reason for avoiding the use of some workers at heights without adequate testing is vertigo. At any height, anyone, even those operating as aircraft pilots, may suffer from vertigo and loss of sense of balance. In the dark or fog, where there is no point or means of reference, the person loses a sense of direction. The result may be an impact or crash. The preventive measures are obvious: only physically and psychologically capable workers should be permitted to work at any elevation, using suitable safeguards or a suitable mechanical device to provide an ability to maintain a safe position.

A few of the other means of avoiding or mitigating the effects of falls include emergency nets, coiled knotted ropes or ladders, fire escapes, and parachutes. Fall protection harnesses attached to restraint systems that arrest falls are available commercially.

IMPACTING OBJECTS

Because a falling object with a force of 600 inch-pounds can cause a skull fracture, it is advisable for any persons where there might be an impact of any force to wear protective headgear, such as a metal or plastic hard hat or helmet.

EFFECTS OF MISSILE IMPACT ON THE CHEST AND HEAD *

Biological Effects Observed	Threshold Velocities (ft/sec) for Missiles of Indicated Weights	
	0.8 lb	0.4 lb
[a] Lung hemorrhages		
Side of impact only (unilateral)	45	80
Impact side and opposite side (bilateral)	110	125
[a] Rib fracture	60	120
[a] Internal lacerations from fractured ribs	90	120
[a] Fatality within 1 hr	155	170
[b] Experimental fracture human skull	15–23 ft/sec range of velocities for 10-lb object (7–15 lb wt range of human adult head)	

[a] Experimental results using dogs.

[b] Computed from data of E. S. Gurdjian, J. E. Webster, and H. L. Lissner, *American Journal of Surgery,* vol. 78, pp. 736, 742, 1949.

* Data AEC Project, Lovelace Foundation for Medical Education and Research, Albuquerque, NM.

FIGURE 16–4 Effects of Missile Impacts on the Chest and Head

EFFECTS OF MISSILES ON HUMAN CADAVERS *

Type of Missile	Mass (g)	Velocity (ft/sec)	Effect on Man
Spherical bullets	8.7	190	Slight skin laceration
	8.7	230	Penetrating wound
	7.4	360	Abrasion and crack of tibia
	7.4	513	Travels through thigh
Bullets	6–10	420–266	Threshold for bone injury
	6–15	751–476	Fractures large bones

*C. S. White and I. G. Bowen, *Comparative Effects Data of Biological Interest,* Lovelace Foundation for Medical Education and Research, Albuquerque, NM, Apr. 1959.

FIGURE 16–5 Effects of Missiles on Human Cadavers

The body can also be impacted by nonfalling solid material such as high-pressure tires or tanks which fail, misslelike fragments from bursting pressurized equipment, windblown solids, debris thrown off by rotating equipment, or other objects. A very common injury is that caused by small rocks or small pieces of metal being thrown by the blades of lawn mowers. These blades have peripheral velocities of between 150 and 300 feet per second. Even a small hard object thrown horizontally by a blade can cause serious injury or damage. Figures 16-4 and 16-5 indicate the effects of missiles on bodies. The figures shown are threshold values; greater masses or higher velocities can cause more severe injuries. A rock or metallic object thrown by a lawn mower's blade would probably have a mass greater than 0.02 pound (10 grams). In addition, rocks and pieces of metal might be sharper and even more penetrating than the spherical missiles used to determine the data in Fig. 16-5.

Certain parts of the body are more susceptible to injury than are others, the eyes being especially critical in this respect. Very small particles thrown from grinding wheels can easily penetrate the very soft tissues that make up the eye. For such operations, therefore, protection is necessary. Safety glasses or safety goggles protect the eyes; face shields will protect both the eyes and other areas of the face.

Impact damage of the body from large masses may not be as penetrating as that from small objects. The results in such cases may be bruises, crushing of tissue, lacerations, or bone injuries or fractures. An impact to the body from the rear by a massive object may not only cause bruises and damage to the spine and ribs but also the "whiplash" injury which is common in rear-end automobile collisions.

Other types of impact injuries include those from:

- The collapse of a structure on a person.
- Impacts between cars, causing rupture of the fuel tank of the car being hit, followed by fire.
- Dropping a heavy object on the foot. Use of safety shoes has done much to alleviate this problem.
- Solid particles blown into the eye by the wind or when compressed air is used to clean work areas.
- A high-pressure flexible hose separating, whipping, and hitting nearby persons (see Chapter 19).
- Persons being crushed when loads dropped on them.
- Collapse of boxes, crates, or similar containers which were stacked too high.
- Being hit by moving equipment or a vehicle.

OTHER ACCELERATION EFFECTS

Figure 16-6 is a checklist showing most of the causes and effects of falls and impacts. Other aspects of acceleration also can be injurious or damaging. The following are very common examples:

HAZARDS CHECKLIST-ACCELERATION

Possible Adverse Effects

Injury to personnel. A person may:
- Be hit by an object set in motion inadvertently
- Hit a hard surface during a sudden start or change in velocity
- Hit a sharp edge or point when startled
- Fall or be thrown backward during sudden forward acceleration
- Lose his or her balance under centrifugal force
- Be thrown against the ceiling of a vehicle in a sudden drop or other falling maneuver
- Fall to the ground or other hard surface
- Be hit by an impacting fragment or missile

Overloading, deformation, and failure of structural members
Deflection of piping
Deflection and bottoming of shock isolated parts and springs
Cracking or breaking of lines or equipment by impact of high-velocity fragments
Breakage of cables, ropes, chains, and pins by sudden overloads
Fracture of brittle materials
Opening or closing of hinged parts, doors, or panels
Seating or unseating of spring-loaded valves or electrical contacts
Shorting of closely spaced electrical parts
Bending of bimetallic strips, thus changing instrument readings and calibration
Pressure surges in liquid systems (water hammer)
Sloshing and loss of liquids from open containers
Loss of fluid pressure

Possible Causes

Acceleration:
- Vehicle, body, or fluid being set into motion or increasing speed
- Outside force applied against an unrestrained body
- Any falling body or dropped object
- Vehicle on a downgrade
- Uncontrolled loss of altitude or height
- Impact by another body
- Turbulence or motion over rough terrain
- Sudden valve opening in a pressure system
- Centrifugal motion
- Sudden reaction by a surprised person

Deceleration:

- Vehicle, body, or fluid decreasing speed or being stopped impact due to:
- Hitting another body, a structure, or the terrain
- A falling body being arrested
- Inadequate shock-absorbing materials or devices
- Sudden closing of a valve in a fluid system with high-velocity flow
- Friction or other resistance to motion

Failure to accelerate or decelerate:

- Inadequate or loss of motive power
- Friction or drag
- Failure of an unlatching or restraining mechanism to release
- Loss, failure, or inadequate braking capacity
- Wet, oily, or other slippery surface

FIGURE 16–6 Hazard Checklists—Acceleration

1. A sudden stoppage of an automobile, with or without a collision, can result in the driver or front passenger being thrown violently against the car's windshield.
2. A sudden start, slowing, or even a circular motion can cause displacement of fluids so that sloshing or spillage results. A liquid in an open tank may overflow its sides if the vehicle on which it is being carried rounds a corner too rapidly, stops suddenly, or starts quickly.

CHECKLIST—ACCELERATION, FALLS, AND IMPACTS

1. Are there structures, equipment, vehicles, or products from which a person might fall and be injured and killed? Have safeguards against falls been provided where such hazards exist?
2. Do in-plant operations include the use of moving vehicles?
   a. Are routes where these vehicles are to operate marked or otherwise restricted?
   b. Have speed limits and other precautionary measures been posted prominently?
   c. Is it a policy that only authorized personnel will operate the vehicles?
   d. Is it a policy that all drivers will be certified to operate specific vehicles?
3. Are vehicles which can be subjected to bumpy roads or to sudden stops or turns provided with seat belts or other means of restraint to prevent injury to the occupants?
4. Are vehicles which might upset equipped with rollover protection for the driver and other occupants?
5. In personnel, carrying vehicles subject to impacts or sudden starts or stops, are cushioning materials used to prevent injury by safeguarding the riders from hitting hard surfaces?
6. In such vehicles, have sharp points, knobs, and other hard protuberances against which personnel can injure themselves by impact been removed or safeguarded?
7. Will the vehicle be able to stop within a reasonable distance if the surface on which it is moving is wet? Does the vehicle have any unreasonably unsafe characteristics on a wet surface?
8. Is the braking surfaces and capability adequate for the vehicle's weight and expected operating speed?
9. Will vehicle, car, or other moving product contain any loose objects or fluid which can be translated (moved out of position) by acceleration, deceleration, or centrifugal force? Are tiedowns or other means or restraint or containment provided?
10. When any liquid is to be moved by a vehicle, could it be lost by sloshing or centrifugal force? Is the container in which the liquid is held sealed, or is it large enough to prevent any loss of liquid?
11. Is there any liquid in a piping system which can cause hydraulic shock (water hammer) by rapid closure of a valve or other sudden stoppage? If this is possible, has an accumulator been provided?
12. Are springs and other shock absorbers on heavy equipment designed to avoid "bottoming"?
13. Can hoists, structural members, or equipment be overloaded by a sudden impact, stoppage, or dynamic load?
14. Are surfaces on which persons will walk so slippery they might fall? Do provisions exist by which such surfaces which become slippery because of leaks, spills, or ice are cleaned up?
15. Is each lifting or lowering device, such as a hoist or elevator, designed to start and stop smoothly to prevent dynamic overloads
16. If the possibility of fragments being thrown off by rotating devices exists, such as particles from grinding wheels, are protective shields provided?
17. Is there a means by which the speed of a rotating, variable-speed device can be kept within safe limits?
18. Is the case in which a high-speed device is rotating strong enough to contain any fragments if the device fails in motion? If not, has the access area into which fragments might be thrown marked and restricted so that personnel or sensitive equipment might be safeguarded if a failure occurs?

FIGURE 16–7 Checklist—Acceleration Falls and Impacts

3. Loose articles can be displaced by a sudden acceleration. This has been very noticeable when there is a sudden drop in an aircraft. When it is due to windshear on a landing aircraft, the results can be a fatal crash.
4. Quick starts and stops of hoists have caused heavily loaded chains and cables to be overstressed and to break. Not only has equipment been damaged as a result in this way, but personnel in the vicinity have also been injured or killed.

Figure 16-7 is a list of items safety engineers can use to develop or check operations in their plants.

BIBLIOGRAPHY

[66] E. M. Roth, W. G. Teichner, and R. L. Craig, *Compendium of Human Responses to the Aerospace Environment* (Albuquerque, NM: Lovelace Foundation for Medical Education and Research for the National Aeronautics and Space Administration, NASA CR-1205).

[67] C. S. White and I. G. Bower, *Comparative Effects Data of Biological Interest* (Albuquerque, NM: Lovelace Foundation for Medical Education and Research, Apr. 1959).

EXERCISES

16.1. On what three major factors does the ability to withstand a fall depend?

16.2. What level of energy can cause a skull to fracture if it is hit by a hard object?

16.3. List five examples of how impact injuries can occur.

16.4. What can a rock thrown by a power mower do to a person's body?

16.5. List some means by which fall injuries can be avoided.

16.6. Provide three examples each of injuries which can result from acceleration, deceleration, or failure to accelerate or decelerate.

16.7. The great number of fatalities and injuries in automobiles has led to the mandatory use of safety equipment for drivers and passengers. Describe some of the required pieces of equipment.

16.8. Prepare five questions, other than for automobiles, for a checklist to be used to review equipment or facilities to determine if the hazards due to acceleration (or deceleration) effects have been adequately controlled.

16.9. List at least two actions in which acceleration is used beneficially.

16.10. Describe some of the effects acceleration might have on a pilot, another worker, or any persons in an aircraft in flight.

CHAPTER 17

Mechanical Injuries

Most of the injuries in industrial plants were originally from mechanical causes. This was especially so in plants with much belt-driven rotating equipment, open-geared power presses, and power hammers. Gradually laws and codes were enacted requiring guards to be provided on points of danger of such equipment, which reduced the number of accidents involving mechanical injuries. They still, however, constitute an extremely common source of injury.

Mechanical injuries that can result from impacts have already been pointed out. Mechanical injuries to the body also include cutting and tearing, shearing, breaking, straining of the body, and combinations of these.

CUTTING AND TEARING

Cutting results when skin or a body appendage comes in contact with a sharp edge. The sharp edge may be fixed, and the body may contact it with enough pressure so it cuts into the skin. If the pressure is great enough, the sharp edge may cut into the underlying muscle and bone. Tearing of the skin may occur when a sharp point or edge first pierces the skin and flesh and then is pulled away violently. In addition, tearing of the skin and flesh can occur when body digits are pulled with great force. For this reason, wearing rings or wrist watches is often prohibited in activities where they might be caught on projections, operating working surfaces, and similar equipment and torn off.

Some of the commonest, although less serious, cutting injuries result from contact with poorly finished surfaces. Sharp edges and burrs in contact with a person can either cut or tear the skin. Legal and industrial standards often lack requirements for manufacturers and engineers to properly eliminate such hazards. Some engineers do require elimination of sharp edges and rough finishes because they are stress raisers—points where part failures can begin. Most manufacturers have also found that poorly finished equipment will not meet their customers' quality standards. The military services, in their human engineering standards, require that sharp corners, projections, edges, and finishes be eliminated on hardware to which personnel have access. In the past few years, automobile companies have adopted similar requirements which have led to recessed door handles, smooth and unbroken arm rests, and other features that avoid cuts or tears to fingers, hands, or other parts of the body.

SHEARING

Shearing will occur when a sharp edge is in a linear motion in a direction vertical to the line of the edge; or it can occur when two objects, one or both of which are in motion, pass close enough to cause a shearing action. Examples of equipment where such actions can occur include powered paper cutters or metal shears; reciprocating mechanisms where there is a close tolerance between a moving part and a fixed surface; and a jamb, where a door closes into a frame.

A common cause of amputation of hands was use of unguarded or badly guarded power shears or cutters. The operator would position the material to be cut and then activate the equipment. The worker would suddenly realize he or she had not positioned the material correctly, would reach in to correct it, and would lose a hand or hands. After installation of two-handed controls was required by OSHA regulations, personnel sometimes bypassed these to increase their production and suffered injuries. There are now even more complex methods of ensuring no other type of interference exists before the shearing action begins, and there is no means to bypass action of the safeguard.

CRUSHING

Crushing of skin, muscle, or a part of the body will occur when it is caught between two hard objects whose separation is progressively reduced until an injury occurs. OSHA standards define "pinch point" as "any point other than the point of operation at which it is possible for any part of the body to be caught between moving parts of a press or auxiliary equipment, or between moving and stationary or another moving part or parts of the press or auxiliary equipment." Actually the definition need not be so restricted to presses and auxiliary equipment. In some instances hazards have been divided into "squeeze points" and "run-in points."

A *squeeze point* is created by two solid objects, at least one of which is in motion. The objects do not have to come into eventual contact with each other, but they cause injury when the lessening distance between them compresses a part of the body. The lessening distance does not have to be a slow process, usually occuring quickly so a part of the person's body is caught. Skin and muscle can be injured in this way between an impacting object and a bone. Prior to passage of the Railway Safety Act, trainmen's loss of fingers by crushing was common as they endeavored to couple railroad cars. Other examples include crushing when a heavy object is dropped on a foot, a misdirected hammer hits a thumb, or a moving vehicle pins a person against a fixed wall or another vehicle.

Run-in points are created when two objects, at least one of which is rotating, are in motion to produce a lessening separation until they touch or almost do so. Common examples include meshing gears, belts running over pulleys, cables on drums, chains on sprockets, rollers on manual-type washing machines, and rolls on rubber mills or paper calendars.

BREAKING

Much of the equipment that can cause crushing of a part of the body that is caught between two hard surfaces can also cause breaking of bones. Breakage will occur if an attempt is made to bend a rigid bone or if a concentrated force impacts a bone that is supported at only one or two points rather than over its entire length. Breakage may take place in material that is fragile and dropped, or unsupported so that vibrations cause oscillations where the object has no resiliency.

MACHINE GUARDS AND SAFETY DEVICES

Industrial physicians have noted that the types of injuries which require treatment because of worker injuries have been changing. Skin cuts, tears and abrasions, or crushed or severed body extremities are becoming fewer compared to other types of injuries. The principal reason for this is the increase in numbers, types, and effectiveness of machine guards and other safety devices.

A guard is a barrier that prevents entry of any part of the body into a hazardous area, such as a cover over a set of gears. A safety device is a machine control which prevents or interrupts the operation if part of the operator's body, such as a hand, is within a hazardous area; or requires its withdrawal prior to machine operation.

An effective guard or safety device must have certain features and meet certain criteria:

1. It must be safe under all conditions. If it fails, ceases to operate, or is opened, the machine will immediately and automatically stop.
2. Access to the danger zone must be prevented while the equipment is operating.
3. It must impose no restrictions, discomforts, or difficulties for the worker.
4. It must automatically move into or be fixed into place.
5. It must be designed for the hazard, the machine, and type of operation which will be present.
6. It must not require delicate adjustment for use or move out of alignment easily.
7. It must be impossible for an operator to bypass or inactivate it without simultaneously inactivating the equipment on which it is mounted.
8. It should require minimum maintenance.
9. It should not itself constitute a hazard.

There are numerous types of guards and other mechanical safety devices; the principal ones are listed next.

GUARDS

Fixed, Total Enclosure

A cover or barrier prevents access to the hazardous equipment so there can be no contact. Solid or wire mesh covers over gears, pulleys, couplings, line shafts, and similar equipment are in this category.

Fixed, with Limited Access

An opening is provided in a fixed guard which permits material to be worked on to be inserted and then removed. The access is too small to admit fingers or hands. This type of guard can be used on presses where small or flat pieces can be inserted. However, the size of the access limits the size of the material which can be processed. If the material being processed becomes jammed, special tools may have to be used, or the guard may have to be removed to correct the problem.

Enclosure with Interlock

When a portion of the guard or the entire guard itself is removed, an interlock shuts off electrical power, disengages mechanical power, or applies a brake so the equipment stops or is otherwise rendered nonhazardous. A common example is the interlock which shuts off power to a washing machine when it is in the "spin" portion of the washing cycle. The disadvantage to many interlocks is that some can be bypassed and therefore rendered ineffective. Figure 12-5 lists some of the many types of interlock devices.

Movable Barrier or Gate

The barrier or gate moves to permit loading or unloading of material in process or which has been processed, such as in a press. The barrier is linked to movement of the press slide. Movement of the press slide can be initiated only when the barrier is in the position of protection.

Optical Sensors

An optical sensor, such as a light source and photocell detector combination, determines whether there is any object in front of the hazardous area. The press will operate only when there is no interruption of the light beam. To ensure proper operation, the light source and detector must be kept clean. The device can fail to operate properly in strong sunlight or other bright light and is therefore usually shaded.

Ultrasonics

Inaudible high-frequency sound senses the presence of an object or body mass in the danger zone. Ultrasonics are not affected by strong light or dirt, but because sound attenuates over distance, the width of protection is limited.

Electrical Field Effect

Antennas around the hazard area create a capacitive field. Any grounded object in the field can be detected. Operators are grounded with wired wrists so the presence of their hands in the work area is sensed. The operation is inhibited until the wired wrists are out of the area of danger. This method can be used over large or irregular-shaped areas.

Two-Hand Control Devices

Two pushbuttons or handles must be used simultaneously by the worker for the equipment to operate. The devices are far enough apart so they cannot be operated with one hand.

Holdout Device

The machine operator's hands can move only in a limited way so they cannot be moved into the hazard zone. Because of the restriction on their movements, operators do not like devices of this type and tend to avoid using them.

Pullout Devices

The machine operator's hands are also restrained by a cable device. In this case the operator can move his or her hands wherever he or she wants until the machine is activated, when the hands must have been moved out of danger in order for the machine to operate. The pullout device is connected to the press slide or die. When this moves, a linkage and cable arrangement pulls the operator's hands and arms out of the danger zone. Operators dislike this type device also, since it tends to restrict their movements. The device requires accurate adjustment of the levers and cables.

Sweeps

When a press is activated, an arm connected to its moving slide swings in front of the danger area, sweeping away the operator's hands. It is not greatly effective with equipment into which the operator must reach. The device itself may injure by hitting the worker's hands, or may knock a standing worker off balance.

Mechanical Feeds

The operator places the part or material, usually a small part, to be processed on a positioning device such as a rotating table. The feeding device then moves the part or material into the point of operation where the hazard exists. After processing, the part is ejected.

Feed Tools

Tools may also be used to place material into or remove it from a press. These tools are used to penetrate guards while preventing a person's fingers or hands from entering the danger zone.

PRECAUTIONARY MEASURES

Certain precautionary measures are common to operation of all mechanical equipment:

- All machine operators should be given instructions on hazards which exist in their equipment and operations, safeguards present, and actions to be taken in case any problem arises.
- Each operator or helper should know the location of any STOP or EMERGENCY button or control on the equipment and periodically ensure it works properly. Supervisors should also ensure workers know the locations and perform the checks. Equipment that has the STOP or EMERGENCY button or control placed so that it is difficult for the operator to access or reach should not be used.
- Equipment with any guard or safety device should be inspected at regular intervals: daily by the operator before beginning to use them; and periodically by the supervisor and maintenance personnel.
- Each operator should be instructed against removing, inactivating, or attempting to circumvent any guard or safety device.
- Repairs, adjustments, and maintenance that require removal of a guard or safety device should be accomplished only by trained and authorized personnel.
- If a machine must be operated by two persons, each should have a control or safety device that prevents injury and premature activation by the other worker.
- Operators of machines in which clothing can be entangled should wear close-fitting clothes and buttoned or short-sleeved shirts or jackets. Neckties, rings, or watches should not be worn. Long hair should be confined in a suitable head covering. Face shields, safety glasses, or goggles should be worn whenever there is a possibility a piece of solid matter might be thrown off by the equipment.
- Where a loading tool is provided to place or remove a part being processed, no attempt should be made to do without it.

Corrective Actions

Figures 17-1 and 17-2 are checklists that will assist safety engineers in reviewing equipment in determining whether or not mechanical hazards exist in their plants. If deficiencies in equipment are found, it is highly desirable that both the plant manager and the equipment supplier be notified. The manufacturer may be asked to either provide a safeguard or recommend one that could be applied. Employers and equipment manufacturers are interested in such knowledge to avoid worker injuries and liability suits. This is not only true for all employers, whose workers' compensation insurance rates might increase, but is increasingly true for equipment manufacturers who want to avoid costly third-party suits. A hazard notification which brings no corrective action puts the manufacturer in an extremely poor position if a liability suit is brought. Manufacturers, therefore, generally will take action or make a recommendation for a safeguard. The chief problem is that the corrective action is often to be done at a cost to the equipment owner.

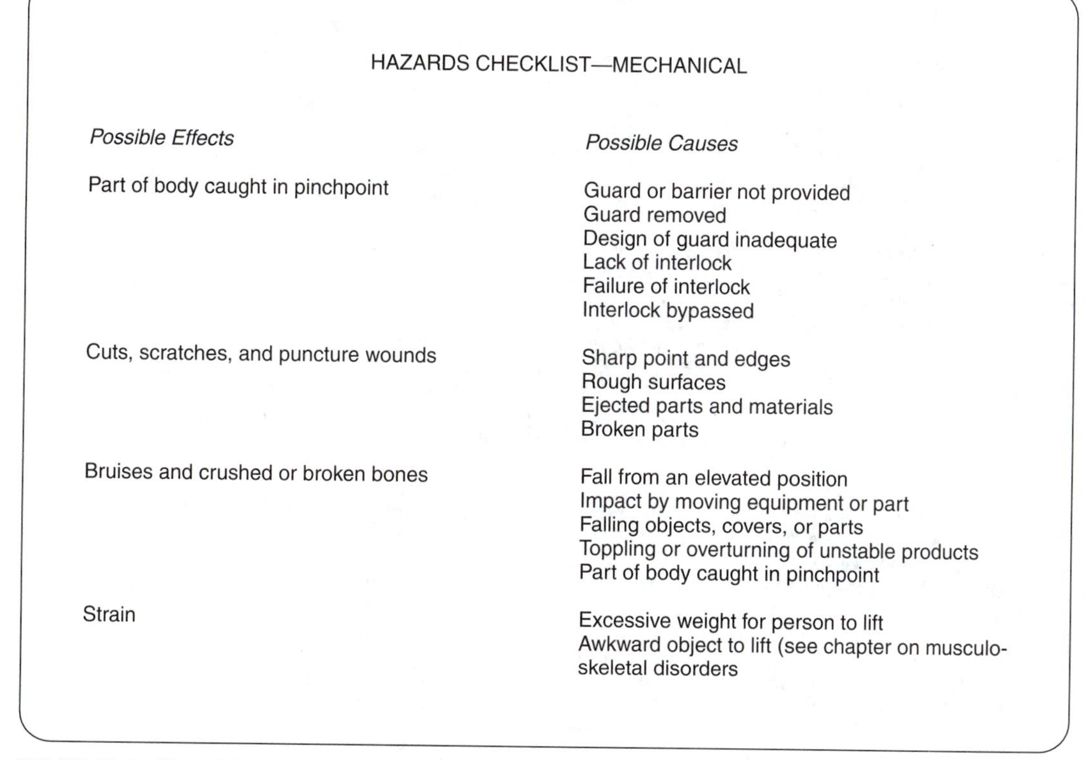
HAZARDS CHECKLIST—MECHANICAL

Possible Effects	*Possible Causes*
Part of body caught in pinchpoint	Guard or barrier not provided Guard removed Design of guard inadequate Lack of interlock Failure of interlock Interlock bypassed
Cuts, scratches, and puncture wounds	Sharp point and edges Rough surfaces Ejected parts and materials Broken parts
Bruises and crushed or broken bones	Fall from an elevated position Impact by moving equipment or part Falling objects, covers, or parts Toppling or overturning of unstable products Part of body caught in pinchpoint
Strain	Excessive weight for person to lift Awkward object to lift (see chapter on musculo-skeletal disorders

FIGURE 17–1 Hazard Checklist—Mechanical

Other organizations that will provide free advice, once a potential hazard is uncovered or thought to exist, include insurance companies, distributors of safety equipment, and, often, state safety agencies.

EXERCISES

17.1. Discuss why personnel in modern plants are less subject to the mechanical injuries that were so common in early industrial plants.

17.2. List the types of mechanical injuries that can occur, an example of how each can be caused, and a corrective measure that can be used to eliminate or minimize possibilities of injury.

17.3. What are the common causes of body strains? How might each type of strain be avoided?

17.4. What are the differences among pinch points, squeeze points, and run-in points?

17.5. List seven characteristics a guard or safety device should have.

17.6. Discuss the various types of guards, indicating advantages and disadvantages of each.

17.7. Discuss the various types of machine safety devices, indicating advantages and disadvantages of each.

CHECKLIST-MECHANICAL HAZARDS

1. Have sharp points, sharp edges, and ragged surfaces that are not required for equipment to function been eliminated and safeguarded?
2. Have guards been provided throughout the plant to keep personnel from injury from moving mechanical parts, such as gears, fans, belts, or other components in motion?
3. Does any equipment have any pinchpoints, rotating elements, or other moving parts which must be guarded?
4. Are ventilation openings small enough to keep fingers away from dangerous locations?
5. Are openings in or around guards small enough to keep persons from inserting fingers into dangerous locations?
6. Is a warning provided against loading cables, chains, and hoists beyond their rated limits? Does the warning on hoists also indicate that loads should not be raised or lowered with quick starts or stops which will put high strains on the cables or chains? Is the equipment posted with the rated load capacities?
7. Has equipment which might be damaged or which might cause damage by impact during operation been provided with bumpers, shock absorbers, springs, or other devices to lessen the effects of impacts?
8. Have critical moving assemblies or other sensitive parts of equipment which could be jammed or damaged by dropped or loose articles been protected by screens, guards, covers, or other barriers?
9. Can large or heavy parts which must be replaced be removed without damaging surrounding components or assemblies?
10. Has adequate space been provided between pieces of equipment to prevent injury because of interference?
11. Have hinged covers or access panels which must be raised to open have means to secure them in their open positions against accidental closures which might cause injury? Are such covers and panels so designed so that they will not protrude where a person might bump into them?
12. If a heavy product or assembly must be lifted frequently, are handles or other handling aids provided? Are lift points provided and labeled?
13. Have personnel been instructed on how to lift a heavy object properly?
14. Have limits been established on weights which might be lifted by one person? Is equipment which should be lifted by more than one person adequately marked?
15. Have lifting devices been provided to assist with heavy or awkward loads?
16. If a product or assembly can be damaged if turned to any but an upright position, is it so marked with a warning and directional arrow? Is every container in which it might be transported similarly marked?
17. Has equipment been designed and tested to ensure it will not overturn or topple easily, especially if it is to contain a hot or otherwise dangerous fluid?
18. Do slide assemblies for drawers in cabinets have limit stops which will prevent them from being pulled out too far inadvertently so that they fall?

FIGURE 17–2 Checklist—Mechanical Hazards

17.8. List five precautionary measures that should be observed by each machine operator.

17.9. List five precautionary measures each supervisor of machine operations should ensure that workers under his or her control should observe.

17.10. If workers believe the machines they are directed to operate are hazardous and they are reluctant to operate them, what should they do?

17.11. Which organizations might be held liable because of any accidental injury caused by faulty equipment?

CHAPTER 18

Work-Related Musculoskeletal Disorders

In the fall of 1995 the Assistant Secretary of Labor for Occupational Safety and Health, Joseph A. Dear, addressed a conference of the American Psychological Association.[68] In that address he described an "emerging hazard," work-related musculoskeletal disorder (WMSD), or occupational overuse disorder (OOD). Musculoskeletal disorders include carpal tunnel syndrome, tenosynovitis, tension neck syndrome, and low back pain. These conditions can be debilitating, "restricting life activities such as driving a car, typing, picking up a child or even writing," said Dear. Occupational disorders of this class (which involve the nerves, tendons, muscles, and supporting structures such as intervertebral discs) have been called by names such as cumulative trauma disorders (CTDs) and repetitive stress injuries (RSIs). WMSDs are musculoskeletal disorders caused or made worse by the work environment.

The Annual Survey of Occupational Injuries and Illnesses is conducted by the Bureau of Labor Statistics. The BLS reports that there are more than 700,000 lost-workday injuries and illnesses related to WMSD. It reports that in 1995, 62 percent of all illness cases were due to disorders related to WMSD. This figure does not include back injuries. In 1994 there were 332,000 cases, compared with 23,900 cases in 1972. This fourteenfold increase is what caused Secretary Dear to describe this as an emerging hazard. In 1993 these injuries cost employers more than $20 billion for 2.73 million workers' compensation claims, not including indirect costs which OSHA estimates may run as high as $100 billion. In 1997 Linda Rosenstock, Director of NIOSH, reported that the volume of NIOSH 800-number calls concerning WMSDs had grown until they were second only to questions about chemical hazards. Through the 1980s an epidemic of these types of injuries began in Australia and spread to Japan, the United States, and Bulgaria. But their occurrence is not new. Ramazzini, in his book *De Morbus Artificum,* prefaced the 1700 edition with these remarks about disease in the workplace:

> That stems mostly I think, from two causes. The first and most potent is the harmful character of the materials that they handle, noxious vapors and very fine particles, inimical to human beings inducing specific diseases. As the second cause I assign certain violent and irregular motions and unnatural postures of the body, by reason of which the natural structure of the living machine is so impaired that serious diseases gradually develop therefrom . . . (quoted by Tichauer, 1975).[69]

NIOSH reports the following risk factors for WMSDs: repetitive, forceful, or prolonged exertions of the hands; frequent or heavy lifting, pushing, pulling, or carrying of heavy objects; prolonged awkward postures; and vibration.[70]

MUSCULOSKELETAL DISORDERS (MSDs): WORK RELATED OR NOT WORK RELATED?

This has been called "the occupational hazard of the 90s." Before it can be acknowledged, the question of its work relatedness must be answered. The *Sacramento Bee* published on April 14, 1994, an article on RSI entitled, "Hand-to-hand combat." It quoted a local hand surgeon: "Saying a person has repetitive motion disorder when in fact they [sic] may have the normal aches and pains of use and age is getting a little too common."

This is a multifactorial disorder. The relative importance of multiple and individual factors in its development is a source of controversy about its work relatedness. To address this question, NIOSH identified 600 studies of the work relatedness of musculoskeletal problems of the neck, upper limbs, or back, and/or occupational and nonoccupational risk factors and analyzed them to evaluate the work relatedness of this disorder. The criteria for work relatedness included measures of: strength of association between exposure to workplace risk facts and MSDs, consistency (the repeated observation of an association), specificity of effect or association (the association of a single risk factor with a specific health effect), temporality (documentation that the cause precedes the effect in time), exposure-response relationship (the relationship of disease occurrence with the intensity, frequency, and/or duration of an exposure), and coherence of evidence (whether an association is consistent with the natural history and biology of disease).[71]

The relationship between workplace factors and the development of MSDs from epidemiologic studies was investigated by addressing certain, but not all, susceptible parts of the body. Certain MSD risk factors for each part were classified into one of the following: **strong evidence** of work relatedness, **evidence** of work relatedness, **insufficient evidence** of work relatedness, and **evidence of no effect**.

For the neck and neck/shoulder, the risk factor of *posture* showed strong evidence of causality, *force* and *repetition* each showed evidence of work-related causality, vibration showed insufficient evidence. For the shoulder MSD, the investigators found evidence of work-related causality for the risk factors of *posture* and *repetition*, and insufficient evidence for force and vibration. They found strong evidence for the work-related causality of elbow MSD for the combination of the risk factors of *repetition, force,* and *posture*. They found evidence that *force* was an individual risk factor for elbow MSD, and insufficient evidence for repetition and posture as individual factors.

For the hand/wrist MSD called carpal tunnel syndrome there was strong evidence for work-relatedness causality involving the combination of *repetition, force, posture,* and *vibration* risk factors. The individual factors of *repetition*, *force*, and *vibration* showed evidence of work-related causality, but there was insufficient evidence for posture as an individual risk factor at work. For hand/wrist tendonitis, there was strong evidence that job tasks that require a combination of the risk factors of *repetition* and *force* are MSD causal, and there was evidence that a single factor of job task *repetition*, *force*, or *posture* is causal of MSD.

For back disorders, *lifting/forceful movement* as well as *whole-body vibration* showed strong evidence that they were each, single causal factors; *awkward posture*, and *heavy physical work* each showed evidence, but static work posture showed insufficient evidence.

The study concluded that "A substantial body of credible epidemiologic research provides strong evidence of an association between MSDs and certain work-related physical factors when there are high levels of exposure and especially in combination with exposure to more than one physical factor. . . ." This evaluative study indicates that MSDs are, indeed, WMSDs—given that the person suffering from the disorder has sufficient exposure at work to certain risk factors.

THE EFFECTS OF WMSDs

The body parts mentioned above are susceptible to WMSD because each involves joints, muscles, and tendons and their related structures, as well as nerves and blood vessels. Any body part, not just those mentioned above, with these features is at risk. Kroemer et al. in 1994 identified the effects of overuse disorders to include "strains," inflammations," and "sprains."[72] As they point out, "A **'strain'** is an injury to a muscle or tendon." Muscles atrophy from lack of blood or neural activity for an extended time. Their fibers can be torn and stretched. Tendons can be torn, scarred, or irritated. **Inflammation** of a tendon or its sheath, Kroemer et al. explain, is a protective response to limit bacterial invasion. A tendon can become roughened from overuse and may irritate its adjacent bursa, contributing to its inflammation. **Sprains** involve the stretching, tearing, or pulling of a ligament. They can result from a single trauma or from repetitive actions.

Kroemer et al. also identify nerve compression and blood-vessel compression as an effect of overuse disorders. Nerve compression arises from pressure on the nerve from either external or internal sources. Repeated or sustained pressure, e.g., from placing of body segments on hard surfaces and sharp edges, or from swelling within the body, can result in nerve impairment and can lead to physiological degradation. A loss or impairment of some normal motor, sensory, and autonomic bodily functions can result. Compressed blood vessels restrict the flow of oxygen to the body parts the vessels supply, as well as the removal of carbon dioxide and metabolic byproducts such as lactic acid. For muscles, the result is not only the potential to atrophy, mentioned above, but also early muscle fatigue and slow recovery from that fatigue. For other body parts served by blood vessels, there can be a similar loss of functioning related to that part or organ when the servicing blood vessels are compressed repeatedly or for an extended time.

WORKER-RELATED FACTORS ASSOCIATED WITH MSDs

Age. A number of personal factors contribute to MSDs. NIOSH reports, in the literature review evaluation cited above, that several studies find age to be an important factor associated with MSDs, while they report that other studies have not found this. They conjecture that a self-elimination process called "survivor bias" could account for the lack of observed relationship between aging and an increased risk for MSD. This bias exists when workers, who experience undesirable on-the-job effects such as MSD, change jobs to escape the problem. The result is that the analyst underestimates the true risk of developing MSDs in the case of the older worker.

Gender. There are studies which show a higher prevalence of MSDs among female workers, and studies which show no difference in gender. Two confounding factors muddy the waters about the role of gender. There are psychological and physiological differences, as well as gender differences in exposure to jobs, that carry a risk of MSD. Women may be more willing to report MSD symptoms and seek treatment than men. Muscles and muscle-fiber, strength, and stature differences exist between the genders. It is well recognized that males tend to be employed in different jobs, than females. Study is needed of those jobs that men and women perform about equally.

Smoking. In general, research shows that smoking is associated with low back pain. It does not find association of smoking with many other MSDs such as tension neck, rotator cuff tendonitis, carpal tunnel syndrome, neck/scapula, or shoulder/upper arm.

Physical Activity. Being inactive physically may increase susceptibility to injury; on the other hand, physical activity may cause MSD injury. MSD symptoms may be reduced through exercise and physical activity, and being in good condition may not protect from MSD. Sports activities outside of the workplace can result in MSD or can contribute to the cumulative effects of work-related exposure, especially where both involve forceful, repetitive movements with awkward postures. The NIOSH epidemiological study of MSD literature summarizes by saying: "... although physical fitness and activity is generally accepted as a way of reducing work-related MSDs, the present epidemiologic literature does not give such a clear indication."

Strength. The NIOSH study of epidemiologic literature shows studies that find a relationship between strength and MSD and those which do not. At issue is the relationship between MSD and a mismatch of physical strength for a job task. This is a concern particularly when the job requires strength that is greater than some predetermined worker's strength-test value. There is support that a relationship exists between back injury and such a mismatch of strength for the job task. Studies of this type have typically been for lifting tasks and have followed workers for a period of no more than one year.

Anthropometrics. Weight, height, weight-to-height ratio, and obesity all are potential risk factors for certain MSDs, particularly carpal tunnel syndrome and lumbar disc herniation.

CARPAL TUNNEL SYNDROME

One of the most common and well-known WMSDs is carpal tunnel syndrome (CTS). NIOSH reports that "Increased awareness of work-related risk factors in the onset of CTS is reflected in the growing number of requests for . . . NIOSH to investigate such suspected problems."[73] NIOSH received three times more such requests in 1992 as compared with 1982. In 1994 the Bureau of Labor Statistics reported 4.8 CTS cases involving lost workdays per 10,000 workers. In 1993 the incidence rate for CTS workers' compensation cases was 31.7 per 10,000 workers. This prevalence may be attributed in part to changes in work tasks. Instead of performing a variety of work tasks on a single job, the worker today may perform a few discrete tasks over and over without a break. In the past, a typist would insert the paper, type a line, throw the carriage, type another line, throw the carriage, and so on until that page was full. Then the typist inserted another page and continued, sometimes changing positions to open a file drawer, etc. Today the word processor or computer operator may work almost continuously at the keyboard with few breaks.

This disorder receives its name from the eight bones in the wrist, called carpals. These, along with certain ligaments, form a tunnellike structure. Human fingers are moved in flexion and extension by tendons that pass through this tunnel. These tendons are attached to the muscles in the forearm. The median nerve, which supplies the sensory cells in the hand, and blood vessels also pass through this tunnel. The flexion and extension of the fingers may cause a thickening of the protective sheaths that surround each of the tendons. These swollen tendon sheaths (tenosynovitis), as well as swollen blood vessels, can apply pressure on the median nerve and cause the symptoms that together make up the carpal tunnel syndrome.

CTS Symptoms. These symptoms are painful tingling in one or both hands, often at night. The pain is frequently enough to disturb sleep. Finger "numbness" sometimes occurs, and fingers are sometimes described as feeling swollen, even though there is little or no swelling apparent. The initial onset of nighttime tingling then progresses to tingling in the daytime. The tingling is reported in the thumb, index, and ring fingers (the median nerve serves the palm, thumb, index, and middle fingers, and the radial side of the ring finger). A weakened grip may follow.

NONOCCUPATIONAL FACTORS OF CTS

Armstrong (1983) cites these factors as frequently reported nonoccupational factors of carpal tunnel syndrome.[74]

- Systemic diseases such as rheumatoid arthritis, acromegaly, gout, diabetes, myxoedema, ganglion formation, and certain forms of cancer.

- Congenital defects including bony protrusions into the carpal tunnel, anomalous muscles extending into or originating in the carpal tunnel, and the shape of the median nerve.
- Acute trauma to the median nerve inside the carpal tunnel which can be produced by a blow to the wrist, laceration, burn, or other acute wrist trauma.
- Pregnancy, oral contraceptives, menopause, and gynecological surgery—uniquely female problems.

These individual factors, and others such as gender and age, may mask or amplify the effects of work-related CTS factors. Women seem to have more susceptibility to CTS than men. Recent studies show that other variables, such as smoking, caffeine, alcohol, and hobbies, also are CTS modifiers.

LOW BACK PAIN

Back disorder may be associated with both occupational and nonoccupational factors. It is common in the general population, particularly low back pain (LBP). General population prevalence has been estimated at 70 percent. Back injuries account for almost 20 percent of all injuries and illnesses in the workplace. It is estimated that costs of these injuries in the United States range from 20 to 50 billion dollars per year. Estimates give a range of 16–25.5 percent of all workers' compensation claims as being for back disorders.

Back Injury Prevention Programs. Most back injury prevention programs found in industry have three general parts. They are: preemployment and job-placement screening, training, and workplace design. The first two approaches are fraught with problems. Preemployment and job-placement strength testing, aerobic testing, and radiographic screening may have merit, but their validity may be difficult to defend. Training may be helpful, but lifting methods are various, and prescribing one way to lift "correctly" and training the employee to lift that way may not fit all job encounters.

Training. Six lifting methods found in the literature illustrate the problem in teaching lifting methods.

1. *The straight-back, bent-knees lift.* This is sometimes called the mechanical or traditional leg lift. It is limited to loads that can be placed between the knees. Changes in spinal curvature may be necessary.
2. *The hip-flex lift.* The trunk is flexed at the hip to enable grasps at floor level. The load is initially lifted by the arms and shoulders. The lift is continued by lowering the hips while simultaneously raising the load above the knees.
3. *The kinetic lift.* The trajectory of the object to be lifted is planned and the body is positioned to most effectively apply force during the object's trajectory, while at the same time avoiding change in spinal curvature.
4. *The dynamic lift.* This is similar to the kinetic lift, except the back is bent. That is, changes in spinal curvature occur.
5. *The stooped lift.* This lift uses a bent-back, knees-slightly-bent position and is called by some the "natural lift" and by others the "free-style" lift (because it is used by those who have not had training in lifting).

6. *The pelvic-tilt lift.* This technique says that however the lift is conducted, the forward tilt of the pelvis is prevented in order to maintain the spinal column in an optimal load-bearing position. This is accomplished by contracting abdominal muscles or posterior leg muscles. That is, one pulls up on the front of the pelvis or pulls down on the back of the pelvis to maintain the spinal curvature in a stable position during lifting.

The traditional lift method is confusing, because humans do not have a "straight back"; therefore, cannot keep a straight back during a lift. The hip-flex method becomes confused with the pelvic-tilt method for many, and the terms "hip-flex" and "pelvic-tilt" are very unclear. Many safety professionals frown upon the stooped lift. The distinction between the kinetic and dynamic lift may be more academic than practical and is confusing to some. Most experts feel that each method has its merits and can be used successfully in situations which take advantage of those merits. The workplace tasks are so variable, involving two-hand saggital lifts, one-hand lifts, two-person lifts, multiple-person lifts, asymmetric lifts, lifts in a single plane, etc., that relying on training alone is not recommended. Figure 18-1 gives general rules for lifting, which might be used in the workplace.

Workplace Design. The most important method in back injury prevention programs is to address workplace design. First, eliminate lifting where possible. That is, eliminate the need to manually raise or lower and carry objects in the workplace. Thus the first step in workplace design to minimize back injuries is to inventory all lifts and eliminate those which are feasible to eliminate. Second, control those lifts which cannot be eliminated. This involves the lift itself, the objects being lifted, the means to lift, and the environment in which the lift is accomplished.

General Rules for Lifting

1. Prepare before lifting. Preparation might include a mental rehearsal of the life (visualization). Long range preparation should include exercises and conditioning.
2. Take a breath before initiating the left. This will tend to distribute the load throughout the trunk of the body, directing the force path distribution over a much wider area than the spine, because the thoracic intercostal muscles will contract by reflex, helping to make the trunk into a better force-bearing, cylinder like member.
3. Maintain mechanical advantage. This includes not only limiting the distance of the center-of-mass of the load from the body as much as possible, but also maintaining joint angles (such as the knees) for optimum muscle leverage. Do not use long reaches.
4. Be sure of your grip and your footing. Sudden changes in the relative location of the center-of-mass, from slips at either coupling can induce sudden forces on the spine. Position the feet carefully.
5. Avoid twisting the body.
6. Avoid, where possible, asymmetric lifts.
7. Test the weight. Do not lift more than can be safely handled.
8. Be careful about spinal displacement. While some flexibility might be maintained, unnecessary bending should be avoided.
9. Do not lift when fatigued.

FIGURE 18–1 General Rules for Lifting

The Lift Itself. If lifting cannot be eliminated, it should be minimized. The lift dimensions, starting height and travel distance, both horizontal and vertical, should be optimized.

The Objects Lifted. The objects to be lifted should be controlled to the extent possible. Awkward packaging should be improved. Rigid frames might be used to support bags and sacks during lifting, or the materials might be transferred to containers more adapted for manual handling. Procurement might be instructed to purchase packaging more amenable to materials handling. The packages should be marked with the location of the center of gravity if the mass is not uniformly distributed. They should have adequate handles, should be of modest dimensions, and should be within the comfort weight range of the foreseeable lifter. They should be designed to be held against the body during lift and carry.

The Means to Lift. Removing the human from the lift may or may not decrease risk. If the means used for lifting is inherently more dangerous than human lifting, then substituting that means may reduce back injury potential but increase exposure to other hazards. There are many ways to lift by machines: exoskeletal machines, robots, cranes, scissor tables, fork lifts, elevators, hoists, etc. Each has its dangers which should be considered carefully for both operational and maintenance activities.

The Lift Environment. The environment in which the lift is accomplished can be a contributing factor to injury frequency and severity. The accessibility of the load for lifting, the space allotted to the human to find a position for lifting, the surface upon which the human must stand during the lift and carry, the lighting, and the ambient temperature are all environmental variables to be considered.

In 1991, NIOSH issued a revised equation for the design and evaluation of manual lifting tasks. This equation uses six factors: horizontal location of the hands, vertical location of the hands, travel distance of the load, asymmetry angle (twisting), lifting frequency, and object coupling (quality of worker's grip on the load) . Using the equation involves calculating values for these factors for a particular lifting and lowering task, thereby generating a recommended weight limit (RWL) for the task. The RWL is the load that almost all healthy employees can lift over a substantial period of time. The methodology involves various tables and the use of other equations. The reader is referred to the *NIOSH Work Practice Guide for Manual Lifting,* found at http://www.osha-slc.gov, for further details.

BACK BELTS

In 1994, NIOSH reviewed and evaluated the scientific literature on the effectiveness of back belts in preventing back injuries related to manual handling job tasks.[75] The conclusion was that there was insufficient evidence to prove the effectiveness of back belts in preventing back injuries related to manual handling job tasks. Since then, Kraus et al. have published an epidemiological study which concludes that the mandatory use of back belts in a chain of large retail hardware stores substantially reduced the rate of low back injuries.[76] The issue needs further study.

There is concern that "risk homeostasis" might be a phenomenon with back belts. Risk homeostasis is a theory that a person may operate at a given risk level regardless of the safeguards and protection provided. Therefore, in the case of back belts and lifting tasks, if a worker feels that she/he is protected by a back belt, that worker may attempt to lift even more with the belt than he/she would without it. Workers should be informed of the questions about the effectiveness of back belts.

ERGONOMICS: A PROGRAM TO CONTROL WMSDs

What is ergonomics? Ergonomics is a vital part of the tool kit of the complete occupational safety professional. It is the application of what is known about worker physical, psychological, and physiological limitations and capabilities to improve human safety, comfort, and productivity in the workplace. It is generally defined as the science of fitting the job to the worker. It is generally accepted that WMSDs, though a multifactorial disorder, can be reduced by establishing an effective ergonomics program. At this writing, OSHA has proposed an Ergonomics Program standard that will require an ergonomics program to be set up for manufacturing, manual handling operations, and other jobs where there are WMSDs.

In 1997 NIOSH published a primer giving the "elements" of an ergonomics program.[77] This primer presents seven steps for evaluating and addressing MSDs in the workplace. They are listed below with slight modification.

Step 1: Look for signs of WMSDs. This involves recognizing signs that may indicate a problem and determining the level of effort required to correct the situation. One source is the company OSHA Form 200 logs; another, workmen's compensation claims. Both these should be regularly reviewed for claims which show WMSD-type effects. Effort should be made to identify worker complaints of undue strain, localized fatigue, discomfort, or pain that does not go away after overnight rest. Job tasks should be analyzed to identify those which require repetitive and forceful exertions; frequent, heavy, or overhead lifts; awkward work positions; or use of vibrating equipment.

The employer should review trade publications, insurance communications, and other references which might indicate risks of WMSDs connected with job operations in the employer's business. Cases of WMSDs found among competitors or in similar businesses should be noticed. Change proposals to increase line speed, retool, or modify jobs to increase individual worker output and overall productivity should analyzed for WMSD risk factors. After recognizing and identifying WMSD risks, the extent of the problem will determine the level of effort required to provide a reasonable prevention effort.

Step 2: Set the stage for action. Ergonomics should be incorporated as part of a company's safety and health program. The principles to control workplace hazards—including hazard identification, case documentation, assessment of control options, and health care management techniques—should include ergonomics problems. As expressed earlier in this book, safety efforts require the involvement and commitment of management and workers. Their involvement, and endorsement of, ergonomics-related efforts must be part of the safety and health program. The program should be designed to cut across organizational units to facilitate recognition, evaluation, and control. Inputs from personnel in safety and hygiene, health care, human resources, engineering, maintenance, human factors/ergonomics should be targeted.

Step 3: Train personnel in order to build in-house expertise. Ergonomics is a recognized discipline in academia. The number of persons having degrees in this area is relatively small. The discipline extends to the Ph.D. level. Their expertise can be tapped to provide training with the aim of enabling managers, supervisors, and employees to identify aspects of job tasks that increase the risk of WMSD. If outside expertise is elected to be used, the instructor should become familiar with company operations, policies, and practices before starting the training program. They should tailor their instruction to the needs of the employers and workers. In-house training materials are available from NIOSH and other groups.

Managers should ask training providers to develop behavioral objectives. These include, but are not limited to, the ability of those who successfully complete the course to:

- Recognize workplace risk factors for WMSDs and understand general methods for control.
- Identify the signs and symptoms of WMSDs that may result from those risk factors.
- Be familiar with company health care procedures.
- Know what the employer is doing to address and control risk factors, and what the employee's role is in them.
- Know reporting procedures for risk factors and WMSDs.
- Demonstrate the way to do a job analysis for identifying risk factors of WMSDs.
- Select ways to implement and evaluate control measures.

Behavioral objectives are objectives that can be measured in some way or another to determine whether or not the trainee has reached an acceptable level of performance. Measures typically include tests of knowledge and skills.

Step 4: Gather and examine evidence of WMSDs. Health indicators to be reviewed include: workers reports of symptoms or risk factors, OSHA logs and other records, results of symptoms surveys and of periodic medical examinations. Identification of risk factors in job tasks should include: screening jobs for risk factors of awkward posture, forceful exertions, repetitive motions, contact with sharp or hard edges or other things that create pressure that can inhibit nerve function and blood flow, and vibration. Other conditions such as cold, work-rest insufficiency, paced work, or new and unfamiliar work should be evaluated. The duration of exposure to risk factors should be examined. Checklists and walk-through surveys are useful. Well-thought-out interviews with workers and supervisors can provide evidence that requires evaluation. A job analysis should be performed which breaks selected jobs down into elements or actions, describes them, and measures and quantifies risk factors inherent in them.

Step 5: Identify effective controls for WMSD-risky tasks, and evaluate these controls to determine effectiveness. Three types of controls are: engineering controls, administration controls, and use of personal protective equipment. The first priority is to eliminate the risk if possible. If this is not feasible, the next priority is to minimize the risk as much as possible. Control for the exposure that remains is to provide and use personal protective equipment. These priorities are accomplished through reducing or eliminating potentially hazardous conditions, using engineering controls, and changing work practices and management policies (i.e., administrative controls).

Engineering controls include design or redesign of the job. This includes the workstation layout, selection and use of tools, and work procedures to take account of the capabilities and limitations of the workforce. Administrative control strategies include the following:

1. Changes to job rules and procedures, such as scheduling more rest breaks, providing personal protective equipment
2. Rotating workers' exposure to jobs that are physically tiring
3. Training workers to recognize ergonomic risk factors and to use techniques to reduce stress and strain while working.

It should be noted that, at the time of this writing, the effectiveness of personal protection equipment to prevent or reduce WMSD is in doubt. Further scientific evidence of effectiveness is needed for various braces, wrist splints, back belts, and other such commonly used devices. Any use of devices should be evaluated carefully. In fact, it is essential to have a program to evaluate any and all control measures implemented, in order to determine their effectiveness to reduce the incidence rate and the severity rate of WMSDs. These measures should be coupled with measures in productivity, quality, job turnover, and absenteeism.

Step 6: Implement health care management. Early detection and prompt treatment can result in timely recovery. Employers, employees, and health care providers are responsible to see that this happens to a reasonable extent. The employer should provide the climate that produces this care through: employee encouragement and training, health care providers' exposure to company jobs and job tasks, support of reasonable job modification and employee accommodation, and ensuring of appropriate employee privacy and confidentiality of medical conditions. The employee should follow workplace safety and health rules and work practice procedures and should report early signs and symptoms of WMSD. Health care providers should gain experience and training in the evaluation and treatment of WMSDs, learn about employee job activities, provide employee privacy and confidentiality, and carefully evaluate symptomatic employees.

Step 7: Maintain a proactive ergonomics program. The ergonomics program should be a continuous program to minimize risk factors for musculoskeletal disorders. It is less costly to build good design into the workplace than to redesign or retrofit later. Therefore ergonomics should be incorporated when planning new work processes and operation or when making changes to existing work processes.

BIBLIOGRAPHY

[68] U. S. Department of Labor, Occupational Safety and Health Administration, Remarks of Joseph A. Dear before the American Psychological Association Conference "Work, Stress, & Health '95," Sept. 14, 1995.

[69] E. R. Tichauer, *Occupational Biomechanics,* Rehabilitation Monograph No. 51, The Center for Safety, School of Continuing Education and Extension Services, New York University, New York, NY 10003, 1975.

[70] NIOSH, "Work-Related Musculoskeletal Disorders," NIOSH Facts, NIOSH Publications Office, May 1997.

[71] National Institute of Occupational Safety and Health, *Musculoskeletal Disorders (MSDs) and Workplace Factors,* Ed. B. P. Bernard, NIOSH Publications Dissemination, Cincinnati, Ohio 45226-1998.

[72] K. Kroemer, H. Kroemer, and K. Kroemer-Elbert, *Ergonomics* (Upper Saddle River, NJ: Prentice-Hall, 1994), 766 pp.

[73] NIOSH, "Carpal Tunnel Syndrome," *NIOSH Facts,* NIOSH Publications Office, June 1997.

[74] Thomas J. Armstrong, *An Ergonomics Guide to Carpal Tunnel Syndrome*, American Industrial Hygiene Association, 1983.

[75] NIOSH, *Workplace Use of Backbelts, Review and Recommendations,* DHHS [NIOSH] Publication No. 94-122, NIOSH Publications Office, 1994.

[76] J. F. Kraus et al., "Reduction of Acute Lower Back Injuries by Use of Back Supports," *International Journal of Occupational and Environmental Health,* Vol. 2, pp. 264–273, 1996.

[77] NIOSH, *Elements of Ergonomics Programs: A Primer Based on Workplace Evaluations of Musculoskeletal Disorders,* DHHS (NIOSH) Publication No. 97-117, March 1997.

EXERCISES

18.1. A NIOSH study concluded that there was a substantial body of evidence of an association between MSDs and certain work-related factors. List some of those factors.

18.2. What are the effects of overuse disorders?

18.3. What worker-related factors are associated with musculoskeletal disorders?

18.4. Name the carpal tunnel syndrome symptoms.

18.5. What are some of the nonoccupational factors of CTS?

18.6. Discuss the prevalence of low back pain.

18.7. What are the three general parts of most back injury prevention programs found in industry? Criticize them.

18.8. List the six lifting methods mentioned in this chapter. Which one is correct?

18.9. To minimize injuries from lifting that cannot be eliminated involves: the lift itself, the object lifted, the means to lift, and the lift environment. Discuss each of these.

18.10. Discuss how some feel that risk homeostasis is a factor in the use of back belts.

18.11. Seven steps for evaluating and addressing MSDs in the workplace are given in the text as part of an industry program using ergonomics. List and discuss each.

CHAPTER 19

Heat and Temperature

High and low temperatures, heat, cold, and the variations of each can all be directly injurious to personnel and damaging to equipment (Fig. 19-1). Effects can be generated, for example, by thermal changes in the environment, which lead to accidents and therefore indirectly to injuries and damage.

EFFECTS ON PERSONNEL

The principal immediate means by which temperature and heat can injure personnel is through burns which can injure the skin and, sometimes, muscles and other tissues below the skin (Fig. 19-2). Skin burns are classified into three degrees of severity.

First-Degree Burns

First-degree burns only cause a redness of the skin, which indicates a mild inflammation, of which the most common is sunburn. This may act as a measure of heat required to produce a skin burn. The rate and severity at which a person will burn depend on the intensity of thermal energy transfer (by radiation, convection, or conduction), the absorptivity of the skin, and the length of exposure.

The occurrence of first-degree burns may vary directly with the time of thermal transfer; for example, the same total amount of heat received by radiation may be more damaging if received in a shorter period of time. Contacts between skin and a hot surface will produce similar types of burns, the severity of injury being dependent on the temperature of the surface contacted and the duration of the contact. First-degree burns may occur with hot surfaces where temperature or contact time are less than those required to cause second-degree burns. Figure 19-3 indicates the effects on the skin in contact with surfaces at different temperatures. Figure 19-4 presents experimental data on burns which resulted from subjecting persons to various heat intensities for slightly more than half a second.

Some persons may become inured to levels of heat and temperature. The skins of some may become tanned, others may grow accustomed to different degrees of

HEAT AND TEMPERATURE

General: heat and temperature are inseparable, temperature being an indicator of the level of heat present. The second law of thermodynamics states that heat will flow from a body at one temperature to one at a lower temperature. The difference in temperature between two points is therefore one indicator of the force which will cause heat to flow.
Temperature differences: heat in a body measurable by temperature is "sensible" heat. The transfer of heat due only to difference in temperature is therefore "sensible heat flow." The mechanics by which these take place are radiation, conduction, and convection. The importance of each depends on numerous factors involved in each individual situation. Radiation is the sole means by which the earth receives heat from the sun. It is of prime concern in other processes, especially where temperatures are very high. Conduction has its greatest importance in matters involving solids, especially metals, and to a lesser extent liquids and gases. Convection is principally concerned with fluids; the relationship with solids is generally at an interface, where conduction also takes place. In some cases convection has been considered a special category of conduction in fluids.
Radiation: any body with a temperature above absolute zero will emit thermal radiation in the form of electromagnetic waves such as visible light or infrared. Like light, it will travel through a vacuum or clear glass, be reflected, and vary in intensity with the square of the distance from the source. Energy radiated is dependent on the temperature and area of the radiating body, not on its mass. The ratio of heat radiated by an actual surface compared to that of a perfect radiating surface ("black body") of the same area and temperature is the emissivity or emittance of the surface.
Conduction: transfer of thermal energy from the molecules in one portion of a substance, or from one substance to another, without physical movement of the substance itself.
Convection: transfer of heat through a fluid by movement of its molecules. The flow of heat may be from or to the surface of a solid, but the medium by which it takes place is a fluid, gas, or liquid. Circulation caused by differences in density within the fluid is known as natural or free convection. Movement of fluid created mechanically by equipment such as pumps or blowers is forced convection.
Multiple flow processes: heat can flow from a surface by more than one means at the same time. In other cases a body may be receiving heat by one means and losing it by another. For example, the sun may be warming a body while air currents cool it.
Heat of reaction: the principal source of heat produced by man is through controlled exothermic chemical reactions, of which two, dissociation and combination, are of special interest. A chemical reaction usually involves consideration of heat of dissociation when the molecule is broken down initially, and the heat of combination (or formation) when new molecules are created.
Heat of combustion: combustion is a complex process involving dissociation and combination. The heat emitted is the summation of all the exothermic and endothermic reactions involved.
Heat of dissociation: in most chemical reactions the first step is dissociation; in many it is the entire reaction. For example, the chemical reaction which results in a nitroglycerin detonation is a dissociation.
Heat of metabolism: life processes involve the use of heat produced by the metabolism of the body as it burns food to release energy. Metabolic heat may create adverse conditions leading to discomfort and errors that produce accidents. It creates problems in cooling in closed environments when a large number of persons are present, or when a person must work in an impermeable protective suit or in a restricted enclosure.
Heat of solution or dilution: some substances are exothermic in the presence of water, giving off heat when no molecular dissociation occurs. Some of these reactions are violent: so much heat is transmitted that the liquids boil and splash. For example, students learning elementary chemistry procedures in laboratories are cautioned to add strong sulfuric acid or strong sodium hydroxide to water, not water to the acid or hydroxide. The wrong procedure could cause the mixture to be thrown on the person doing the pouring.
Electric heat: heat is produced in electrical systems principally by resistance of a conductor to flow of current.
Mechanical heat: the gas laws are equations that indicate the relationships among pressure, volume, and temperature of a gas. When any of these is changed, work is done that involves either heat flow,

FIGURE 19–1 Heat and Temperature

change in temperature, or both. An isothermal process is one which involves flow of heat but not change in temperature (for example, during boiling or condensation). An isentropic processes involves no heat flow from or to the substance, but may involve changes in pressure, volume, and temperature.
Friction: conversion of mechanical energy to heat generally by continuous motion. It may occur by solid-to-solid contact, and also by contact between a solid and a fluid.
Latent heat: heat which must be absorbed when a substance changes from solid to liquid or liquid to gas, or emitted during changes from gas to liquid to solid. At any given pressure, the temperature at which a change of phase of any substance occurs and the latent heat that is required, will remain the same. For example, at sea level (14.7 psia), approximately 970 Btu are required to evaporate 1 pound of water. This will occur at 212°F. To freeze 1 pound of water, 143.5 Btu must be removed from water that is at 32°F.

FIGURE 19–1 Continued

bearable temperature. The effects of cold on skin and body tissues may produce effects similar to those of heat, with severities and effects varying in the same way.

Second-Degree Burns

Second-degree burns are much more serious than are first-degree burns. Blisters of the skin will form, and in severe cases, fluid will collect under the skin. The skin beneath the blisters is extremely sensitive, red in color, and exudes considerable amounts of fluid. Second-degree burns are sometimes much more painful than are third-degree burns, since the nerve endings may be exposed; in third-degree burns the endings are deadened. Broken blisters will expose the body to possible sources of infection. The most common cause of second-degree burns, like that of first-degree burns, is solar radiation.

Third-Degree Burns

Although the damage is more serious, the destruction of the nerve endings in third-degree burns can cause less pain than lesser burns that leave them exposed.

In third-degree burns, the skin, subcutaneous tissue, red blood cells, capillaries, and sometimes muscle are destroyed. Burned skin may be white, light gray, or even charred black. Third-degree burns from moist heat, such as from steam scalds or immersion in very hot fluid, are usually white, in folds, and sometimes with separated outer layers. Those from dry heat, such as fires, are charred and blackened. The areas around a third-degree burn may also contain burns of lesser degrees. Third-degree burns due to immersions in scalding liquids are much more damaging than are spills of hot liquid on the skin. Spilled liquids are generally limited to film depth, exposure of fluid is usually less, and the evaporation of the hot liquid provides a degree of cooling.

Burns caused by direct contact between the skin and a high-temperature substance may be hundreds of times more damaging than heat flow by radiation or convection of air because of tissue destruction, loss of plasma from the body, and disturbance of the body's fluid and chemical balance. Destruction of capillaries can

THE SKIN

The skin protects underlying structures from damage and invasion by harmful organisms; acts as a sense organ; helps in regulating body temperature; stores fat, water, and other substances important to the body; and excretes water and oils. The skin of an adult has an area of about 20 square feet. Its thickness varies from about 1 millimeter over the eyelid to about 2–3 millimeters on the back, palms, and soles.

It consists of two main parts, the epidermis (corneum) and dermis (corium). The epidermis is the outer portion, composed of a number of layers of cells. Overall, the epidermis is comparatively thin, but its thickness varies considerably from one part of the body to another, also being thickest on the palms of the hands, the soles of the feet, and other locations subject to rubbing or pressure. It is free of blood vessels. The outer portion of the epidermis is made up of several layers of different kinds of lifeless and scale like cells. These are sloughed off constantly and replaced by cells from below.

The inner layer of the epidermis performs a number of functions. It undergoes frequent cell division, which replaces those cells which are compressed to form the outer layer of epidermis. Inpocketings of cells from the inner layer of the epidermis extend deep into the dermis to make up the hair follicles, oil and sweat glands, and nails. The inner epidermal cells also contain the pigment, which gives the skin its principal coloration. (The outer layer is normally translucent, with a yellowish tinge.)

The dermis (corium) is much thicker than the epidermis. It holds the glands and hair follicles, which are ingrowths from the inner layer of the epidermis, blood and lymph vessels, fat cells, sense organs, and nerves. The fat acts as a cushion against impact injury, assists in preventing excessive loss of body heat, acts as a food reservoir, and guards delicate structures by absorbing outside shocks.

Hair is present over most of the body. Each consists of a *shaft,* which projects beyond the surface, and a root embedded in the skin. The root and two layers of cells form *the hair follicle,* and consist of tightly packed dead cells. There are a countless number of these follicles. Each is associated with an oil (sebaceous) gland, which secretes a film of oil to keep the skin from drying out and cracking. The oil also keeps the hair pliable and moist.

The subcutaneous layer below the dermis contains the sweat glands. There are approximately 2.5 million sweat glands all over the body. They are most numerous on the palms of the hands, soles of the feet, forehead, and in the armpits. The sweat glands pass perspiration to the surface of the skin where it evaporates and cools the skin and body. Perspiration is over 99 percent water, with traces of protein and some sodium chloride. Moisture which passes through the sweat glands is brought there through the blood vessels in the dermis. The amount of moisture increases as the flow of blood increases due to a warm environment, physical exertion, stress, or illness.

The color of the skin is determined principally by pigments in the second layer of the epidermis. The overlying layer is almost transparent. The blood vessels near the surface also vary the color shade of the skin. When capillaries are contracted (vasoconstricted), as when affected by contact with a cold surface, the skin appears pale. When the skin is warmed, the capillaries expand (vasodilation), blood flow increases, and the skin appears redder.

FIGURE 19–2 The Skin

permit gangrene to set in, because blood cannot flow to any remaining tissues. Facial burns often result in injury to the upper respiratory tract, usually with swelling which can cause obstruction to breathing.

Freezing of skin and tissue can produce effects similar to those of third-degree burns, destruction of capillaries that carry blood, and gangrene. Cryogenic burns, caused by extremely cold fluids or metal, act in the same way; because of extremely low temperatures, the intensity is greater than with usual freezing. A cryogenic gas in

Severity Level	Characteristics
0	No burn
1+	Reddening erythema
2+	Patchy white coagulation; the epidermis can be pushed aside and removed rather easily. The underlying dermis is pink.
3+	Solid white coagulation; the epidermis can be removed but not as readily. The underlying dermis is grayish white.
4+	Steam blebs
5+	Carbonization

FIGURE 19–3 Burn Classification by Perkins

EFFECTS ON SKIN IN CONTACT WITH SURFACES AT DIFFERENT TEMPERATURES*

Temperature (°F)	*Sensation or Effect*
212	Second-degree burn on 15-second contact
180	Second-degree burn on 30-second contact
160	Second-degree burn on 60-second contact
140	Pain: tissue damage (burns)
120	Pain: "burning heat"
91± 4	Warm; "neutral" (physiological zero)
54	Cool
37	"Cool heat"
32	Pain
Below 32	Pain; tissue damage (freezing)

*R. F Chaillet et al., *Human Factors Engineering Design Standard for Missile Systems and Related Equipment,* U.S. Army Human Engineering Laboratories, AD 623-731, Sept. 1965.

FIGURE 19–4 Effects on Skin in Contact with Surfaces at Different Temperatures

contact with a body part, which is usually warm, generates a layer of gas which momentarily acts as protection to the skin. Direct contact between metal and skin may result in the two adhering to each other.

CLASSIFICATION OF BURN SEVERITIES

Severities of burns have been classified in a number of ways, of which two follow. Blocker[76] has presented one classification of burn injuries according to criticality:

1. Critical burns
   a. Second-degree burns exceeding 30 percent of the body surface.
   b. Third-degree burns exceeding 10 percent of the body surface.
   c. Burns complicated by respiratory-tract injury, major soft-tissue injury, and fractures.
   d. Electrical burns.
   e. Third-degree burns involving critical areas, for example, the hands, face, or feet.
2. Moderate burns
   a. Superficial second-degree burns of more than 15 percent of the body surface.
   b. Deep second-degree burns of 15 to 30 percent of the body surface.
   c. Third-degree burns of less than 10 percent, excluding the hands, face, and feet.
3. Minor burns
   a. First-degree burns.
   b. Second-degree burns of less than 15 percent of the body.
   c. Third-degree burns of less than 2 percent of the body surface.

A second system of rating burns using six levels of severity was developed at the University of Rochester.[77] As listed in Fig. 19-5, burns of 2+ or worse are considered

THERMAL ENERGIES APPLIED OVER 540 Msec
REQUIRED FOR SKIN BURNS*

Degree of Burn	*Thermal Energy*	
	White Subjects (cal/cm^2)	Black Subjects (cal/cm^2)
First degree	3.2	Not stated
Second degree	3.9	1.8-2.9
Third degree	4.8	3.3-3.7

*C. S. White and I. G. Bowen, *Comparative Effects Data of Biological Interest,* Lovelace Foundation for Medical Education and Research, Albuquerque, NM, Apr. 1959.

FIGURE 19–5 Thermal Energies Applied over 540 msec Required for Skin Burns

incapacitating. Those of levels between 1, 2, and 3 are mild if the area diameter is less than 0.40 inch; intermediate, between 0.40 and 0.57 inch; and severe if more than 0.57 inch. Depth of burn increases with exposure time.

Tolerance to Burns

Turnbow[78] indicates that a person's ability to survive exposure to heat is governed by two factors: (1) tolerance to pain, and (2) heat-exposure level at which second-degree burns begin. Both of these are personal factors. The report goes on to quote studies indicating that persons will suffer "unbearable pain" when skin temperature is between 108° and 113° F; and that at skin temperature of 111° or higher, injury will result because the rate of cell destruction is greater than that of repair. The conditions described must be examined carefully, especially the qualification of "normal persons." A case in point is that Japanese baths generally consist of water heated to 110°F ± 2° to 3°F. Experience has shown that by gradual acclimatization, Westerners also become accustomed to such bath temperatures.

BURNS TO THE EYE

Welding and new process techniques can cause damage to the eye, which is more sensitive than the skin. Some of these new processes use strong electromagnetic radiation, such as ovens and other drying processes, radars, and lasers. The injuries that can result are principally thermal in nature. The sources and effects of these types of radiation are presented in Chapter 27.

OTHER TEMPERATURE EFFECTS ON PERSONNEL

Normal body temperature is approximately 98.6°F. Heat production by metabolic oxidation of food warms the body and is balanced by heat losses. Where there is an imbalance, the net difference between production and loss governs body temperature. In a cold environment, metabolism must be increased to match losses; in warm surroundings the body must lose any excess or store it. The relationship between the various factors may be expressed by

$$S = M - E \pm R \pm C \pm D - K - F$$

where
S = heat excess or deficiency
M = heat produced by body metabolism
E = heat loss by evaporation
R = heat loss or gain by radiation
C = heat loss or gain by convection
D = heat loss or gain by conduction
K = heat expended in physical exertion
F = heat loss due to respiration, excretion, urination, etc.

Metabolism always involves production of heat. An individual has an energy input which depends on the calorie content of the food he or she eats. Any excess over losses is stored in the body as fat. If energy from body stores or other sources is not available, body exertion must or will be reduced or the person will lose body fat.

During activity the body's metabolic rate increases. A hot environment will increase heat gains by radiation, convection, or conduction. A humid environment will lessen heat loss from the body by evaporation. The body will attempt to balance any heat gain by perspiring to increase cooling by evaporation. If the loss of heat does not balance the gain, the body will suffer.

HIGH TEMPERATURES

Continued exposure to high temperatures and humidity (Fig. 19-6) or to hot sun is a common cause of heat cramps, heat exhaustion, or heat stroke. The same degree of exposure may produce different effects, depending on the susceptibility of the person exposed. Heat cramps are an initial form of heat exhaustion, but both differ from heat stroke. As the name implies, heat cramps involve muscular pains and spasms, cold sweating, and vomiting.

Heat evaporation is caused by excessive perspiring, because of a hot environment or strenuous physical exertion. The ability of the body to carry heat to the skin is inadequate, and certain critical areas are deprived of blood. The skin may initially be red, but as blood is transferred to more critical areas, it will turn white as the blood vessels constrict. A person suffering from heat exhaustion may lose the ability to stand erect. If the person has fainted, he or she will generally regain consciousness after the body is lowered from an upright position. The person will be weak, dizzy, and suffer from nausea, generally because of the loss of salt from the body while perspiring. People's abilities to avoid heat exhaustion vary. Where exposure to high-temperature atmospheres cannot be avoided, personnel must add large amounts of salt to their diets.

Heat stroke is more serious than heat cramps or exhaustion. The patient becomes dizzy, weak, and irritable. Vision may be blurred with a reddish haze. The patient will have a temperature of 105°F or more, with correspondingly hot skin, but it will be dry, with no perspiration. Breathing may resemble snoring. The patient may vomit, suffer convulsions, and lose consciousness.

Temperature and Performance

Numerous investigators have studied the effects of temperature on performance. In almost all instances, there is agreement that stresses generated by high temperatures degrade performance. However, opinions differ on the degree and point at which degradation begins. Surveys of pertinent literature on the subject indicate that numerous factors will affect heat stress. In most cases these factors will be additionally detrimental; in others, they mitigate heat's adverse effects. These factors have been classified[79] broadly into five categories, which indicate that the effects of heat will depend on the following factors:

1. Intensity of the heat. (It should be understood that the term "heat" in this specific connotation implies conditions of high temperature and humidity which produce discomfort.)

MOISTURE AND HUMIDITY

General: when water is present as a liquid, molecules leave its surface and move into the immediately adjacent space. Here they continue to move about, some returning to the liquid surface, others diffusing farther away. When the number of molecules returning is the same as those leaving, the liquid-vapor mixture is said to be in equilibrium. The space immediately above the surface is then saturated with water vapor. The force exerted by the molecules of water at and near the surface to leave the liquid is called the vapor pressure of the liquid. To become a gas, molecules in the liquid must overcome the forces that attract them to other molecules. In doing so, they absorb energy, or latent heat of vaporization. The rate at which molecules leave the liquid and the vapor pressure are governed by temperature alone, increasing as temperature increases, decreasing as temperature is reduced. When the temperature of a saturated space is reduced, more moisture will condense than evaporate until equilibrium is again reached. The temperature at which saturation occurs and condensation begins is the dew point, or saturation temperature.

Absolute humidity: weight of moisture in a cubic foot of space. At any specific temperature, the vapor pressure and absolute humidity will be the same at sea level as at any other altitude. The pressure over the liquid has no bearing on humidity.

Relative humidity: ratio (as a percentage) between the actual pressure of water vapor present to the saturation vapor pressure at the same temperature. A simpler way to compute relative humidity is to divide the weight of water gas present in 1 cubic foot of space by the weight of water that would be present at the saturation temperature.

Measuring humidity: the psychrometer is the most common instrument for measurements of humidity. Two similar thermometers are used. The bulb of one is encased in a wet, porous cloth sleeve. The moisture in the cloth evaporates if the air tested is not saturated. Evaporation of water requires that heat be absorbed from its surroundings. This cools and contracts the liquid in the thermometer, lowering the reading. The result is the wet-bulb temperature. The ambient, or dry-bulb temperature, is read from the other thermometer. The two temperatures are used with a psychometric table or chart to determine relative and absolute humidities, dew point, and other conditions of an air-water mixture.

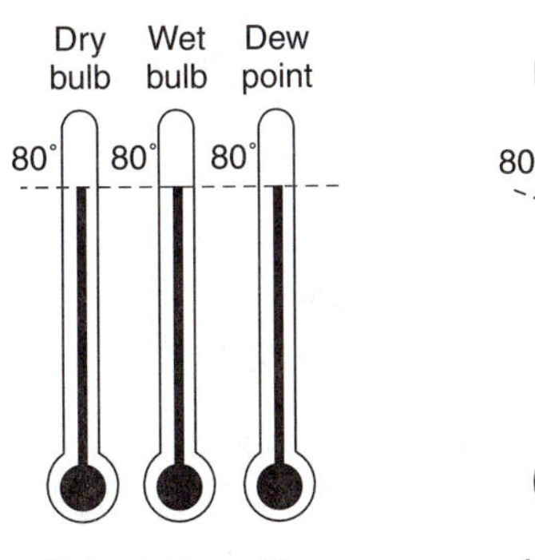

Saturated condition

Dry bulb Wet bulb Dew point

80° 73° 70°

Unsaturated condition

Effective temperature: comfort index which combines the effects on a body of temperature, humidity, and air movement. It is not a temperature measurable with instruments. It is equivalent to the comfort a person generally feels (there are always exceptions) in a saturated atmosphere with the same dry-bulb temperature, and with a specific movement of air. A person should be equally comfortable

FIGURE 19–6 Moisture and Humidity Continued on page 300

under different conditions, provided the effective temperature remains approximately the same.

Wet-bulb globe-temperature index (WBGT): effective temperature includes no provision for gain or loss of heat by radiation. (In an enclosure, the effective temperature could remain the same, but the person would feel cold if the walls were cold, warm if they were warm or hot. These effects are frequently noticeable in a room maintained at constant temperature by air condition. A person sitting near a window or badly insulated wall will be warm on a hot, summer day and cold on a frigid, wintry day.) To compensate for the radiation effect, a globe thermometer is used instead of a dry-bulb thermometer. (The globe thermometer is a black-painted copper sphere, 6 inches in diameter, into which a thermometer is inserted. It therefore measures temperatures which include radiant heat effects.) Personnel exposures can then be calculated by one of two formulas:

Indoors or outdoors, with no sun load: WBGT = 0.7 WB + 0.3 GT
Outdoors with solar load: WBGT = 0.7 WB + 0.2 GT + 0.1 DB

If it is necessary to determine the average exposure of a person over a long period of time when the WBGT varied, a time-weighted average can be obtained:

$$\text{Average WBGT} = \frac{\text{WBGT}_1 \times t_1 + \text{WBGT}_2 \times t_2 + \ldots + \text{WBGT}_n \times t_n}{t_1 + t_2 + \ldots + t_n}$$

Exposures should not exceed those in the following chart:

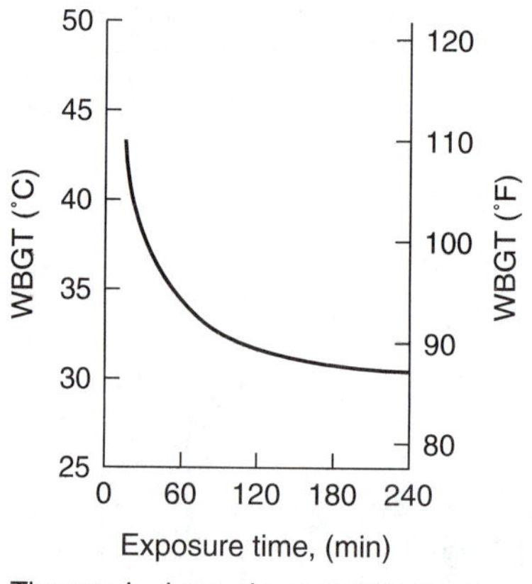

The graph shows the upper limits of exposure for unimpaired mental performance.

FIGURE 19–6 Continued

2. Duration of the exposure period.
3. Tasks involved.
4. Persons performing the tasks.
5. Presence of other stresses.

High humidity may cause psychological and physiological stresses in personnel, especially at high temperatures. Personnel working in such atmospheres become irritable, suffer fatigue more easily, have less ability to concentrate, and make numerous errors. Some suffer from headaches caused by sinus trouble, prickly heat, and fungus infections. The amount of work that can be accomplished at high temperatures decreases as humidity increases. The length of time protective clothing impermeable to moisture can be worn decreases as the humidity builds up inside the clothing. Provisions must therefore be made to minimize the presence of moisture inside garments where temperatures are elevated.

Figure 25-3 indicates some degrees of comfort and discomfort in various environments. Tests have shown that in general personnel will be comfortable in temperatures of 70° to 80°F even when the humidity is as high as 70 percent. The exact amount will vary with the person, the clothing worn, the activity, the wind velocity, and numerous other factors. At a temperature of 85°F and a humidity of 90 percent or more, a person may begin to feel uncomfortable. At 92°F or higher with a humidity of 90 percent or more, there may be increases in body temperature, pulse, and breathing rates, accompanied by heat exhaustion. The body temperature increases because evaporation and heat loss from the skin are inadequate to dissipate the heat produced by the body. At temperatures of 95°F or more, evaporation of perspiration constitutes almost the entire means by which the skin is cooled to maintain normal body temperature.

High relative humidity means there will be little or no evaporation, because a high percentage of moisture is already present. Certain environments fall in the latter category, especially when they are not fitted with air-conditioning equipment or when it fails: windowless industrial buildings, protective suits of impermeable material, steam process plants, and laundries.

A report prepared by the British Navy[80] on conditions during World War II describes the effects on personnel in non-air-conditioned submarines. During one patrol in the Indian Ocean, the submarine ran submerged for 13 hours. The wet-bulb temperature was 92° to 93°F. There were four cases of heat stroke, all personnel developed prickly heat, and 10 men had swollen ankles from excess body fluids. It was estimated that personnel efficiency was 60 percent of normal.

In addition, the following statement was made about the effect of high temperature and humidity on the crews of surface ships in the same region:

> There was little doubt that efficiency was reduced amongst the other members of the naval community who were adversely affected by the warmth, but it was not always easy to prove. There was, however, considerable circumstantial evidence. Men decoding signals in cipher offices, accountants, store officers, and others engaged in sedentary mental work, described how they were unable to think in their hot offices, and either felt sleepy or made unreasonable mistakes. Others remarked how mistakes were frequently made without the knowledge of the person making the mistake. Officers in charge of gunnery control

> positions noticed that the crews of transmitting stations and calculating positions made more individual mistakes, and showed poorer team work, than in home waters. Engineer crews noticed that the workmanship of their skilled technicians was of a poorer quality than the work turned out in cooler water, and jobs took longer to complete. The winding on electrical armatures in electrical shops took longer than in home waters although more men were occupied on this work. Heavier aircraft, with bombs weighing 500 to 600 pounds each, took half as long again in the tropics as in temperate waters. The bomb magazines were very hot in certain ships; and 126°F dry-bulb was recorded in a magazine in a fleet carrier in the Red Sea. In addition it was common belief that everyone was below par.

After humidity, the next most significant factor in increasing heat stress is duration of exposure. Short-term exposures of less than one hour to ambient dry-bulb temperatures of 160°, 200°, and 235°F caused no significant impairment of performance by persons doing additional tests. However, one hour is probably the upper limit of time during which high temperature will not degrade performance. Even these data cannot be applied universally, since they are based on results for specific tasks. In one case the task involved learning, which may have increased the tolerance to heat. The data for experienced subjects would probably be similar, but their levels would differ, depending on the other stresses involved. The maximum level for reliable human performance would be an effective temperature of 85°F.

Acclimatization is another major factor. Short exposure to high temperature generally will not affect performance; longer exposures may result in degradation. However, it has been found that continued exposure to adverse conditions, such as high temperature, enables persons to withstand heat successfully.

The length of time for acclimatization varies. The ability to withstand thermal stresses depends on the individual; some persons require little or no adjustment for a changed condition. Again, response to acclimatization is subject to individual differences in adaptation and motivation.

Ability to withstand high temperatures without impairment of physical or mental performance appears to be affected by such other adverse factors as the following.

- A disagreeable odor (see Fig. 25-4) may make a person irritable at any time. An odor that can be tolerated at low temperatures not only is frequently offensive at high temperatures but reduces the tolerance to heat. In addition, high temperatures often cause perspiring and additional odors, and an increased evaporation of liquids and moisture brings out still more offensive odors and lower tolerance to ambient temperatures.
- Fatigue and lack of sleep both also reduce tolerance to high temperatures. Conversely, warm and moderately elevated temperatures make people sleepy, so they tend to doze off. Such conditions have caused innumerable accidents when personnel controlling moving vehicles have fallen asleep. In less severe cases, vigilance is reduced.
- Worry, frustration, and nervousness not only reduce the ability of a person to withstand heat stress, but may themselves be further increased by high temperatures. This may continue to increase the cycle and a person's inability to withstand high temperatures, causing the person to be more nervous, irritable, and

frustrated. The person perspires more, the ability to withstand stress drops, and the tendencies to and probabilities of errors and accidents increase.

- Smoke from cigarettes, dust, pollen to which a person is allergic, and other particulate matter also reduce the level of thermal stress tolerance. Occupants of rooms polluted with cigarette smoke frequently have increased sensitivity to high temperatures, requiring lower ambient temperatures for the same degree of comfort than in uncontaminated air. Many companies now forbid smoking at all or limit smoking to certain areas such as rest rooms.
- Temporary illnesses, such as cold, motion sickness, or high acceleration, will reduce tolerance to high temperatures. Persons with fevers may find even moderate temperatures objectionable.

Effects of Cold

Cold is generally not a problem in industrial plants, the exception being in those activities where cryogenic or highly refrigerated activities are conducted. In such places, the chief hazard from the effects of cold is contact between the skin and a cold metal surface. The cold burns of a freezing or cryogenic liquid have already been pointed out.

Outdoor activities, such as construction during winter months in frigid climates, can result in other types of cold injury. MacFarlane[81] has listed three types of cold injury that can occur, especially in the body's extremities.

The first is chilblains, a relatively mild form of tissue damage. Poor circulation, as a result of cold, causes damage to the tissue of the extremities. Local itching and swelling characterize this form of injury.

The main examples of wet cold syndrome, a second type of injury, are the well-known trench foot and immersion foot. These result from exposures below 53°F (12°C) for several days. Moisture from cold and sweat contributes to the pathogenesis of these disorders. Feet and legs become cold, pale, and numb, and they cease to sweat. After vasoconstriction, vasodilation takes place and the feet become red and swollen. Nerve injury is frequent. Loss of sensitivity and anesthesia persist for weeks, even after the feet have been warmed. Blood vessels are damaged, and plasma and red cells leak into tissue spaces.

Frostbite results from prolonged and severe vasoconstriction at temperatures below 32°F (0°C). In mild cases, tissue is not necessarily frozen. In more severe cases, ice penetrates the tissue, causing its death, and very often, gangrene sets in if circulation has been severely reduced. On warming of the injured extremities, vasodilation and swelling occur. Considerable pain and damage to the liver, kidneys, and adrenals are sometimes noted.

The mechanisms of the body make adjustments in an attempt to maintain normal body temperature. Under conditions of extreme cold, especially where wind is present, this may be impossible. Injuries will then occur. Severe cold injuries such as those of deep frostbite are generally irreversible, gangrene often setting in so that amputation of the affected extremities is required.

In cool air the body loses heat principally by radiation from exposed skin surfaces and a small amount of convection or conduction. The rate of heat loss from a warm body increases with movement of air across the exposed skin. This will produce a cooling effect and chilling. The heat lost from skin exposed to a 10-mile-per-hour wind when the ambient temperature is 10°F will be the same as that from skin in still air at −9°F. These equivalent temperatures, known as chill factors, are indicated in Fig. 19-7.

The wind chill factor is an indication of relative heat loss only; it does not indicate that tissue will freeze because of the wind. Freezing will not occur unless the temperature is 32°F or lower. For example, the wind chill factor for a 15-mile-per-hour wind at an ambient temperature of 40°F is 22°. Water or blood will not freeze, but the body will attempt to take corrective action to compensate for heat loss. Circulation will increase in the affected area. The body may attempt to provide more heat by shivering. If a body is exposed to a temperature of 32°F or less, wind chill may cause frostbite to occur sooner than it normally would.

The means to mitigate heat-loss problems resulting from wind chill is obvious: obtain shelter of some sort to break the wind. An impermeable cover, a tent or lean-to, the cab of a vehicle, coverage with a blanket of leaves or newspaper, or even a hole in a snowbank will provide protection.

COOLING POWER OF WIND ON EXPOSED FLESH EXPRESSED AS AN EQUIVALENT TEMPERATURE
(under calm conditions)

Estimated wind speed (mph)	Actual Thermometer Reading (°F)											
	50	40	30	20	10	0	−10	−20	−30	−40	−50	−60
	Equivalent Temperature (°F)											
Calm	50	40	30	20	10	0	−10	−20	−30	−40	−50	−60
5	48	37	27	16	6	−5	−15	−26	−36	−47	−57	−68
10	40	28	16	4	−9	−24	−33	−46	−58	−70	−83	−95
15	36	22	9	−5	−18	−32	−45	−58	−72	−85	−99	−112
20	32	18	4	−10	−25	−39	−53	−67	−82	−96	−110	−124
25	30	16	0	−15	−29	−44	−59	−74	−88	−104	−118	−133
30	28	13	−2	−18	−33	−48	−63	−79	−94	−109	−125	−140
35	27	11	−4	−20	−35	−51	−67	−82	−98	−113	−129	−145
40	26	10	−6	−21	−37	−53	−69	−85	−100	−116	−132	−148
Wind speeds greater than 40 mph have little added effect.	Little danger (for properly clothed person). Maximum danger of false sense of security.				Increasing danger. Danger from freezing of exposed flesh.			Great danger.				
Trenchfoot and immersion foot may occur at any point on this chart.												

Source: NAVMED Bulletin 5052-29.

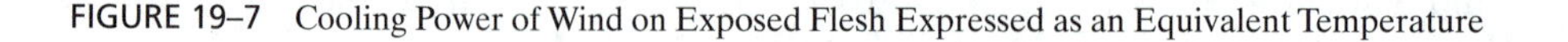

FIGURE 19–7 Cooling Power of Wind on Exposed Flesh Expressed as an Equivalent Temperature

One aspect of cryogenics that might involve hazards to workers is the announcement of superconductivity of electricity at relatively high temperatures. Relatively high temperatures means above the boiling point of nitrogen at – 321° F. The expected great demand for low temperatures may greatly boost the need and use of cheap nitrogen. The major problem with nitrogen has not been its cryogenic cold but the failure of persons to realize that, even at only a few degrees above its boiling point and higher, the odorless and colorless nitrogen has resulted in fatalities. Common use of nitrogen evaporating in closed spaces may prove lethal to workers.

ADDITIONAL EFFECTS

High temperatures, low temperatures, and temperature variations can all cause damage directly or generate other conditions that can result in damage. Almost every industrial plant or operation involved with any chemical process can have safety problems deriving from temperature and heat.

Susceptibility to Fires

Fires and their causes are discussed in Chapter 22. One of the worst effects of elevated temperatures is the increased susceptibility to fire. If the temperature is high enough or the volatiles in the organic material are reactive, a fire may start spontaneously. Dry vegetation and oily rags are highly susceptible to ignition this way. Thermal radiation from flames, molten metal, or other high-temperature sources can cause charring of materials such as wood, paper, and cloth. Charring can also occur when such materials are in contact with a high-temperature source such as a steam line, hot electronic equipment, or an overheated bearing.

Spills of liquid fuels are more hazardous in warm and hot weather than in cold. Not only will the liquid evaporate faster to form large volumes of flammable fuel-air mixtures, but less powerful initiation sources will be required to ignite them. (Spilled gasoline will be less of a hazard during the winter, if out-of-doors, than during the summer.)

Reaction Rates

Rates of chemical reactions will increase with increases in temperature. Hot fuel-air mixtures may explode (see Chapter 23) when ignited, rather than burn as they would when cold. Rates at which other chemical reactions will occur increase with increases in temperature. A chemical process may go out of control if temperature is not maintained within safe limits. The reaction may become so violent that an explosion results. High temperature can cause normally nonreactive material to react. For example, trichlorethylene was considered nonflammable until reports were received of cases in which it exploded. Investigation revealed that dip tanks which held trichlorethylene were being welded when explosions occurred. The tanks had been inadequately cleaned, and the temperature of the welding arc was high enough to cause the trichlorethylene to react by decomposing. Decomposition of trichlorethylene is a violent reaction. (Trichlorethylene was widely used for cleaning all types of materials, from clothing to metals, until it was found to be carcinogenic.)

Because of the increase in reactivity with temperature, reactive materials should be stored in low-temperature spaces. This is especially true of materials such as explosives, which should be stored in magazines where no heat is provided. If explosives are to be used in the field, such as in construction work, they should not be left where solar radiation can raise their temperature.

Other Problems

Increased temperature can generate other problems which lead to failures and accidents.

- *Corrosion:* corrosion at low temperatures may be negligible; at elevated temperatures the corrosion reaction may be rapid. A corroded part can then weaken and fail at a critical instant.
- *Field composition:* Hydraulic fluids may decompose more rapidly, generating contaminants which could plug flow orifices so that equipment fails.
- *Reliability:* reliability of electronic devices is degraded by heat and increased temperature, so that part and equipment failures are more frequent.
- *Increased gas pressure:* increased temperatures will raise the pressure of gas in a confined space (see Chapter 20). If the increases in temperature and heat are great enough, the container in which the gas is confined might rupture. Even a small rise in temperature of a cryogenic liquid could produce a sharp increase in vapor and container pressure, so that the container bursts.
- *Freezing of water:* freezing of water lines may cause bursting of pipes and other closed vessels in which expansion is restricted. Ice or other congealed liquids can immobilize parts designed to move or to restrict the distance they can move. Accumulation of ice can prevent a body from moving along a fixed surface or from separating. The pressure of freezing ice has pushed steel bridges off their foundations. The added weight of accumulated ice could cause the collapse of a structure.
- *Liquid expansion and overflow:* an increase in its temperature could cause a liquid to expand, creating problems. The liquid in a tank completely filled could expand so that it overflows. Such an overflowing flammable liquid would then generate a severe fire hazard. Ullage in closed tanks is provided to lessen the effects of liquid expansion and pressure with temperature.
- *Effects on metals:* an increase in temperature will generally reduce the strength of most common metals. It will also cause most metals to expand, so that there are dimensional changes. If a warmed part is restrained from expanding, it will buckle or burst its restraints. Failure to understand this is a common cause of deformation, damage, failures, and collapse of welded structures. Internal restraints and stresses imposed during high-temperature manufacturing processes, not extreme loads in use, have caused splitting and failures of long, heavy steel girders.

- *Loss of ductility:* reduced temperatures can cause loss of ductility of metals and increase brittleness so that they fail. Brittle failures have caused steel structures such as bridges to collapse, ships and heavy equipment to break up, and gas transmission lines to crack.

Figures 19-8 and 19-9 are checklists to assist safety engineers in making reviews of their plants to determine whether hazards from heat and temperature exist.

Protection from Thermal Extremes

Protection from human exposure to thermal extreme that can result in personal harm or impaired performance can be provided at the thermal source, the pathway to the endangered host, and the host. When the heat or cold is generated by equipment or processes, the *source* of discomfort can be altered in some cases to produce less heat or cold, at less than extreme temperatures. Replacement of the offending source by one less offensive is sometimes feasible. By far, the most common methods of remedy are at the pathway to the host and at the host.

Air conditioning is the most common remedy. The environment along the *pathway* between the host and the source is artificially heated or cooled and/or air is circulated. Of course, this is more feasible for indoor applications. Another *pathway* protection is the use of structure and materials to provide a protective barrier between the source and the host. That barrier could be a thermal insulation material, or a reflective material, or it could be as simple as a roof to provide shade. Contact with hot or cold surfaces can be prevented by providing mechanical barriers. Warning signs can be used in the pathway.

At the *host*, the person in danger, many protective interventions are possible:

- Conduct medical screening to select workers based on health and physical fitness
- Provide acclimatization by graded work and exposure to temperature extremes
- Administer appropriate work-rest cycles to limit exposure
- Provide ample hot or cold liquid, food, salt intake
- Provide or ensure adequate clothing
- Provide protective clothing such as reflective apron, cooling vest, cooling suit, cooling helmet
- Provide "spot" heating or cooling through sources directed at the host
- Monitor workers during sustained exposure to temperature extremes
- Train the worker to recognize symptoms of extreme exposure and to know what action to take

In occupations where they must work in severe cold, workers spend time and energy in self-preservation, and there are many impediments to productivity.[81] Some workers, particularly chemical workers, must perform critical jobs wearing full-body protective suits that increase heat stress.[82] The safety professional should have a thermal stress management program ready to apply as needed.

CHECKLIST— HEAT AND TEMPERATURE

1. Are there sources of heat in the plant having temperatures high enough to cause burns, with which a person might come in contact inadvertently? Can these sources be isolated so that such contacts will not occur?

2. If heated equipment or products must be handled, are they of a temperature which will permit continuous contact?

3. Is any material present which would melt, warp, lose strength, or catch fire because of a high-temperature heat source?

4. Will operating temperatures cause damage to paints or other finishes?

5. Is there any material which will become sensitive or highly reactive because of elevated temperatures?

6. Is there any place where an organic material might char because of contact with a high-temperature source such as a high-pressure steam line?

7. Are containers of pressurized gases with high vapor pressures kept away from or shielded from sources of heat?

8. Are pressurized containers, such as gas cylinders, shielded from the direct rays of the sun?

9. Is a warning provided on small pressurized containers not to throw them in a fire, even when empty?

10. Are persons who must work in enclosed spaces protected against high environmental temperatures which could create tendencies for them to produce errors leading them to accidents?

11. Are persons who must operate a moving vehicle in cold weather protected from a wind chill?

12. Is any cryogenic fluid used in the plant which requires protection against "cold" burns?

13. Is any cold metal surface present which might contact a person and cause his skin to adhere and freeze?

14. Is there any material which will become brittle and break under stress when subjected to a low temperature and result in an accident?

15. Is there a necessity to provide fuels, coolants, or lubricants which are capable of operating at either an extraordinarily high or extraordinarily low temperature?

16. Is it necessary to provide freeze plugs or other safety devices which will prevent damage to a water system if the water freezes?

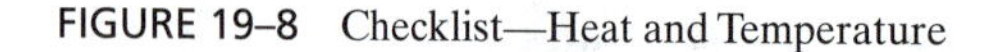

FIGURE 19–8 Checklist—Heat and Temperature

HAZARDS CHECKLIST— HEAT AND TEMPERATURE

Possible Effects

High temperature:
- Burns to personnel
- Reduced personnel efficiency and errors
- Heat cramps, strokes, and exhaustion
- Reduced relative humidity
- Ignition of combustibles
- Charring of organic materials
- Reduced strength of metals and other materials
- Melting of metals and thermoplastics
- Distortion and warping of parts
- Weakening of soldered seams
- Peeling of finishes, blistering of paint
- Expansion causing binding or loosening of parts
- Decreased viscosity of lubricants
- Increased evaporation and leakage of liquids (fuels, lubricants, toxic liquids)
- Increased gas diffusion
- Increased reactivity
- Breakdown of chemical compounds
- Premature operation of thermally activated devices
- Increased electrical resistance
- Opening or closing electrical contacts due to expansion
- Changes in other electrical characteristics

Possible Causes

High temperature:
- Generation or absorption of heat from:
  - Heat-engine operation
  - Fire or explosion
  - Other exothermic chemical reaction
  - Electrical heating
  - Solar heating
  - Aerodynamic or other vehicular friction
  - Friction between moving parts
  - Internal friction due to repeated bending or other work process such as repeated impacts
  - Gas compression
  - Biological or physiological processes
  - Welding, soldering, brazing, or metal cutting
  - Hot climate or weather
  - Organic decay processes
  - Nuclear reaction
  - Immersion in hot fluid
  - Lack of insulation from thermal sources
  - Inadequate heat-dissipation capacity or cooling-system failure
  - Hot spots due to coolant fluid circulation being obstructed

FIGURE 19–9 Hazards Checklist—Heat and Temperature Continued on page 310

Low temperature:
- Frostbite or cryogenic burns
- Icing of operating equipment
- Freezing of liquids
- Condensation of moisture and other vapors
- Reduced viscosity of liquids
- Gelling of oils and lubricants
- Reduced reaction rates
- Increased brittleness of metals
- Loss of flexibility of plastics and organic materials
- Contraction effects, especially opening of cracks in metals
- Jamming or loosening of moving parts due to contraction
- Delayed ignition in furnaces and combustion chambers
- Combustion instability in engines
- Changes in electrical characteristics

Temperature variations:
- Dimensional changes, especially in metals
- Cycling fatigue of metals
- Pressure changes in confined gases and liquids
- Variations in stresses

Low temperature:
- Loss of heat because of:
  - Mechanical cooling or refrigerating processes
  - Heat loss by radiation, conduction, or convection
  - Cold climate or weather
  - Endothermic reactions
  - Rapid evaporation
  - Immersion in cold fusion
  - Presence of cryogenic liquid
  - Exposure to heat sink
  - Gas expansion
  - Joule-Thomson effect

Temperature variations:
- Stopping and starting of heat engines and other powered equipment
  - Diurnal heating and cooling
  - Gain and loss of heat due to changes in radiation, conduction, or convection

FIGURE 19–9 Continued

BIBLIOGRAPHY

[76] T. G. Blocker, Jr., *Studies on Burns and Wound Healing* (Austin: University of Texas, 1965).

[77] J. B. Perkins et al., *The Relationship of Time and Intensity of Applied Thermal Energy to the Severity of Burns* (Rochester, N.Y.: University of Rochester, Dec. 1952).

[78] U. W. Turnbull et al., *Crash Survival Design Guide*, Flight Safety Foundation, U.S. Army Aviation Material Laboratories, Technical Report 67-22, July 1967.

[79] J. F. Wing, *A Review of the Effects of High Ambient Temperature on Mental Performance,* Aerospace Medical Research Laboratories, AD-624-144, Sept. 1965.

[80] F. P. Ellis, *Personnel Research in the Royal Navy, 1939–1945,* Royal Navy Research Committee, London, March 1950.

[81] W. V. MacFarlane, "General Physiology Mechanisms of Acclimatization," in *Medical Biometeorology,* S. W. Troup, ed. (New York: Elsevier Publishing Co., 1963), pp. 372–417.

[82] C. Brian Malley, "Cold Stress Revisited," *Professional Safety,* American Society of Safety Engineers, 1992, pp. 21–23.

[83] J. S. O'Connor II and Kim Querrey, "Heat Stress and Chemical Workers: Minimizing the Risk," *Professional Safety,* American Society of Safety Engineers, 1993, pp. 35–38.

EXERCISES

19.1. What is the difference between heat and temperature?

19.2. Describe the three principal means by which heat can be transferred.

19.3. What is sensible heat? Latent heat?

19.4. Describe first-, second-, and third-degree burns.

19.5. What effects can high temperatures have on a body?

19.6. How do temperature and humidity affect performance?

19.7. Describe five factors that can affect tolerance to high temperature.

19.8. What is effective temperature? WBGT? What is the principal difference between the two?

19.9. What are some of the consequences of exposure to low temperatures?

19.10. What is "wind chill"? Give an example.

19.11. Describe the effect of temperature on reaction rates. How can an increase in reaction rates be dangerous?

19.12. Describe five other effects that temperature can generate.

19.13. Prepare three additional checklist questions that can be used with those listed in Fig. 19-8.

CHAPTER 20

Pressure Hazards

On April 27, 1865, the side-wheeler steamer *Sultana* carried more than 2,000 Union soldiers, far more than her normal capacity, up the Mississippi. Many of the soldiers were eagerly hurrying home after being released from Southern prison camps. Quick repairs had been made to the vessel's leaky boilers at Vicksburg and again at Memphis. A few miles north of Memphis the boilers blew up, with an explosion heard miles away. It tore the *Sultana* apart, hurling men and parts of the vessels hundreds of feet. It was estimated that 1,600 to 1,700 died either from the explosion or from drowning. (No accurate determination of the number of men killed was possible because no passenger count had been made.) The pressures at which the Sultana's boiler normally operated, and even the pressure at which it ruptured so violently, would be considered low compared with pressures in common use today.

It is not necessary to have much pressure to have conditions where serious injuries and damage can occur. It is commonly and mistakenly believed that injury and damage will result only from high pressures. Actually, there is no agreement on the term "high pressure," beyond the fact that it is greater than normal atmospheric pressure.

A hurricane wind can drive a straw endwise into a tree or wooden telephone. Yet a 70-mile-per-hour wind exerts a pressure of only one-tenth of a pound per square inch (0.1 psi) (see Fig. 20-1); a wind of 120 miles per hour has a dynamic pressure of only 0.25 psi.

The American Gas Association indicates that a high-pressure gas distribution line is one which operates at a pressure of more than 2 psi. The American Society of Mechanical Engineers (ASME) rates only those boilers which operate at more than 15 psi as high-pressure boilers. OSHA standards state: "High-pressure cylinders mean those with a marked service pressure of 900 psi or greater." The military services and related industries have categorized above-atmosphere pressures as

Low pressure—1 atmosphere to 500 psia.

Medium pressure—500 to 3,000 psia.

High pressure—2,000 to 10,000 psia.

Ultrahigh pressure—above 10,000 psia.

Pressure: A physical force against a fluid. The unit force magnitude equals the unit pressure times the unit area the force covers. In the United States, the most common indication of pressure is in pounds per square inch (psi). Other units of pressure: pressures are often indicated in metric terminology, feet or inches of water, inches of mercury, atmospheres, and millibars of atmospheric pressure.

Standard water pressure: in the United States, one cubic foot of water weighs 62.4 pounds, which works out to be 4.33 pounds per square inch.

Standard atmospheric pressure: is 14.7 pounds per square inch.

Absolute pressure (psia): is pressure measured from the point at which no particles of any fluid exist to create a pressure. Absolute pressure is that of atmospheric pressure plus that indicated on a gage.

Gage pressure (psig): is that shown on a meter.

Static pressure: is pressure when the fluid is quiescent and the force it exerts is only that due to the gravitational weight of the liquid. Dynamic pressure is pressure exerted due to the kinetic-force movement of a fluid.

Water pressure: is often designated in heights of water above atmospheric. Such usages are generally expressed in feet or inches of water. Vacuum is the measure of pressure less than that of the standard atmosphere.

Stored pressure energy: the expansive energy contained in a fluid.

Pump: a device to increase the pressure of a liquid. A blower does the same for a gas.

Pressure regulator: a device to maintain a constant pressure or flow rate from a source whose pressure must be limited and which might change.

Accumulator: a device used to dampen pulsations in a fluid system.

Pressure relief valve: a device which permits discharge of fluid from a system if it exceeds a set value.

Rupture disk: a thin membrane which prevents flow in a fluid system until the membrane breaks because its designed rating is exceeded, permitting discharge of the fluid.

Ullage: the amount by which a cylinder falls short of being full, usually the amount of gas left to prevent any excessive increase because of a temperature rise.

FIGURE 20–1 Pressure Definitions

The term "high pressure" can therefore be almost any level prescribed for the equipment or system in use. For accident prevention purposes, any pressure system must be regarded as hazardous. Hazards lie both in the pressure level and in the total energy involved. Here, again, the hurricane can be used as an example: its relatively low pressure can exert a force against a building that can devastate the structure.

The term "high pressure" will therefore be used in this chapter in a comparative sense of any pressure greater than that of a standard atmosphere, but without any set quantitative value. Some fundamental precautionary measures to be used with pressure systems are given later in Fig. 20-4.

Pressure-Vessel Rupture

When the expansive force of a fluid inside a container exceeds the container's strength, it will fail by rupture. A slow rupture may also occur by popping of rivets or by the opening of a crack which provides passage for leakage of fluid. When bursting is rapid and violent, the result will be destruction of the container—frequently with fragmentation, and sometimes with generation of a shock wave. If personnel are in the vicinity, injuries could result from impacts (as described in Chapter 13) from fragments. A shock wave produced by a rupture can produce blast effects similar to and as damaging as those generated by detonations (see Chapter 23). The rupture of a pressure vessel occurs when the total expansive force acting to cause the rupture exceeds the vessel's strength.

The process by which a boiler rupture occurs can be described as follows. When heat is applied to water in the boiler, its temperature increases to the boiling point, causing it to evaporate, and the steam to exert pressure. Generally the steam leaves the boiler and is replaced by new supplies of water. When the input of heat from the boiler equals that removed by steam flow, equilibrium is reached and pressure remains constant. If steam flow output is prevented or restricted so it is inadequate to remove all the excess heat supplied, the temperature and pressure in the boiler will increase. If inadequate supplies of fresh water are not provided, any water vapor can turn to dry gas and then increase in pressure. If a safety device is not provided or is inadequate to limit gas pressure to a safe value, the strength of the boiler might be exceeded, causing it to fail. (A boiler can also fail at normal pressures if its strength has been degraded in any way.)

Boilers are therefore equipped with safety valves, which permit pressures to be relieved if they exceed set values (Fig. 20-2). Low points in boilers are provided with fusible plugs. During normal operations the plugs are covered with water, which keeps them relatively cool. If the water level drops too far, the plugs become uncovered, are no longer protected by the water, get hot, and melt. This opens another path for relief of steam pressure.

UNFIRED PRESSURE VESSELS

On January 15, 1919, in Boston, Massachusetts, pressure destroyed a 50-foot-high, 90-foot-diameter tank full of molasses that was used for the fermentation of alcohol. The rupture loosed 2.3 million gallons of molasses into the streets, immediately killing a milkman in a horse-drawn wagon. The flood then knocked down track supports of an elevated railway, so that several persons riding there fell to their deaths. It took months to clean up the molasses. The plant owners were ordered to pay $1 million in damage claims.

Pressure vessels do not have to be fired to be hazardous. Inputs of heat can occur in other ways. The heat of the sun on outdoor pressure vessels is a common example. Portable cylinders, some of which at room temperature contain gases at pressures up to 2,000 psi when filled, should be stored only in shaded areas. (The vapor pressure of carbon dioxide is 835 psi at 70°F and 2,530 psi at 140°F; nitrous oxide is 745 psi at 70°F and 2,450 at 140°F.) Pressure vessels inside buildings should not be located near sources of

PRESSURE-RELIEVING DEVICES

Type	*Use*	*Method of Operation*
Safety valves	Gas or vapor	Safety valves are frequently called "pop" valves because they pop full open when a preset pressure is exceeded.
Spring-loaded		Opens when the total force, due to static pressure upstream of the valve, exceeds the preset force of a spring acting on the opposite side of a valve plug.
Weight-lever		The force on the plug closure is exerted by a lever on which weights are mounted. Combinations of lever-arm distances and weights provide a simple and reliable means of setting these devices. Their use is generally restricted to stream boilers operating at less than 100 psig.
Solenoid		Transducers sense pressure in the vessel being protected. If the established limit is exceeded, a signal is sent to energize a solenoid which opens the valve. Electrical power is required; if power fails, the device becomes inoperative. For this reason, use of solenoid-actuated safety valves is confined to noncritical service.
Pilot		This is a two-stage valve which overcomes the disadvantages of a large, single-stage direct-acting valve. A small pilot piston controls the area above the poppet of the valve. When the pilot opens, the pressure above the main poppet is reduced so that it rises off its seat. The valve has better control and its stability is better than a single-stage valve of similar capacity.
Relief valves	Liquids	Relief valves do not pop full open when a preset valve is exceeded, but open slightly and then open further as the pressure increases.
Spring-loaded		Similar to the spring-loaded safety valve, except that it is for liquid use and does not open fully when first activated.
Power-actuated		Movements to open or close are modulated and controlled entirely by an auxiliary source: electric, air, steam, or hydraulic power activated by a pressure sensor upstream of the valve.
Safety-relief valves	Liquid, gas, vapor	Can be used as either a safety valve or a relief valve, depending on the application.
Vacuum breaker	Air	Action is similar to that of a spring-loaded safety or relief valve, except that the atmospheric pressure outside the vessel, being higher than that inside, forces it open.

FIGURE 20–2 Pressure-Relieving Devices Continued on page 316

Frangible disk	Liquid, gas, vapor	A thin piece of metal incorporated into a pressure system to prevent damage that could result from an excessively high pressure peak. Disks can be flat, prebulged in either direction, or corrugated in the form of a bellows. They rupture when a stress imposed by pressure exceeds the strength of the disk; pressure forces the disk against a knife-edge or point, or it is exploded by a small explosive charge initiated when a pressure sensor indicates the pressure level of the protected vessel has exceeded a specified level. The disadvantage of the disk is that the system, once opened, empties and is inoperative until the passage created is closed by replacement of the disk is often used in conjunction with a safety or relief valve, which provides the primary relieving capability. If this is inadequate, the disk will rupture. Imperfections in manufacture, installation, or caused by corrosion can cause the disk to fail prematurely.
Fusible plugs	Liquid, gas, vapor	The plug is a fitting filled with a alloy that melts at a predetermined temperature. To protect a boiler the plug is located in the boiler shell at a point that is normally under water, which keeps it cool. If the water level drops below that of the plug, the water's cooling effect is lost, the temperature of the plug increases, the alloy melts, and steam passes through, relieving the pressure. The plug must be replaced to refill the boiler. Fusible plugs are also used in compressed-gas cylinders to prevent their bursting violently of they are involved in a fire. Similar usages are made on other pressure vessels.

FIGURE 20–2 Continued

heat, such as radiators, boilers, or furnaces. Vessels containing cryogenic liquids can absorb heat from normal environments that could cause boiling of liquids and very high pressures. Gas compression can cause an increase in temperature and very high pressures.

Aerosol cans and other pressurized containers might be thrown into a fire. Warnings on cans caution against disposing of cans in this way. Formerly, these cans would explode violently, but aerosol cans are now made with a low-melting-point alloy which acts as a fusible plug (see Fig. 20-2).

Excessive Pressures from Nonthermal Sources

A vessel or container can be overpressurized so that it fails. The simplest example is the bursting of a child's balloon. An automobile tire may normally be filled to a pressure of 30 psi. However, if it is filled from a tank which holds compressed air at 100 psi, there is the possibility the tire could be pressurized to a point at which it fails.

The *Safety Journal* (Vol. 29, No. 3) reported 11 cases where beer kegs became pressurized with carbon dioxide gas and exploded, causing eight deaths and much property damage. Although the system was a low-pressure one, many of the explosions were due to the lack of pressure-release devices, dirt in the regulator, or improper use of the regulating valve. Practically all pressure-vessel failures occur at flaws in the material where stresses are concentrated. If the flaw is serious enough, failure may occur at or below the normal operating pressure. If the vessel is overpressurized, it will fail at a weak point, where a flaw exists.

In September 1969 a high-pressure natural gas line near Houston, Texas, became blocked internally. The compressors kept packing gas into the line, raising the pressure until the line failed.

Another common source of injury is through pressure-gauge failure. In such instances, the thin-walled bourdon tube or bellows inside the case fails under pressure, generally because of metal fatigue or corrosion. Unless the case is equipped with a blowout back, the face of the gauge will rupture, propelling fragments toward the normal location of a person reading the gauge. A blowout back does not prevent failures but it ensures that no fragments will be propelled toward viewers. Boiler furnaces have blowout panels in case there is an explosion due to delayed ignition of unburned fuel gases.

DISCHARGES FROM SAFETY VALVES

To lessen the possibility of a rupture because of overpressurization, safety valves are often provided. Possible discharges from such valves should be conducted to locations where they constitute no danger, especially if the fluid discharged is very hot, flammable, toxic, or corrosive. On September 25, 1916, a tank car filled with gasoline was left in the freight yard at Ardmore, Oklahoma, where it remained for two days. The second day was sunny and hot. The gasoline vaporized, creating pressure in the tank so that the safety valve opened. Gasoline vapor escaped for several hours, blanketing the neighborhood and filling buildings with a combustible mixture. Ignition occurred, followed by an explosion. The flame was described as "fire bounding along the ground in streaks," with reports of clothing being ignited as far away as 350 feet from the tank car. All buildings within 400 feet were completely wrecked. Gasoline vapors spread through the streets so that buildings 1,200 feet away were damaged.

In 1973, a safety valve on a pressurized vessel at the El Segundo, California, plant of the Standard Oil Company opened, releasing oil over the surrounding neighborhood. Houses, cars, structures, vehicles, and the landscape far from the plant were contaminated by the windborne oil.

DYNAMIC PRESSURE HAZARDS

The pressures in cylinders of compressed air, oxygen, or carbon dioxide are over 2,000 psig when the cylinders are full. A large cylinder, such as that used for oxygen for oxyacetylene welding, will weigh slightly more than 200 pounds. The force or thrust generated by gas flowing through the opening left when a valve breaks off a cylinder can be 50 to 20 times greater than the cylinder weight. This can be compared to the propulsion system of a rocket or guided missile, which generally produces a thrust only 2 to 3 times the weight of the vehicle being propelled.

Spectacular accidents have occurred when such charged cylinders were dropped or struck so that the valve broke off. The cylinder took off, in some cases smashing through buildings and rows of vehicles, creating the tremendous havoc that a heavy steel projectile traveling at high speeds can generate. Computations indicate that if a valve is broken off, a cylinder filled to 2,500 psig can reach a velocity of 50 feet per second in 1/10 of a second. Safeguards to prevent accidents with pressure cylinders are presented in Fig. 20-4.

Whipping of Hoses and Lines

Whipping of flexible hoses can also generate injury and damage. In one instance, the end fitting of a compressed-air line was not tightened adequately when the line was connected. The line separated when it was pressurized. It then began whipping about until it hit and killed a worker by crushing his skull. Such types of accidents are not restricted to pressurized gas lines but may occur with water hoses as well. A whipping line of any kind can tear through and break bones, metal, or anything else with which it comes in contact. All high-pressure lines and hoses should be restrained from possible whipping by being weighted with sand bags at short intervals, chained, clamped, or restricted by all of these means. Rigid lines should be preferred to flexible hoses, but if the latter must be used, they should be kept as short as possible. Rigid lines should also be secured, especially at bends and fittings, since an accidental disconnection could cause even these lines to whip. Workers should be instructed never to attempt to grab and restrain a whipping line. The valve that controls flow to the line should be shut.

Other Effects

Failure to realize the consequences of flow from a compressed gas system can generate many injuries. No pressurized systems should be worked on. Each pressurized vessel or line should be considered hazardous until all pressure has been released. This can be done by checking a gauge directly connected to the container or line (after making sure that the gauge works and the valve to the gauge is open), by opening a test cock (without standing in front of it), or by noting the equipment or line is already open to the atmosphere.

An airman in a launch facility for a ballistic missile system was working on a nitrogen gas line that was pressurized to almost 6,000 psi. He failed to determine whether the line had been depressurized. He loosened the bolts of a flange, which separated slightly. A very thin sheet of compressed gas shot out, cutting into his leg like a knife.

Personnel have been warned that such effects could occur. If a pressurized line is suspected, no attempt should be made to use fingers to probe for a leak. They might be cut by the gas. A bit of cloth on a stick, a soap-and-water solution, or sprays specifically made for the purpose can be used with safety.

It has already been pointed out that a hurricane wind, with a pressure of only 0.25 psi, can drive a straw endwise into a wooden pole. Dirt, debris, fillings, and other particles can be blown by compressed gas, which would probably have a far higher pressure, into an eye or through the skin. OSHA standards permit compressed air to be used for cleaning purposes if its pressure is less than 30 psi (some states permit even less), if there is effective guarding against chips, and if personal protective equipment is used. The harm which can be done by indiscriminate use of compressed air at these pressures can be considerable. It should be general practice that compressed air never be used for cleaning clothes or anything near a person's body. The safeguards on compressed-air guns supposedly prevent chips, dirt, and other small particles from blowing back to the user. However, these particles can ricochet off equipment or walls back to the user, be driven by the jet into the skin, hit other nearby persons, or contaminate the environment.

Cases have been cited in which compressed air entered the circulatory system through cuts in the skin. A case cited by the Navy[85] illustrates this:

> After washing some machine parts in a cleaning solvent, a mechanic held them in his hand and blew compressed air over them to dry them. He had a small cut on one finger. A little bit later, the man complained that he felt he was going to explode. Somebody took him to a hospital, where his trouble was diagnosed as air bubbles in his bloodstream. The cause was the compressed air striking the cut and entering his circulatory system. He recovered, but other people have died from this kind of injury.

A similar case was described in the TWA Safety Bulletin:

> In a Massachusetts plant, a woodworker, covered with sawdust, held the compressed air nozzle 12 inches away from the palm of his left hand and opened the valve. Before realizing what was happening, his hand swelled up to the size of a grapefruit. Severe shooting pains started from his finger tips to his shoulder. The most excruciating pain was in his head, as though the top of his head were coming off. When found, he was actually holding down the top of his head with one hand .... (Apparently the man recovered.)

It should also be a firm company rule that no person will direct a compressed air at any other persons, especially in horseplay. Even the restricted pressures permitted by OSHA can cause fatal injuries when applied to any body opening.

WATER HAMMER

Water hammer is caused by a sudden stoppage of liquid flow so that a shock effect occurs, which can cause the rupture of a line. The mass of liquid had momentum. If the flow is terminated abruptly by closing a valve at the downstream end of a line, the momentum of the liquid is transformed into a shock wave (water hammer) which is transmitted back upstream.

The shock is transmitted back through the liquid because liquids are practically incompressible. The energy shock involved may be adequate to break fittings and lines, especially if they are made of brittle materials which do not stand shock well. To avoid damage to liquid lines, the use of quick-closing valves should be avoided. If they must be used, the shock can be alleviated by a suitable air chamber or accumulator connected to the lines slightly upstream of the valve. The air in the chamber provides a cushion which can be compressed to absorb the energy of the water hammer.

A water-hammer problem can be recognized by its noise. It provides a warning that the problem exists. Failure to take remedial action may result in the violent rupture of a line, often with fragmentation that could cause injury to nearby personnel and equipment, or leaks at joints and fittings.

Ruptures during startups of steam lines are not uncommon. When a piece of steam equipment has been shut down, residual steam will condense. If it is not drained properly, the condensate remains as a liquid. When the steam is turned back on again, it may propel a slug of condensate ahead until it violently hits a bend, causing noise and possibly damage. Compressed-gas systems may accelerate dirt or debris inside a line to high speeds when the line is pressurized. Here again, the solid matter can cause rupture of a line and possible damage to downstream equipment.

NEGATIVE PRESSURE (VACUUMS)

The negative difference between atmospheric and below-atmospheric pressure or vacuum can be as damaging as a positive one, even when both are positive pressures in the absolute sense. Unintended vacuums or negative pressures can be extremely damaging, because structures involved may not be built to withstand reversed stresses.

In one instance, a large intercontinental ballistic missile was shipped in an aircraft from the high-altitude Denver location where it had been manufactured to the sea-level base where it was to be launched. The propellant tanks were sealed to prevent entrance of moisture or contaminants. There the internal pressure of the seal-in air was that of the atmosphere at Denver. When the plant landed at the sea-level field, the ambient pressure was much higher. The difference in the two pressures caused the tanks to collapse, because they had not been built to sustain external loads.

Much of the damage done by high winds during hurricanes and tornadoes is due to negative pressures. Most building are designed to take positive loads but not to resist negative pressures. Such negative pressures might be generated on the lee side when winds pass over. Although the actual difference is very small, the area over which the total negative pressure will act is very large, so a considerable force is involved. For example, a roof on a small house may be 1,500 square feet. If the difference in pressure is only 0.05 psi, the force tending to tear off the roof equals $1500 \times 144 \times 0.05$, or 10,800 pounds (equivalent to 5 tons of weight).

Condensation of vapors is another source of vacuum pressure which could cause collapse of closed containers. A liquid occupies far less space than does the same weight of its vapor. A vapor which cools, such as steam, will liquefy, leaving water in the space which it occupied, and the partial pressure will decrease greatly. Unless the vessel is designed to sustain the load imposed by the difference between the outside and inside pressures, or unless a vacuum breaker is provided, collapse may occur.

TESTING OF PRESSURE SYSTEMS

Each pressure system should be tested prior to use, and pressure vessels should be tested periodically after that to determine their adequacy for continued service. Wherever possible, hydrostatic testing, using water and not a gas such as steam or compressed air, should be used. If the vessel being tested fails suddenly, the rapid expansion of gas might cause the rupture to be violent, possibly generating a blast wave with injury or damage. Hydrostatic testing, using water as a fluid, has two major advantages. Leaks created by pressurization of a vessel can be detected easily. The test can then be interrupted or continued with increased care. Because fluids expand little, in case of a vessel's rupture, no shock wave will be generated. Although the potential energy of the compressed water may be converted to kinetic energy, any fragments will be propelled for only small distances (far less than in a gas rupture).

Tests are usually conducted to proof pressures. A proof pressure test will generate stresses which should not exceed the yield strength of the container or its metal. Proof pressures are usually 1.5 or 5/3 times the maximum expected operating pressure (under nonshock conditions) of the vessel. Strain measurements indicate whether any permanent deformation remains after the pressure is released. Proof pressure tests may call for pressure to be increased to the required value, held for a specific time while it is being inspected, and then released and reinspected. Sometimes this cycle may have to be repeated.

Less common is the burst pressure test. When such a test is conducted, the vessel may leak because it will be stressed beyond its yield point and be permanently deforment, but it will not rupture. A vessel subjected to a burst pressure test cannot be reused for the function for which it was made. Burst pressure tests are therefore used only to test prototypes or a sampling of a large number being produced, to verify the adequacy of calculations and manufacture.

Nondestructive testing methods are also used to inspect pressure vessels. Dye penetrant, magnetic particle, radiography, and ultrasonics are some methods used. Many of these tests are used in conjunction with pressure testing, sometimes being made before and again after being subjected to the proof pressure.

If a flaw is located, it often becomes necessary to determine what corrective action should be taken. So many factors are involved in such a determination that only one recommendation can be made: Follow the recommendation of the inspector who made the tests.

LEAKS

Leaks are another possible hazard source. Possible causes of leakage can be categorized as follows:

- *Poorly designed systems or connections:* the possibility of leakage exists wherever there is a connection in a system containing a fluid. It is therefore advisable to keep the number of connections to a minimum. Elimination of short pipe segments and the use of welded lines are beneficial in this respect.
- *Separable connection:* where necessary, these should be selected with care, since certain types have greater tendencies to leak than others. The numerous designs of flanged, screwed, and mated fittings offer a broad variety suitable for any type of fluid, hazard involved, proposed system use, speed of separation, and cost.
- *Fluid contained:* the type of fluid in the system may influence the leakage hazard. Gases have a greater tendency to leak from or into a pressurized container than do liquids, but a liquid will be a greater problem where weight alone causes a leakage. Leakage will be less with higher-velocity fluids than with a fluid of low viscosity.

- *Inadequately fitted or tightened parts:* cross-threading and lack of adequate engagement of screwed parts are in this category. Bolts on flanged fittings or nuts on mated connections which are not tightened may permit separation of surfaces when pressure is applied. Lack of compression seals, gaskets, and packings will permit liquids and gas to leak. Failure to close valves adequately will allow fluids to flow through.
- *Fittings loosened by vibration:* this situation is especially likely where connections are in lines not secured against movement. Vibrations transmitted will not only loosen screwed connections but also lessen tension on nuts and bolts holding together flanges and other types of connections, open valves, and cause damage to seals, packings, and gaskets.
- *Cracks and holes:* these are caused by structural failure. Overpressurization can cause stresses in container walls, which can crack and open under tension caused by the pressure. Cracks may be generated where a sharp object scratches a metal surface. A hole can be caused by an impact by a sharp object on a container wall. A hole may be produced where the side of a hose or pipe vibrates against a hard surface.
- *Porosity:* metals may be porous or contain defective matter in which they were cast or welded, permitting the slow passage of fluid. Seals, packing, or gaskets may be materials that allow passage of fluid. Nonmetallic sheet materials used for diaphragms or containers may be permeable. Materials suitable as sealants for pressurized systems may be unsuitable for vacuum systems.
- *Corrosion:* corrosion can create holes which can extend entirely through the metal of a container. Corrosion can roughen surfaces so that passages exist where mating metal surfaces should create barriers to flow of fluid. Corrosion can weaken metals so that cracking and failures create openings. It can permit the failure of nuts, bolts, and other devices which hold connectors together.
- *Worn parts:* parts can wear out or be damaged during use. Disassembly and reassembly of parts can cause stripping of threads, cuts in gaskets, and damage to mating surfaces. Flexing of rubber and plastic hoses by changes in pressure, kinking and unkinking, or dragging them on rough surfaces will wear them out and cause them to leak. Continued exposure of nonmetallic materials to solar radiation can cause their deterioration and failure.
- *Interference:* dirt, contamination, or other solid materials can prevent close contact between faces of flanges, threads, gaskets, or other mating surfaces of connectors. Valves may not be able to seat properly in the closed position, allowing continued flow of fluid.
- *Overpressurization:* excessive pressures can cause overstressing of the container, possibly causing distortion, cracking, or separation, leading to leakage. It can increase the rate to flow through a hole so that formerly permissible leakage becomes unbearable.
- *Temperature:* higher temperatures may cause the loosening of connections where dissimilar metals are joined because of the difference in expansion. Increases in temperature will reduce viscosity so that systems may begin to leak. Organic sealants may begin to ooze out. Cold can cause organic materials to crack to

permit flow of pressurized gases to areas where they may have adverse effects (as happened to one of the solid propellant boosters for the space vehicle *Challenger*).

- *Operator errors:* valves, drains, or other closure devices may not have been closed adequately to stop flow, or they may have been opened in error. Containers may have been overfilled so that a liquid reached a level where leakage or overflow occurs.

EFFECTS OF LEAKAGE

One of the chief effects of leakage is the release of fluids which are flammable, toxic, radioactive, corrosive, injurious, or damaging in other ways. The degree of hazard involved with each of these would be one factor in the determination of the amount of leakage permissible.

- *Materials contamination:* fluid leakage from a tank, hose, or other container can cause damage of materials with which the fluid comes in contact, even without creating other hazards. Leaks of oil or colored fluid may make fabrics or foodstuffs unusable. Contaminants in a container may degrade the purity of its contents so that it cannot be used—for example, oil leaking into a tank of gasoline.
- *Loss of stores:* leakage of containers can result in loss of material required for other operations or for economic purposes. Lubricants leaking from a bulk storage tank might create a fire hazard. Loss of oil by leakage from an engine may cause a lack of lubrication and consequent engine failure due to increased friction, wear, and heat.
- *Loss of system fluids:* certain items of equipment, such as hydraulic actuating devices, are dependent on the presence of an adequate supply of fluid. Depletion or loss of the fluid can make the system less effective by decreasing the response or by causing it to fail entirely.
- *Loss of system pressure:* some pressurized systems may not be able to maintain the level of pressure necessary for their operation. Pneumatic and vacuum systems fail or are less effective if leaks are present.
- *Electrical hazard:* water or other conductive fluids leaking into electrical connectors or onto live electrical equipment can cause short circuits. Water leaking onto floors near electrical equipment can increase the shock hazard to a person standing there. Rain leaking into a car's engine can prevent it from operating.
- *Temperature:* leakage of cold liquid on a hot surface can cause it to contract and crack; hot glass is especially susceptible. The presence of fluids leaked onto hot operating equipment can reduce its effectiveness by increasing heat loss. Leakage of warm air into refrigerated equipment may ruin food and increase operating costs. On cold equipment, the fluid may freeze and cause blockage of parts' movement or of heat flow.
- *Moisture:* leakage of moisture into an enclosed space may increase humidity to a point at which it becomes harmful to personnel, material, and equipment. Insulation wet from leakage will be degraded and lose its effectiveness.

LEAK DETECTION

Detection of massive leaks is generally comparatively easy, actual flow or its effects often being readily detectable. Jets of leaking liquids can be seen, splashes or pools of liquid can be noted, and losses can be measured. Containers can be equipped with liquid-level indicators or pressure devices to measure conditions inside the container, which will indicate losses and of leakage taking place. A container and its contents can be weighed to determine the extent of any leakage.

Discharges of gases can sometimes be noted by the whistling noises produced. Care should be taken to avoid probing with fingers for very high-velocity leaks from very high-pressure systems. Such thin, knifelike jets can be powerful enough to cut through skin and bone. A simple method for searching for invisible gas leaks is to use cloth streamers on a stick as indicators. The streamers are passed over a suspected area; any gas leak will cause the ribbons to flutter.

Very small leaks are more difficult to locate, but detectors are available for the purpose. Some are used for any type of gas or liquid; others have specific uses only. Soap solutions and bubbles are simple, cheap, and generally effective for small gas leaks. Initially, the unaided ear was used. Now, extremely small leaks can easily be detected by sonics and vibrations through electronically amplified devices.

Another method in common use is the addition of an odorant (Fig. 25-4). Hazardous gases that lack odors are given one which is easily detectable. Other types of equipment usable as leak detectors (oxygen, combustible gas, infrared, ultraviolet, clathrate, and others) are indicated in Fig. 25-5. Leakage of a radioactive gas can be detected by a suitable detector of radioactivity.

A problem in determining the point of leakage is that pressure can cause temporary enlargement though which leakage occurs; but when the container empties and flow stops, it may be difficult to find where the leak took place. Pressure in a closed container may create stresses in its walls which expand gaps between gaskets or seals and metal surfaces, holes, or cracks. As pressure is reduced in the container by loss of fluid, any gaps, holes, or cracks may decrease in size or even disappear. Under such conditions, where the opening and leakage are small, it may be necessary to keep the system pressurized to determine where the leak is taking place. Only the minimal pressure necessary to determine any leakage should be used, since the small existing fault may lead to a complete rupture or failure of the container.

DYSBARISM AND DECOMPRESSION SICKNESS

Dysbarism is the collective name given to physical disturbances in the body caused by variations in pressure. Many of these disturbances result from, and are particularly damaging because of, rapid pressure changes. Some have been known by other names, although produced by the same two basic causes: (1) release from solution of gas in the blood and (2) expansion of free gas in the body cavities.

The amount of gas that will remain in solution in any liquid at any temperature is dependent on the partial pressure of the gas (Henry's law). This variation with pressure is especially true in the blood, where its temperature is almost constant. Gases in the bloodstream include oxygen, carbon dioxide, nitrogen, and other inerts (see Fig. 24-1). Oxygen and carbon dioxide are chemically bound to the red cells, oxygen more so than carbon dioxide. Nitrogen has the highest partial pressure of the gases present, but it is present in solution, not in chemical combination. Nitrogen is the chief contributor to dysbarism.

The partial pressure of nitrogen at sea level is 570 torr (mmHg). At pressures greater than that, an increased amount of nitrogen will enter solution; as pressures decrease, it will leave the blood. The rate at which it leaves the solution governs the severity of any problem that may arise. If the change in pressure takes place slowly, the free nitrogen has an opportunity to leave the body with no damage. As an underwater swimmer proceeds deeper, more and more of the gas breathed goes into solution in the blood. In this case, the increased gas in solution is not only nitrogen but also oxygen, which is in excess to that which goes into chemical combination in the blood cells. Ascent of the swimmer must be slow to permit the gas freed because of the decreasing pressure to be eliminated from the body.

Analyses of pressure effects indicate that obese or older persons are more prone to decompression sickness than are thinner and younger people. This is attributed to the greater amount of fat present in the obese or older person, in which nitrogen can dissolve. Fat will dissolve five times as much nitrogen as would the plasma of the blood. The central nervous system may also be disturbed by rapid and severe pressure changes, which may affect the person's ability to coordinate movements and to maintain equilibrium. The person may suffer from vertigo and disorientation, and sensory perception may be reduced. The person may become dizzy, suffer from headaches, and in extreme cases even suffer partial paralysis. These changes do not often result from changes in altitude, but they may occur in deep underwater swimming or when working under very high pressures. The pressure at a depth of 32 feet of water equals the entire height of the atmosphere, and swimmers have gone as deep as 700 feet. For such swimmers, changes in depth must be made with care if damage to the body is to be avoided.

Bends is an ailment involving pain in the joints, muscles, or bones (which results in the limb affected being held in a bent position from which it is difficult to straighten, hence the name), generally considered to be caused by the formation of gas bubbles. Bubbles will form when the rate of gas release from the blood and tissue is greater than its rate of elimination from the body, generally through the respiratory passages and lungs. Studies of symptoms indicate that the principal locations of bends are the knees, shoulders, elbows, and wrists. The reason for the localization in these areas is not yet clear; it has been theorized it may be due to the positions of the joints, restriction of circulation, deposits of fat, and other restrictions.

Effects similar to those of bends have been given other names: caisson disease and decompression sickness. Caisson disease was the earliest to become a problem, occurring to workers under air pressures much greater than 1 atmosphere. When the pressure was rapidly reduced, such as when workers went too quickly from a pressurized to an unpressurized environment, they suffered pains and symptoms similar to bends. In addition, they might have skin itch or rash, visual defects, difficulty in breathing, or paralysis. The first pressurized caisson was used in England in 1851, but it was in an operation after that the first man died of caisson disease after working under a pressure of 40 psi (88 feet). Of 600 men working on the foundations of the St. Louis Bridge over the Mississippi, 119 suffered from the disease, 14 of whom died.[85]

Decompression sickness is a general term which includes bends (due to a rapid ascent from great depths of water), caisson disease (due to a rapid decrease from high air pressure to 1 atmosphere), and aeroembolism (due to a rapid decrease in pressure from altitude). It differs slightly from dysbarism, which also includes the effects caused by differences in pressures between gases trapped in body cavities and barometric pressure.

Chokes are believed to be due to expansion in the lungs and respiratory passages because of the pressure of gas released. This may cause difficulty in breathing, coughing, and pains in the chest which may vary from a tightness to a burning sensation. Skin disturbances include rashes and discoloration. Rashes occur principally on the abdomen, thighs, and chest, where there are considerable amounts of fatty tissue. Rashes are usually accompanied by pain and itching and a coloration which varies from light red to purple-red. This is believed to be due to dilation of blood vessels in the skin by the gas released from the blood. In especially severe cases, such as may occur from caisson disease or ascents from great depths of water, bleeding into the tissues may result. Rashes and mottling are regarded as signs of danger. Other effects are itching and prickling of the skin, sensations of hot or cold, or release of tissue fluids immediately under the skin.

Aeroembolism concerns the same problem; however, it is induced by rapid changes in pressure where the barometric pressure is one atmosphere or less. It first became apparent with the use of aircraft capable of rapid ascents to high altitudes. Aeroembolism involves one factor which may cause it to be more damaging than that of caisson disease. At higher altitudes, not only will gases physically come out of solution, but oxygen will be released from its combination within the red cells. Ebullism, or boiling of body fluids, occurs at altitudes of 63,000 feet or more, when pressure inside the body is less than 5 torr.

Gas in Body Cavities

Free gas within the body cavities also may cause pain and injury to persons subjected to rapid changes in pressure. One of the commonest symptoms is the "stuffed" feeling or pain in the ears caused by blockage of the Eustachian tube. This tube connects the middle ear with the pharynx at the back of the mouth. Blockage prevents equalization of the air pressure on the opposite sides of the ear drum, causing distention of the ear drum and pain in the ear. If the blockage is partial, movement of the jaws may be adequate to clear the tube and relieve the pain. Temporary but complete blockage may occur when a person has a cold or sinus trouble. Rapid changes in pressure during an illness may be dangerous, and such experiences as flying, underwater swimming, or activities under elevated pressures should be avoided. A decrease in altitude from 8,000 to 3,000 feet causes an increase in barometric pressure of 2.25 psi. Swimming to a depth of a little more than 5 feet will produce the same effect.

Pressure changes can also cause pain in the teeth. It may be a dull ache or a sharp, severe pain. In some cases, fluid in the mouth may enter under fillings as altitude changes, causing the pain. It is also believed that other instances were "referred" pains (pains which seemed to originate in the teeth, but were actually due to sinus troubles).

Expansion of gas in accordance with Boyle's law in abdominal cavities, such as the stomach, colon, and bowel, may cause "gas pains." This is almost always due to an increase in altitude and a decrease in barometric pressure. When bloating and pains do occur, they generally appear during ascent or soon after. The gas is mostly nitrogen from air which was swallowed, and from changes in food in the digestive system, with some diffusion of oxygen and carbon dioxide from the bloodstream. Pains often disappear immediately with passing of wind or belching.

Effects on Bone

Nitrogen released in the arteries may block the flow of blood to the bones. Unless the block is removed, the portions of bone affected may begin to deteriorate from lack of nourishment and oxygen. This effect is known as aseptic necrosis. If the block is not removed, a new blood passage will form to the bone to feed the cells there. However, sudden shocks and pressures before the bones heal can cause pieces to break off. The problem is especially critical when it involves a joint, so that permanent crippling results. Aseptic necrosis itself is not painful, so symptoms may appear only after six months to a year after repeated attacks of bends.

Decompression Injury Prevention

Decompression sickness (or one of the variants mentioned) is normally not a problem with atmospheres of less than 18 psig. OSHA Regulations for Construction require that preventive measures be undertaken for persons who have been subjected to pressures of 12 psig or more. The time required for decompression from this and higher pressures depends on the pressure and the duration of the exposure. The OSHA regulations include tables and other information which must be followed for decompression without adverse effects.

Soon after the reasons for decompression sickness were postulated, it was found that the only remedy was to resubject the victim to high air pressure. Each construction site involving operation in compressed air at pressures greater than 12 psig must have completely equipped locks for compression and decompression through which crews can enter and leave the work area, and also a medical chamber for emergencies. If any worker appears to have symptoms of decompression sickness, the worker is put into the medical chamber and subjected once again to the pressure at which he or she had been working. Decompression is then carried out very slowly, since bubbles once formed are difficult to dispose of. The recompression will, however, reduce the bubbles' size so the effects are mitigated.

Each compressed-air worker must be provided with an identification badge which the worker is to wear at all times. The badge must have the employee's name, address of the medical chamber, telephone number of the physician in charge, and instructions that in case of an emergency of unknown or doubtful cause or illness, the worker can be rushed to the medical chamber.

COMPRESSED-GAS CYLINDERS

The widespread use of compressed gases in portable cylinders makes it advisable that safety engineers be familiar with their characteristics, hazards, safety features, and precautionary measures which should be observed in their use. The properties of most of the gases generally transported and handled in portable cylinders are listed in Fig. 20-3.

The gases drawn from cylinders are stored therein in one of three ways: permanent gases, liquified gases, and dissolved gases.

Permanent Gases

Permanent gases have boiling points of –150°F or lower. At room temperatures they cannot be liquefied, no matter how high the pressure. Oxygen, nitrogen, and helium are examples of permanent gases.

Liquefied Gases

Liquefied gases liquefy at temperatures of –130°F or higher at 1 atmosphere, but can be liquefied and maintained as liquids at higher pressures. Such gases include propane, chlorine, and butane. Carbon dioxide is also in this category but becomes a solid rather than a liquid. Opening the valve of a cylinder in which a liquefied gas is stored permits gas to vaporize and flow out and additional liquid to evaporate. The pressure of a gas in the cylinder will depend on the vapor pressure of the liquid, which, in turn, depends on the liquid's temperature.

Dissolved Gases

Only one common gas, acetylene, is used as a dissolved gas in cylinders. Acetylene will decompose and explode violently at pressures between 15 and 16 psig. Acetylene cylinders are therefore filled with calcium silicate, an inert porous material, which is then saturated with acetone. Acetylene is then pumped into the cylinder, where it goes into solution in the acetone. Acetone will hold 35 times its own volume of acetylene. The maximum pressure for such cylinders is 250 psig at 70°F. The acetylene should not be drawn from the cylinder at a pressure greater than 15 psig.

Sizes and Volumes

Compressed-gas cylinders vary in size, the smallest being 2 inches O.D. by 15 inches long. The size most commonly seen in industrial plants is 51 inches high with an outside diameter of $8\frac{1}{2}$ or 9 inches. Cylinders of nitrogen and acetylene have larger diameters.

The volume, weight, and pressure of gas loaded into a cylinder varies with its characteristics. The pressure at which a permanent-gas cylinder may be filled is marked on the shoulder of the cylinder. A marking of D.O.T. 3A-2000 indicates that the cylinder, produced in accordance with D.O.T. Specification 3A, is designed for a filling pressure of 2,000 psig at 70°F. Cylinders with nonflammable gases such as nitrogen and helium may be overfilled to a pressure of 10 percent greater. The pressure in a permanent-gas cylinder will decrease in proportion to the amount of gas used. The

PRESSURIZED-GAS CHARACTERISTICS

Gas	*Formula*	*Boiling Point (˚F)*	*Cylinder Pressure (psig)*	*Physical State in Cylinder*	*Specific Volume (ft^3/lb)*	*Flammability L.F.L–U.F.L (%)*	*Toxicity*	*Odor*	*D.O.T. Label*	*Other Hazards*
Acetylene	C_2H_2	Sublimes	250	In Solution	14.7	2.5–81	Slightly narcotic	Garlic like	Red gas	
Air		–312.7	2,640 max.	Gas	13.3		Nontoxic	None	Green	High pressure
Allene	C_3H_4	–30.1	102	Liquid	9.6	2.1	Probably anesthetic	Sweet	Red gas	
Ammonia	NH_3	–28.0	114	Liquid	22.6	15–28	TLV, 50 ppm; dangerous through inhalation and body contact	Pungent, irritating	Green	
Argon	Ar	–302.6	1,875 max.	Gas	9.7	–	Suffocation only	None	Green	High pressure
Arsine	AsH_3		90	Gas	5.0	Limits unknown	Highly toxic; TLV, 0.05 ppm	Garlic like	Poison and red gas	
1.3-Butadiene	C_4H_6	24.06	22	Liquid	6.9	2–11.6	Slightly anesthetic; TLV, 1,000 ppm	Mildly aromatic	Red gas	
Butane	C_4H_{10}	31.1	16.3	Liquid	6.4	1.9–8.5	Slightly anesthetic	Faintly disagre-able	Red gas	
1-Butene	C_4H_8	20.7	24	Liquid	6.7	1.6–9.3	Slightly anesthetic	Slightly aromatic	Red gas	
Carbon dioxide	CO_2	–109.3	830	Liquid	8.8	–	Suffocation; TLV, 5,000 ppm	None	Green	
Carbon Monoxide	CO	–312.7	1,500 max.	Gas	13.8	12.5–75	Chemical asphyxiant; TLV, 50 ppm	None	Red gas	
Carbon sulfide	COS	–58.4	160	Liquid	6.5	11.9–28.5		Rotten eggs	Red gas	
Chlorine	Cl_2	–30.1	85	Liquid	5.4	–	TLV, 1 ppm; by inhalation + contact	Pungent, irritating	Green	Supports combustion

FIGURE 20–3 Pressurized-Gas Characteristics Continued on page 330

PRESSURIZED-GAS CHARACTERISTICS

Gas	*Formula*	*Boiling Point (°F)*	*Cylinder Pressure (psig)*	*Physical State in Cylinder*	*Specific Volume (ft^3/lb)*	*Flammability L.F.L–U.F.L (%)*	*Toxicity*	*Odor*	*D.O.T. Label*	*Other Hazards*
Chlorine trifluoride	ClF_3	53.15	6	Liquid	4.2	–	TLV, 0.1 ppm; by inhalation + contact	Sweet, irritating	White	Highly reactive
Cyanogen	$(CN)_2$		60	Liquid	7.4	6–32	TLV, 10 ppm	Pungent; almond like	Poison gas	
Cyclopropane	C_3H_6	–27.15	75	Liquid	9.2	2.4–10.4	Anesthetic; TLV, 400 ppm	Ethereal	Red gas	
Dimethyl ether	$(CH_3)_2O$	–12.68	60	Liquid	8.4	3.4–18.0	Anesthetic	Slightly ethereal	Red gas	
Ethane	C_2H_6	–127.53	543	Liquid	12.8	3.0–12.5	Suffocation only	None	Red gas	
Ethyl chloride	C_2H_5Cl	54.3	5	Liquid	6.0	3.8–15.4	TLV, 1,000 ppm	Pungent, ether like	Red gas	
Ethylene	C_2H_4	–154.7	1,200	Gas	13.8	2.7–34	Anesthetic	Sweet	Red gas	High pressure
Ethylene oxide	C_2H_4O	51.3	35 max.	Liquid	7.7	3.0–100	TLV, 50 ppm	Irritating	Red	
Fluorine	F_2	–306.6	300	Gas	10.2	–	TLV, 0.1 ppm	Sharp, penetrating	Red gas	Highly oxidizing; no safety relief device
Helium	He	–452.0	2,120 max.	Gas	96.7	–	Suffocation	None	Green	High pressure
Hydrogen	H_2	–423.0	2,400 max.	Gas	192.0	4.0–7.5	Suffocation	None	Red gas	High pressure
Hydrogen Chloride	HCL	–121.0	613	Liquid	10.9	–	TLV, 5 ppm; by inhalation and contact	Pungent, suffocating	Green	
Hydrogen fluoride	HF	67.1	1	Liquid	17	–	TLV, 3 ppm; inhalation and contact	Sharp, penetrating	Corrosive	Highly irritating
Hydrogen sulfide	H_2S	–75.3	252	Liquid	11.2	4.3–45	TLV, 10 ppm	Rotten eggs	Red	

FIGURE 20–3 Continued

PRESSURIZED-GAS CHARACTERISTICS

Gas	Formula	Boiling Point (°F)	Cylinder Pressure (psig)	Physical State in Cylinder	Specific Volume (ft^3/lb)	Flammability L.F.L–U.F.L (%)	Toxicity	Odor	D.O.T. Label	Other Hazards
Isobutane	iso-C_4H_{10}	10.9	31	Liquid	6.5	1.8–8.4	Slightly anesthetic	Faintly sweet	Red gas	
Isobutylene	C_4H_8	19.6	24	Liquid	6.7	1.8–9.6	Unknown, probably anesthetic	Faintly Coal gas	Red gas	
Krypton	Kr	–243.2	590 max.	Gas	4.6	–	Suffocation	None	Green	High pressure
Methane	CH_4	–258.9	602 max.	Gas	23.7	5.0–15.0	Suffocation	None	Red gas	High pressure
Methyl bromide	CH_3Br	38.2	13	Liquid	4.1	10.0–15.0	TLV, 20 ppm; inhalation and contact	Chloroform like	Poison	
Methyl Chloride	CH_3Cl	–11.36	59	Liquid	7.6	7.0–17.4	TLV, 100 ppm	Faintly sweet	Red gas	
Methyl-mercaptan	CH_3SH	42.7	11	Liquid	7.5	3.9–21.8	TLV, 10 ppm	Extremely disagreeable	Red gas	No safety relief device
Monomethyl-amine	CH_3NH_2	20.6	29	Liquid	12.1	4.2–20.8	TLV, 10 ppm	Strong ammoniacal	Red gas	No safety relief device
Natural gas	Mixture		2,400 max.	Gas	24.0	3.8–17	Suffocation	Depends on odorant	Red gas	High pressure
Neon	Ne	–410.6	2,070 max.	Gas	19.2	–	Suffocation	None	Green	High pressure
Nitric oxide	No	–241.1	500	Gas	12.9	–	TLV, 25 ppm	Slightly irritating	Poison gas	Highly reactive
Nitrogen	N_2	–320.46	2,640 max.	Gas	13.8	–	Suffocation	None	Green	High pressure
Nitrogen dioxide	NO_2	70.1	0	Liquid	4.7	–	TLV, 5 ppm	Slightly irritating	Poison gas	Highly reactive
Nitrous oxide	N_2O	–129.1	745	Liquid	8.7	–	Anesthetic	Sweet	Green	High oxidizing
Oxygen	O_2	–297.4	2,640 max.	Gas	12.1	–	–	None	Green	High oxidizing; high pressure

FIGURE 20–3 Continued

PRESSURIZED-GAS CHARACTERISTICS

Gas	Formula	Boiling Point (°F)	Cylinder Pressure (psig)	Physical State in Cylinder	Specific Volume (ft^3/lb)	Flammability L.F.L–U.F.L (%)	Toxicity	Odor	D.O.T. Label	Other Hazards
Phosgene	$COCl_2$	45.6	11	Liquid	3.9	–	TLV, 0.1 ppm	Musty hay	Poison gas	
Propane	C_3H_8	–43.8	110	Liquid	8.7	2.2–9.5	TLV, 1,000 ppm	Faintly disagreeable	Red gas	
Propylene	C_3H_6	–53.9	137	Liquid	9.4	2.4–10.3	Suffocation	Faintly sweet	Red gas	
Sulfur dioxide	SO_2	14	34	Liquid	5.9	–	TLV, 5 ppm	Pungent, irritating	Green	
Sulfur hexafluoride	SF_6	–59.4	310	Liquid	2.5	–	Suffocation	None	Green	
Trimehthylamine	$(CH_3)_3N$	37.17	13	Liquid	6.0	2.0–12	TLV, 10 ppm; inhalation and contact	Fishy	Red gas	No safety relief device
Vinyl chloride	C_3H_3Cl	7.0	34	Liquid	6.2	4.0–22		Pleasant	Red gas	Carcinogen
Vinyl fluoride (inhibited)	C_2H_3F		355	Liquid	8.4	Unknown	Unknown; probably anesthetic	Sweet	Red gas	
Vinylmethyl ether (inhibited)	C_3H_8O	42.8	28	Liquid	6.7	Unknown	Unknown; probably anesthetic	Sweet	Red gas	
Xenon	Xe	–162.4	503 max.	Gas	2.9	–	Suffocation	None	Green	High pressure

*TLV of vinyl chloride was 500 ppm until it was determined vinyl chloride is a carcinogen. The matter is being studied at this time to determine whether a maximum should be set or no exposure permitted

FIGURE 20–3 Continued

PRECAUTIONARY MEASURES FOR USE WITH
COMPRESSED-GAS CYLINDERS

1. Compressed gas-cylinders should be operated and handled only by personnel who have been instructed in proper procedures for their use and in the hazards involved.

2. Personnel using the contents of a compressed-gas cylinder should be familiar with the properties of the contents, the hazards involved, and precautionary and emergency measures to be taken for those hazards.

3. Cylinders should not be banged, dropped, or permitted to strike each other or against other hard surfaces.

4. Cylinders should be secured by chain to a fixed support to prevent them from being dropped or from falling over. The cylinder valve should never be opened unless the cylinder is secured, since the thrust from the gas might cause the cylinder to fall.

5. Cylinders should not be dragged, slid, or rolled. Small cylinders may be carried by one man; larger ones by two men or by a suitable truck on which the cylinder can be secured firmly.

6. Cylinders should be protected from anything that will cut, gouge, or damage the metal and reduce the strength of the cylinder. No one should bang the cylinder with a hard object to determine how full it is.

7. Protective caps should be kept on the cylinders wherever the cylinders are not in use, or are to be transported from one place to another.

8. The cylinder valve should be kept closed wherever the cylinder is not in actual immediate use.

9. Cylinders should be protected against heat, which would increase the gas temperature and pressure. Outdoors they should be stored in shaded locations and not where direct rays of the sun could hit them. They should not be stored near other sources of heat, such as boilers, furnaces, radiators, or hot process equipment.

10. Cylinders should not be stored near sources of ignition or near flammable materials such as oil, gasoline, or wastes. Cylinders containing flammable gases should not be stored near cylinders containing oxygen or other oxidizers. Inside a building, there should be a separation of at least 20 feet between oxygen and fuel-gas cylinders unless there is a fire-resistive partition between the two. Cylinders should not be permitted to come in contact with electrical circuits.

11. No tampering of safety relief devices should be permitted, and no attempt should be made by the user to remove, repair, or modify cylinders, valves, or safety relief devices. In case of any problems, the cylinder should be capped and returned to the supplier.

12. The user should examine the label and markings on the cylinder prior to connection to make certain that it contains the gas he intends to use. Any cylinder whose content is not positively identifiable by markings should be returned to the supplier.

13. The cylinder valve should be opened slowly and the valve and fittings watched closely for signs of leaks. If the valve leaks, it should be closed, and the cylinder stored in a location where leakage will not constitute a hazard, marked as leaking, and returned to the supplier.

FIGURE 20–4 Precautionary Measures for Use with Compressed-Gas Cylinders Continued on page 334

14. If a valve sticks, never hit it with a hard object to loosen it. Return the cylinder to the supplier.

15. Acetylene cylinders should always be kept in an upright position -to avoid loss of the acetone in which the acetylene is dissolved.

16. Acetylene should not be used at pressures exceeding 15 psi.

17. Oil, grease, or other combustible material should never be used to lubricate or clean valves, regulators, gauges, or fittings on cylinders holding oxygen or other oxidizer.

18. Before an attempt is made to remove a regulator from a cylinder, make certain the regulator is depressurized by closing the cylinder valve and releasing all pressure from the regulator.

19. When a toxic or highly reactive gas is to be used, a cylinder holding the smallest amount necessary for the operation should be used.

20. Cylinders should not be kept in unventilated enclosures such as cabinets or lockers.

21. Fuel-gas cylinders with total capacities exceeding 2,000 standard cubic feet or 300 pounds of liquefied petroleum gas should not be kept inside a building except when in use or ready for immediate use.

FIGURE 20–4 Continued

cylinder should be considered empty when it is down to 25 psig. Empty cylinders should be marked with a tag or label or the letters "MT" on the side of the cylinder.

Since acetylene should not be used at a pressure greater than 15 psig, the rule considering a cylinder empty at 25 psig does not hold. Acetylene, especially for welding, is generally used at a pressure of 5 to 8 psig. A cylinder can be considered empty when it no longer provides an adequate supply of gas.

Cylinders loaded with a liquefied gas are not filled completely; a small vapor space is left for expansion (ullage). This is accomplished by prescribing the "filling density" that is permitted. The filling density is the ratio (expressed as a percentage) between the weight of the liquid to be loaded and weight of water which would totally fill the cylinder. A liquefied-gas cylinder should also be considered empty when its pressure reaches 25 psig. Figure 20-4 lists precautions.

Markings

Each cylinder must be marked with its chemical name to indicate its contents. The name must be applied by a means not readily removable; it may be stenciled or stamped or printed on a label. If the cylinder is to be shipped by an interstate carrier, it must also carry a pertinent D.O.T. label. In addition, other labels are frequently added with a warning of the hazard involved and precautionary measures that should be taken.

HIGH-PRESSURE SYSTEMS (PNEUMATIC AND HYDRAULIC)

1. All personnel who maintain, repair, or operate pressure equipment should be familiar with their hazards and the precautionary measures which must be observed.

2. Only qualified and authorized personnel should install, maintain, repair, adjust, or operate pressure system equipment, especially the safety devices involved.

3. Personnel assigned to work on pressure systems should be familiar with the locations and purposes of control valves, and with procedures for emergency depressurization.

4. Pressurized equipment and lines should be depressurized before any repairs are attempted or any effort made to loosen or open any parts. Pressure vessels and lines should be considered hazardous until it has been absolutely determined that all pressure has been released. Personnel should themselves ensure that a system is depressurized either by checking a gauge connected to the immediate line or equipment, by opening a test cock, or by noting that a disconnection in a line already exists. A pressurized line should not be bled by loosening a fitting.

5. All pressure-system components should be marked to indicate their rated pressure and direction of flow.

6. Face shields or goggles should be worn by all persons working on or with pressure systems.

7. Compressed air at pressures greater than 30 psig should not be used to clean filings, shavings, or other solid particles from work areas or from equipment unless protective equipment is worn. Compressed air should not be used to clean clothing or parts of the body.

8. Compressed air should never be directed at another person or used in horseplay.

9. Pressure systems being installed or repaired should be kept clean and free of dirt or debris, which might be accelerated to high speeds in the lines and cause internal damage to piping or equipment.

10. Rigid lines should be used where possible, instead of flexible hose. Lines should be secured on both sides of a bend, at suitable intervals along straight runs, and near connection fittings.

11. Where a flexible hose is necessary, it should be as short as possible. It should be adequately clamped (preferably) or chained to a secure fixture of strength adequate to restrain the hose in case it breaks. A long hose should be weighted down with 100-pound lead weights (preferably) or sandbags, or otherwise secured along its length, at 6-foot intervals. The entire length of the hose should be contained by a structural system separate from the hose, but in no way interfering with its normal flexibility.

12. Hose should not be used at a pressure greater than its manufacturer's rating. The hose should be checked to determine whether it is intended for use at the pressure desired. Hoses and fittings should meet appropriate industry or governmental standards. When not in use, hose should be plugged or capped and stored in a designated location where it will not be subject to deterioration.

13. No pressure line should be used as a step.

14. Pressure vessels, hoses, lines, and other equipment should be tested periodically and at the pressures indicated in prescribing regulations, codes, or standards.

15. No system requiring a pressure regulator, pressure-reducing valve, safety valve, or other relieving device should be activated unless it is in place and in operable condition. Only qualified and authorized personnel should change the settings on these valves and regulators. Where changes are made, the valves should then be tested to be certain that they are operating at the desired settings.

16. Each high-pressure vessel and any line which could be closed off should be equipped with a suitable relieving device. There should be no shutoff valve between the vessel or line and the relieving device. The relieving device must be sized to permit flow which will keep pressure to a safe level.

17. Safety and relief valves should be tested at prescribed intervals. Test levers should be left unsecured so that the valve will operate freely. Leaking valves should be replaced. The discharge from a pressure-relieving device must be sized to permit flow which will keep pressure to a safe level.

18. Valves should be checked periodically to ensure that they will work adequately and safely under the pressures which will exist in the systems in which they are installed. Valves should not be installed in an inverted position, since they might fill with debris and be impossible to close.

FIGURE 20–5 High-Pressure Systems (Pneumatic and Hydraulic) Continued on page 336

19. Valves should be used only in the manner for which they are designed. Gate valves should be used only in the fully opened or fully closed positions.

20. Where possible, valves should be installed in fail-safe configurations so that any valve failure will result in the safest possible condition for the system.

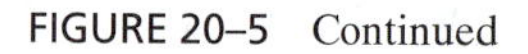

FIGURE 20–5 Continued

Cylinders are often color-coded. Unfortunately, there is no industry standard for the colors of the various gases in the cylinders for industrial usage. Cylinders for medical use are color-coded green. All cylinders used by the Department of Defense are color-coded using a system of basic colors and stripes; the basic colors indicate the type of hazard the gases in the cylinder represent, and the stripes indicate the gas.

Safety Devices

All D.O.T.-approved cylinders must have safety devices except hoses that are (1) 12 inches or less in length, exclusive of the neck, and with an outside diameter of $4\frac{1}{2}$ inches or less unless they are charged to 1,800 psig or more; (2) filled with fluorine, or other poisonous gas or liquid; or (3) charged with nonliquefied gas under 300 psig.

Some types of pressure relieving devices used on compressed-gas cylinders are safety relief valves, rupture (frangible) disks, and fusible metals. Frangible disks are set to burst at pressures far above those of the gases contained in the cylinder but below the pressure at which the cylinders must be hydrostatically tested periodically. Sometimes a combination of frangible disk and fusible metal is also used. The disk will function only after a temperature is reached at which the fusible metal melts. The problem with the frangible disk/melting alloy combination is that a temperature might be reached which is too low to melt the alloy but high enough to cause the cylinder to burst. Any cylinder might also rupture if intense localized heating occurs, weakening the cylinder so that it fails before any pressure relief device is actuated. See Fig. 20-5 for precautionary measures to be used with high-pressure systems. Figures 20-6 and 20-7 are checklists which can be used to alert safety engineers to potential pressure problems.

BIBLIOGRAPHY

[84] *Life Line*, U.S. Navy Safety Center, Norfolk, VA, Nov./Dec. 1973.

[85] D. Hunter, *The Diseases of Occupations* (Boston: Little, Brown, and Company, 1994).

HAZARDS CHECKLIST–PRESSURE

Possible Effects

High pressure:
- Injury:
  - Eye or skin damage due to blown dirt or other solid particles
  - Whipping hoses hit personnel
  - Lung, ear, and other body damage by overpressurization
  - Cutting by thin, high-pressure jets
- Container ruptured (internal pressure) or crushed (external pressure):
  - Blast effects
  - Fragments of ruptured container blown about

Leakage:
- Leaks in lines and equipment designed for lower pressures
- Blowout of seals and gaskets
- Release of toxic, corrosive, flammable, odorous, or high-temperature fluids
- Loss of system fluids
- Early fuel exhaustion
- Loss of system pressure
- Loss of lubricants
- Contamination and degradation of materials
- Slippery of surfaces
- Short-circuiting of electrical circuits and equipment]
- Displacement of air or other gas by liquid
- Vibration and noise

Possible Causes

High pressure:
- Overpressurization:
  - Connection to system with excessively high pressure
  - Regulator failure
  - Heated gases in closed containers
  - Heating fluids with high vapor pressures
  - Water hammer (hydraulic shock)
  - Deep submersion
  - High acceleration of liquid system
  - Warming cryogenic liquid in a closed or inadequately ventilated system
  - Excessively high combustion rate for boiler, evaporator, or other fired vessel
  - Pressure relief failure:
    - No pressure relief valve or vent
    - Faulty pressure relief valve or vent
    - Relief inadequately sized
  - Failure at normal pressure:
    - Deteriorated pressure vessel or lines
    - Inadequate connection
    - Failure or improper release of connectors
    - Inadequate restraining devices

Leakage:
- Reservoir losses:
  - Overfilling of container
  - Erroneously open drain or connection
- Inadequately fitted or tightened parts
- Worn parts and connections
- Fittings loosened by vibration
- Cracks caused by structural failure
- Porosity or other weld defect
- Contact surfaces inadequately finished or dirty
- Wrong type of gasket or seal
- Cuts in seals, gaskets, or hoses
- Hose holes caused by wear, kinking, or deterioration
- Hole torn by impact

FIGURE 20–6 Hazards Checklist—Pressure Continued on page 338

Blowout of seals and gaskets
Permanent deformation of metal containers
Excessively rapid motion of hydraulically or pneumatically activated equipment
Unsecured container propelled about by escaping gas

Low pressure:
- System inoperable
- Implosion of pressure vessel
- Inadequate air for respiration
- Physiological damage (atelectasis)

Pressure changes:
- Compressive heating
- Joule-Thomson cooling
- Physiological disturbances (cramps, the bends)
- Condensation of moisture

Low pressure:
- Compressor or pump failure
- Condensation or cooling of gas in a closed system
- Decrease in gas volume due to combustion in a closed system
- Inadequate design against implosion forces
- Increased altitude

Pressure changes:
- High gas compression
- Rapid expansion of gas
- Rapid change of altitude
- Rapid rise toward surface from underwater
- Explosive decompression

FIGURE 20–6 Continued

CHECKLIST—PRESSURE HAZARDS

1. Has each pressure vessel been designed, manufactured, tested, and installed in accordance with the applicable code?
2. Has each pressure vessel been proof-pressure tested after manufacture and periodically in the plant?
3. Are all the lines, fittings, and hoses rated for the pressures they must withstand? Have they been pressure tested?
4. Where tires must be inflated to high pressures, is the operator protected in case of a rupture?
5. Are flex hoses and their connections secured to prevent whipping if there is a failure?
6. Is any pressure vessel or system located near, or can it be subjected to, an unintended high heat input which will raise the pressure to a dangerous level?
7. Is it possible to accidentally connect a pressure vessel or system to a source of pressure greater than that for which the system, product, or any component was designed?
8. Does each container or line which might be overpressurized have a relief valve, vent, or burst diaphragm to protect it?
9. Will the exhaust from each relieving device be conducted away for disposal? Are toxic or flammable fluids either prevented from being exhausted or disposed of properly?
10. Is there a program for periodically testing each pressure device, relief device, and regulator?
11. Is each container which holds a flammable, toxic, corrosive, or otherwise dangerous fluid suitably identified and marked? Has each such container or line been analyzed to determine the effects of leakage? Could a leak of a flammable fluid, gas or liquid, come in contact with a source of ignition?
12. Are there instructions to ensure that any leaked or spilled fluid should be cleaned up immediately?
13. Should the system, assembly, line, or vessel be marked with a warning that it is to be depressurized before any work is started on it? Have maintenance personnel been instructed regarding this requirement? Is there a means to depressurize each pressure system to permit work on it?
14. Are personnel who work with compressed air or other pressurized fluid required to wear face shields?
15. Has "horseplay" with compressed air been forbidden?
16. Does any line with rapidly moving fluid have a quick-closing valve or other shutoff device which could result in water hammer and a shock wave?
17. Is there a possibility of a closed pressure-vessel collapse because of condensation of steam or other gas, decrease in altitude, or excessive operation of a vacuum pump? Does such a vessel have a vacuum relief valve?
18. Are large vacuum tubes which might implode if damaged adequately protected?
19. Do direct-reading pressure gauges have shatterproof glass or plastic faces, and blowout plugs at the rear?
20. Is flexible hose protected against chafing, twisting, or other damage?
21. If there is an accumulator, does it have a warning indicating the maximum operating pressure and other operating instructions necessary to avoid injury?

FIGURE 20–7 Checklist—Pressure Hazards

EXERCISES

20.1. Give three definitions of high pressure. Why should there be different levels for different applications?

20.2. What could cause a pressure vessel to rupture?

20.3. Explain the difference between proof pressure and burst pressure.

20.4. What is the principal difference between a safety valve and a relief valve?

20.5. What precautions should be taken in the event of discharges from safety valves?

20.6. Why must pressurized gases be secured carefully?

20.7. Why must a pressurized system be deenergized before being worked on? How could this be done?

20.8. What causes water hammer, what are its effects, and how can it be avoided or eliminated?

20.9. List seven causes of leakage.

20.10. List seven possible adverse effects of leakage.

20.11. What are the relationships among dysbarism, decompression sickness, bends, chokes, and gas pains?

20.12. Discuss what must be done to prevent decompression sickness. What should be done if a person shows symptoms of decompression sickness?

20.13. What are the three means by which gases are maintained in gas cylinders?

20.14. What types of relieving devices are used on gas cylinders?

20.15. List ten precautionary measures to be taken with gas cylinders.

20.16. Prepare six questions to be added to the checklist in Fig. 20-7.

CHAPTER 21

Electrical Hazards

The use of electricity and electrical equipment and appliances is so common that most persons fail to appreciate the hazards involved. These hazards can be divided into five principal categories: (1) shock to personnel, (2) ignition of combustible (or explosive) materials, (3) overheating and damage to equipment or burns to personnel, (4) electrical explosions, and (5) inadvertent activation of equipment. Electromagnetic effects and hazards are discussed in Chapter 27.

SHOCK

Electrical shock is a sudden and accidental stimulation of the body's nervous system by an electrical current. Current will flow through the body when it becomes part of an electrical circuit which has a potential difference adequate to overcome the body's resistance to current flow.

Although potential difference determines whether the resistance will be overcome, the damaging factor and the chief source of injury and death in electrical shock is current flow. The effects produced by various flows of a 60-hertz alternating current may be:

1. *1 milliampere (0.001 ampere):* shock becomes perceptible. Normally at this magnitude its chief effect is involuntary reflex action. The reaction could cause the affected person to make an inadvertent motion by which he or she might be injured. The sudden uncontrolled motion can cause a person to jerk violently, lose balance and fall, bump his or her head, or otherwise produce an injury worse than the shock itself.
2. *5 to 25 milliamperes (0.005 to 0.025 amperes):* this intensity is adequate to cause an adult to lose control of the muscles affected. A victim may "freeze" to the conductor with which he or she comes in contact, losing the ability to release the grasp voluntarily. "Let-go" current indicates the highest amperage at which a person still in contact with a conductor can still control muscles, even though stimulated. Let-go currents for 60-hertz circuits are approximately 9 milliamperes for men and 6 for women.

3. *25 to 75 milliamperes (0.25 to 0.75 amperes):* current in this range can be very painful and injurious. Prolonged contact may produce collapse, unconsciousness, and death as paralysis of the respiratory muscles causes asphyxiation. Death generally results if the paralysis lasts more than 3 minutes.
4. *75 to 300 milliamperes (0.075 to 0.300 amperes):* current of this magnitude that lasts a quarter of a second or more can be almost immediately fatal. This is due to "ventricular fibrillation," in which the muscle fibers of an individual's heart (ventricles) twitch, or fibrillate, in self-perpetuating waves instead of contracting in a coordinated beat. The rhythm of the heart is disturbed, it fails to function, blood circulation stops, and a lack of oxygen to the brain and tissue results. Alternating currents are more dangerous in this respect than are direct currents. Alternating current lasting one heart cycle or more will cause fibrillation in almost every case if the amperage is great enough. Direct current will produce fibrillation only if applied during a short, specific, vulnerable instant of a heart's cycle. The heart rarely recovers by itself from ventricular fibrillation. However, a strong, short-duration countershock can stop fibrillation and reestablish normal heart rhythm. The countershock excites all the muscle fibers at the same time; they are then quiescent for an instant; after which they may resume their normal beat.
5. *2.5 or more amperes:* a current of this magnitude will clamp (stop) the heart as long as it flows. Blood pressure falls as circulation stops. Beating of the heart and circulation of the blood will usually resume when the current is cut off. Unconsciousness generally occurs. High voltages frequently produce respiratory paralysis, but resuscitation applied immediately may succeed. Alternating currents of this magnitude will cause burns to both the skin and internal organs. Skin burns due to arcing will be especially damaging if contact areas with the conductor are small. Such burns reduce skin resistance, permitting higher current flow with shocks of greater severity than through an intact skin. Overheating of the body and burns to internal organs may also occur.

OTHER FACTORS

Shock intensity and effects also depend on the current path, the frequency, and the duration. Relatively large currents can pass from one leg to the other with only contact burns. A similar current from arm to arm or arm to leg may clamp the heart or paralyze the respiratory muscles. The effects of any specific amperage vary with the frequency. Neither alternating currents with potentials as low as 18 volts nor direct currents up to 140 volts have ever been fatal. Direct currents three to five times those of alternating current are generally required to produce the same effects. This is based on usual conditions. There has been much concern regarding possible electrocutions in hospitals at far lower currents during operations, especially those such as open-heart surgery.

Currents with frequencies of 20 to 100 hertz are the most hazardous. A 60-hertz current, ordinarily used in most commercial systems, is especially hazardous because it is close to the frequency at which the greatest possibility of ventricular fibrillation exists. Fibrillation is less probable at frequencies above 100 hertz, the probability decreasing inversely with the increase in frequency. High-frequency currents are less hazardous from this standpoint, since they flow along the surface of, rather than through, the conductor or body. Frequencies greater than 2,000 hertz will cause severe burns but have less internal effect than do lower frequencies for any specific amperage.

The current that flows through a body depends on the resistance of the body and any additional resistance between it and the earth. The resistance of the human body to current flow is contained almost entirely in the skin. The skin consists of two layers (see Fig 19-2). The outer one, composed of dead, scaly cells, has a high resistance when dry. Dry, clean, and unbroken, it has an electrical resistance of 100,000 to 600,000 ohms, depending on its thickness. Wet or broken, its resistance may be 500 ohms or less. This is due to the fact that the current can then pass to the inner skin layer, which has less resistance. This lower resistance results from the body fluids present, which make it moist and conductive. Skin considered dry varies appreciably with actual conditions. A person with a naturally dry skin may have a body resistance ten times that of a person with a moist skin. The resistance of a person who has washed and dried his or her hands may be only one-half that before the person washed them, even if they had been clean. Dirt on the skin increases its resistance.

Current passing through the body is also dependent on any other resistance it may encounter, which includes the internal resistance of the body's tendons, muscles, and blood. Internal body resistance is comparatively low, averaging 300 ohms (with a maximum of 500 ohms) for current flow from head to foot. The path of the current through the body not only governs the resistance it will encounter, but also the severity of the injury it will inflict. Paths through the heart or trunk are much more dangerous than one from leg to leg. Currents through the heart, respiratory muscles, or brain are the most critical.

When skin is dry, resistance is so high it may be adequate to protect a person from a mild shock, but having adequately dry skin is extremely unusual. When the skin becomes even imperceptibly wet, a person is in danger of getting a shock of great severity. A worker who is perspiring or in a moist atmosphere is in much greater danger of a fatal injury than when the skin is comparatively dry. Anyone performing physical labor perspires, even though it may be only slightly. It must therefore be assumed that the body is always wet, resistance is low, and the possibility of shocks high. Burns which puncture the skin reduce the body's resistance even more.

Differences in resistance under wet and dry conditions can be great, resulting in extreme changes in the tendency for a person to be shocked. An accidental contact of a finger when dry involves a resistance of 400,000 ohms but only of 15,000 ohms when wet. Using Ohm's law, the current that would pass when contact is made with a 120-volt circuit would be

$$\begin{aligned}\text{current} &= I = \text{voltage/resistance} = E/R \\ \text{dry} &= I = 120/400{,}000 = 0.3 \text{ mA} \\ \text{wet} &= I = 120/15{,}000 = 8 \text{ mA}\end{aligned}$$

CAUSES OF SHOCK

There are five principal ways in which a person can be shocked:

1. Contact with a normally bare energized conductor.
2. Contact with an energized conductor on which the insulation has deteriorated or has been damaged so that it has lost its protective value.
3. Equipment failure which causes an open or short-circuit.
4. Static electricity discharge.
5. Lightning strike.

Bare Conductors

Accidents are frequent in which persons are electrocuted because of lack of care near energized bare conductors. One of the commonest causes of such fatalities is contact with a live overhead electrical line. This often happens in the construction industry, when the boom of a crane, the raised end of a dump truck, a scaffold being erected, or a high metal ladder hits an energized power line. California reported that in one year 36 workers lost their lives through electrical contacts, of which 17 fatalities occurred when equipment or tools accidentally touched high-voltage overhead lines. Eleven of these fatalities involved hoisting apparatus.

The operator may not be affected unless attempting to leave the vehicle. For example, the 34-foot boom of a mobile rig was being positioned over a pump house to pull up a pump motor. When a well-puller operated the boom controls from the rear of the rig, the boom contacted a 12-kilovolt overhead line and he was killed instantly. When he heard the noise of the contact, the truck driver stepped from the cab of his truck to investigate and was hit by the full 12-kilovolt charge when his foot touched the ground. He was rushed to a hospital, where doctors amputated both legs, but he died a few hours later from massive burns and internal bleeding.

Persons working on rooftops, making repairs or installing TV antennas, have also been electrocuted when they hit bare high-voltage lines.

Although safety codes require enclosures of high-voltage equipment, failure to secure the enclosures against unauthorized entry has led to many fatalities. Another example from California can be cited:

> An 18-year-old janitor opened the bolted door giving access to the rear of a 2,400-volt main service cubicle, which he entered. A Z-shaped bus bar about three feet above the floor extended diagonally across the rear of the cubicle. The worker brushed against the bus bar and was electrocuted. The day before he had been found asleep behind the switch board. His work assignment did not include cleaning electrical equipment, and he had no reason to enter the area.

The entrance to a cubicle or other enclosure in which there is hazardous high-voltage equipment should be locked. Keys should be in the possession of the persons in charge and those with authorized entry such as qualified electricians and security personnel. Small enclosure accesses (panel, drawer, or door) should either be locked or be provided with an interlock (see Chapter 12) which will deactivate the equipment if the enclosure is opened. It is generally advisable to provide additional shielding around bus bars and other live conductors to prevent contact even after the enclosure is opened. Warnings should be placed both on the access to the hazardous area and on the equipment to alert personnel concerning the hazard.

Working on live high-voltage lines has been a common cause of accidental electrocutions, and such practices should be discouraged. If it absolutely must be done, it should be attempted only by qualified workers fully aware of the hazards involved, properly equipped and safeguarded, and only with the buddy system in use.

Accidents have occurred where a circuit was opened and an electrician began work. Another person, wanting to use a piece of equipment and not knowing work was being done, reenergized the circuit.

Electrical systems shut down for repair, maintenance, or modification should be locked out after being deenergized. The switch which energizes the equipment should be deenergized and opened. An attempt should be made to operate the equipment on this circuit. A warning tag should be placed on the handle of the opened switch which controls the circuit, after it is secured with a key or combination lock to prevent inadvertent activation. The person who will do the work on the line should be the only one to have the lock's key or combination. If more than one person is to work on the circuit, each worker should have an individual lock on the switch. The supervisor in charge may add a lock. Each lock remains in place until the person who put it there is finished working. No attempt should be made by any other person to remove the lock.

Circuits which contain capacitors store electrical charges with very high potentials. When the power to the system is turned off, unless grounded, the capacitors may remain charged. An unwary person who touches the capacitor or the circuit to which it is connected can be shocked as the current passes through the body to the ground. Although voltage may be very high, the currents involved are generally quite low. The shock may therefore be more surprising than painful and is usually not as injurious as a shock from an active circuit. In addition, a capacitive discharge occurs once and is nonrecurring.

Deenergizing by grounding eliminates the possibility of a person being shocked by touching a charged capacitor. To avoid capacitive shocks, safe design practice is to provide resistance grounds to deenergize the system. The resistance is high enough to prevent imposed charges from draining off, allowing the circuit to perform properly at its designed potential. On the other hand, resistance to ground is low enough to discharge any capacitance within a few seconds after the system is deactivated.

The National Electrical Code, which is concerned with large capacitors, requires the residual voltage of a capacitor which operates at 600 volts or less to drop to 50 volts or less within 1 minute after the capacitor is disconnected from the source of supply; and in 5 minutes for capacitors rated at more than 600 volts (see Chapter 5). Discharge, where required, must be accomplished either through a permanent connection or one which takes place automatically when the capacitor is removed from the electrical supply.

When electronic equipment with capacitive circuits is to be tested, personnel would do well to wait and then discharge the circuit with a grounding rod. The National Electrical Code, which is concerned with larger industrial equipment, prohibits the use of manual means of connecting a discharge circuit.

ELECTRICAL INSULATION FAILURES

Lines and equipment can be electrically insulated, but should the insulation be defective because of deterioration or damage, a person could be shocked. There are numerous ways in which deterioration or damage can occur.

One theory of insulation breakdown is that materials are not uniform throughout: localized areas may therefore have different specific heats and resistances. Heat due to power loss in the conductor causes the temperature in high-specific-heat areas to increase more than in others and to deteriorate and fail first.

In another theory, electrons from the conductor or quanta of energy from external radiation destroy chemical bonds in the insulation. This permits a flow of electrons so that a conductive path is created and the electrical resistance of the material is reduced.

Other causes and effects of insulation degradation might be:

Heat and elevated temperatures: flow of current always results in production of heat and increased temperatures. Even at moderate temperatures, heat causes a slow but gradual breakdown of some polymers.

Moisture and humidity: this is dependent on the absorptive properties of the insulation material. Moisture provides paths through and over insulation, increasing its conductivity. Many insulations absorb little moisture, whereas others, such as nylon, absorb as much as 8 percent. Numerous fatalities have occurred when electrical cords were used in damp locations. The reduced resistance caused the user to become the medium through which a circuit to the ground was established.

Oxidation: this may be due to the presence of oxygen, ozone, or other oxidizers in the atmosphere. Ozone is a special problem in enclosed spaces in which rotating equipment, such as motors or generators, operates. Sparking and arcing by such equipment are prolific producers of ozone. Ozone itself is a much more reactive and unstable form of oxygen. Although present to a much lesser extent in the atmosphere, it may create much more damage than will oxygen. It is produced naturally in small amounts by electrical discharges such as lightning strikes and by ultraviolet radiation of oxygen. Unlike oxygen, the percentage of ozone varies from place to place on the earth's surface and in its atmosphere. Its concentration is affected by the presence of smoke, smog, and other particulate matter in the air, altitude, latitude, local weather conditions, season, and amount of solar radiation received. The effects on insulation are therefore governed by location and protection.

Radiation: ultraviolet and nuclear radiation all have the ability to degrade the properties of insulation, especially polymers. Photochemical processes initiated by solar radiation cause the breakdown of polymers: natural or synthetic rubber, vinyl chloride, and vinyledene chloride, during which hydrogen chloride gas is produced. The acidic hydrogen chloride then causes further reactions and breakdown, crazing and cracking, and degradation of desirable physical and chemical properties.

Chemical incompatibility: incompatibility with acids, lubricants, salt spray, acid rain, or other materials can cause chemical breakdowns of insulation. In some instances, chemical reactions may result not only from external sources but also from decomposition products from the action of radiation on polymers or ozone produced on the conductor by coronas. Figure 8-1 points out the problem of acid rain in the Northeast and Canada.

Mechanical damage: installation and use of conductors is a common cause of damage to insulation and therefore must be done with care. Abrasion, crushing, cutting, flexing, and crimping are frequently found to have ruined coverings of wires and cables. Abrasion may occur when a wire is pulled through an opening that is too small, when it is dragged along a rough surface, or when too many conductors are located in a conduit. Vibration is a frequent cause of abrasion damage. The wire is located against a vibrating surface whose motion gradually wears into its covering. Unintentional cutting of insulation may take place when wire is pulled over a sharp edge. A heavy part or vehicle rolling over a wire located on a hard surface may crush it. A wire caught between two sharp surfaces, such as between a metal door and its frame, may be crimped and damaged. Bending a wire, especially in cold weather, may cause the insulation to crack.

High voltages: high voltages can cause sparking or corona effects (Fig. 21-1) which create holes in insulation or chemically degrade it. Insulation benefits against voltages at lower levels are reduced. Sparking will occur when the dielectric strength of the insulation is exceeded and is especially prevalent where weakness or damage has been created by other causes. Coronas will produce nitrous oxide, which reacts with moisture to form an acid that attacks the insulation.

Biological factors: some insulations act as nutrients for living organisms. Rats, other rodents, and insects may eat organic materials, cutting through or weakening them so they fail under load. Fungi may form on organic insulations, especially in moist atmospheres. Growth generally occurs where there is also contamination. The effects of fungi have shown to be deleterious; however, the exact degree of damage to such parasitic growth alone is difficult to determine because of the additional presence and effects of moisture and contamination. Plasticizers used in manufacture of plastic insulations which contain natural or fatty oils are especially susceptible to growth of fungi.

Pressure: vacuum may cause outgassing of volatile materials from organic insulators, reducing their resistivity. Outgassing will cause gaps in the material, dimensional changes, and weight loss.

EQUIPMENT FAILURES

In addition to insulation failure, other portions of equipment could fail so that a person will be shocked. A broken, energized power line falls on an automobile. The owner unknowingly grasps a handle to open the door and becomes a path to ground. Similarly, faults in electrical equipment, especially portable tools, cause the housing to

SPARKS, ARCS, AND CORONA

A *spark* is a sharp, rapid, heavy discharge of electrons. It may be a single discharge which completely exhausts the energy in an electrical system, or it may be a series, as the energy required for a discharge is replenished. Initial contact between conductors is not necessary for sparking as it is in arcing. A spark will therefore occur and recur each time an ionized path is completed between two conductors or a conductor and the earth.

An *arc* is a sustained stream of electrons across a gap created when two metal surfaces that were in contact or in close proximity separate while current is flowing. Although they are frequently called break sparks, the word "arc" is more generally used to indicate current flow across a separating gap. As the distance between the contacts increases, the arc grows in length until separation is so great that the voltage across the two is no longer adequate to sustain it. The arc then breaks down, and the flow of electrons halts. Much less energy is required to maintain an arc across a lengthened gap than to initiate one over the same distance. An arc will remain as long as power remains on the system; even brief interruption of current or a decrease in potential may kill it. It may also be extinguished if the gap becomes too great for it to span, or if its path is lengthened by displacement by wind or a magnetic field. Unlike a spark, an arc will not re-form unless the gap is closed again or the voltage across the gap exceeds the breakdown potential.

The length to which an arc will be drawn depends on a number of factors. As mentioned previously, voltage and current across the gap are of prime importance. Current required to maintain an arc over a specified distance also depends on the metal of which the contacts are made; those with low melting points require less current.

The dielectric constant of the medium separating the two contacts also determines the resistance to the electron stream. An increase in air pressure increases its dielectric value and reduces the tendency to arc or spark. For this reason, some electrical systems in which arcing or sparking may be particularly undesirable, such as in potentially flammable atmospheres, are pressurized.

An ionized gas is extremely conductive compared to one that is not; energy is required to produce ionization in a nonionized gas before a spark will pass. The rate at which ionization occurs depends (in addition to the factors already mentioned), on the nature of the surface from which it takes place. When a surface is rough, ionization takes place more rapidly from projections, where the intensity of current per unit area is greater, than from a smooth surface. When a surface is smooth and extensive, an arc may "wander," moving as the arc bows. Wandering is also governed by the type of metal of which the conductors are made.

Arcs may occur when an electrical switch is opened, a current-carrying wire is broken, or a relay opens. Arcs are produced intentionally in certain types of equipment, such as electric welders, for their heat-generating capabilities. The same heat, produced accidentally in arcs and sparks, can be extremely damaging. Combined with frictional heat, they cause the breakdown of carbon brushes on rotating electrical equipment, such as motors and generators. Arcs damage the surfaces of metal contacts, such as relays, which open and close frequently. Arcing causes pitting of one contact and buildup of the other with the metal removed from the first. When an arc forms, its diameter decreases, and a small, localized area on one contact grows hotter and hotter until metallic particles are emitted. These minute amounts of metal are carried along with the stream of electrons to the other contact. The metal deposited in the new location is less dense than in its original place and occupies more space. The gap between the contacts is gradually reduced until it is closed.

The *corona* effect occurs in high-voltage circuits when breakdown of a gap between two conductors is only partial. The potential exceeds a specific level, so gas ionization occurs but is less than that needed to produce sparking. Corona discharge starts more readily at those electrode points or protuberances which are negative than at those which are positive. Ions form a cloud, which begins to drift to the opposite electrode, establishing a low-grade continuous flow. This is the corona effect. It may produce a faint glow in the dark, a hissing sound, and electromagnetic interference. In severe cases the glow which results around a projecting surface may be so bright that it is called "St. Elmo's fire."* In any case, should potential difference between two conductors increase, a complete breakdown of dielectric and a spark may result.

FIGURE 21–1 Sparks, Arcs, and Corona

Arcs, sparks, and coronas all cause breakdown of insulators, ruining their usefulness as nonconductors. Ionization produces nitrous oxide from atmospheric nitrogen. The oxide and moisture form acid, which affects insulation. The destructive effect of corona discharge is less apparent and takes longer than does a spark. However, its long-term deleterious effects are damaging, chiefly because they weaken insulation badly without apparent signs, so that sparks may eventually occur when least expected.

Arcs, sparks, and coronas also generate electrical "noise," which reduces the effectiveness of electronic equipment. However, the worst effect of arcs and sparks is the release of energy, which may cause the ignition of combustibles. This may involve not only flammable gases, but solid materials that may be present close to where the sparks occurs. Even sparks produced by static electricity frequently contain more than enough energy to cause fires. Energized power circuits are therefore more hazardous if sparks or arcs occur.

* F. B. Silsbee, *Static Electricity*, National Bureau of Standards Circular C438, June 10, 1942, states: "It is probable that such a discharge caused the destruction of the dirigible airship *Hindenburg* in 1937."

FIGURE 21–1 Continued

be energized. The appliance may have a broken connection or wrongly replaced internal wiring which touches the tool or appliance housing. When the device is turned on, the user receives a shock when he or she touches the housing. (Newer models have improved by double-insulation systems which reduce such possibilities to almost zero. However, even double-insulated types are hazardous when wet, unless the insulation is of a special type. Electrical products totally immersed in water can still be fatal.) If the tool is not grounded and the person who touches the equipment is not insulated, the person may form the connection to ground through which current will pass (see Fig. 21-2).

To minimize this problem with power tools, three-wire systems were used in which one wire in a plug provided a connection to a grounded outlet. Unfortunately, in some instances, especially in homes, the outlets were not grounded properly, so the ground wire of the tool was useless. In other instances, older tools may not have had three-wire connections. Because of this, portable-tool manufacturers now use double insulation.

OSHA requirements stipulate that the frames and all exposed non-current-carrying metal parts of portable electrical machinery operating at more than 90 volts to ground will be grounded. The ground will be through a separate ground wire and polarized plug and receptacle.

OTHER SHOCK PROTECTION

In addition to the measures already indicated, other means of protecting personnel against shock can be provided in the design and operating procedures of electrical and electronic systems. The most common of these can be summarized as follows:

GROUNDING AND BONDING

The earth acts as an infinite store from which electrons can be drawn or to which they can return. Any undesirable excess or deficiency can be eliminated by providing a path from where it exists to earth. Positive ions in a system can then be neutralized by gaining electrons; electrons can be conducted to earth. In some countries this is called "earthing". In the United States the term "grounding" is preferred, and the earth or path to it is a "ground." In some devices such as electronic equipment, a massive metallic body acts as the reservoir of electrons and ions in place of the earth.

Grounds can be designed and installed into a system or they can be accidental. Unless noted otherwise, we use the word "ground" here to indicate a design ground. Installed grounds are basically safety mechanisms to prevent (1) overloading of circuits and equipment which would destroy them or shorten their lives, (2) shock to personnel, and (3) arcing or sparking that might act as an ignition source.

Grounds may protect a system, equipment, or personnel. Certain designs used on high-voltage transmission lines are sophisticated types which follow the standards set by the American Institute of Electrical Engineers (AIEE) or other codes. The grounding systems and standards of the National Electrical Code (NEC), which apply to buildings and related facilities, are more common. Some of the terminology used here is from the NEC and may therefore be familiar; other expressions and material are from other sources.

System grounds: are found in electric circuits and are designed to protect conductors of a transmission, distribution, or wiring system. The phrase "voltage to ground" is often used in electrical codes. It indicates the maximum voltage in a grounded circuit measured between the grounded wire and a wire that is not grounded. Where a ground is not used, voltage to ground indicates the maximum voltage between any two wires. The wire that connects the circuit to earth is the "grounding wire" or "ground"; the wire to which it is connected is the "grounded wire."

Three-wire systems can be used to illustrate the principles of grounding. In such systems, current generally flows along two wires and the third is neutral. In distribution systems for building and related facilities, the neutral wire is always the one grounded when grounding is installed. High-voltage transmission lines sometimes ground all three wires, but this is less common. The types of grounding systems that have been used on transmission lines include:

1. *Solid grounds.* The neutral wire is grounded without any impedance which might restrict current flow.
2. *Resistance grounds.* The neutral wire is connected to ground through a high resistance at a transformer.
3. *Reactance grounds.* The neutral wire is connected to ground through an impedance which is principally reactance.
4. *Capacitance grounds.* Each line of a circuit is connected to a capacitor; the other side of each capacitor is grounded.
5. *Resonant grounds.* This is a tuned, parallel system which uses capacitance grounds and a ground from a transformer neutral through an induction coil.

Solid grounds are the most commonly used, especially in interior electrical systems of buildings. Resistance and capacitance grounds are designed into most electronic equipment. These types of grounds involve circuitry comparable to two-wire systems in which it is necessary to maintain potentials within prescribed limits.

One purpose of grounding the neutral in a three-wire system is to activate over-current protection devices before damage is done when a fault occurs. Should one of the two wires which normally carries current be broken or accidentally grounded, current will flow through the neutral, through the installed ground, and back to the power source. This short circuit will open the protection devices and deenergize the affected portion of the system.

Where the neutral is not grounded, accidental grounding of one of the other wires will cause an increase in voltage to ground of the remaining system. The definitions of "voltage to ground" for grounded and ungrounded systems will illustrate this point. According to these, a 220-volt three-wire grounded neutral system will have a voltage to ground of 110 volts. Where there is no ground, the voltage to ground is 220 volts, the maximum voltage between any two wires. The excessively high voltages may cause burnout of equipment, burning or breakdown of insulation, arcing and sparking, and shock to personnel

FIGURE 21–2 Grounding and Bonding

in contact with metal energized through the breaks.

Other possibilities exist by which an excessively high voltage can be produced which would create similar hazards if the system is not grounded. A fault in a step-down transformer could result in applying the distribution system potential, or part of it if greater than normal, to a building wiring system. An accidental connection between the two systems would produce the same result. Where grounds existed, the overcurrent protection devices would deenergize and safeguard the system.

Equipment grounds: may be used on the metal parts of a wiring system, such as the conduit, armor, switch boxes, and connected apparatus other than the wire, cable, or other circuit components. They may also be provided for equipment such as metal tables and cabinets which may come in contact with an energized circuit or source of electrical charges. Equipment on which undesirable charges may be induced or generated should also be grounded.

Metal of electrical equipment may come in contact with an energized circuit whose insulation is deteriorated or cut, or through which arcing can take place. A person may then touch the metal surface inadvertently, receiving a shock. The degree of shock would depend on whether the equipment was grounded. If it was not, the person in contact with the metal would act as a ground, the current passing through his body. If the equipment was grounded, the person might or might not receive a shock. If current did pass through his body, the amount would be inversely proportional to the resistance of his body compared to that of the equipment ground. If the resistance of his body was high enough, no current would pass.

Bonding: ensures that all major parts of a piece of equipment are linked to provide a continuous path to ground. A bond is a mechanical connection which provides a low-resistance path to current flow between two surfaces which are physically separated or may become separated. A bond can be permanent, such as one in which the connection is welded or brazed to the two surfaces, or it may be semipermanent, being bolted or clamped where required.

Where permanent types are used, the parts themselves can be joined and narrow gaps filled with weld or brazing metal. Where separation is wider, a strip of metal can be welded or brazed at both ends across the gap. Bonds connecting a vibrating part to another part that may or may not vibrate should be of a flexible material which will not fail under vibration. Corrosion because of the joining of dissimilar metals may cause the electrical resistance across the bond to increase. This is especially noticeable in humid or corrosive atmospheres. The types of metal for the bond and its fastenings must therefore be selected with care.

Grounding and bonding requirements: grounds and bonds should:

- Be permanent wherever possible.
- Have ample capacity to conduct any possible current flow. (Note: A ground should not be designed to normally be part of a current-carrying circuit.)
- Have as low impedance as possible.
- Be continuous and, wherever possible, made directly to the basic structure rather than through other bonded parts.
- Be secured so that vibration, expansion, contraction, or other movement will not break the connection or loosen it so that the resistance varies.
- Have connections located in protected areas and where accessible for inspection or replacement.
- Not impede movement of movable components.
- Not be compression-fastened through nonmetallic materials.
- Not have dissimilar metals in contact or be selected to minimize corrosion.

Exceptions: Grounding is not advantageous in all cases; some electrical systems are safer ungrounded. Leea has pointed out: Some electrical systems (necessarily of limited extent), must be left ungrounded for safety reasons. For example, the electrical system of a hospital operating room is purposely ungrounded because a spark from an insulation failure would otherwise ignite the anesthesia-permeated atmosphere. When ungrounded, an insulation failure "to ground" produces no current flow, and hence no spark, no ignition and no explosion. Electric blasting caps present a similar condition; a short-circuit current returning through the earth could fire the caps if their two connecting wires touched the earth more than a few inches apart.[a]

Other exceptions include two-wire battery-supply systems for power circuit breakers, or voltages to ground that exceed 300 volts. In such ungrounded installations, detectors are necessary which will indicate when a ground exists.

[a] R. H. Lee, "Electrical Grounding: Safe or Hazardous?" *Chemical Engineering*, July 28, 1969, p. 162.

FIGURE 21–2 Continued

- *Insulation:* insulating parts of electrical equipment which a person will contact routinely or accidentally during operation of the system is advisable. Insulated knobs, dials, handles, and buttons on controls, switches, drawers, and meters are such items. Rheostats and potentiometer control shafts can be coupled to nonconductive rods and knobs.
- *Personnel can be insulated:* mats of rubber or other nonconductive material can be provided on which to stand when electrical equipment is being worked on, repaired, or maintained. Rubber gloves and nonconductive shoes should be worn when work must be performed on circuits that might be energized. Care must be taken to ensure the dielectric strength of the material is such that protective equipment has not been degraded by damage or deterioration.
- *Interlocks (Fig. 12-5):* where an enclosure is breached, the circuit is broken automatically and the system deenergized. Because enclosures are frequently opened for maintenance purposes, during which "hot" circuits must be checked, interlock switches must be operable deliberately when the access panel is open. Such switches should be of a type which reinstitutes the safety function when the enclosure is closed again.
- *Isolation:* electrical equipment, especially high-voltage types, should be isolated to keep unauthorized or untrained personnel from approaching too close. Large transformers with exposed terminals can be located in vaults or fenced enclosures to which only unauthorized persons have access. Panel boards, generators, large motors, batteries, bus bars, and other electrical equipment which might be hazardous should be enclosed or provided with guards to prevent accidental contacts.
- *Marking:* points of access to hazardous electrical equipment should be marked with suitable warnings. Some organizations require the insides of panels to circuit breakers, fuse boxes, and similar equipment be painted with colors and designs that would immediately indicate they were open.
- *Warning devices:* a suitable warning device may be connected to electrical equipment to indicate when it is energized. This may be a light, steady or flashing; a suitably colored indicator; an "on-off" sign; or an audible signal.
- *Ground-fault circuit interrupter (GFCI):* the National Electrical Code defines a ground-fault circuit interrupter as "a device whose function is to interrupt the electrical circuit to the load when a fault current to ground exceeds some predetermined value that is less than that required to operate the over-current protective device of the supply circuit." The GFCI was developed and intended for protection of personnel against shock. All outdoor 120-volt single-phase 15- and 20-ampere receptacle outlets on residential properties and bathroom outlets must have approved ground-fault circuit protection to comply with the National Electrical Code. It may be provided anywhere else and will provide additional protection against line-to-ground shock hazards.

Circuit breakers and fuses will open only under comparatively large currents which could prove fatal to personnel. The GFCI is sensitive to very small currents, far lower than those that will cause injury. The time for the GFCI to detect these small currents is short enough to open the circuit before a person can be electrocuted (see Fig. 21-3).

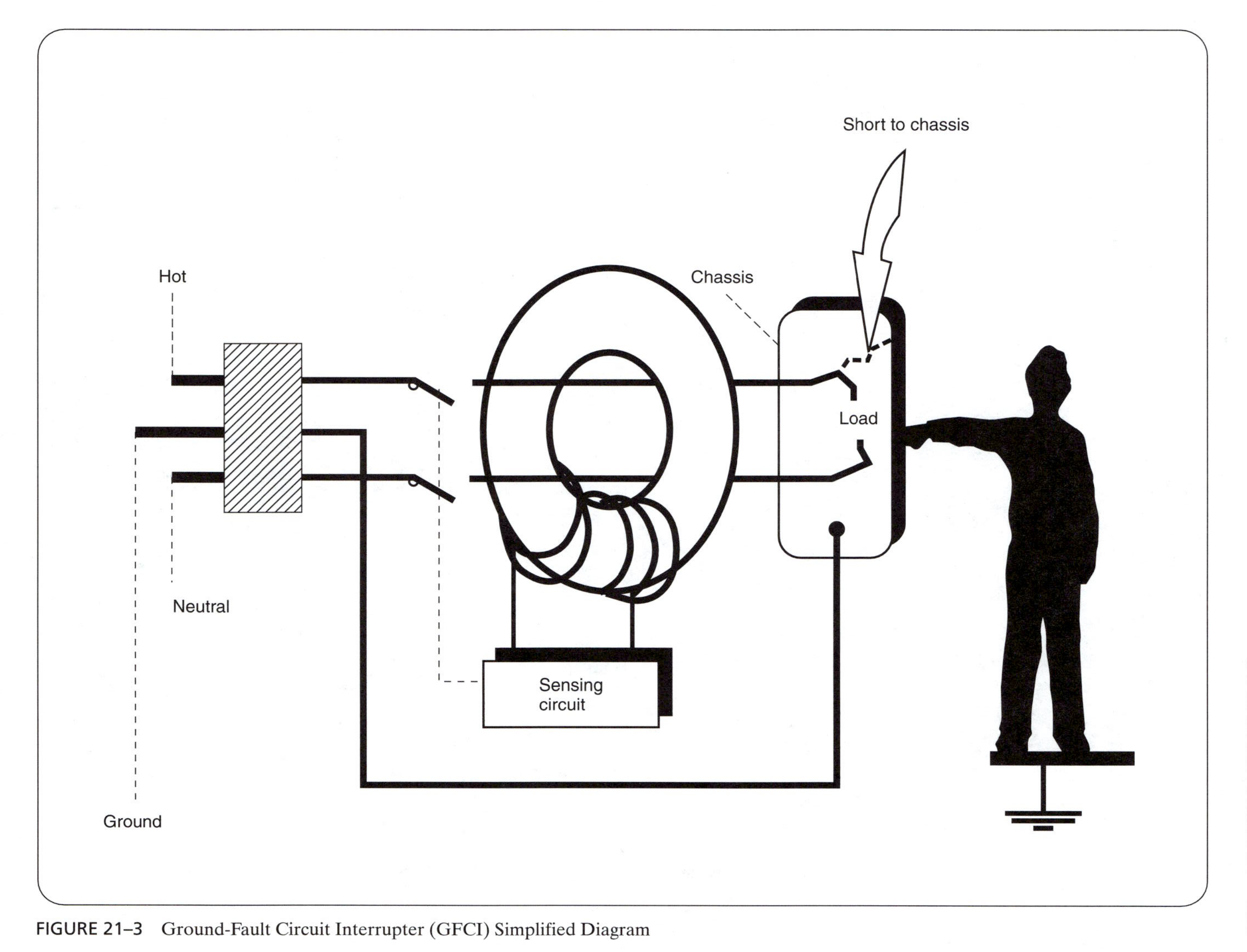

FIGURE 21–3 Ground-Fault Circuit Interrupter (GFCI) Simplified Diagram

The GFCI monitors current flow in a circuit as it operates normally. If current greater than the level for which the GFCI is set passes to ground by any route other than the one designed into the system, an imbalance in current flow is created and sensed, and a circuit breaker opens immediately. Thus, if a condition exists where a conductor is grounded directly, current flow will be so great that the circuit breaker will open even when a GFCI is installed. The GFCI provides protection if the activated circuit is inadvertently grounded through a high-resistance path such as a person's body.

In Fig. 21-3, current passes along the hot wire, load, and neutral path, through a sensing coil (toroidal coil), shaped like a doughnut in the diagram. If there is a short to the equipment chassis, the path of least resistance will flow along the hot wire and the system ground wire. This creates an imbalance in the hot-neutral path, which is sensed by the sensing circuit/coil. The protected circuit is immediately opened by the GFCI (shown in Fig. 21-3 by the open switches on the hot and neutral wires). Even if there is a path to ground, which for some reason is through the human body, the GFCI will trip in a fraction of a second at currents well below the dangerous threshold.

The GFCI will provide protection only in a ground-fault situation (from an energized line to ground). There are two types of ground faults: hot to ground and grounded neutral. It will not provide protection when a person forms the path between two lines of a circuit without a ground contact. However, most electrical accidents are of the line-to-ground type. When a typical GFCI is used in a two-wire system, without a system ground wire, the GFCI will trip when there is a flow to ground. However, this flow could be through the person who connects to the fault. That person's safety then depends upon the GFCI. A properly installed three-wire system and GFCI provides a redundancy that is important.

STATIC ELECTRICITY

The various ways in which static electricity can be generated (see Fig. 21-4) keeps increasing as new material is created and put into use. Static electricity discharges, since they are of a capacitive nature, also involve high potentials and low current. In many cases they involve only a single discharge; however, there are many types of static electricity generators which can caused repeated sparking.

There are ways of controlling the static electricity problem; the best specific means to do so depends on the individual case.

- *Selection of suitable materials:* avoidance of materials which generate static electricity is often the simplest method. For example, where charge generation on a person's body could create problems, the person should be instructed to wear cotton clothing rather than clothing of wool, nylon, or other synthetic fibers.
- *Making a material suitable:* a coating that can be sprayed on a surface to make it conductive frequently can reduce or eliminate the static electricity problem. Sprays are available for applications on materials such as plastic car seats, carpeting, and thin sheeting.
- *Bonding and grounding:* the various surfaces on which charges could accumulate are bonded to provide a path by which neutralization can occur. Grounding may

STATIC ELECTRICITY

Static electricity consists of an excess or deficiency of electrons on the surface of a material, Unfortunately, the manifestation that static electricity is present is often an electrical discharge which eliminates either the excess or the deficiency, When two surfaces are in close proximity, electrons will be present on one side of the interface, and a second layer with charges of opposite polarity will form parallel and close to the first, on the opposing face. As long as the surfaces remain in close proximity the charges oppose each other, so that for all external effects, the combination itself is electrically neutral. The magnitude of the charges that accumulate will depend basically on the nature of the material and the area and geometry of contact of the surfaces. (A surface that has a deficiency of only one electron in 100,000 atoms is very strongly charged.) Friction is not necessary to generate static electricity. However, friction increases the release of electrons and the production of ionized particles.

Because generation of static electricity depends on the areas in contact, being a surface phenomenon, flow of a fluid through a screen, filter, or similar, device is especially productive of static charges, The Shell Oil company carried out an investigation to determine the cause of sparking inside a fuel tank being filled with filtered jet fuel. Sparks up to 2 feet long were generated when the filter was used, even though the lines were grounded. When no filter was used, no sparking occurred.

Relaxation: the process by which electrons leave negatively charged regions and flow to ground or to positively charged ions when charged surfaces are separated. Relaxation time is related to the ease with which electrons can flow through a material (conductivity), or its reciprocal (resistivity). Conductive surfaces have little tendency to retain static charges once they are grounded.

Whenever charged surfaces are separated rapidly the electrons may be prevented from moving to neutralize opposing charges if the conductivity of the material through which the electrons must pass is low. Resistance to electron movement by a metal is extremely low and neutralization occurs easily. Static electricity is therefore rarely a problem when both surfaces are metal. With other substances, such as insulators, electron movement will be impeded; after separation, excess electrons will remain on the insulator's surface.

Conductivity of a sheet of material will be increased by any moisture that may be present. A study [a] of the possibility of generating electrical potentials on garment fabrics great enough to initiate detonation of sensitive explosives and flammable gas mixtures indicates the following effects of humidity at 75 F.

65% relative humidity	Potentials produced were inadequate to detonate the explosives or ignite the flammable mixture.
35% relative humidity	Nylon and wool, nylon and cotton, and nylon and wool-nylon contacts produced potentials greater than 2,650 volts, enough to ignite the more sensitive materials.
20% relative humidity	Higher potentials were produced by combinations given above; high potentials also from wool and cotton contact.
Less than 20% relative humidity	Dangerous voltages produced on the body, even with cotton.

Charges tend to leak into the gas which separates one surface from another. When the gas is ionized or has a high electrical conductivity, the surface charge will leak off and be lost at any protuberance, such as a piece of dirt or sharp point. Leakage into air, especially humid air, increases with the sharpness of the protuberance.

Occurrence of static electricity: extremely high voltages for experimental work are produced intentionally by running a broad nonconductive belt rapidly over a stationary surface and collecting the charges generated. High potentials are produced similarly, if less desirably, during production of sheets of paper, cloth, plastics, fiberglass, and other nonconductive materials; by flow of nonconductive fluids; and by interactions between other nonconductive surfaces.

[a] F.L. Hermack, *Static Electricity in Fibrous Materials*, National Bureau of Standards Report 4455, Dec. 1955.

FIGURE 21–4 Static Electricity Continued on page 356

One of the commonest and most familiar means of charge generation is by a person walking across a wool or nylon carpet, especially when scuffing his shoes. Done briskly, an individual can accumulate charges with a potential difference as high as 10,000 volts between his body and the ground – enough to produce an arc in air over $\frac{1}{8}$ inch. Nearing a conductive, grounded surface or one which accepts electrons will permit a rapid discharge which gives the charged individual a shock. Touching another person at lower potential may give shocks to both as the charge passes from one to the other.

Multilayered clothing of wool and nylon is another common generator of static electricity. Movement of the wearer creates friction between the various layers and accelerates production of static charges. Should the wearer approach close enough to a grounding device, a long spark may be produced when discharge occurs.

Wool is an excellent generator of static charges. However, woolen socks or underwear may accumulate moisture as the wearer perspires, reducing the tendency to produce and accumulate static charges. The energy and the potential involved with such items depend on how long the garments have been worn.

A study of charge accumulations in out-of-door arctic environments states[b] that the calculated stored energy on clothing generally remained at less than 0.011 millijoule. When a person removed his outer garment, static energy increased to 0.04 millijoule. If the same person entered a warmed enclosure with low relative humidity and removed his outer garment, a potential between his body and the ground of 8,000 volts, with stored energy of 3.9 millijoules, would be produced. This combination of voltage and energy is more than adequate to provide a spark that could ignite primary explosives or flammable mixtures of common fuels with air.

Large sheets of plastic will produce extremely high accumulations of static charges and therefore energetic discharges. In 1964, removal of a plastic cover from the third stage of a NASA Delta rocket caused the motor to ignite. Two men were killed and three others critically burned by the motor exhaust.

Other examples of problems with frictional electrification involving solids occur in, but are not limited to (1) mining and milling industries, where organic and metallic dusts often ignite and expiede from the sparks generated; (2) ordnance plants, where even small static charges can initiate violent explosions; (3) paper cloth manufacturing and processing plants, where the great charges and potentials generated can shock personnel badly, ignite fibers, cause pitting of bearings, and damage electrical equipment; and (4) materials-transporting systems, where high voltages are generated by conveyor belts, solid materials blown through ducts,or the tires of vehicles moving over a road surface.

Fluids: the mechanisms by which static charges are produced with fluids can be categorized as:

1. Friction between a flowing liquid and a solid surface such as the walls of a pipe, hose, or tank, or a filter.
2. Friction between a fluid flowing in contact with another liquid.
3. Spraying of a liquid into air.
4. Settling of suspended foreign particles, solid or liquid, or upward movement of bubbles through a liquid.
5. Movement of rain, snow, or hail through moist air.

[b] J. H. Veghte and W. W. Millard, *Accumulation of Static Electricity on Arctic Clothing,* Arctic Aeromedical Laboratory, AD-412-781, May 1963.

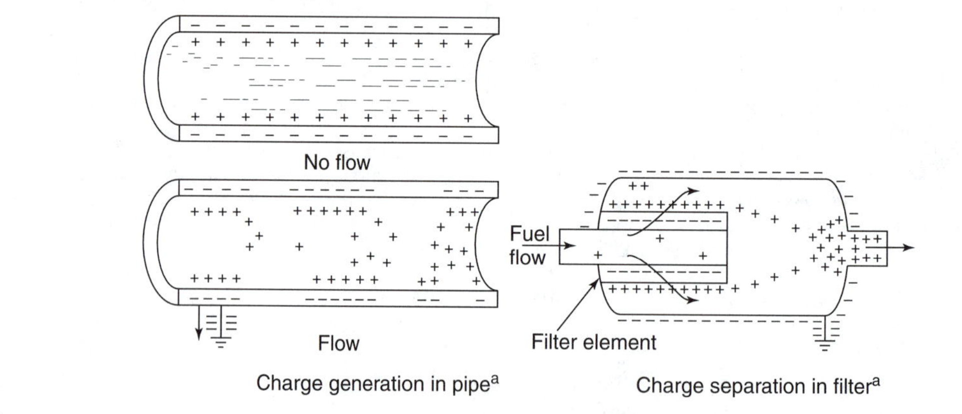

Charge generation in pipe[a] Charge separation in filter[a]

FIGURE 21–4 Continued

When a fluid, such as diesel oil, flows through a pipe, the liquid becomes charged because of its comparatively low conductivity. Charges move with the liquid, accumulating as it flows. This moving accumulation is known as the "streaming current". It may enter a tank with the fuel, sometimes in extremely dangerous amounts. Once the liquid enters a tank, the charges may require hours to dissipate, the period depending on the relaxation time of the fluid and the material of which the tank is made. Generation of such accumulations by combustible fluids is especially hazardous since discharges can be ignition sources which cause fires involving the very liquids that produced the charges.

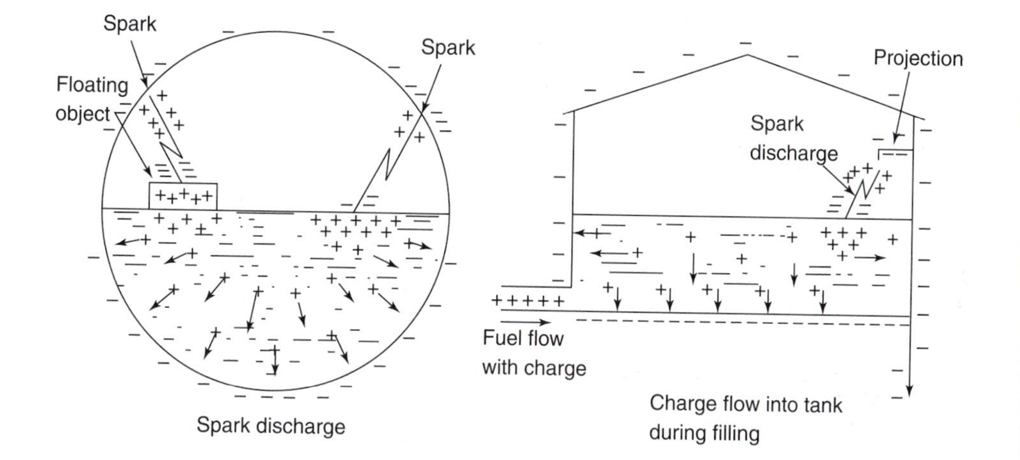

Spray electrification: charged particles are generated by the rapid separation of double layers during a spray process. Spray electrification differs from flowing liquid charging in that all the charges, negative and positive, remain on the droplets when they are produced. The very fine particles retain negative charges, and the larger droplets are positive. Settling of the larger droplets leaves the atmosphere negatively charged. The possibility of sparking between charged fuel in a tank and the tank's walls and roof is enhanced by the ionized condition of the mist in the tank atmosphere. Sloshing of fuel if the tank is aboard a moving vehicle may produce additional charges.

Filling a fuel tank may therefore be a hazardous operation[c]. The hazard increases with the rate at which the tank is filled and its size. A metallic object on the surface, such as a float, will aggravate the situation even more by acting as one plate of a condenser and collecting electrical charges from the fuel. The tank sides and roof act as the other plate of the condenser. Potential increases with the distance between the two; when it becomes great enough, a spark may occur. Since the atmosphere in the tank is occupied with fuel vapor containing ionized particles, the potential required to produce a spark is much less than that required in dry air.

[c] Explosions of large tankers at sea have been attributed to static electricity that developed during cleaning operations. Lt. Cdr. H. D. Williams, U.S. Coast Guard, in "Dangerous and Exotic Cargoes", *U.S. Naval Institute Proceedings,* Naval Review Issue, 1973, describes this problem, which is very briefly summarized here: "Benzene left on the tank bottom vaporizes and produces a blanket of vapor which at normal ambient conditions is in the explosive range... Droplets formed by water sprayed on the tank walls (to wash them down) create static charges in the water vapor. Any ungrounded object passing through this atmosphere will collect a static charge which will generate a spark as it nears a ground."

FIGURE 21–4 Continued

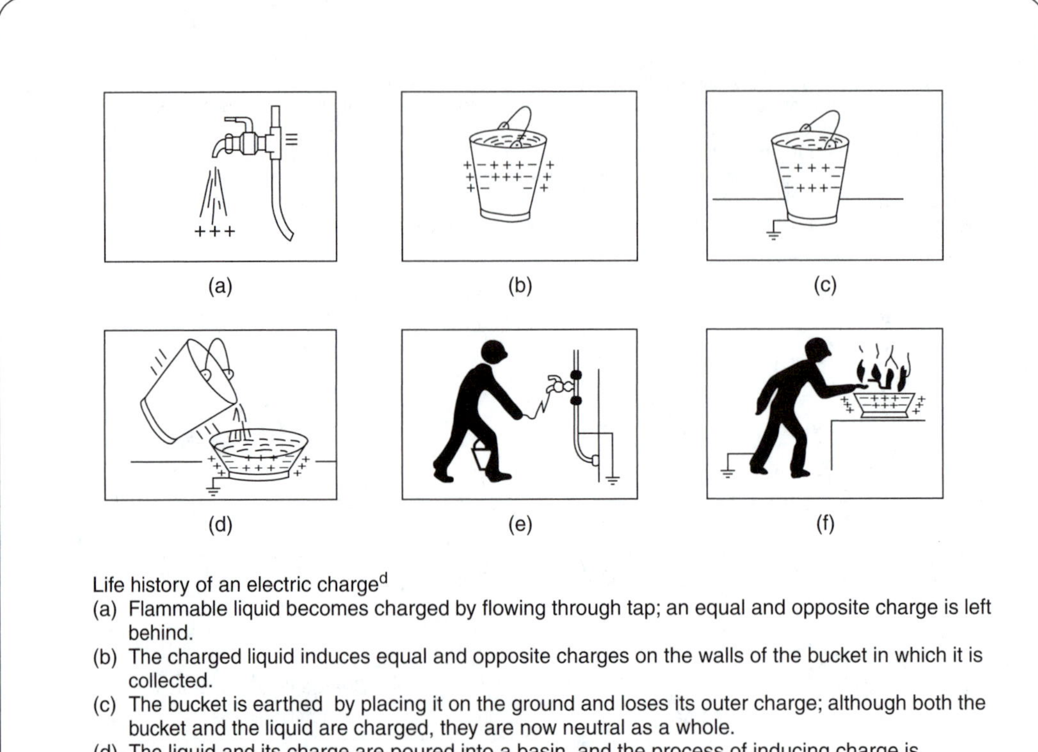

Life history of an electric charge[d]

(a) Flammable liquid becomes charged by flowing through tap; an equal and opposite charge is left behind.
(b) The charged liquid induces equal and opposite charges on the walls of the bucket in which it is collected.
(c) The bucket is earthed by placing it on the ground and loses its outer charge; although both the bucket and the liquid are charged, they are now neutral as a whole.
(d) The liquid and its charge are poured into a basin, and the process of inducing charge is repeated; the exterior charge on the bucket is left behind.
(e) A man in rubber-soled shoes shares the charge with the bucket and a spark passes from his hand to an earthed fuel tap, causing a fire.
(f) Another man without rubber–soled shoes (who is 'earthed') puts his hand near the surface of the charged liquid and a spark passes, causing another fire; this has, in fact, caused many accidents, and earthing the basin makes very little difference.

[d] W.F. Cooper, "Electrical Safety in Industry," *IEEE Reviews*, Vol. 117, Aug. 1970.

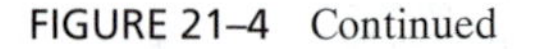

FIGURE 21–4 Continued

be to earth or to any other reservoir of charges to which electrons can be discharged or from which they can flow. Grounding bars on doors leading to hazardous areas, and ground-stats which personnel must wear for hazardous operations, are examples. *Note:* Gasoline trucks were formerly equipped with drag chains to bleed off static charges. It was found, however, that they were ineffective during dry weather because of the low conductivity of the road surface, and they were unnecessary when the road was wet. Their use is no longer recommended.

- *Electrostatic neutralizers:* there are three major types. The radioactive neutralizer is a source of radium or polonium that emits alpha particles (see Chapter 27) bearing two unit charges of positive electricity. These charges neutralize the negative charges on the material being processed. Radioactive neutralizers are harmful to human tissue and must therefore be installed only in accordance with the manufacturer's instructions and with existing health regulations. The high-voltage neutralizer produces a very high potential in air near the surface being treated. The atoms of gas in the air become ionized and release electrons. The air then has both positive and negative ions, which combine with the electrons and ions produced by the material being treated. A high-voltage neutralizer could cause persons injury from shock. The induction neutralizer functions by producing potentials of polarity opposite to that causing static electrification. Induction neutralizers can be built to any size necessary. Sizes impracticable for ionizing or high-voltage neutralizers would probably be feasible with induction neutralizers.
- *Humidification:* Raising the relative humidity above 65 percent permits static charges to leak off and dissipate. As indicated in Fig. 21-4, this is because of the conductive film of moisture that forms on the surface, which would otherwise be nonconductive.

LIGHTNING

Lightning is a massive, natural discharge of static electricity involving very high potentials and high current flow. Lightning follows the path of least resistance to earth; lightning arresters, rods, and grounds provide such paths. As mentioned in Chapter 25, because of the heights of their masts, ships at one time were extremely vulnerable to lightning strikes.

As protective devices, lightning rods are placed so their upper ends are higher than any nearby structure. Grounds are low-resistance paths to provide easy passage of current to earth. This protects electrical circuits and all metal equipment in a building or structure from direct passage of lightning. The secondary effect of lightning, induction in any nearby conductor, can be absorbed better if the system and equipment have grounds and overload protection. If lightning protection systems were not available, the sudden and massive surge of current would tend to burn out most circuits through which it would flow.

Lightning warning, prevention, and protection have changed considerably in recent years. Facilities which are sensitive to lightning can install lightning warning systems. These are designed to give real-time warnings of lightning activity within a 100-nautical-mile radius of a direction-finding antenna. Lightning strokes have a unique signature which is received by the antenna and identified by the warning system computer, which records and displays the peak rate of flashes per minute and the number of strokes per last hour. The display can show the distance and location of the strikes and color-code the intensity on a scale such as, "occasional, moderate, and intense" or "yellow alert" (storm within one-half hour), "amber alert" (storm within minutes), and "red alert" (imminent danger).

Lightning prevention includes practical engineering using tall towers with multiple-point discharge arrays. Thousands of pointed rods are spread over the protected area like an umbrella. During electrical activity, a protective corona diverts the lightning. These rods do not draw lightning as does a lightning rod, but instead the design goal is to dissipate cloud charge. More exotic prevention techniques such as chaff seeding of clouds and rockets fired into charged clouds are reserved for very special needs. The chaff seeding of millions of thin aluminum-coated nylon fibers to depolarize clouds is claimed to reduce the number of strikes from a cloud by as much as 75 percent. Rockets tethered to earth by wire are fired into clouds to create a path to ground that triggers lightning—theoretically neutralizing the cloud.

Research has shown that the pointed lightning rod may shift the lightning bolt to the roof it is supposed to protect. This shift can occur if there is a rapid blue-green corona build-up on the rod. Some lightning protection systems use an array of blunt rods so that the corona builds up slowly and over a wide area. It is said that, for these systems, any lightning bolt is likely to hit away from the protected building altogether. During an electrical storm, natural or manufactured objects, and even a person, may act as a ground if it or the person is higher than the surroundings. Even a person in a boat or open field is tall enough to present a path of lower resistance than the same height of air. Persons in such situations have been electrocuted in this way. A tree is another natural ground for lightning. Alone in a field, it may provide protection against rain, but it is more hazardous when it comes to lightning strikes.

Certain types of terrain and structures are more subject to lightning strikes than are others, again because of the paths they provide. Such structures require that protection be provided to prevent damage; others have inherent protection. All-metal buildings provide large inductive areas that protect its inhabitants. Tall buildings with metal skeletons also give protection. However, unless specific grounds are provided, the current flow from the lightning may follow a random path to earth, damaging circuitry and equipment, and shocking personnel in contact with metal surfaces.

General precautions can be described for persons in an electrical storm to avoid being struck by lightning. If a metal structure or building, or one protected by lightning rods and grounds, is available, it should be used as a shelter. If a metal building is not protected, it is not advisable to remain in contact with metal surfaces or electrical equipment. Outdoors, a person in a field or boat on a body of water distant from high points should lie down. Protection in open areas should be sought in ravines and ground depressions if there is no danger of flooding. In a wooded area, protection from rain and lightning should be sought under a low tree, away from the tallest in the area. Metal fences should be avoided, since they may act as grounds. Any person in contact with one at such times could be shocked.

Most occurrences of lightning strikes have occurred to elevated structures where the electricity has passed to the earth. Many strikes never touch the ground, as the bolts pass from cloud to cloud. This has been common in aircraft flying through storm areas. In addition, a most unusual occurrence took place on June 11, 1987, at Wallops, VA, when lightning ignited three small NASA rockets to be used to position space satellites. Although normal protection had been provided, the protective system of surge suppressors and lightning rods had been overwhelmed.

IGNITION OF COMBUSTIBLE MATERIALS

The commonest means by which combustibles are ignited electrically are by a spark, arc, or corona (Fig. 21-1) through a flammable mixture. Ignition by contact with an electrically heated surface generally requires the expenditure of considerable power to cause the heating. A spark or arc, on the other hand, may involve comparatively little energy, but is discharged rapidly in a limited space where it is adequate to cause the ignition of an extremely small volume of flammable mixture. The energy released from this combustion is enough to cause further propagation of fire. A spark or arc is the passage of a stream of electrons across a gap between two conductors or a conductor and the earth. When a gas such as air is the dielectric between two oppositely charged points, it will undergo progressive ionization. The gas closest to the electrodes will ionize first and, if the strength of the electrical field is strong enough, ionization will continue progressively over the entire gap. When breakdown of the dielectric is complete, a spark will cross the gap.

CONTAINMENT OF DISCHARGES

It is advisable to eliminate all electrical circuits and equipment from hazardous areas in which flammable atmospheres might exist. By suitable design, equipment can sometimes be located outside and wiring routed around such areas. However, in many cases these measures may not be possible or economical. Suitable provisions must therefore be made to eliminate the possibility or minimize the effects of electrical sparking, arcing, or coronas. This can be done in numerous ways, sometimes for this reason alone, and sometimes as the result of a design adopted for other reasons and benefits.

INHERENTLY SAFE DEVICES

In some instances it may be possible to use devices whose operating power levels are low enough to make them inherently safe even in dangerous locations. Miniaturized equipment, semiconductor components, and printed circuits may not only save weight but may require currents and voltages far less than those necessary to ignite flammable mixtures.

Embedment, Encapsulation, and Potting

Units can be encapsulated or embedded in a suitable material, such as a plastic solid or foam. Encapsulation is the process for encasing a part or an assembly of discrete parts within a protective material that is generally not over 100 mils thick and does not require a mold or container. Embedment uses a material generally thicker than 100 mils, whose thickness may vary. Two principal methods of embedment are in use. In the first, the equipment assembly is fabricated, tested, and placed in a mold. A suitable material is then cast around it. In the second method, the equipment is fitted into a hollowed-out container, covered, and sealed in place. In potting, the protective material bonds to the container to form an integral unit.

Materials used for encapsulation and embedment are generally poor conductors of heat. Electrical equipment with considerable power loss may suffer from high temperatures which result. Encapsulation or embedment, however, has numerous advantages. The plastic used is nonflammable, thick enough to prevent sparks from reaching the surrounding atmosphere, and thermally nonconductive, so its outer surface is never hot enough to ignite any gas. The container keeps the device clean, dry, and free from corrosive materials. It provides excellent support against vibration and impact. The result is prevention of possible unauthorized tampering or adjustment of components.

Hermetic Sealing

Hermetic sealing was originally developed to prevent entrance into a container from the atmosphere of moisture, dirt, combustible gases, or other corrosive and degrading materials. Its advantages are similar to those of encapsulation. Heat removal is less of a problem, since the operating device can be cooled in various ways. The heat generated can radiate to the inside walls of the sealing container, through which it is conducted and from which it is removed. When the container holds a gas, convective flow of the gas can assist in carrying heat to the container walls. When a liquid is used, it carries the heat away, but radiation transfer is generally eliminated.

Hermetic sealing is especially useful for equipment that may be subjected to and affected by changes in altitude, or that must be submerged in a liquid, dust, or vacuum environment. The container can be filled with a fluid—gas or liquid—which is either inert or of thermal capacity suitable for transport of the heat to be removed. To prevent overpressurization due to expansion caused by the heat produced during operation, the container may be sealed when hot. The fluid is added, brought to a temperature higher than the expected maximum operating temperature, and sealed. The container must be strong enough to withstand the partial vacuum that will result when the container and its contents are cooled. Sealing must be complete; otherwise the fluid will be lost and effectiveness of containment destroyed.

Liquid Filling

Liquid-filled containers have been used for years for cooling and protecting electrical equipment such as large transformers and circuit breakers. With transformers, the principal requirement of the liquid is to act as a coolant to remove the heat generated by power losses. A secondary advantage in this case is the greater dielectric strength of the liquid compared to that of air. Not only is loss insulation required, but sparking and arcing that might result from a broken wire or broken insulation is eliminated.

With circuit breakers, the chief purpose is arc suppression. Opening a breaker when a current is flowing in a very high voltage system would require an extremely wide, impractical gap. An arc would span the gap between contacts in such circuits if air were the dielectric. When a liquid, such as silicone oil, is used instead, the arc can be suppressed quickly. Cooling and arc suppression reduce the deterioration and damage to contacts. Pitting and roughening are minimized. The possibility of a flammable atmosphere is eliminated. Long arcs that might change their paths to nearby grounded metal equipment are avoided.

That oil-filled electrical equipment can explode was shown when in 1973 a transformer in a bank of three exploded during a final check after being installed in a new office building in Chicago. The transformers had been operating for a few days when the explosion, possibly due to a short-circuit in one, took place. One of the four employees working on the transformers was killed; the others were hospitalized with burns.

Explosion-Proof Equipment

The National Electrical Code defines "explosion-proof" as: "Enclosed in a case which is capable of withstanding an explosion of a gas or vapor which may occur within it, and of preventing the ignition of the specified gas or vapor surrounding the enclosure by sparks, flashes or explosion of the gas or vapor within, and it must operate at such an external temperature that a surrounding flammable atmosphere will not be ignited thereby."

An explosion inside an explosion-proof device may appear to be an anomaly. It occurs when flammable gas leaks into an enclosing case and is ignited by a spark or arc. Leakage may take place through the joints necessary to connect equipment and conduit. To work freely, the rotating shaft of a motor or lever of a switch must have clearance through which flammable gas may pass. Gas may also enter enclosures by which components of electrical systems are joined, as ambient pressure and temperature change.

To confine an internal explosion, joints must be made "flame-tight." In addition, the case must be able to withstand four times the maximum pressure that could be developed by a hydrocarbon explosion.

Flame will not carry through a passage long in proportion to its width and diameter. Flame-tight joints have ground surfaces with extremely close clearances. If an explosion occurs, the distance a flame front must travel to leave the enclosure is comparatively long. Gas products will be cooled below the ignition temperature of any flammable mixture before it reaches the outside of the case. Screwed fittings produce the same result if at least five full, finely machined threads are engaged. Explosion-proof systems therefore use threaded pipe conduit. Seals must also be provided to prevent gas or flame passing from one part of a system to another. Installation and repair of explosion-proof equipment and lines must be done with care (Fig. 21-5).

Another approved means of providing explosion-proof wiring to connect equipment in a hazardous location is the use of mineral-insulated, metal-sheathed cable. Refractory, noncombustible mineral insulates the electrical conductors and fills the liquid- and gas-tight tubing in which it is enclosed. The mineral generally is compressed magnesium oxide; the tubing is copper. Appropriate fittings are required for connections and terminations to prevent sparking at these points. This type of cable may be used in all hazardous locations.

To operate explosion-proof equipment "at such external temperatures that a surrounding flammable gas will not be ignited thereby" means that heat-producing equipment needs to be cooled suitably. External case temperatures must be comparatively low. An explosion-proof motor draws in cool air, circulates it past the hot inner case, and then discharges it. Lamps used in explosion-proof lighting fixtures must not

CARE OF EXPLOSION-PROOF EQUIPMENT

1. Explosion-proof equipment selected for use in hazardous atmospheres must be designed for the specific gas or vapor that might be encountered. This is because of the differences in flash points, explosion pressures, and ignition temperatures of the hazardous fuel.

2. Explosion-proof systems should be installed and maintained only by authorized electricians.

3. In making modifications or repairs, the same standards should be observed as for a new installation.

4. Circuits should be deenergized before joints or threaded connections are disassembled, or lighting fixtures opened for relamping. All enclosures should be reassembled tightly before the circuits are reenergized. Ensure that screwed connections have at least five full threads engaged. Flat ground joints must be pulled up tight.

5. The flat ground joints should be protected against mechanical injury and foreign material which might prevent a close fit. Hammers or prying tools should not be allowed to damage these surfaces; grease, dirt, paint, and other particulate matter should be removed carefully before the joints are reassembled. Files or abrasive materials should not be used to remove accumulated corrosion. If corrosion cannot be removed with solvents, the parts should be replaced.

6. Where sealing compound has been removed for modification or repair, the system will be resealed according to original standards when reassembling. The sealing compound must have a melting point of not less than 200°F and must not be affected by the liquid or gas whose presence constitutes the hazard.

7. In storing explosion-proof equipment, covers should be assembled to their mating bodies. A film of light oil or lubricant of a type recommended by the equipment manufacturer should be applied to the body and cover-joint surfaces. The cover-attachment screws intended to hold explosion-proof joints together should be firm and tight when the unit is assembled. Only bolts and screws provided by the equipment manufacturer should be used.

8. Threaded joints should be tightened adequately to prevent loosening due to vibration. Clearance between joints, shafts, and bearings should be maintained within the tolerance prescribed by the manufacturer.

9. Explosion-proof lighting fixtures should be marked for the maximum wattage lamps that may be used. This maximum should not be exceeded.

FIGURE 21–5 Care of Explosion-Proof Equipment

exceed the maximum wattage marked. This is necessary to keep the surface temperature of the fixture safely below the ignition temperature of the flammable gas.

Pressurization

Where approved explosion-proof equipment is not commercially available or economical, it may be possible to use pressurized systems for electrical equipment in hazardous locations. Explosion-proof equipment is generally preferable and more economical for individual units. Where a large item or area or number of units is to be contained in one enclosure, pressurization may be more suitable.

Pressurization involves use of air or inert gas at pressures above the ambient atmosphere. The elevated, positive pressure prevents entrance of flammable gas into an enclosure where electrical equipment might act as an ignition source. When air is used, it is blown through at a pressure of 0.1 inch or more of water. The clean air also purges the system of any contaminant gas that might enter, eliminating accumulations which could constitute a hazard. An inert gas as the pressurant provides an internal atmosphere lacking in the oxidizer which a flammable substance needs to burn. With either air or an inert gas, the enclosure is provided with a pressure switch which prevents energizing the system before it is safe or if pressurization fails.

Pressurization is necessary or desirable for some types of electrical equipment which must operate at high altitudes. The lower pressure reduces the dielectric effect of air so that coronas and sparking may occur at voltages lower than those at sea level.

Isolation

Isolation can be employed two ways: all sources of electrical power can be located in places sealed off from the hazardous environment or the hazardous environment can be enclosed and sealed off from any ignition source.

The National Electric Code categorized locations and atmospheres according to their degree of fire hazard into classes, divisions, and groups. The NEC provides greater detail on the various factors that affect and constitute the hazard, especially with commonly used, highly dangerous gases and liquids, and on safe installation practices.

To ensure that electrical equipment is available which will be safe in hazardous atmospheres, the National Board of Fire Underwriters sponsors the Underwriters' Laboratories to make tests in suitable environments. The Laboratories determine whether devices, equipment, and materials for electrical systems are suitable for use in the various NEC categories. It is not an enforcement agency. It provides labels for equipment of which it approves, which indicate to the user that UL criteria have been met for specific NEC categories. The laboratory also provides an examination service by which labels are not issued but equipment is placed on a list indicating it has been approved.

HEATING AND OVERHEATING

Use of electrical power results in the production of heat, either by design or as an unavoidable and unproductive loss, because of the resistance to flow of a conductor. The electrical energy converted into heat is equal to the square of the current times the resistance (I^2R) for a direct-current system.

One of the principal effects of electrical heating, especially overheating, is to cause accidental fires. It can (1) raise a flammable mixture to a temperature where it ignites easily; (2) raise the mixture to its ignition temperature where it ignites; (3) cause insulation, wood, and other organic materials to melt, char, or burn; (4) cause rapid evaporation of liquid fuels so that flammable concentrations are created; and (5) cause breakdown of noncombustible compounds. For the mechanisms by which ignition and combustion take place, see Chapter 22.

In addition, overheating can cause burnouts of operating equipment, raising its temperature so high that the equipment fails and sometimes ignites. The damage that results from overheating will depend to a certain extent on the rapidity with which it occurs. Protection can be provided in a circuit which will deenergize the system if it occurs at a rate slow enough to permit the protective device to operate properly.

However, a condition could exist—for example, failure or blockage of cooling equipment—which causes the item that should be protected to overheat and be damaged even when a normal amount of current is flowing.

Another adverse effect of electrical overheating is that exposed metal surfaces may become so hot that unwary persons who touch them may get burned (see Chapter 18). Requirements for military electronic equipment stipulate that temperatures of exposed parts, including the equipment enclosure, will not exceed 140°F (60°C) and front panels and operating controls will not exceed 120°F (49°C).

Electrical Explosions

A conductor which is inadequate, either because of its size or material, to pass a very heavy current may explode. Rapid overheating from overcurrents due to short circuits or current surges can cause switches, fuses, or circuit breakers to explode violently. The heat generated by the resistance to flow can cause the metal of the conductor to melt and vaporize extremely rapidly. This effect is used purposely in certain electroexplosive devices, but when it occurs unintentionally, it can be extremely damaging. The overcurrent, which may prove to be excessively heavy, can be due to short-circuits, transient surges due to startups, or lightning strikes.

Electrical explosions can be generated in other ways. Oil-filled breakers or transformers may contain contaminants such as free water. The heavy current can cause high heating of these contaminants so that they vaporize and create high pressures that will rupture the containers in which the oil and electrical equipment are held. Capacitors will not only explode because of heavy currents, but the polarized electronic type will also do so if they are subjected to current and voltage of the wrong polarity. This can occur after installation or repair work if the capacitor is installed incorrectly or if leads are reversed in a direct-current system. In 1979 the Federal Aviation Agency (FAA) ordered all lithium-sulfur dioxide batteries removed from U.S. registered aircraft because of a rash of incidents involving exploding batteries.

CIRCUIT AND EQUIPMENT PROTECTION

Almost all circuit protection is to guard against overloads of current. Protective devices ensure that current flow does not produce heat which causes temperature to rise to dangerous levels. Circuits are protected by fuses and circuit breakers, and specific pieces of equipment by cutouts. All these break the path along which current flows and deenergize the circuit when an overload condition arises.

Fuses

Fuses contain metals that conduct rated currents without overheating but that melt when a current above a designated value is imposed. The conductor in the fuse is a zinc alloy, either in wire or ribbon form. One or more portions with a reduced cross section are located near the middle of the conductor. When the circuit is overloaded, heat melts the narrow portion of the fuse link, opening the circuit before other damage can occur.

There are differences in types of fuses, not in modes of operation. The three available basic types are screw-in plugs, ferrules, and knife-edge cartridges. The National Board of Fire Underwriters specifies that fuses will be adequate to carry 110 percent of nominal rated current indefinitely, but will open when 150 percent of nominal rating exists for a prescribed length of time. Between these percentages, whether a fuse will blow depends on the length of time of the overload. A lesser tolerance might cause the fuse links to melt during power surges when the circuits are first activated. Time-lag fuses are designed to carry such short-time overloads which would not damage the system. This type of overload is common: motors draw greater than normal current when starting. A regular fuse would blow if a motor was started under load. The length of time to blow a fuse is therefore made proportional to the amperage of the current that might flow. If time-delay fuses were not used, it would be necessary to install fuses of higher capacity to carry the surges. This would reduce the protection afforded the circuit during continuous operation.

Temperature affects electrical resistance of a metal. The link of a very hot fuse will open more easily than normal. A fuse which is colder than the conductor in the circuit it is supposed to protect will have less resistance. A greater overload current will be necessary to cause it to open. In some instances it may not open under a load which might damage the circuit and equipment. The protective device should therefore be located where ambient conditions provide approximately equable temperatures.

Circuit Breakers

Three types of circuit breakers are in general use: magnetic, thermal, and combinations of both. The magnetic type uses a solenoid, or coil, which encloses a metal plunger connected to a latch or tripping device. With an overload, the allowable amperage is exceeded and the magnetic force produced in the coil pulls the plunger into the solenoid. To close the circuit again, the plunger and latch are repositioned, either manually or automatically, after a suitable interval. A circuit breaker will trip whenever or as soon as its preset limit is exceeded. It differs in this respect from a fuse, whose limit can be exceeded to a much greater degree without being blown.

Thermal breakers use a bimetallic actuator. An overload causes heating and bending of an element whose movement releases the latch. A thermal trip opens on any prolonged overload and is not affected by minor, short-lived power surges. Some circuit breakers have magnetic and thermal elements in series: the magnetic trip guards against heavy overloads and short-circuits; the thermal trip protects against light, sustained overloads.

UNIT PROTECTION

Fuses and circuit breakers are intended primarily for protection of complete circuits, including wiring and attached equipment. Protective devices have also been built into or provided for separate pieces of critical equipment, such as motors. Overcurrents that would not open or damage a fuse or circuit breaker might damage equipment that is more heat sensitive. Also, during parallel operations of equipment, one unit may be overloaded while others are operating at less than normal loads. The total current would be less than that required to blow a fuse or circuit breaker; the current through the overloaded piece might damage it.

Thermal and magnetic relays (cutouts) are used for protection of individual pieces of equipment. In the thermal type, a bimetallic disk holds two sets of contacts closed. The current flows through the disk and contacts and then through the equipment to be protected, such as a motor winding. Excessive current will raise the temperature of the disk until its distortion causes separation of the contacts and opens the circuit. When it cools, the disk snaps back and the contacts touch again, closing the circuit. Other thermal cutouts have bimetallic elements that have difficult configurations but operate on the same principle.

A second type of thermal device uses an expanding metal. In one, current passes through a heater element inside a manganese tube. The tube expands when current heats the element. Overcurrent causes expansion of the tube to such an extent that it forces a lever to trip which opens the circuit.

Another group of thermal devices uses solder. One end of a metal shaft holds a spring-operated ratchet, the other end is set in solid solder in a chamber surrounded by a coil. The solder remains solid as long as normal current flows through the coil. The added heat and higher temperature of an overcurrent cause the solder to melt, releasing the ratchet and opening the circuit.

Magnetic-type cutouts are similar to those of circuit breakers. Solenoids produce magnetic forces under overloads which cause relays to open and break the circuit. Thermomagnetic types incorporate the principles used in both types of relays.

Resets

Devices can be designed with manual or automatic resets. In the manual type, a reset button must be pushed to close the circuit. Manual resets for relays and circuit breakers have an important advantage over automatic types; they call attention to the fact that the circuit has opened. Manual types generally require resetting of the tripping mechanisms and not the actuator. For example, in solder-type thermal devices, the solder cools and solidifies after the current flow is stopped. The spring that trips the lever which opens the contacts must be reset.

Automatic devices depend on the fact that when a circuit is opened, current flow is cut off and the system cools and returns to its original condition before it was activated. Fuses are not in this category, since their elements are destroyed and must be replaced. In magnetic types, a decrease in current causes release of a plunger, permitting closing of contacts and current flow. Bimetallic thermal devices, distorting or expanding, also return to their original configurations and can be made automatic, depending on the type of tripping mechanism. Automatics are frequently provided with warning lights which indicate to an operator that the system has been opened.

WHY AN OPEN CIRCUIT?

The reason that a protective device opens in a circuit should be determined or known before the device is reset. An occurrence more than once is an indication that an overload or short-circuit exists. Remedial measures should be taken.

ACCIDENT PREVENTION MEASURES FOR ELECTRICAL HAZARDS

1. Only fully qualified and authorized electrical technicians should make electrical connections or work on installation or repair of electrical equipment.
2. Work on energized electrical circuits should be avoided if possible. If work must be done on energized circuits, the buddy system should be used, employing two or more electricians provided with suitable protective equipment. Insulated tools should be used, as well as rubber gloves, mats, and blankets to provide insulation from the ground and from conductors connected to the ground.
3. A bare conductor should not be touched without adequate protection until after the system has been deenergized and tested. Only designated electrical personnel or their supervisor should deenergize the system. The circuit should be tagged and locked out at the point at which it has been broken to prevent its being turned on inadvertently.
4. The fact that a circuit has been deenergized should be verified at a terminal board or switch box before a connection is broken.
5. All personnel who will be affected when the power is turned off should be notified beforehand as to when and how long the outage will take place be. Before the system is energized again, personnel working on lines or others who might be endangered should be informed.
6. Interlocks should be rendered inoperative by removal, modification, or destruction.
7. The voltages and frequencies of circuits to be worked on should be ascertained before any work is done, so that necessary precautions may be taken.
8. Ratings of overload protection (such as fuses and circuit breakers) should be checked to determine whether they adequately protect the line's maximum current-carrying capacity.
9. Cords to electric tools and other portable equipment should be checked before using, and replaced or repaired if defective. All such tools, equipment, and extension cords should be grounded.
10. Ground connections should not be made to gas or steam pipes, an electrical conduit system, dry pipe sprinkler system, or the air terminals of a lightning protector. All grounding cables and connections temporarily removed any reason should be replaced as soon as possible.
11. Water should not be used on electrical equipment fires. When possible, electrical equipment should be deenergized before fire fighting.
12. All uninsulated conductors (such as bus bars on panel boards and switchboards, or high-voltage equipment connections in accessible locations) should be in enclosed, protected areas. In building interiors they may be in equipment rooms or fenced areas. Outdoors, high-voltage equipment should be in vaults or enclosed by security fences. Doors and gates should be locked, with the keys in the possession of electrical technicians or other responsible personnel familiar with the hazards.
13. Extreme care should be exercised when construction, transportation, or handling equipment having long booms or extensive heights is operated near high-voltage lines or equipment. If feasible, lines or equipment should be deenergized temporarily during such operations.
14. Personnel should avoid working on electrical circuits or equipment while clothing or shoes are wet, or while hands or feet are immersed in water. Wet areas on which personnel must stand should be covered with dry wooden boards or rubber matting.
15. Personnel should not wear rings or watches, or carry keys, lighters, or similar metallic objects while working on electrical systems or in strong microwave radiation fields.
16. Electronic equipment should be relieved of capacitive charges before work on such equipment is begun. Test equipment and panel boards should be grounded. Working personnel should stand on rubber floor mats or nonconductive floors.
17. Tape measures used to measure distances in the vicinity of energized equipment should not be of metal or contain any metallic reinforcement.
18. Electrical equipment that cannot be kept out of locations which might have flammable or explosive atmospheres should be explosion-proof or intrinsically safe.
19. Grounding devices should be engineered for the particular site and installed to ensure maximum protection. Doors of buildings, magazines, or rooms in which highly flammable or explosive materials are located should be provided with grounded push bars or plates. Shoes and floors should be conductive types.

FIGURE 21–6 Accident Prevention Measures for Electrical Hazards Continued on page 370

20. Trucks, storage tanks, missiles and explosives trailers and dollies, aircraft, and any other equipment and structures on which electric charges might accumulate should be grounded. Normally, however, 10 ohms of resistance should not be exceeded when grounding is for lightning protection.
21. Lightning protection should be provided on elevated structures and buildings.
22. Fuses, circuit breakers, or thermal cutouts should be used for overcurrent protection. When a protective device deenergizes a circuit, the reason for the action should be determined and corrective action taken where necessary.
23. A fuse with a capacity greater than that prescribed for the circuit should not be used as a replacement. Jumpers should not be used across fuse terminals to keep current flowing in a circuit.
24. Lamps used in any circuit or equipment should not exceed the specified voltage for that fixture. Lamps should be screwed firmly into their sockets.

FIGURE 21–6 Continued

A replacement fuse should not be of capacity greater than that prescribed for the circuit. Jumpers across bridge terminals should never be used to bridge a circuit.

Inadvertent Activation

Inadvertent activation of electrical equipment can cause injuries, fatalities, damage, or other problems. Personnel have been injured or killed when they crawled inside to repair or clean electrically operated equipment. The equipment would start unintendedly, and an injury or fatality would result. The precautionary measures to preclude such occurrences are the same as those indicated to prevent mistaken activation of an electrical circuit: (1) opening the circuit, (2) locking it open, (3) ensuring that only the person working on the system can unlock it, and (4) trying to activate the system when the circuit is locked open.

Personnel have been injured after accidentally hitting a start button or switch, activating the equipment they were using before they were ready. Persons who were changing the bits on drills or drill presses have unintentionally hit their shoulder or arm while still holding the chuck key. The results were torn, broken, or severed fingers. Precautionary measures to prevent such unwanted activation would be to ensure start buttons were located so distant that the problem could not arise. Also, such buttons should be recessed to avoid unintentional activation. Even safer are switches with safety covers that must first be raised to activate a critical circuit.

The heat from fires at upper levels of high-rise buildings has activated elevators, which then automatically moved toward the floors on which the fires were taking place. Any person aboard the elevator would be carried into the fire zone rather than to safety.

Figure 21-6 indicates some accident prevention measures against electrical hazards, while Figs. 21-7 and 21-8 are checklists which may help to alert personnel against potential problems.

HAZARDS CHECKLIST—ELECTRICAL

Possible Effect	*Possible Cause*
Shock injury	Accidental contact with live circuit through: Touching bare conductor Inadequate insulation Cutting through insulation Deteriorated insulation Defective assembly of electrical tool or appliance Erroneous connection Lightning strike
Thermal effects: Burns Degradation of performance Overloading and burnout of equipment Ignition of combustibles Melting of soldered connections Degraded reliability Softening and melting of plastics Circuit breakers, fuses, and cutouts opening deactivating equipment	High I^2R losses Inadequate cooling Overloads Short-circuits caused by: Inadequate or deteriorated insulation Erroneous connection Bare conductors touching Dirt, contamination, or moisture Corrosion Excessive or loose particles of solder or cut wire Bent connector pins Improper wiring Improper mating of connectors
Arcing and sparking causes: Ignition of combustibles Buildup and welding of contacts Surface damage to metals Interference with electrical equipment operation Electrical noise and cross talk	Lightning strike Gaseous gap between conductors caused by: Loose connection Opening of switches, relays, circuit breakers and similar devices Electric arc welding Lack of bonding or grounding Deteriorated or inadequate insulation

FIGURE 21–7 Hazard Checklist—Electrical Continued on page 372

Inadvertent activation of the product or a device: Untimely equipment starts Endangering personnel working on or in equipment supposedly inoperative	Lightning strike Stray current from: "Sneak" circuit Cross-connection Personnel error Misapplied test-equipment power Static electricity discharge Coupling
Electrical system failure, making: System inoperative in hazardous situation Safety equipment inoperative Release of holding devices Detection and warning devices inoperative Interruption of communications	Malfunction caused by: Power-source failure Power surge opening fuse or circuit breaker Component failure System overloading Short-circuit Operator error Lightning strike
Explosion of: Batteries Circuit breakers, transformers, and similar equipment Capacitors	Short-circuiting Presence of liquid or its contaminants which disassociate violently when current passes through. Lightning strike

FIGURE 21–7 Continued

CHECKLIST—ELECTRICAL HAZARDS

1. Are any extraordinarily high voltage or amperage levels used which would require special safeguards? Have those safeguards been provided?
2. Are all items which should be electrically grounded, grounded adequately? Are the grounds tested periodically?
3. Is there any location where a live circuit is not insulated? Is adequate protection provided to keep personnel from contacting such circuits? Is protection also provided where insulation might have deteriorated to the point where it should be replaced?
4. Is there any surface, other than a heating element, hot enough to burn a person or ignite a material?
5. Are the voltage and amperage high enough to cause arcing or sparking which could cause ignition of a flammable gas or combustible material?
6. Are there any points, such as motor brushes or open circuit breakers, where arcing or sparking can occur close to any fuel?
7. Are there any places where lint, grease, or other flammable material which can be ignited can accumulate?
8. Is there any possibility of inadvertently activating a piece of equipment while a person is in a position to be injured? Is there a means of cutting off power to a piece of equipment while it is being worked on or replaced? Is there a means by which the cut-off power can be locked out?
9. Are fuses, circuit breakers, and cutouts sized to protect the circuits and equipment they are supposed to protect?
10. Are fuses and circuit breakers in a readily accessible and safe location? Are accesses to them kept clear?
11. Are all the electrical installations and systems in the plant in accordance with the requirements of the OSHA standards and of the National Electric Code?
12. Is there at least one emergency button on each piece of hazardous equipment? Can the hazard which might be considered the basis for the need for the button be eliminated or controlled by a better method? Can the button be easily reached by anyone who might have to use it?
13. Is an interlock which removes power provided on the access to any equipment interior where a person could receive a fatal shock?
14. Are wires and cables protected against chafing, pinching, cutting, or other hazards which could damage the insulation, so a person could get a shock, or which could cut the metal conductor?
15. Are the locations of underground cables marked so that they will not be cut by excavating equipment?
16. Are wires, cables, and conduits adequately secured to the structures along which they pass or to the chasses of the equipment on which they are installed?
17. Are wires and cables kept off floors over which vehicles must pass? If they must be on the floor, are they adequately protected against damage?
18. Are batteries firmly secured where they are to be used? Are battery compartments ventilated so that hydrogen cannot accumulate during charging?
19. Where batteries are to be used, is the location marked with the polarity, voltage, and type(s) of battery to be used?
20. Where batteries may be "jumpered" for engine starts, are they posted with instruction indicating the proper way it is to be done and the precautionary measures to be taken?
21. Are materials and equipment which can generate static electricity grounded to prevent accumulations of static charges?

FIGURE 21–8 Checklist—Electrical Hazards

EXERCISES

21.1. Cite some hazards involved in the use of electrical equipment.

21.2. What is "let-go" current? What amperage range would this include?

21.3. What is ventricular fibrillation? What range of current could cause ventricular fibrillation? Is it more likely to occur with alternating or with direct current? Can the person with ventricular fibrillation recover without outside assistance?

21.4. Why is a hot, sweaty person in more danger from shock when working on electrical equipment than a person whose skin is dry?

21.5. Why should electrical equipment that is being shut down for repair, maintenance, or modification be locked out after being deenergized? Describe how it is done.

21.6. Give five factors that can cause insulation failure.

21.7. Describe encapsulation, embedment, and potting and their uses. How has use of solid circuitry lessened the possibility of electrical shock or circuit damage?

21.8. Explain the differences between sparks, arcs, and coronas.

21.9. Why and how are electrical interlocks used?

21.10. Give four factors that affect the generation or retention of static charges.

21.11. Tell which of the following could produce high static-electricity potentials: wood, paper, copper, plastic sheet, aluminum, fiberglass, dry air, flowing gasoline, nylon, a rubber conveyor belt.

21.12. Describe four means by which problems with static electricity can be reduced.

21.13. How can effects of lightning strikes be lessened? What should you do if caught in the open during a thunderstorm?

21.14. Describe five ways by which problems of electrical charges can be eliminated or reduced.

21.15. Why is circuit or electrical protection provided? How do fuses work? How do circuit breakers work? What are cutouts and how do they work?

CHAPTER 22

Fires and Fire Suppression

It has been said that for a fire to start requires a fuel (Fig. 22-1), an oxidizer, and a source of ignition. This is an oversimplification; the process is more complex. First, both the fuel and oxidizer must be in suitable proportions, in intimate contact with each other, and in proper modes to enter into a reaction. The means by which permanent gases, liquids, and solids reach states where they will burn are similar in certain respects and far different in others.

FUELS

Materials that will burn are innumerable. The following list therefore indicates major categories of combustibles, giving some examples. Some substances listed may fall into more than one category.

Normally Flammable Materials

Normally flammable materials include:

1. Fuels for heating, internal combustion engines, welding, and rocket engines.
2. Solvents and cleaning agents.
3. Lubricants.
4. Coatings such as paint, lacquers, or waxes.
5. Industrial process chemicals.
6. Refrigerants such as ammonia and methyl chloride.
7. Insecticides.
8. Plastics and polymers.
9. Hydraulic fluids and hoses.
10. Vegetation and wood products.
11. Paper products.
12. Cloth and other fiber materials.

COMBUSTION AND FIRE

Fuel: substance that acts as a reducing agent, giving up electrons to an oxidizer in a chemical combination. It may be an element, such as carbon, hydrogen, or magnesium; a single compound, such as carbon monoxide (CO) or methane (CH_4); a complex compound, such as wood or rubber; or mixtures of these.

Oxidizer: substance that acquires electrons from a reducing agent (fuel) in a chemical reaction. It may be an element such as fluorine, oxygen, or chlorine; a compound that will readily release fluorine, oxygen, chlorine, or other elemental oxidizer, such as hydrogen peroxide (H_2O_2), potassium hypochlorite ($KClO_3$), or lead dioxide (PbO_2); a strong acid, such as nitric (HNO_3), hydrofluoric (HF), or sulfuric (H_2SO_4), or compound that will release negative radicals, such as sodium nitrate ($NaNO_3$) or nitrogen tetroxide (N_2O_4).

Note: In some instances, certain elements considered fuels may act as oxidizers, and vice versa. Sulfur will burn with oxygen to form sulfur dioxide (SO_2), but hydrogen will burn with sulfur to form hydrogen sulfide (H_2S). Hydrogen sulfide itself is a fuel and will burn in air. Oxygen is an extremely powerful oxidizer. but will give up electrons to fluorine, the most powerful oxidizer in existence, to form oxygen difluoride (OF_2).

Fire: rapid oxidation-reduction reaction which results in the production of heat and, generally, visible light.

Flammability limits: lowest and highest percentages by volume of fuel gas to air at 1 atmosphere that will burn. The difference between the two limits is the *flammability range*. When the amount of fuel present is too little to permit a self-sustaining reaction, the mixture is said to be "too lean." It is below the lower flammability limit (LFL). When the fuel is so plentiful that there is an inadequacy of oxidizer, the mixture is "too rich": above the upper flammability limit.

Flammability limits have sometimes been used interchangeably with "explosive limits." That is, the LFL and the lower explosive limit (LEL), and the UFL and upper explosive limit (UEL), are considered synonymous. Technically, this common usage is not correct. Hydrogen in air will burn when it reaches at least 4.0 percent of the mixture by volume; it will explode when it is 18.3–50 percent. Methane, propane, and butane are burned as fuels for cooking and heating with few problems. However, these gases have generated disasters when concentrations exceeding their lower explosive safety limits were ignited.

From a safety standpoint, lower flammability limits are of much greater interest than upper limits, since they indicate the lowest concentrations at which combustion will begin. However, certain substances, such as acetylene, hydrazine, ethylene oxide, and n\propyl nitrate, have upper flammability limits of 100 percent. This means, in effect, that they will burn with no oxidizer, such as the oxygen in the air, present. Certain of these were considered for use as rocket propellants. Since no oxygen was needed, there would be a savings in weight. For this reason they were known as "monopropellants."

Flash point: lowest temperature at which a liquid fuel will give off enough vapor to form a momentarily ignitable mixture with air. A fuel will not burn as a liquid. In the presence of its liquid, the vapor pressure of a gas is governed by its temperature. There is a definite temperature for any liquid at which it will provide enough gas, by vaporization, to approach the lower flammability limit in air. If an outside source of ignition is then used, these vapors will burn in a momentary flame and then die out. This temperature is the "flash point." Vapor pressure at this temperature is inadequate to sustain continuous burning after ignition, so the reaction stops as soon as the available combustible gas is consumed.

Flash points are determined in either closed or open testers. The Tagliabue (TAG) closed tester is generally used for liquids whose flash points are less than 175°F. The Pensky-Martens Tester is primarily for fuel oils with flash points between 150°F and 230°F. The Tagliabue and Cleveland Open Cup Testers are mostly for flammable liquids in transportation. They are preferred for high-flash-point liquids.

FIGURE 22–1 Combustion and Fire

Closed-cup flash points will generally be 10–20 percent lower than open-cup results for the same liquids.

Fire point: if the liquid is heated higher than the flash point, the higher temperature will cause vapors to be produced rapidly enough to permit continuous burning after ignition by an open flame or spark. The lowest temperature at which continuous burning occurs is the fire or ignition point. (With solids, this is also known as the kindling temperature.) The fire point is always higher than the flash point; but the difference between the two varies, since vapor pressure depends on the substance in question.

Flash point versus fire point: the flash point is of more interest in safety than the fire point, since it indicates that the lower limit of flammability is being approached, and the level of hazard. Both are approximations determined under stipulated conditions in the laboratory. Actual field conditions may and will differ. Also, a high-flash-point liquid heated until its vapor pressure is equal to that of a more volatile liquid at a lower temperature constitutes an equal hazard.

Stoichiometric mixture: A stoichiometric mixture is one that contains exactly those amounts of fuel and oxidizer which will react completely so that neither fuel or oxidizer remains.

Exothermic reaction: reaction in which heat is emitted as a result.

Endothermic reaction: reaction in which heat is absorbed.

Autoignition temperature (AIT) or spontaneous ignition temperature (SIT): the autoignition temperature is the lowest temperature at which a flammable mixture will burn without application of an outside spark or flame, and continue to burn without further application of heat. (Fire point is the lowest temperature at which a liquid produces enough gas to sustain continuous combustion after being ignited by a spark or open flame.)

FIGURE 22–1 Continued

13. Rubber products.
14. Metals such as sodium, potassium, cesium, rubidium; metal dusts, powders, fibers, ribbons, or fine wires.

Materials—Both Flammable and Nonflammable

Some materials are nonflammable or of low flammability in air but will burn in the presence of a strong oxidizer, high oxygen concentration, very high temperature, or strong ignition source. Some of the more common substances in this category are:

1. Halogenated hydrocarbons such as tricholoroethylene.
2. Foam and silicone rubbers.
3. Plastics and polymers, plastic-coated fabrics, flame or fire retardants, wire insulation.
4. Metals in massive form, such as magnesium, aluminum, titanium, or zirconium.
5. Powered metals, such as magnesium, aluminum, titanium, and iron.
6. Specially compounded hydraulic fluids and lubricants.
7. Sealants, packings, O-rings, diaphragms, and valve seats.

Products of Other Reactions or Processes

Products of other reactions or processes that act as fuels include:

1. Carbon monoxide produced by incomplete combustion of organic compounds or other carbonaceous materials.
2. Hydrogen released during charging of storage batteries.
3. Hydrogen released by decomposition of water on very hot surfaces.
4. Combustible gases released by distillation of organic materials.
5. Hydrogen released by reaction between the metals themselves of sodium, potassium, or lithium and water.

OXIDIZERS

Our commonest oxidizer is, of course, the oxygen in air. Most of our terminology relating to fires and to tests indicating the hazard levels of fuels is predicated on atmospheric oxygen as the oxidizer. Modern industrial processes now employ or produce oxidizers that may be far more hazardous than air, which is diluted oxygen.

Many substances considered nonflammable or of very low combustibility in air will burn in oxidizer-rich atmospheres. Methylene chloride will not burn in air but will burn in oxygen-rich or nitrogen tetroxide atmosphere. Trichloroethylene and other halogenated hydrocarbons, considered to be of very low combustibility, will form explosive mixtures with nitrogen tetroxide, requiring only a suitable activation source to explode.

Flare-type combustion is the burning of fuel in an oxidizer-rich atmosphere causing a higher, more intense temperature and rate than would occur in a fuel-air mixture. This type of combustion may also occur with such substances as solid propellants, where a source of air is not required for combustion so that blanketing a fire will not cause its extinguishment.

Even slightly pressurized air, such as from bellows, will increase combustion of fuels. Pure oxygen will, of course, be a stronger oxidizer than will air, of which it comprises only 21 percent. A small oxygen leak can raise the oxygen level in air into which it is leaking to a point where a very dangerous fire hazard could exist. For this reason, oxygen cylinders, even when considered empty, are not to be stored near a combustible gas or near combustible materials. Flares of oxygen-fed acetylene are used as welding and cutting torches.

Fluorine is the only element which is a stronger oxidizer than oxygen (which it will replace even in water), under certain conditions even combining with inert gases. Other elements and compounds are slightly less energetic than oxygen in a reaction. Some are of such a nature they may not only initiate reactions more readily than will oxygen, and much more rapidly than will air, but they will cause certain fuels to ignite without an outside source of ignition. A few other strong oxidizers include chlorine and other halogens and halogen compounds, nitrates, nitrites, peroxides, and strong acids (sulfuric, hydrofluoric, hydrochloric, and so on). These oxidizers should be handled with care to avoid their coming in contact with a fuel.

GASES

When the fuel and oxidizer are both gases, the mixture of the two must be within flammable limits to ignite. Flammability limits (see Fig. 22-2) and flammability ranges should be considered to be only relatively approximate, not fixed, values. Many factors can affect, extend, or constrain these limits. There may be differences in the exact proportions of the mixture; its temperature; the type, shape, magnitude, and direction of the ignition source; the type and concentration of oxidizer; in some cases, mixture pressure; and other factors.

In any gaseous mixture, the exact concentration of fuel and oxidizer at any minute volume is uncertain. A sample of gas as indicator or a calculation may indicate a mixture is too lean to burn, but there may be pockets where measurements are not made where the fuel concentration is within the flammable range. Equations exist from which an average fuel-air ratio can be calculated if a specific amount of fuel, such as a solvent, is spilled in a room of known volume. The calculated fuel-air ratio may be below the flammability limit; however, the difference in density between the vaporized solvent and air may cause stratification, so the flammability limit is exceeded below a certain level. There have also been cases where confined spaces, such as tanks, have been monitored for flammable gases and found to be generally safe. However, flammable gases lighter than air were trapped under roofs or ceilings and ignited inadvertently.

Gas-air mixtures outside normal flammability limits can be ignited if the ignition energy, such as from a welding or cutting torch, is very high. The differences in limits with different temperatures are shown for these combustible gases in Figs. 22-2, 22-3, and 22-4.

A mixture that would not ignite at low energy levels might do so at high energy inputs. A 0.2-millijoule spark will not ignite a 79°F butane-air mixture under any circumstances. A 0.5-millijoule spark will ignite the mixture if it is within narrow flammability limits. When the spark is 10 millijoules or more, the flammability range is much wider. Substances that might not ignite under normal circumstances and might be considered of low flammability will ignite readily under the heat of a welding arc or flame.

Flammability limits are normally predicated on mixtures of fuel and air. Changes in percentages of oxygen present in air will change the percentage of fuel that could be present in a flammable mixture. This in effect will change the flammability limits and range.

FLAMMABLE AND COMBUSTIBLE LIQUIDS

The words "flammable" and "combustible" may appear to be synonymous, but they have different connotations when related to liquids. Any liquid having a flash point less than a stipulated temperature is considered "flammable"; above that point, "combustible." Hazard ratings by flash point differ from organization to organization, as shown in Fig. 22-5.

It should be emphasized here that a liquid which is less than at the "flammable" temperature but still "combustible" remains a dangerous material. It will ignite easily if

FLAMMABILITY LIMITS OF COMBUSTIBLE GASES AND VAPORS*

Compound	Formula	LFL (% by vol. in air)	UFL (% by vol. in air)
Acetates			
Ethyl acetate (acetic acid, ethyl ester[a])	$CH_3COOC_2H_5$	2.2	11.0
Amyl acetate (acetic acid, amyl ester[a])	$CH_3COOC_5H_{11}$	1.0	7.1
Vinyl acetate (acetic acid, ethanyl ester[a])	$CH_3COOCH{:}CH_2$	2.6	–
Alcohols			
Methanol[a] (methyl alcohol)	CH_3OH	6.7	36
Ethanol[a] (ethyl alcohol)	CH_3CH_2OH	3.3	19
Propanol[a] (propyl alcohol)	$CH_3CH_2CH_2OH$	2.2	14
Isoproponal[a] (isopropyl alcohol)	$CH_3CHOHCH_3$	2.0	11.8
1-Butanol[a] (butyl alcohol)	$CH_3CH_2CH_2CH_2OH$	1.7	12.0
Cyclohexanol	$CH_2CH_2CH_2CH_2CH_2CHOH$	1.2	–
Cyclohexanol, 1-methyl[a]	$CH_3CH_5CH_2CH_2CH_2CH_2CHOH$	1.0	–
Ethers			
Ethyl ether (ethoxyethane[a])	$CH_3CH_2OCH_2CH_3$	1.9	36
Methyl ether (methoxymethane[a])	CH_3OCH_3	3.4	27
Methyl ethyl ether (methoxyethane[a])	$CH_3OC_2H_5$	2.2	–
Hydocarbons			
Methane[a]	CH_4	5.0	15
Ethane[a]	CH_3CH_3	3.0	12.4
Ethene[a] (ethylene)	$CH_2{:}CH_2$	2.7	36
Propane[a]	$CH_3CH_2CH_3$	2.1	9.5
Butane[a]	$CH_3CH_2CH_2CH_3$	1.8	8.4
Butene-1[a] (butylene)	$CH_3CH_2CH{:}CH_2$	1.6	10
Butadiene 1-3[a]	$CH_2{:}CHCH{:}CH_2$	2.0	12.0
n-Pentane	$CH_3CH_2CH_2CH_2CH_3$	1.4	7.8
Hexane[a]	$CH_3CH_2CH_2CH_2CH_2CH_3$	1.2	7.4
Heptane[a]	$CH_3CH_2CH_2CH_2CH_2CH_2CH_3$	1.05	6.7
Acetylene (ethyne[a])	$HC{:}CH$	2.5	100
Propylene (propene[a])	$CH_3CH{:}CH_2$	2.4	11
Aliphatic hydrocarbons	Mixed	0.8–5.3	5–14
Paraffins C_5 and up	Mixed	1.2–3.1	–
Gasoline (Av. US Prem. Grade)	Unleaded	1.2–1.8	7.1–8
Jet fuel (JP-4)		1.3	8
Naphtha (petroleum naphtha)		0.8	5
Turpentine		0.7	–
Cyclic compounds			
Toluene (methylbeneze[a])	$C_6H_5CH_3$	1.2	7.1
Xylene (benzenedimethyl[a])	$C_6H_4(CH_3)_2$	1.1	6.4
Styrene (ethenylbenzene[a])	$C_6H_5CH{:}CH_2$	1.1	6.1
Cyclopropane[a] (trimethylene)	$CH_2CH_2CH_2$	2.4	10.4
Cylcohexane[a] (hexahydrobenzene)	$CH_2CH_2CH_2CH_2CH_2CH_2$	1.3	7.8
Cycloheptane[a] (heptamethylene)	$CH_2CH_2CH_2CH_2CH_2CH_2CH_2$	1.1	6.7
Methyl cyclohexane, methyl (hexahydrotoluene)	$CH_2CH_2CH_2CH_2CH_2CH_2CH_2$	1.1	6.7
Ketones			
Acetone (2-propanone[a])	CH_3COCH_3	2.6	13

FIGURE 22–2 Flammability Limits of Combustible Gases and Vapors

Methyl ethyl ketone (2-butanone[a])	$CH_3COC_2H_5$	1.4	10
Pinacolone (2-butanone,3-3-dimethyl)	$CH_3COC(CH_3)_3$	1.4	–
Acetic acid (ethanoic acid[a])	CH_3COOH	5.4	–
Acetaldehyde (ethanal[a])	CH_3CHO	4.0	60
Ethylene oxide (ethane, 1,2-epoxy[a])	CH_2OCH_2	3.6	100
Propylene oxide (propane, 1,2-epoxy[a])	CH_3CHCH_2O	2.8	37
Monochlorides			
Methyl chloride (methene, chloro[a])	CH_3Cl	7	–
Vinyl chloride (ethene, chloro[a])	CH_2CHCl	3.6	33
Ethyl chloride (ethane, chloro[a])	CH_3CH_2Cl	3.8	–
Other compounds			
Carbon monoxide	CO	12.5	74
Hydrogen sulfide	H_2S	4.0	44
Ammonia	NH_3	15	28
Hydrogen	H_2	4.0	75

[a] Official name, International Union of Chemist.

[a] M. G. Zabetakis, *Flammability Characteristics of Combustible Gases and vapors,* Bulletin 627, U.S. Bureau of Mines, Washington, D. C., 1965

IGNITION LIMITS IN DEPENDENCE ON THE TEMPERATURE

	Temperature (deg. C)				
	17	*100*	*200*	*300*	*400*
Lower limit, $\%H_2$	9.4	8.8	7.9	7.1	6.3
Upper limit, $\%H_2$	71.5	73.5	76	79	81.5
Lower limit, $\%CO_2$	16.3	14.8	13.5	–	11.4
Upper limit, $\%CO_2$	70.0	71.5	73.0	–	77.5
Lower limit, $\%CH_4$	6.3	5.95	5.5	–	4.8
Upper limit, $\%CH_4$	12.9	13.7	14.6	–	16.6

* W. Jost and H. O. Croft, *Explosion Process in Gases*, New York, McGraw-Hill Book Company, 1946.

FIGURE 22–2 Continued

subjected to a high-energy ignition source, if heated to a temperature above its flash point, or if it is in the form of a spray or mist.

Liquid Fuels

Liquids will not burn as such but will first change phase to gas. Reactions then will proceed under the same conditions as for gases, provided the liquid vaporizes fast enough to produce gas in quantity adequate to sustain the fire. After a fire has been initiated, its flame emits heat. The rate at which the liquid-vapor will burn is then dependent on the rate at which heat flows back from the flame to the liquid, causing it to evaporate,

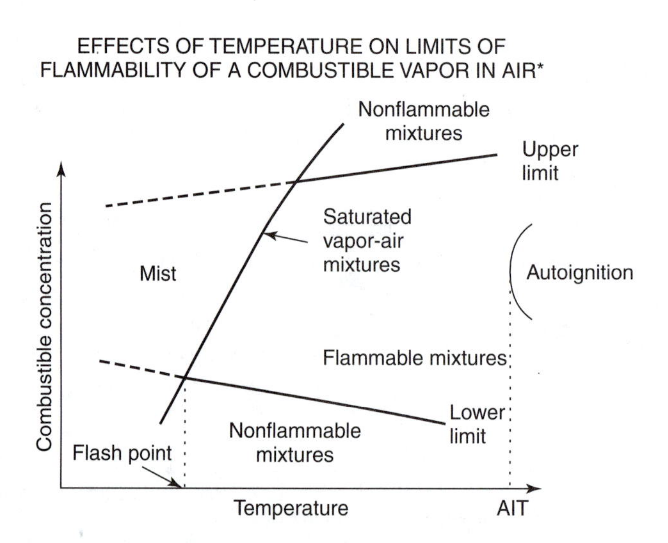

* R.W. Van Dolah, M.G. Zabetakis, D.S. Burgess, and G.S. Scott, *Review of Fire and Explosion Hazards of Flight Vehicle combustibles,* ASD-TR-61-278, U.S. Bureau of Mines, Washington, D.C., Apr. 1961 (ASTIA AD-262989).

FIGURE 22–3 Effects of Temperature on Limits of Flammability of a Combustible Vapor in Air

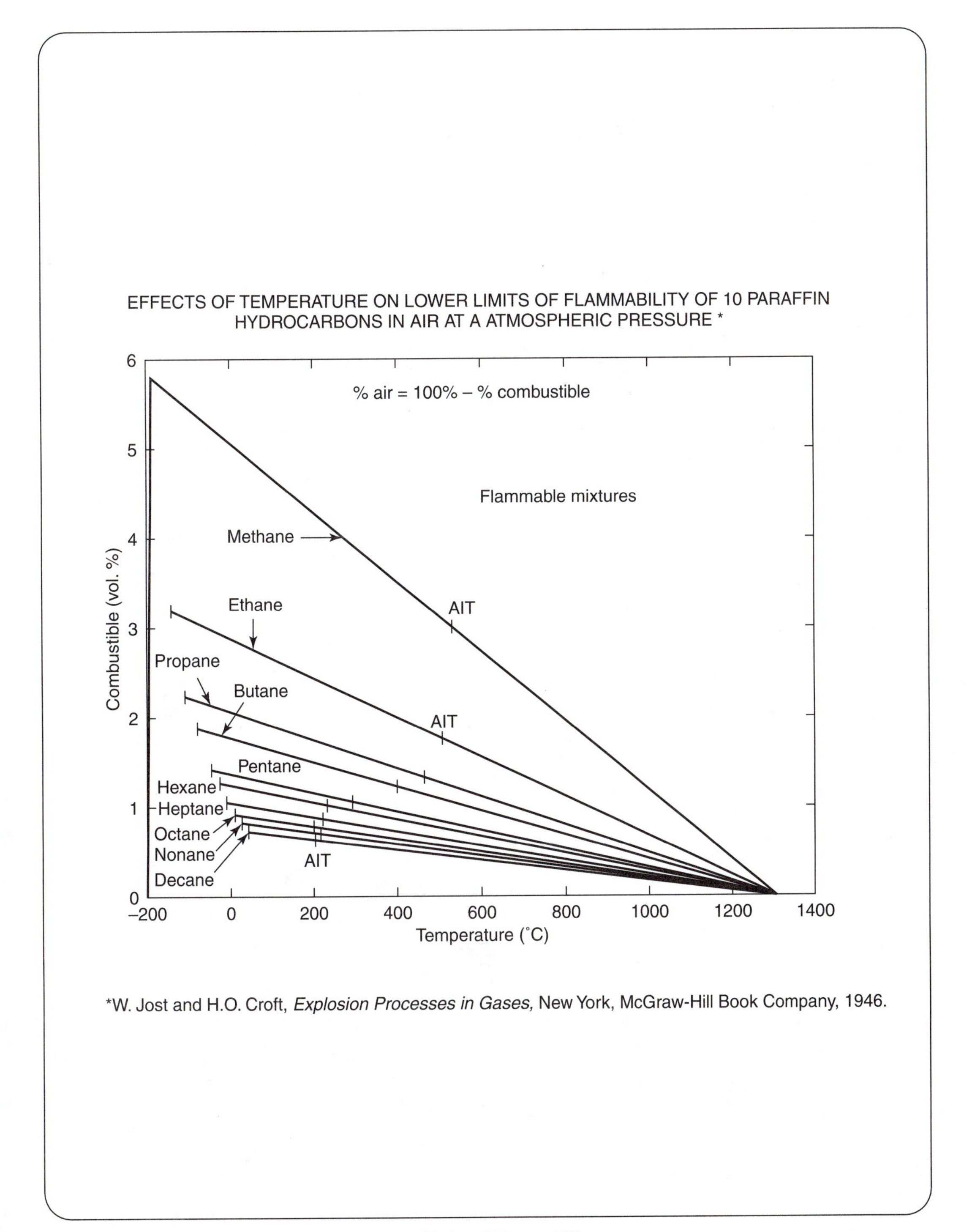

FIGURE 22–4 Effects of Temperature on Lower Limits of Flammability

HAZARD RATINGS OF LIQUID FUELS

OSHA Standards

Criteria	Hazard Class
Flammable Liquids Flash point below 73°F Boiling point below 100°F	1A
Flash point below 73°F Boiling point 100°F or above	1B
Flash point 73°F or above Boiling point 100°F or above	1C
Flash point 100°F and below 140°F	II
Combustible Liquids Flash point 140°F and less than 200°F	IIIA
Flash point 200°F or more	IIIB

National Fire Protection Association (NFPA)

Flash Point (°F)	Hazard
20 or less	High (Class I)
Above 20 to 70	Moderate (Class II)
Above 70 to 200	Slight (Class III)

Department of Transportation

Criteria	Hazard Class
Flash point 80°F or less	High (Class I)
Flash point below 80°F Boiling point below 125°F	Special group I
Flash point below 80°F Boiling point 125°F or more	II
Flash point 80°F to 200°F	II
Flash point above 200°F	III

Underwriters' Laboratories Relative Scale

Class	Rating
Ether	100
Gasoline	90-100
Alcohol (ethyl)	60-70
Kerosene	30-40
Paraffin oil	10-20

FIGURE 22–5 Hazard Ratings of Liquid Fuels

and on the rate at which the fuel and oxidizer combine to produce a flame. This process involves a number of factors, principally:

1. The temperature, brightness, and emissivity of the flame.
2. The heat required to vaporize the liquid (latent heat of vaporization).
3. The heat required to raise the temperature of the liquid to the point at which it will vaporize.
4. The rate at which the liquid will absorb heat.
5. The extent of the exposed liquid surface.
6. Wind velocity over the liquid in the form of a spray or mist.

Pool Burning

A fire may be initiated from a spill of flammable liquid or from a vat containing such a liquid. It may then be of interest to know how rapidly such a pool of liquid will burn after ignition.

Burning rates of liquid from a pool can be considered either mass or linear. Mass burning rate indicates the weight of liquid burning in pounds per minute. The linear burning rate is the speed at which specific depths of fuel are consumed. The linear burning rate depends on the amount of liquid vaporized per unit time. This, in turn, is dependent on the heat flow to the liquid surface.

Sprays and Mists

Sprays and mists are means by which extremely rapid vaporization can be produced. This can be done intentionally in furnaces and in internal-combustion engines, accidentally from leaks or other causes. When the spray or mist particle sizes are less than 10 micrometers, the mixture will act as a pure gas mixture, with ignition temperature far below that of a liquid's flash point (Fig. 22-3). Reduction in droplet size in a spray will also reduce the lower flammability limit.

Leakage from hoses and containers can occur either as sprays or as streams. Sprays can be produced from high-pressure leaks through very small openings. Jets of leaking fluids can turn into sprays and mists as they hit solid objects. Liquids inside a tank will do so as they hit its wall. This can happen when the tank is in a moving vehicle, such as a truck, aircraft, or automobile. This sloshing will produce the droplets and mists over the liquid, making them easier to ignite.

Not only will fuel sprays and mists ignite more easily than will a large mass of liquid, but ignition may occur at temperatures far below the flash point, which is based on evaporation from the surface of a liquid mass. Mists will ignite at temperatures at which vapor pressures from which masses of liquids are inadequate to produce flammable vapor-air mixtures. This increases the fire hazard. However, the hazard is still less than with a flammable gas mixture, because a spray consists of very fine drops still in the liquid phase. To burn, the drop must first vaporize, which requires that the heat of vaporization be provided. An ignition source must supply this added amount of heat, which can be 20 to 30 times greater than for gases, for the spray to burn.

Demonstrations during lectures on fire prevention may include one spectacular example of flammability of liquid sprays. The demonstrator will select an aerosol spray can and point to the label, which may say something like: "The liquid in this can is not rated as flammable." He or she will then light a long match and generate a spray from the can so that it hits the flame. The spray will ignite and be consumed in one rapid puff. The need for care near sprays, such as during painting operations, becomes readily understandable.

FLAMMABLE SOLIDS

Solids burn in heterogeneous reactions because solid and gaseous phases are always involved, and sometimes liquid. A very few solids, such as naphthalene, sublime on being heated and change from solid to gas without passing through the liquid phase. They may then burn like any other gaseous fuel. After the igniting, the heat will cause rapid sublimination of additional fuel, accelerating the process. Some subliming solids even have flash points, like flammable liquids.

Some solids melt, vaporize as do ordinary liquids, and then burn in the same way most liquids eventually do. Any metal with a melting point lower than its flame temperature falls into that category.

A simple element, such as carbon and many metals, having a melting point higher than the flame temperature undergoes a slow oxidation process which creates flammable products. These products then burn with an added oxidizer. More complex fuels, such as coal and flammable compounds, may involve either vaporization, slow oxidation, or decomposition to form combustible gases, which then burn.

Fuel-oxidizer mixtures or compounds, such as in solid propellants, create both flammable and oxidizing gases on being heated. The oxidizer is generally produced by decomposition of a compound.

Other than those solids which become gases simply by change of phase, solids undergo reactions by which they also become gases, or oxidize when an oxidant contacts a metal surface. The oxidation reaction can vary from one that is relatively slow to a more rapid one as temperature increases. Generally this is the process metals and other solids in the last three categories follow. Even many of the metals that melt will form oxides, which then become gases. All metals but platinum and gold will ignite under suitable conditions.

Carbon reacts with oxygen in air to form carbon monoxide and carbon dioxide. Any dioxide formed will contact additional carbon, causing it to react again to produce more carbon monoxide. The flame that results from this incomplete combustion gradually diffuses and mixes with additional air. Carbon monoxide is highly combustible. The reaction then proceeds to completion, most of the carbon dioxide formed leaving the combustion space; the remainder once again goes through the process of forming carbon monoxide. The heat from the hot flame radiates back to the carbon, continuing the reaction. The heat may cause expansion and cracking of solid surfaces which receive the heat of radiation. Cracking increases the surface open to reaction and accelerates combustion until equilibrium is reached.

Coal is a complex mixture of substances, its constituents varying with the grade of coal. All contain free carbon and volatile matter in different amounts. Coal will burn more rapidly than will pure carbon because of the volatiles it contains. The amount of volatiles, and thus the ease of ignition and the combustion rate, depend on the grade of coal. (Anthracite, lowest in volatiles, is hardest to ignite; subbituminous and lignite are the easiest.) The volatiles—combustible gases trapped between the solid particles, consisting in part of trapped hydrocarbon gases and sulfur or its compounds—will burn immediately as they are released from the exposed coal. Sulfur freed from the coal will melt, evaporate, and burn. Heat of combustion of the volatile matter will then initiate combustion of the harder-to-ignite carbon, which burns as just described.

Most structural metals in large masses, blocks, thick sheets, or bars will not burn as such. The reactions on the surfaces of such massive materials are limited, and heat is conducted away too quickly for the metal to be raised to the ignition point. Although they expand, they will not crack as does coal.

The ability of a metal fuel to ignite depends on the intimacy of contact between the solid fuel surface and the oxidizer. Increasing the contact area will reduce the ignition temperature and energy required. In some cases, rough, fractured surfaces may provide adequate exposure to strong oxidizers for combustion to begin. Reducing the size of particles, which increases areas of exposure, will have the same effect with solids as for liquids. However, the heat a drop receives by radiation increases its evaporation rate, but a particle of metal either melts, then evaporates, or forms a coating of oxide, or both. The energy required to ignite a cloud of metal dust is primarily a function of particle size and, where sizes differ in a mixture, on the size of the smallest particles. Other factors that affect ignitability are the specific heat of the particles (since they absorb heat as their temperature is raised), the rate at which ignition energy is applied, the presence of inert solids, and whether the particles are in clouds or settled. When certain dusts burn, the reaction speeds may be great enough to be classed as explosions. Metals have been milled so fine they will ignite on exposure to air.

After dusts, the most combustible forms are thin wires and foils, where combustibility varies with thickness. Ribbons are harder to ignite, while bulkier metals may not even burn. For the more massive forms of ignitable metals, energy adequate for ignition of dusts may result only in slow oxidation.

IGNITION

Ignition consists of energizing the molecules of fuel and oxidizer in a mixture so they collide with each other with sufficient velocity and force to initiate a reaction (see Fig. 22-6). Activation energy for a reaction is generally a form of heat; however, light and mechanical energy may also cause disassociation of molecules. Photons of light travel at extremely high velocities, and although they are extremely small particles, their energies are great enough to cause initial reactions between gas mixtures of hydrogen and chlorine. These will not react in darkness at room temperature but will ignite violently when exposed to sunlight. Shock also can provide the additional energy to initiate such a reaction in a sensitive substance like acetylene or hydrazine. Most ignition

CHAIN REACTIONS

A combustion reaction begins when fuel and oxidizer molecules collide with enough force to break apart into fragments. These fragments may be atoms or free radicals. Initially, their velocities will be so high from the impacts that the fragments will be unable to recombine when they collide again with other fragments. They will, however, continue to collide with other fragments and molecules, using up their energy, until they reach a lower kinetic-energy state, as they will combine with another fragment when they collide rather then bounce apart or fragment further. When such a combination occurs, heat is given off.

Any mixture of reacting gases will be made up of molecules of the original gases, atoms, radicals, and both simpler and more complex molecules produced by the various reactions that occur. When an entire series of reactions can reproduce itself so as to be sell-sustaining after initial activation, it is called a chain reaction.

Chain reactions can be either branched or unbranched. The commonest example of a branched-type chain is combustion of hydrogen and oxygen. Even this comparatively simple reaction does not take place in a single step. To date, approximately 14 different intermediate steps have been identified.

The combustion of methane, a simple compound, undergoes a more complex reaction:

	Reactants		Products	
Steps in chain mechanism	$2CH_4$ + ignitor	$\longrightarrow$	$2H + 2CH_3$	
	$2H + 2O_2$	$\longrightarrow$	$2OH + 2O$	(1)
	$CH_4 + OH$	$\longrightarrow$	$H_2O + CH_3$	(2)
	$O_2 + CH_3$	$\longrightarrow$	$CH_20 + OH$	(3)
	$CH_2O + O$	$\longrightarrow$	$HCO + OH$	(4)
	HCO	$\longrightarrow$	$H + CO$	(5)
	$CO + OH$	$\longrightarrow$	$CO_2 + H$	(6)
	$2OH$	$\longrightarrow$	$H_2O + O$	(7)
	$2O$	$\longrightarrow$	O_2	(8)
Overall reaction	$CH_4 + 2O_2$	$\longrightarrow$	$2H_2O + CO_2$	

Branching produces a large number of chains, so numerous reactions take place simultaneously. Unless there is interference, the number of chains increases geometrically. If the overall speed with which each reaction takes place is high enough, the process may turn into an explosion. The combustion rate is accelerated by the creation of large numbers of chains in the presence of high temperature and strong ignition sources. In a hydrogen-oxygen mixture, the change from rapid burning to explosion is limited only by the concentrations of the two gases. In air, nitrogen interferes with and slows the branching process. Carbon dioxide is a more effective suppressant than nitrogen, because it not only dilutes a flammable mixture but may enter the reaction, absorbing active particles so they cannot continue the chains.

A break in a chain can take place in other ways. The hydrogen or oxygen can be depleted. A foreign material may act as a catalyst to cause new types of reactions to take place which terminate the chains. The atoms, particles, or molecules may hit the walls of their container so that their velocities are reduced until they are no longer effective should collisions occur. A solid surface cooler than the hot gases will reduce the energy of the particles so that the number of activated molecules will be fewer than are required to continue the chain. Metal may adsorb particles, removing them from the chain.

The energy output of the overall reaction depends on the intermediate steps, no matter the sequence through which the reactions proceed. Some of the steps, such as the initiating reaction, are endothermic (absorb heat), others exothermic (emit heat). Even if the chain is not completed, the energy produced will be the net total of the heat emitted or absorbed by all the intermediate steps that have taken place. If the sum of these results is the release of heat, the reaction will continue; if the heat is absorbed or less than that required to continue the process, the reactions will be quenched and stop. The rate at which the overall reaction between a fuel and oxidizer supposedly takes place is actually governed by the speed of reaction of the slowest intermediate step.

Chain reactions also occur in decomposition, polymerization, and halogenation. To a safety engineer, understanding that chain reactions do occur in a chemical process is necessary to comprehend how deflagrations and detonations occur and how to avoid them, the actions of catalysts, and, most important, how and when certain fire extinguishants, such as the halogenated hydrocarbons, work.

FIGURE 22–6 Chain Reactions

sources can be divided into a number of categories: open flames, electrical arcs or sparks, hot surfaces, mechanical or electrical sparks, spontaneous ignition, adiabatic compression, or catalytic action.

The configuration of the energy source can have a great effect on the energy required to ignite a flammable mixture. Far less energy is required for ignition of a point source, such as a flame or spark, than with a diffuse source, such as a hot surface, because of the higher intensity of the energy in a more limited volume of flammable mixture.

Electrical Arcs and Sparks

Ignition of flammable mixtures from electrical arcs or sparks (see Fig. 21-1) is a common cause of accidental fires. These can be produced by:

1. Sparking of electric motors, generators, or other electrical rotating equipment.
2. Arcing between contacts as switches and relays open to break the flow of current.
3. Arcs at points of broken, inadequate, or failed insulation.
4. Discharges of static electricity accumulated on ungrounded surfaces.
5. Lightning strikes.
6. Discharge of charged capacitors through a gas.
7. Accidental contacts during a short-circuit.
8. Poor contacts between conductors, such as poorly fitted light bulbs and their sockets.
9. Broken or cut conductors carrying electrical currents.
10. Arcs intentionally created during electric welding.

Hot Surfaces

Numerous examples exist of hot surfaces from which flammable mixtures can be ignited. In many instances, ignition may not take place immediately. Organic materials in contact with hot surfaces may brown, char, and ignite after a long time as volatiles are emitted which then reach their autoignition temperatures. In the same way, heat can cause decomposition of heavy oils into less complex but more reactive molecules, which then ignite as the temperature of the heated surface increases. Some of these hot surfaces are:

1. Electrical heaters or hot plates.
2. Exposed, hot parts of operating engines or compressors, especially exhaust manifolds.
3. Overheated wiring, motors, or other electrical equipment.
4. Boilers and steam surfaces, stacks, and chimneys.
5. Steam radiators, lines, and equipment.
6. Burning cigarettes.

7. Metals heated by friction, such as brake drums and bearings.
8. Surfaces heated by radiation from the sun or from fires.
9. Metals being welded.
10. Hot industrial process equipment.

Mechanical and Chemical Sparks

Metallic and chemical sparks (see Fig. 22-7) can be generated by:

1. Dropping or banging of steel tools, chains, or equipment parts on other steel or other extremely hard surfaces.

MECHANICAL AND CHEMICAL SPARKS

Mechanical sparks: produced by friction between certain metals or other hard substances. Small particles are torn from a larger mass; the energy expended in the tearing process then causes them to be heated to incandescence. To spark, a metal must have three basic properties: it must be of such composition that the energy to create particles will also be enough to heat them to high temperatures. Lead, zinc, and tin are too soft to spark. "Nonsparking" tools are made of a steel alloy containing beryllium-copper. Tools of this material are soft, so they will rarely spark. Unfortunately, they deform easily and slip when force is applied.

A metal that sparks also oxidizes and burns easily. Temperatures of such sparks are generally at or about the burning temperature of the metal. Metals with this property are those which are highly electropositive. Cerium burns easily in air; it is used in an alloy which has the necessary hardness and fracture properties for "flints" in cigarette lighters to produce sparks. The third property affecting sparking characteristics is specific heat. For the same energy involved, a metal of low specific heat will reach a higher temperature. Beryllium and titanium have the combination of properties which causes them to present the greatest sparking hazards of common metals.

Nonmetallic sparks can be produced in a similar way with carbon. This frequently happens with the hard carbon brushes used in electric motors and generators. The constant rubbing of the brushes over commutator bars causes breaking away of small particles already heated by passage of electric current and friction. The breaking away of the already hot particles permits them to reach the point of incandescence easily.

Chemical sparks: particles of incandescent carbon or organic materials such as wood or paper. The carbon is produced by the incomplete combustion of hydrocarbon fuels, generally deposited as soot on cold metal surfaces. A built-up deposit may be heated to incandescence again, break off, and be carried away by convection currents.

Combinations: mechanical sparking becomes a particular problem when there is impact between aluminum and iron oxide. The oxide may be present in the form of rust on steel, or as a protective coating. The reaction that occurs between the two is the same as that in thermite welding. The impact force provides the energy required to initiate the reaction:

$$2Al + FeO_3 \longrightarrow Al_2O_3 + 2Fe + heat$$

Metals treated with strong oxidizers, such as concentrated acids, will also show increased tendency to spark and ignite under impact or friction. This is evidently because of the reduction in activation energy required when an oxidizer stronger than air is present.

FIGURE 22–7 Mechanical and Chemical Sparks

2. Friction of moving metal parts against other metals or extremely hard surfaces; for example, braked steel wheels skidding on rails or rocks.
3. A rotating metal body striking a hard fixed surface or metal, or a substance caught between rotating and fixed metal surfaces.
4. Abrasive wheels or disks used for metal cleaning or grinding.
5. Scattering of incandescent particles from wood or paper fires.
6. Spatter and slag created during welding or gas cutting.
7. Aluminum hitting iron oxide.
8. Exhausts with very hot carbon particles from internal combustion engines.

Spontaneous Combustion

Large stockpiles of garbage or low-grade bituminous or subbituminous coal, lignite, and peat, must be guarded against spontaneous heating and ignition. Some recommended procedures are to store them under water or layers of earth, coat the surface of the pile with a sealant to prevent entrance of air, limit heights of piles, monitor temperatures at the bottom of piles, and break piles apart regularly to release accumulated heat.

Other solids which heat and ignite spontaneously are organic materials such as cotton, wood shavings, hay, and masses of refuse. In most of these, adsorption of atmospheric gases and bacterial processes provide the first exothermic reactions. The decomposition of organic compounds may form additional carbon, which adsorbs and reacts with more oxygen, causing emission of more heat.

Unsaturated oils and fats are another group of substances that may heat or ignite spontaneously. In the presence of air, these oils and fats combine slowly with oxygen to become fully saturated. As soon as the heat from the slow reaction causes the temperature to rise high enough, the mass may ignite.

Bulk oils have little tendency to ignite spontaneously. The heat generated will raise the temperature of the entire mass only slightly because of convection within the liquid. Danger arises when oil remains on fibrous materials, such as cotton rags or wastes, exposing large oxidizable surfaces to air. A single rag will rarely ignite from this cause, because the heat generated is lost rapidly. In a pile, however, the rags themselves act as insulation to keep the heat from being dissipated. The heat accumulated from the numerous simultaneous reactions can be adequate to cause local ignition, from which fire spreads rapidly. The length of time for a fire to start spontaneously with oil-soaked materials varies from hours to days, depending on the amount and type of oil, the quantity of rags or waste, the temperature to which it is subjected, and the availability of air. This type of ignition is the reason for warnings to keep oily rags and waste in closed containers and to dispose of them frequently.

Insulation on a steam line or other hot surface that has been contaminated by lubricating or fuel oil should be replaced for this reason, since it represents a fire hazard. The oil may not flame at room temperature without an outside source of ignition, or at the temperature of the hot surface. However, the heat may cause the oil to oxidize and react exothermally. The additional heat generated by the rise in temperature may be adequate to cause localized ignition of the oil.

Hypergols

When a fuel and oxidizer react so rapidly on being mixed at room temperature that combustion starts immediately without an outside ignition source, it is a hypergolic reaction. The term hypergolic was originated for use with rocket propellants; these ignited as soon as the hypergol mixed with oxygen. This ignition was intentional, but similar reactions have been known throughout the history of chemistry which have caused accidental fires.

In such a reaction, the heat initially emitted will produce combustion with no other ignition source required. Nitric acid, for example, in highly concentrated form is hypergolic with numerous fuels. Under such conditions, the possibility that such propellants might mix accidentally constitutes a considerable hazard. Such a reaction ignition took place in the silo of a Titan II missile when a bit of nitrogen tetroxide (oxidizer) accidentally came in contact with unsymmetrical dimethyl hydrazine (fuel). Because the two substances were hypergolic, they burned rapidly in the confined space. The pressure created by the combustion gases was adequate to throw parts of the heavy concrete doors of the silo hundreds of feet away.

Pyrophors

A special category of hypergolic fuel includes those substances that will react rapidly not only with oxygen in high concentrations, but even with the oxygen portion of air. These fuels are pyrophoric. Such fuels and reactions have also been known for a long time. White phosphorus removed from water will burst into flame as it dries. Gases such as silicon and boron hydride will also ignite spontaneously in air. Some of the most hazardous pyrophors are the iron sulfides: FeS_2, FeS, and FeS_2S_3. FeS_2 is found in iron ore and coal in small amounts as iron pyrites. The others are produced during industrial processing of materials containing sulfur in ferrous vessels. During production of sulfuric acid or petroleum fractions, there is generally little hazard while the iron sulfides are submerged in process liquid or otherwise kept wet. As soon as the sulfide dries, however, it will react with the oxygen in air in an exothermic reaction which produces so much heat that even a small amount of sulfides is dangerous. FeS_2 will also react with oxygen in the presence of small amounts of moisture to produce iron sulfate, sulfuric acid, and heat. A reaction of this type in petroleum-processing equipment may ignite hydrocarbons present, the hot combustion products sometimes causing pressures adequate to rupture vessels violently.

Adiabatic Compression

Adiabatic compression of a gas will increase its temperature. In some instances this may be desirable; in others, detrimental. Diesel engines compress flammable mixtures to the temperature at which they ignite and burn. In air compressors the heat of compression must be removed to reduce the temperature of the air. Explosions have been caused by the ignition of hydrocarbons in hot air leaving the compressor. Generally these hydrocarbons are lubricating oils or their residues. Compressors are especially hazardous in this respect, since they involve high concentrations of oxidizer, flammables, and ignition source. Because of the very high temperatures present, combustion will be so rapid that an explosion might result.

Adiabatic compression of gas with increase in temperature can occur under other conditions. Air entrained as bubbles in a pipe full of fuel can be compressed if there is sudden stoppage of flow. The pressures that result from such stoppages can be considerable.

Radiation

Radiation may be either a direct or indirect source of fire ignition. Sunlight can be concentrated intentionally or accidentally by a lens or curved reflector to cause ignition of combustible materials. Solar reflectors provide some of the highest temperatures available without the use of nuclear devices. In some countries short of fuel, small reflectors are being used to concentrate heat in amounts adequate for cooking purposes. Less efficient concentrators of solar energy may still constitute sources strong enough to cause fires. Flames, industrial heating furnaces, highly incandescent metals, and glowing solid combustibles can also radiate enough energy to ignite flammable nearby materials. Lasers (see Chapter 27) can generate beams whose intensities may cause combustibles to ignite.

Catalytic Action

A catalyst is frequently defined as "a material that will accelerate a reaction without itself being consumed." It appears from this that the catalyst does not take part in the reaction. This is not so. A catalyst is involved in the overall reaction between two other substances by interposing other reactions in which it takes part and which require less activation energy than will a direct reaction between the other substances. In many instances these intermediate reactions are not well understood. However, the first intermediate step is one that generally involves an exothermic reaction between one of the normal reactants with an atom or radical of the catalyst. The initial reaction requires less activation energy than would the two normal reactants if the catalyst was not available. The second intermediate reaction then restores to its original the catalyst from its modified form. An example of a two-step sequence illustrates the ability of copper oxide to accelerate oxidation reactions of organic compounds

$$2\,CuO + R \longrightarrow RO + Cu_2O, \text{ where } R = \text{organic compound}$$

$$Cu_2O + O \longrightarrow 2\,CuO$$

Some other metals whose oxides act in this way are lead, chromium, manganese, mercury, silver, vanadium, molybdenum, and iron. Hydrazine must not be permitted accidentally to contact iron rust, mercuric oxide, lead oxide, chromic acid, or similar substances. Potassium cuprocyanide also causes decomposition of hydrazine.

In some instances, a mixture of catalysts will increase the effect beyond that produced by all of them separately. Hopcalite, used for oxidation of carbon monoxide in some gas masks and air-regenerating systems, consists of 50 percent manganese oxide, 30 percent silver oxide, 15 percent chromic oxide, and 5 percent silver oxide. This mixture is more efficient than is a similar type of catalyst which uses only the oxides of manganese and copper.

The reactivity of a catalyst can be defined by the rate of reaction or by the heat liberated during a time period, and with comparable reactants. Each catalyst has a limit beyond which its effect increases no further. The activity of manganese naphthenate is a maximum when present in a concentration of 0.8 percent. Addition of more catalyst produces no further increase in reaction rate.

IGNITION SOURCES

Statistics on the types of equipment involved in fires and initiating factors might be of interest. A survey of fire statistics made by the National Fire Control Data Center of the U.S. Department of Commerce[88] analyzed the causes (as it lists causes) of fires which occurred in two states, California and Ohio. Figure 22-8 presents data from a number of tables in the report of the survey combined into one list, which might be of interest to occupational safety personnel.

IGNITION DELAY

The time required for a flammable mixture to ignite depends on its composition, temperature, pressure, and the energy of the ignition source. Generally, an entire mixture is not brought to its autoignition temperature all at once. Local heating occurs so that a small volume is first brought to the autoignition temperature and ignites. The localized reaction creates the flame, which propagates through the remaining mixture. The lower the mixture's temperature, the longer the time lag, or ignition delay, before a sustained flame appears.

A mechanical spark entering a combustible mixture heats a small volume of gas while itself being cooled. If that small volume reaches the ignition temperature before the spark is cooled, ignition takes place. Otherwise, there will be no fire, and the particle will cool. A spark, therefore, constitutes a greater hazard with a warm or hot mixture than a cool one, for two reasons: (1) less energy and time are required to ignite the mixture by raising it to its autoignition temperature, and (2) there will be less cooling effect on the spark.

Walls surrounding a mixture can act either as a heat source to cause its ignition or as a heat sink to delay ignition. Both of these may constitute hazards. The hazard of a heat source which could raise a flammable mixture to its autoignition point is evident. The danger of a cold surface is not so apparent. However, a cold surface can act as an "inhibitor," a substance whose action is diametrically opposed to that of a catalyst. An inhibitor slows or stops any tendency to burn. Cold furnace walls or boiler tubes can reduce fuel temperature to such a degree that normal ignition will be delayed while fuel vapors accumulate. The fuel may then ignite, but the accumulation's burning almost all at once may produce a minor explosion due to excessive buildup of the resulting burned gases.

CAUSES OF FIRES*

Sorting sequence	"Cause" category[1]	Definition	Residential fire causes–percent	Nonresidential causes–% basic ind.	Nonresidential causes–% manufacturing
1	Exposure	Caused by heat spreading from another hostile fire.	3	6	4
2	Natural source	Caused by sun's heat, spontaneous ignition, or chemical, lightning or static discharge.	0.9	3	6
3	Incedendiary/suspicious	Fire deliberately set or suspicious circumstances.	11	7	10
4	Explosives, fireworks	Self-evident; explosives used as incendiary devices included in category 3.	0.7	7	14
5	Smoking	Cigarettes, cigars, pipes as heat of ignition	13	1.9	5
6	Children playing	Includes all fires caused by children playing with any materials contained in categories below.	5	0.8	1.2
7	Heating systems	Includes central heating, fixed and portable local heating units, fireplaces and chimneys, water heaters as source of heat.	13	5	5
8	Cooking equipment	Includes stoves, ovens, fixed and portable warming units, deep fat fryers, open fried grills as source of heat.	18	0.9	3
9	Air conditioning, refrigeration	Includes dehumidifiers and water cooling devices as well as all air conditioning and refrigeration equipment as source of heat.	0.7	0.8	0.9
10	Electrical distribution	Includes wiring, transformers, meter boxes, power switching gear, outlets, cords, plugs, lighting fixtures as source of heat.	7	45	7
11	Appliances	Includes TV's, radios, phonographs, dryers, washing machines, vacuum cleaners, separate motors, hand tools, electric blankets, irons, electric razors, can openers as heat source.	7	2	8
12	Gas	Material first ignited was a gas; natural, LP, manufactured, anesthetic, acetylene, other gas.	0.3	0.4	0.2
13	Flammable, combustible liquid[2]	Material first ignited was flammable liquid; gasoline, ethyl alcohol, ethyl ether, acetone, jet fuel, turpentine, kerosene, diesel fuel, cooking oil, lubricating oil, etc.	0.9	4	7
14	Open flame, spark (heat from)	Includes torches, candles, matches, lighters, open fire, backfire from internal combustion engine as source of heat.	5	4	7
15	Other equipment	Includes special equipment (radar, X-ray, computer, telephone, transmitters, vending machine, office machine, pumps, printing press), processing equipment (furnace, kiln, other industrial machines), service, maintenance equipment (incinerator, elevator).	0.4	7	14
16	Other heat	Includes all other fire caused by heat from fuel-powered objects, heat from electrical equipment arcing or overloaded, and heat from hot objects not covered by above groups.	2	2	4
17	Unknown	Cause of fire undetermined or not reported.	10	9	15

1 "Cause" as used here is a shorthand notion for what is sometimes a complex chain of events leading to a fire.
2 Note that incendiary fires involving flammable liquids are covered in category 3, not 13.

* California 1975 and Ohio 1976 experiences combined.

FIGURE 22–8 Causes of Fires

In some oil-burning furnaces or kerosene stoves, cold fuel, air, or metal surfaces have led to failures to ignite even when a strong ignition source, such as a flame, was used. The fuel vapors increased steadily. Repeated attempts at ignition finally warmed the fuel enough to permit its ignition. The overpressures generated by the burning masses in such cases have been known to blow boiler furnaces apart. For this reason, such furnaces are equipped with blowout panels. If an explosion in the furnace occurs, pressure will be released before boiler tubes, drum, or walls can be damaged.

The principle of inhibition by a metal surface to reduce the ignition possibilities of flammable mixtures was known even before the mechanism was fully understood. The miner's lamp invented by Humphrey Davy used a wire mesh to absorb and dissipate heat from a burning candle. This reduced the possibility of igniting flammable gases in a mine. The cooler metal surfaces between which a hot gas must pass to leave an explosion-proof electrical fixture provide the cooling effect necessary to prevent ignition of a flammable mixture outside the fixture.

EFFECTS OF FIRE ON PERSONNEL

The National Fire Protection Association indicates that a majority of the fatalities from fires are caused by suffocation or inhalation of smoke or fire gases, not burns.

In one nine-month period, 1,382 of 1,803 injured firemen (77 percent) were smoke-inhalation victims. Fifty-three men died as a result of a fire, involving less than 90 gallons of hydraulic fluid, in a Titan II missile launch silo, almost all of carbon monoxide poisoning.

A fire in any closed structure will have two immediate toxic effects: production of carbon monoxide and depletion of the air present. Combustion in an enclosed space is generally incomplete, and partially burned gases (especially carbon monoxide), decomposition products, soot, and smoke will be generated. Most of these are described in Fig. 22-9. Other terms regarding toxicity of these combustion products are also discussed in Chapter 24.

Temperature and Heat

The effects of temperature and heat are presented in Chapter 19. A standard time-temperature curve, which indicates the rate at which temperature in an enclosed space will increase during a fire, is a straight line between zero and five minutes, although temperatures are quite unstable during that period and a curve to represent all conditions is difficult to provide. These first few minutes are of prime importance, since it is the time when a fire can sometimes be caught before it creates extensive damage. In addition, it is the period when personnel have an opportunity to and must escape before the temperature reaches the generally acceptable tolerance limit of 150° to 160°F.

Burns

Burns can result from direct contact with burning materials, hot surfaces, flame, soot, smoke, or burned gases, or from radiation from these. The various degrees of burns are described in Chapter 19. Burns of the skin and upper respiratory tract by soot and

Not only is the mechanism of combustion complex, but so is the number of toxic products that result. Most of those produced, the types of which depend on the original fuel involved and with air as the oxidizer, are listed below. Figure 25-1 and Chapter 24 also are pertinent. A few of these toxic gases produced by means other than combustion have also caused fatal consequences.

Carbon monoxide: Carbon monoxide (CO) has been the commonest killer because of the frequency with which it occurs and the rapidity with which it reaches lethal concentrations in a fire. It will be produced by the incomplete combustion of every organic compound that burns (or becomes and then burns) any form of carbon. Deaths from carbon monoxide have resulted from malfunctioning furnaces and fuel-burning heaters, and from exhaust fumes from motor vehicle operations. These deaths often take place because of operation in cold weather when windows are closed, because CO is odorless and colorless. The effects of carbon monoxide on the body are described in Chapter 24. It may be sufficient to repeat here that 1.28 percent by volume CO in air will be fatal in 1 to 3 minutes; 0.64 percent in 10 to 15 minutes; 0.32 percent in 30 minutes to 1 hour; and 0.16 percent within 2 hours. Anything over 0.05 percent is considered dangerous.

Carbon monoxide itself is a fuel. Produced in large amounts in an enclosed space, it can create a flammable atmosphere when additional oxygen, such as air, is introduced. Fire fighters who have created openings into closed areas where fires existed have been blown back out again. The carbon monoxide, heated to high temperatures, mixed with air and burned rapidly enough to generate overpressures adequate to hurl back them back and to burst windows, doors, and walls. A hot fire in the open air or a ventilated space may be less hazardous to personnel, especially to fire fighters, than one in an enclosed space. Where air is available, combustion is complete and little carbon monoxide is left to provide a highly toxic and flammable atmosphere.

Carbon dioxide: Fires in air lead to depletion of oxygen and production of carbon dioxide (CO_2) during complete combustion of carbon or organic materials. Comparisons of percentages of oxygen and carbon dioxide show that as the first decreases, the second increases in approximately the same amount. A small increase in carbon dioxide will cause an increased respiratory rate; a large one will increase the rate until the demand of the body cannot be satisfied adequately. Respiratory collapse will then occur. Carbon dioxide in extremely large amounts also acts as an asphyxiant and fire extinguishant. Concentrations greater than 5.0 percent in the environment should be considered dangerous, not because of imminent effects but because of their indication of abnormality from usual conditions.

Hydrogen cyanide: Although hydrogen cyanide (HCN) produced in fires of organics is much more toxic than carbon monoxide, the volume produced is usually less. A concentration of 100 ppm may be fatal in 30 to 60 minutes. It is produced during the combustion in air of chlorinated hydrocarbons, plastics, leather, rubber, silk, wool, or wood. HCN, like CO, is lighter than air and therefore constitutes a much greater hazard in an enclosed space than in the open air.

Phosgene: Phosgene ($COCl_2$) is also produced during decomposition or combustion of chlorinated hydrocarbons, such as carbon tetrachloride, Freon, or ethylene dichloride. It is toxic and dangerous in extremely small dosages. Twenty-five ppm can be fatal in 30 to 60 minutes. Its toxicity is indicated by the fact that during World War I it was used intentionally as a poison gas.

Hydrogen chloride: Hydrogen chloride (HC_2) is produced during combustion of materials containing chlorine. It is not as toxic as hydrogen cyanide or phosgene but is dangerous during long exposures.

Hydrogen sulfide: Hydrogen sulfide (H_2S) is a highly toxic gas whose presence makes itself readily apparent by its odor of rotten eggs. Released naturally in excavation, it also can be produced from combustion processes involving sulfur. If reactions that normally end up as sulfur dioxide do not go to completion, hydrogen sulfide may result, such as for rubber products or untreated petroleum.

FIGURE 22–9 Combustion Gases Continued on page 398

Concentration of 400 to 700 ppm (0.04 to 0.07 percent) can be fatal in 30 to 60 minutes. The odor of rotten eggs warns of its presence; however, it desensitizes the organ of smell so the ability to detect its presence is lost rapidly. Hydrogen sulfide is flammable, so that, like carbon monoxide, it will burn when exposed to additional air. Because of its offensive odor, unlike carbon monoxide, it is generally not allowed to decumulate in amounts great enough to create an explosive atmosphere.

Sulfur dioxide: Sulfur dioxide (SO_2) results from complete combustion of sulfur. It is a toxic gas with an irritating odor which makes it detectable in concentrations of 3 ppm and is considered to have an initial danger level of 5 to 10 ppm. Only 140 ppm is considered possibly lethal in 30 to 60 minutes. (See the comment on deaths at Donora, Pennsylvania, in Chapter 25.) Sulfur dioxide is extremely irritating, since it combines with water to form sulfuric acid. An unprotected person is generally unable to withstand breathing a lethal mixture because of the irritation and coughing it produces. Figure 8-1 points out that extensive combustion of fuels causes development of sulfuric acid in acid rain.

Oxides of nitrogen: These include of nitrogen oxide (also found in acid rain), nitrogen dioxide, and nitrogen tetroxide. Nitrous oxide (N_2O) is found naturally in air concentrations of 0.5 ppm, created at high altitudes by the combination of nitrogen and air or atomic oxygen. Nitric oxide (NO) and nitrogen dioxide (NO_2) are compounds produced as products of the reaction when a fuel in burned in air.

Combinations of these nitrogen oxides and of water can produce nitric acid (HNO_3), which can then react with metals to form nitrates. An exposure of 100 ppm (0.01 percent) for 30 minutes or more can be fatal. A health hazard is considered to exist when the concentration of nitrogen oxide is 10 ppm or more.

Ammonia: Ammonia (NH_3) is produced when wood, wool, silk, a refrigerant, or other compound containing nitrogen and hydrogen burns in air. It is not as toxic as most other gases described here and is easily recognized by its sharp odor. However, even small amounts irritate the eyes and respiratory tract.

Acrolein: Acrolein is a highly toxic material which results from heating of fatty materials to 600 F and from combustion of some paints or woods. Exposures of 30 minutes or more to 150 to 240 ppm can be fatal. Acrolein is an irritant that affects the eyes and respiratory system. The burning sensation it produces in the eyes causes tearing. This may reduce the ability of an affected person to make a successful escape from a fire.

Metal fumes: Fires involving electronic equipment may cause melting and vaporization of solder, liberating fumes of lead, tin, and sometimes antimony.

Soot and smoke: Soot and smoke, although they are different, have similar effects on the body. Soot is unburned carbon produced by the decomposition of organic materials in the absence of enough oxygen for combustion, or by impinging of a flame on a cold surface. Smoke is a mixture of particles of soot, ash, and solid combustion products. Hot soot or smoke can cause burns to the skin. It can burn and irritate the upper respiratory tract (see Fig. 24-1) and cause tearing of the eyes. Internal inflammation can produce heavy exudation (edema) of tissue fluid from the membranes lining the respiratory passages, which may then block the flow of air. This may be compounded by vapors from extinguishants to cause more physiological problems. These gases may create problems requiring inhalation of increased amounts of air.

Another important effect of soot or smoke is to decrease visibility, which may hamper efforts of rescuers, fire fighters, and persons attempting to leave a burning structure. Smoke produced by combustion of wood products is generally gray-black; of fabrics, brown; of paper or straw, whitish-yellow; and of hydrocarbons, such as oil or acetylene, deep black.

FIGURE 22–9 Continued

smoke have already been mentioned. Hot gases will do the same. Radiation can be extremely damaging, its severity depending on the type of flame. Although hydrogen burns with the release of the most heat per pound of any fuel, its emissivity is so low that it will radiate comparatively little heat. Gasoline and jet-fuel flames have emissivities of about 1; the colorless hydrogen flame, about 0.07. The flame is so hard to distinguish that when pure hydrogen is burning, it is difficult to detect. Hydrogen leaks can be ignited easily, but personnel may not be aware there is a fire. If a worker must walk into an area where a hydrogen fire might exist, it is frequently advisable the worker carry in front a long, easily ignitable object, such as a broom.

FIRE DETECTION SYSTEMS

Fire detectors note abnormal environmental conditions such as the presence of smoke or an increase in temperature, light intensity, or total radiation. Some gas-detector principles are noted in Fig. 25-5, but additional explanations are provided here for some types of fire detection.

Thermal-Expansion Detectors

Thermal-expansion detectors use either a bimetallic element or confined fluid. The bimetallic type is one made of two metals with different rates of expansion. When heated, one metal will expand more than the other, causing the element to bend, either closing or opening an electrical circuit. This will occur when the element reaches a predetermined temperature. In the confined-fluid type, the abnormally high temperature produced by a fire causes the liquid or gas to expand and operate a pressure switch when a preset activation point is reached. This type of device provides a signal when a small mass of fluid is heated to a high temperature or a large mass is heated moderately.

Thermosensitive Devices

Thermoelectric devices use low-heat-capacity thermocouples. When heated, they generate voltages sufficient to activate sensitive relays. Their operation depends on differences in temperature between an exposed thermocouple joint heated by a fire and an unheated or "cold" junction. Thermocouple detectors are based on rate of temperature rise, not on absolute temperature, and depend on the time for the cold junctions to be heated by current flow. Thermocouple sensing elements are connected in series. Rapid heating of one element to a high temperature or of all elements to a moderate temperature will indicate a fire.

Thermoconductive Detectors

Thermoconductive detectors each use two conductors separated by special insulating materials. Heating the insulation at a specific rate or to a specific temperature causes its electric resistance to decrease rapidly. Current will then flow between the conductors to signal a fire. The device operates on heat flow, but an insulator can be used whose resistance decreases rapidly at a fixed temperature.

Radiant-Energy Detectors

Infrared radiant-energy detectors use photoelectric cells to detect changes in infrared energy radiated by burning materials. They generally give fire warnings when rapid fluctuations occur in radiation intensities, such as the presence of flames. When radiation changes take place repeatedly within a selected range of frequencies, a fire signal is given.

Ultraviolet (UV) detectors use either a solid-state device or a gas-filled tube to detect the presence of fire. Unlike infrared detectors, these are not considered to be sensitive to sunlight (or to artificial light).

Light-Interference Detectors

Light-interference detectors operate by attenuation of visible light passing through a column of smoke or heated gas. There are various types. In one, a fixed-intensity light beam aimed at a photoelectric cell causes constant current flow. This keeps a relay open when conditions are normal. When light intensity is reduced by smoke or other interference, the current decreases and the relay closes to provide a warning signal.

In another type of light-interference detector, the light beam does not hit the photoelectric cell under normal conditions, but passes near it. In the presence of smoke, interference by solid particles reflects and scatters the beam so that the light hits the photoelectric cell. A warning signal is produced when a preset signal strength, generated by a specific amount of combustion products, is exceeded. This type of device is used in popular home fire detectors.

Ionization Detectors

A very small amount of radioactive material in a detector's smoke chamber ionizes the air to make it electrically conductive. As a result, a weak electric current can be made to flow through it and the detector's circuit. Any particles that enter the smoke chamber interfere with this current flow. If the interference is great enough, the current flow drops below a preset level or stops, so that an alarm or buzzer sounds.

FIRE CLASSIFICATIONS

Accidental fires are divided into classes:

Class A. Class A fires involve solids, such as coal, wood, paper, and refuse, which char or produce glowing embers. Any rapid combustion is due to volatile materials liberated by heating.

Class B. Class B fires involve gases and liquid that must be vaporized for combustion to occur. In many cases, liquids in this fire class will float on water and continue to burn as long as they are in contact with air.

Class C. Class C fires are either Class A or Class B fires but also involve electrical equipment or materials near electrically powered equipment.

Class D. Fire involving magnesium, aluminum, titanium, zirconium, or other easily oxidized metals are included in Class D. Combustion temperatures and energy are high compared to those of hydrocarbon or wood fires.

Special Categories. This class includes fires involving extremely active oxidizers or fuel mixtures, such as flammables with oxygen, nitric acid, or hydrogen peroxide, or fires involving solid missile propellants.

FIRE SUPPRESSION

Fires can be suppressed by various means:

Isolation

Isolation of fuel from the oxidizer is a common means of extinguishing fires and can be accomplished in several ways, some very easily. Where the fire is fed with fuel from a fluid source such as a ruptured line, it may be possible to stop the flow of fuel by closing an upstream valve. Blanketing with a covering of inert solid, thickened water, heavy nonflammable gas, or foam is a valuable means. A film of inert liquid or powder with a specific gravity less than that of a burning liquid can be floated on its surface. A noncombustible material such as an asbestos blanket, a metal cap put onto a container in which a fire exists, or water flooding a shallow vessel is also effective.

Increasing the volume of inert gas in air will dilute the oxygen a fuel requires to burn. Diluting a gaseous fuel may make the mixture so lean that it will be below the lower flammability limit. Similarly, diluting a liquid fuel with a noncombustible liquid will reduce the volume of flammable vapors produced so that the fire is no longer self-sustaining.

Cooling Combustibles

When a fire exists, either the flames or the combustible may be cooled to a temperature below the point for a fire to continue. Of the two, cooling the combustible is much more effective. This can be done by a coolant's absorbing sensible heat, or more effectively, by its absorbing latent heat as water is vaporized. Blowing away a flame will keep the heat of reaction from reaching the surface of a flammable liquid or solid. A cold quenching device or surface may absorb enough heat to prevent propagation either through a gas mixture or back to a solid or liquid.

Reaction Inhibition

Combustion involves numerous chain reactions (see Fig. 22-6). Breaking a chain by removing a necessary reactant so that the reaction cannot propagate will kill a fire. This can be done by:

Catalyzing a reaction so that it moves in a direction it would ordinarily not take.

Using an inhibitor to force the process along a chain which will lead to endothermic rather than exothermic reactions, causing suppression of the fire.

Using an inhibitor to deactivate excited particles; for example, electrical charges can be provided to positive radicals or removed from negatively charged ions.

Halogenated hydrocarbons inhibit chain reactions. They are fuels, generally with high ignition points, which decompose into radicals and ions. These inhibit normal chain production and quickly suppress a fire. However, they are effective only against hydrocarbons burning in air. They have no effect where other oxidizers or other types of fuel are involved.

Water

Water is by far the commonest fire extinguishant. It has limitations in certain situations but generally is available, low in cost, simple to use, and effective. The amount of water required to suppress flames generally increases with the mixture and flame temperature. Most fires involve Class A combustibles, and with these materials water is most effective. As a spray or with additives for specific purposes, it can be used for other classes of fires.

Streams. The principal effect of a solid stream of water is to cool the burning fuel below its ignition temperature, using lower-sensible-heat effects. Very rarely does water on a solid stream provide a blanketing effect. The few occasions on which it can do so are generally when the fire is in a container or depression that can be flooded. Even here, the cooling effect of the water may be more instrumental in extinguishing the fire than the blanketing action. It is estimated that only 5 to 10 percent of the water in a solid stream is actually effective in fire extinguishment. A great advantage of a solid stream is that it can be used from comparatively long ranges, the distance depending on the available water pressure. Streams of water are usually not recommended for fires involving magnesium or other metals because of the violent and explosive reactions that result. (In some cases this might be advantageous by shattering the burning metal into fragments that can be cooled more easily.)

Firefighters sometimes use large amounts of water to flush away spills of gasoline or other deleterious materials into drains and sewers, especially in industrial plants. However, personnel in many sewer districts indicate this technique is recommended only for small spills of less than 50 gallons. Large ones should be contained, covered with "light water" (later in this chapter), and the spilled gasoline collected with skimmers, absorbents, or other means.

Spray and Fog. A spray or fog increases the effectiveness of water as a coolant. A solid stream cools because the portion of water that contacts the burning material absorbs sensible heat. A spray or fog cools by absorbing heat of vaporization and by breaking chain reactions. The heat-absorbing effects of streams and sprays can be compared as follows: Assume the water has an initial temperature of 40°F. In a stream it may be raised to a temperature of 80°F before it runs off. One pound will absorb 40 Btu. A pound of spray will be heated to 212°F, absorbing 172 Btu of sensible heat, and then it evaporates, absorbing an additional 970 Btu in latent heat, or a total of 1,142 Btu.

The steam produced will have two additional effects. Steam acts as a diluent, which reduces the concentration of oxygen and therefore the combustion rate. Steam may also combine with carbon in hydrocarbon chain reactions to produce carbon monoxide and hydrogen. These will not burn in an atmosphere whose oxygen concentration has been reduced by steam.

The effectiveness of sprays and fogs permits them to be used on all classes of fires. However, the principal disadvantage—that they be applied at close range—restricts their use with metallic fires, where explosions may occur. Nor can they be used with foam extinguishants, whose effectiveness they destroy.

Steam. Steam's principal effect as a diluent of oxygen in air is negated to a great extent by the fact that it keeps the temperature fairly high. On the other hand, vaporization of the hot water does have a cooling effect. Its inhibiting effect is not present in an organic fire.

Thickening Agents. Thickening agents reduce one of the principal disadvantages of water streams: runoff. Increasing the time that water remains on a burning surface will increase the water's effectiveness, not only as a coolant but also in protecting flammables which it coats. Water has been thickened to syruplike consistency with methyl cellulose, clays, or gum. In some types of thickeners, sodium and calcium borate are used, which create mixtures so thick they are "slurries." Slurries adhere to surfaces with which they come in contact. In the presence of heat, a reaction occurs to break down the agent, creating a fire extinguishing and retarding coating. Borate slurries have been used successfully in drops from aircraft on brush and forest fires.

Salts. Salts are added to water for three reasons. Calcium chloride and lithium chloride depress the freezing point to –40°F or less, permitting the water solution to be used in extremely cold weather. Potassium carbonate is deposited from water solutions to form a coating of salt on burning Class A materials, blanketing them from the air they need to burn. Some salts are active inhibitors or produce gases, such as carbon dioxide, which are inhibitors. Salts which have been considered for this purpose are sodium fluoride, chloride, and bromide; potassium sulfate and bicarbonate, sodium carbonate and bicarbonate; and copper chloride. Salt solutions should not be used on fires involving strong oxidizers, with which they may form explosive reaction products.

Detergents. The action of detergents to water sprays and fog will reduce the surface tension of the water. The size of the droplets will be reduced, increasing the vaporization, the cooling effect, and the ability of the water to penetrate Class A combustibles. In addition, the detergent will decrease the surface tension of many hydrocarbon fuels, increasing the effectiveness of the water in cooling fires in which they are involved.

Gas Extinguishants

The use of gas extinguishants is not as common as that of water, but in some instances is more effective, especially in enclosed spaces. Chief among these gas extinguishants are:

Carbon Dioxide. Carbon dioxide is the most widely used gas firefighting agent. It acts as extinguishant in four ways: as a coolant, by blanketing, by reducing oxygen concentration, and as a combustion inhibitor. Discharged from a container, the liquid carbon dioxide vaporizes and cools as it expands. The vaporing effect due to expansion reduces the gas temperature so far that "snow" is formed. The snow and cold gas are effective as coolants to bring the burning materials below the ignition temperature. When large amounts of carbon dioxide are available, a fire may be blanketed, so there is separation from the oxygen in air that a fire requires. Where carbon dioxide for blanketing is not available, it may act as a diluent to decrease the concentration of oxygen so that combustion is reduced or halted.

Carbon dioxide is also effective as an inhibitor in chain reactions. Hydrogen produced in the chain reacts with carbon dioxide to form the monoxide and water. The water and free carbon create additional carbon monoxide and hydrogen. Both reactions are endothermic, absorb large amounts of heat, and reduce the temperature of the combustibles. Although much carbon monoxide is produced, the decreased concentration of oxygen temporarily reduces the probability of its ignition. However, if the monoxide is permitted to accumulate, it may later be ignited by a hot surface as air becomes available. The inhibiting effect of carbon dioxide makes it more effective than nitrogen or other inert gases. For a fire to be suppressed with nitrogen as a diluent, the concentration of oxygen in air must be reduced to less than that when carbon dioxide is present.

Nitrogen. Nitrogen, as well as other similar inert gases except carbon dioxide, has fire-suppressant capabilities only as a diluent to reduce the concentrations of oxygen in air to less than that necessary for combustion. As already mentioned, an oxygen concentration that will not permit combustion in a carbon dioxide atmosphere may support combustion in the presence of a diluent, such as nitrogen, which does not inhibit chain reactions. Actually, nitrogen may prove harmful, because at high temperatures it combines with carbon to form cyanogen $(CN)_2$ and with oxygen to create nitrogen peroxide and other oxides of nitrogen. These gases, although small in amounts, are extremely toxic and have been injurious and fatal to persons who have inhaled products of combustion in air.

Halogenated Hydrocarbons. Extinguishants of the halogenated hydrocarbon type act solely by inhibiting chain reactions. The effectiveness of such extinguishants depends on which halogen is used and its concentration. The less reactive the halogen, the more effective the compound as an extinguishant. Thus the order of effectiveness from best to poorest would be iodine, bromine, chlorine, and fluorine.

The disadvantage is that certain of them, even in unreacted states, are highly toxic. Methyl iodide is very highly effective as an extinguishant in hydrocarbon-air fires. Unfortunately, its extreme toxicity eliminates it from consideration for widespread use. Halogenated extinguishants are also extremely expensive, generally limiting their use to situations where less costly agents are inadequate. Halogenated compounds are unsuitable as inhibitors of fires involving oxidizers other than oxygen. In addition, high temperatures decompose complex halogenated hydrocarbons into simpler molecules of fuel and oxidizer, which may burn. On a very hot surface, these halogenated agents may themselves become hazards. Effort has therefore been expended on the development of halogenated fire extinguishants which have thermal stability temperatures (minimum decomposition temperatures) of at least 500°F.

Halons 1301, 2402, and 1211 have a special disadvantage. Each has been identified as a stratospheric ozone-depletion agent. An international agreement, the Montreal Protocol on Substances that Deplete the Ozone Layer, requires a complete phaseout of the production of these halogens by the year 2000, except where there is an essential use for which no adequate alternatives are available. Research and development programs for alternative agents are being undertaken.

Halon 1211 is the most generally used halogenated hydrocarbon. (Halon numbers are used only with halogenated hydrocarbons. The first digit on the left indicates number of carbon atoms in the basic molecule, which will always be present. The second digit indicates the number of fluorine atoms; the third, fourth, and fifth digits the chlorine, and bromine atoms; iodine is not used.) Halon 1011 is bromochloromethane (CH_2BrCl); Halon 1301, bromotrifluoromethane (CBF_3); Halon 1202, dibromodifluoromethane (CBr_2F_2); Halon 1211, bromochlorodifluoromethane ($CBrClF_2$), and Halon 2402, dibromotetrafluoroethan ($C_2F_4Br_2$). Carbon tetrachloride was once used extensively as a firefighting agent. However, it produced large amounts of phosgene in a flame, and the adverse effects of its toxicity more than overshadowed its effectiveness.

Foams

Foams suppress fires by cooling and blanketing, sealing off the burning fuel from the surrounding atmosphere. Firefighting foams are classified as either mechanical or chemical. They gradually break down under the action of heat or water. At high temperatures caused by fires, the breakdown rate is greater than at normal temperatures. This affects the density of the foam that can be used. Foams are not suitable for fires involving gaseous fuels, such as propane or butane, or materials that react violently with water.

Mechanical foam is produced by mixing a solution of water and foam concentrate with air or inert gas in a foam maker or special pump. A solution of 3 to 6 percent concentration in water is used. The concentrate may be a "protein" or "synthetic" type made of a suitable chemical compound. Since air itself is an oxidizer, consideration has lately been given to using inert gases, such as the exhaust of internal-combustion engines. In air foam, the ratio of air to water solution depends on the generating equipment used. In fires involving liquid fuels in compartmented spaces, 3 to 1 is generally used; ratios of between 5 to 11 are used in crash rescue operations.

Mechanical foams are also classified according to their expansion ratios. This is the volume of foam produced per unit volume of solution. Commonest are low-expansion foams, which produce 7 to 12 cubic feet of foam per each cubic foot of solution. High-expansion foams have ratios of between 16 and 18; and superhighs, up to 1,000. Low-expansion foams are generally of the protein type: high-expansion types, nonprotein; and superhighs use a suitable detergent.

Heat causes foam to break down. Therefore the expansion ratio must be adapted to the type of fire being fought. A low ratio will be more effective, but less economical. Stability of a foam is sometimes measured by its "quarter drainage time." Standard test apparatus and procedures determine the length of time for 25 percent of a sample of foam to break down into its liquid solution. Additives sometimes improve breakdown resistance for specific types of fires. Protein concentrates sometimes have additives which make them more suitable for use on alcohol or similar fires. Care must be taken to ensure that other extinguishants used concurrently with foam do not cause the foam to break down more rapidly than usual. Dry chemical powders, incompatible foam concentrates, and streams of water frequently cause this to happen.

Mechanical foams for liquid hydrocarbon fires usually have a 4-to-1 expansion ratio and 4-minute draining time. For volatile fuels such as gasoline, coverage is 0.4 to 0.5 gpm per square foot of burning surface; half that amount can be applied in tanks containing low-volatility fuel oils.[87]

Reignition of the flammable material may occur after the foam breaks down. Newly developed foam solutions therefore sometimes contain agents which cause the foam to produce a film on the surface of the fuel after breakdown. This film separates the fuel from air and prevents reignition.

Chemical foams are not as extensively used as mechanical foams, being approved for Class B fires almost exclusively. They are generated by mixing two chemicals, generally sodium bicarbonate and aluminum sulfate (or their solutions), a foaming agent, and water. The chemical reaction produces carbon dioxide, which causes expansion of the solution into foam. Chemical foams are usually comparatively thick, with expansion ratios of 9 or 10 to 1. They break down less rapidly than mechanical foams; however, their stabilities depend to a great degree on a water or solution temperature, with 60° to 85°F as the optimum range.

Light Water

The Navy developed an aqueous film-forming foam (AFFF), "light water," which it uses in conjunction with Purple-K (potassium bicarbonate) powder to replace protein foam. Light water is a mixture of complex fluorocarbons. It retards combustion by forming a vapor-securing film on the surface of flammable liquids. The Navy found the combination of light water and Purple-K powder to be the most effective agent combination available for fighting oil and engine fuel spills.

Solid Extinguishants

A common instruction for combating Class B fires, such as those involving automobile oil or grease, is to cover them with dirt or sand. Either of these blankets the fuel, separating it from the air it needs for combustion. One advantage of any solid is its ability to remain on a burning surface without flowing off.

Dirt or sand may serve in emergencies, but other solids are more effective. Solid firefighting agents may provide all the mechanisms for extinguishment but are less effective than liquids on Class A fires. For metal fires, cooling by solids is not of importance. Liquids are better in this respect, because the heat of vaporization absorbs more heat than does sensible heating of solids. Since solid extinguishants reduce temperatures only slightly, their effectivity on metal fires depends principally on the impermeability of the cover they produce. However, breaking of the solid cover permits reignition of the fire. For this reason, solids are frequently used in conjunction with other types of extinguishants, which may not themselves be adequate to suppress the metal fire but will prevent it from restarting.

Sodium and potassium bicarbonate are the principal solid extinguishants in use for liquid fuels. Both solids in water decompose to carbon dioxide and simpler salts. The effects of carbon dioxide and water in blanketing, cooling, and inhibiting combustion have been mentioned previously. In addition, the sodium and potassium themselves are powerful inhibitors which interfere with the chain reaction involving carbon. Both displace carbon in the reactions which take place in the flame. Of the two, potassium in the carbonate provides greater interference and is twice as effective as is the sodium.

Other Fire Extinguishants

The unsuitability of water, water solutions, inert gases, halogenated hydrocarbons, and foams on metal fires has lead to the development of other extinguishants. Special substances have been formulated for use on fires involving specific metals. One of these is trimetoxyboroxine (TMB). It is a colorless liquid, generally used in a methyl alcohol solution. The methanol ignites on a magnesium fire and the TMB decomposes to form a molten boric oxide coating on the metal. This blankets the burning metal from air. The TMB-alcohol solution is flammable; therefore, it is necessary that care be taken to avoid spread of the fire to nonmetallic materials. Besides for magnesium, TMB can also be used on zirconium or titanium fires.

Numerous other proprietary extinguishants are manufactured for specific types of metal fires. Met-L-X, Met-L-Kyl, and Lith-X are other dry-powder agents which have been used.

Unsafe Suppressants

The use of certain suppressants under wrong conditions may be hazardous. For example, water streams on certain fires are dangerous. They should not be used on very hot or burning magnesium because of the violent reactions. The water decomposes to release hydrogen, which will burn explosively as it diffuses into air. Other materials with which water will burn are listed next. These materials need not be burning to react; however, firefighters must be careful with their use of water in the vicinity of these materials.

WATER-REACTIVE CHEMICALS		
Acetyl bromide	Lithium	Sodium
Acetyle chloride	Phosphorus oxychloride	Sodium amide
Aluminum borohydride	Phosphorus trichloride	Sodium hydride
Calcium	Potassium	Sodium hydroxide, solid
Calcium oxide	Potassium peroxide	Sodium peroxide
Diborane	Potassium hydroxide, solid	Sulfur chloride
Dimethyl sulfate	Rubidium	

Use of carbon dioxide on hot steam lines may be dangerous; the very cold carbon dioxide can cause carbon steel to fracture so that the steam in the line is released.

Certain extinguishing materials may be conductive and therefore dangerous when used on electrical fires or materials near high-voltage electrical equipment. Tests on 550-volt ac live equipment with calcium chloride and foam caused arcing. When a stream of water hit a metal plate, current was conducted over the following:

Water with calcium chloride	6 feet
Water with soda acid	4.5 feet
Foam	5 inches
Water	2 inches

Magnesium will also burn in carbon dioxide. Carbon dioxide will therefore have no effect on magnesium fires, and its use will create an additional hazard. Titanium and zirconium burn in carbon dioxide and nitrogen when ignited under specific conditions.

Extinguishants of exotic fuels and for specific purposes must therefore be selected with care to ensure they are proper and adequate for the service required and will not increase fire rates and intensities.

EXTINGUISHING SYSTEMS

Extinguishers can be portable, mobile, or fixed automatic systems. Each has advantages. Fixed installations must be custom made, usually for specific installations and types of material to be protected. Portable and mobile systems are, of course, flexible, can be changed easily, and can employ equipment that is in very wide use.

Fixed Automatic Systems

Time is critical when an accidental fire occurs. It has therefore been found highly advantageous to use fire extinguishing systems that can sense the presence of a fire and then take prompt action to extinguish it. Such systems are especially valuable where fires can spread rapidly, where personnel may be present only infrequently—which are especially dangerous—where materials are located that could produce highly toxic gases, or in high-rise or other buildings difficult for firefighters to reach.

The study by the National Fire Data Center[88] also indicated the results of an analysis of the effectiveness of automatic sprinklers in structural fires in Ohio. Generally, the use of sprinklers resulted in substantial reductions of fire losses. The average amounts of the savings depended on the type of property involved. Losses in manufacturing and basic industry plants were 2.5 and 2 times greater in plants without sprinklers than in those that did have them. In buildings used for educational purposes the ratio was far more significant. In this 1978 report, the average damage per fire in a sprinklered building was only $70; in an unsprinklered building it was $9,572, or 136.7 times as great.

Sprinkler systems using water are the most common and most important of fire extinguishing devices. Sprinkler heads may be closed or open. When a fire takes place, a fusible element in the head melts, permitting flow to the sprinklers. The heads are temperature rated, with the ratings depending on the maximum normal temperature near the ceiling of the room. The heads are color coded according to their operating temperatures.

In the wet pipe system, the piping up to the sprinkler heads is filled with water. Flow begins from the head as soon as the fusible element breaks. The constant presence of the water makes this type of system usable only in climates where there is no danger of freezing: in warm climates or in heated areas. Otherwise, freezing of the water might burst the lines.

The dry pipe system uses air under pressure between the sprinkler heads and a special dry pipe valve. When a sprinkler opens, the air pressure drops, water pressure forces the dry pipe valve open, and water flows through the lines and out the sprinkler head. Because of the time required for water to flow through the lines, reaction time after the fusible link melts is greater than in a wet pipe system. This arrangement can be used in cold or outdoor locations.

The deluge system generally delivers greater amounts of water than other systems and is therefore used in high-hazard areas. Sprinkler heads are always open, with air at atmospheric pressure back to the deluge valve. The deluge valve is opened by a suitable detection system or by an operator. A preaction deluge system using closed heads and dry pipes is filled with air at either atmospheric pressure or a pressure slightly higher. It is activated by a separate fire detection system. A high-speed deluge system with low-pressure water in the piping is also available where rapid response is required.

Carbon dioxide systems, generally used for local applications where water or foam might prove too damaging, dispense the gas when the system is activated by a fire detection device. They must be used in conjunction with an alarm system to alert personnel that the carbon dioxide hazard of asphyxiation exists. Foam can also be applied through piped systems, generally to protect flammable liquid tanks and vats. Water is mixed with a foam solution or powder when the system is activated by a fire detection device. The mixture forms the foam to cover the fire and surrounding areas. Most recent are piped systems to dispense a halogenated hydrocarbon, generally Halon 1211 or 1301. The advantages of this type of system are many: where gas is effective, it acts rapidly, causing none of the damage of water or foam, and it is operable in cold weather. Halon 1301 is safe to humans in concentrations up to 10 percent by volume in air for exposures up to 20 minutes. (Above 900°F it decomposes to give off hydrogen fluoride, hydrogen bromide, and free bromine.) The principal limitation is that halogenated hydrocarbons act only on fires involving organic materials.

A recent development in sprinkler systems is the "quick-response" sprinkler. This is basically a standard spray sprinkler with a fast-response operating mechanism. This design came from the recent market in residential sprinklers. A more sensitive heat-responsive element replaces the actuating mechanism of standard spray sprinklers. NFPA standards permit the use of quick-response sprinklers throughout buildings of light- and ordinary-hazard occupancy.

The worst problem with sprinkler systems is that personnel forget to reopen the valves they use to shut off the system when they want to perform maintenance or repair. A commentator pointed out: "In the period 1970–1975, there were 471 failures involving protection systems out of a total of 3,105 recorded fire events. Fires in which

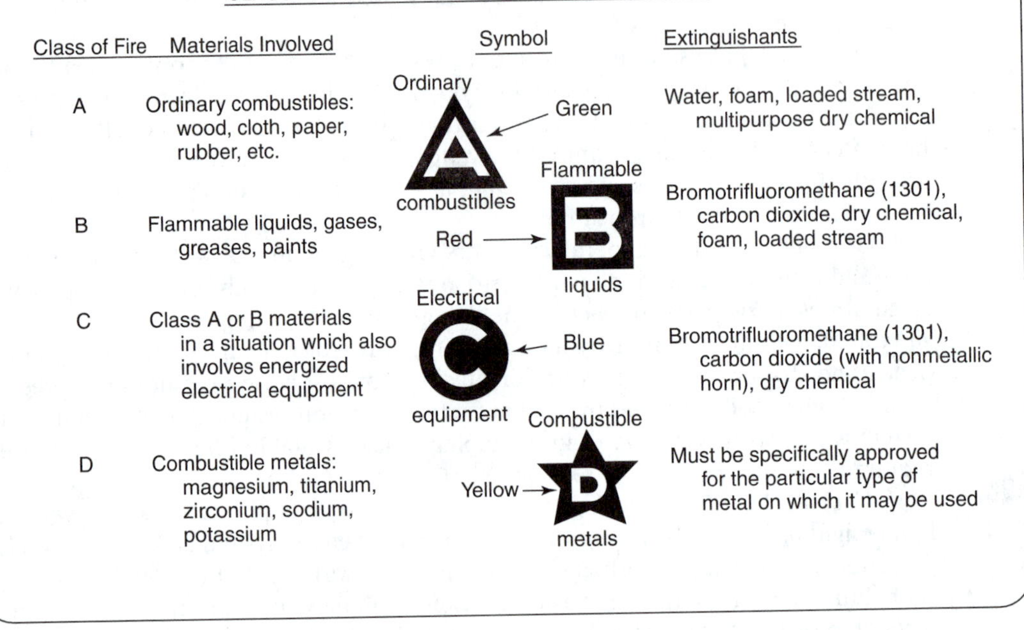

RATING PORTABLE FIRE EXTINGUISHERS

Portable (and mobile) fire extinguishers are rated in two principal ways: type of fire for which they are suitable, and extinguishing capacity.

CLASSIFICATION ACCORDING TO TYPE OF FIRE

Class of Fire	Materials Involved	Symbol	Extinguishants
A	Ordinary combustibles: wood, cloth, paper, rubber, etc.	Ordinary combustibles (A), Green	Water, foam, loaded stream, multipurpose dry chemical
B	Flammable liquids, gases, greases, paints	Flammable liquids (B), Red	Bromotrifluoromethane (1301), carbon dioxide, dry chemical, foam, loaded stream
C	Class A or B materials in a situation which also involves energized electrical equipment	Electrical equipment (C), Blue	Bromotrifluoromethane (1301), carbon dioxide (with nonmetallic horn), dry chemical
D	Combustible metals: magnesium, titanium, zirconium, sodium, potassium	Combustible metals (D), Yellow	Must be specifically approved for the particular type of metal on which it may be used

FIGURE 22–10 Rating Portable Fire Extinguishers

sprinkler control valves were shut alone produced over 15 percent of the total loss sustained during the period."[88] In addition, the National Fire Protection Research Foundation conducted a study to determine means to lessen sprinkler reaction response times.

Portable Extinguishers

Portable extinguishers are described in Figs. 22-10 and 22-11. Portables might be usable for from only 8 seconds to about 1 to 1.5 minutes. Depending on the type, they might be of Class A, B, C, or any combination of these.

Extinguishers should be placed only after careful consideration of the fire hazards present and marked with the types of fire against which they would be effective and should be used. Personnel who might be called upon to operate these extinguishers should be familiar with both their capabilities and limitations. They should not be located where they could not be usable because they might be blocked, too high, or too heavy. Personnel should make no attempt to use them for purposes for which they

CLASSIFICATION ACCORDING TO CAPACITY

Portable (and mobile) extinguishers are capacity-rated according to the area for which they are considered capable of providing protection. For example, at one time a portable extinguisher designated "1A" would constitute one Unit Of extinguishing capacity for a Class A fire. A 1A unit would, supposedly, protect an area of 2,500 square feet of light-hazard occupancy and 1,250 square feet for ordinary hazard. These square footages were based on the assumption that a fire starting at some point within such areas could be controlled initially by a rated extinguisher. (They did not indicate that the extinguisher could control a fire the size of the square footage indicated.) To ensure that fires could be fought in their incipient stages, maximum distances are stipulated so that little time is lost in obtaining and using extinguishers.

Occupancies of areas that could be involved in fires are defined as:

Light hazard: areas where fires that occur would be only of minimum severity because of the relatively small amount of combustible material present, such as in offices, telephone exchanges, classrooms, and assembly halls.

Ordinary hazard: includes light-manufacturing buildings, parking garages, and warehouses not classified as extra hazard, where fires that occur would be of average severity.

Extra hazard: includes facilities such as paper and textile mills or processing plants; aircraft servicing shops, paint, dipping, and other flammable-liquid-handling areas, where fires that occur would be extrasevere because of the quantity or nature of combustible material present.

At one time it was also considered that units rated 2A, 3A, and 4A could be used to provide protection for areas 2, 3, and 4 times that of a 1A unit. However, similar fires could start in different-sized areas. In some instances protection was governed by the rapidity with which an extinguisher could be obtained and used, and by the capacity. (There have been numerous cases where control of small fires was lost because of inadequate extinguisher capacity or because a fire flared up again after an extinguisher had been emptied.) OSHA standards have therefore stipulated coverages as follows.

Basic minimum extinguisher rating for area specified	Maximum travel distances to extinguishers (ft)	Areas to be protected per extinguisher		
		Light-hazard occupancy (ft^2)	Ordinary-hazard occupancy (ft^2)	Extra-hazard occupancy (ft^2)
1A	75	3,000	Note 1	Note 1
2A	75	6,000	3,000	Note 1
3A	75	9,000	4,500	3,000
4A	75	11,250	6,000	4,000
6A	75	11,250	9,000	6,000

Note 1: The protection requirements specified may be fulfilled by several extinguishers of lower ratings for ordinary or extrahazard occupancies.

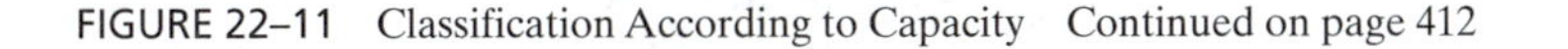

FIGURE 22–11 Classification According to Capacity Continued on page 412

Extinguishers for Class B fires were formerly rated in two ways. One unit of extinguishing capacity (1B) was supposedly adequate for each 625 square feet of floor area to be protected agains a Class B fire, or against 1 square foot of flammableliquid surface in a deep tank. The deep-tank criterion is still used in the OSHA Standards; however, the term "deep-tank" has been changed to "appreciable depth," which is defined as "a depth of a liquid greater than one-quarter inch." The OSHA standards do not set area protection requirements for Class B fires as they do for Class A, but are limited to maximum travel distances:

Type of hazard	Basic minimum extinguisher rating	Maximum travel distance to extinguishers (ft)
Light	4B	50
Ordinary	8B	50
Extra	12B	50

Extinguishing capabilities are determined by test in accordance with criteria set by the Underwriters' Laboratories. Capabilities depend not only on the size of the extinguisher, but principally on the extinguishant used. Thus similarsized extinguishers can have different ratings, depending on the principal material used and any additives the manufacturer provides to improve his product. (Ratings of extinguishers can be obtained from manufacturers or their representatives.) Thus an extinguisher may have a rating of 2A, 4B, or C, indicating capacities against Class A and B fires. C ratings lack numerical values, since they indicate only that they may be used against Class A or B fires which also involve energized electrical equipment or wiring.

To ensure that extinguishers with capabilities suitable for the classes of fires that might arise are located in the hazard areas and used as intended, they should be marked prominently. Fire-class decals are available at nominal cost from organizations such as the National Fire Protection Association.

FIGURE 22–11 Continued

might not be suitable. A regular inspection and maintenance program must be undertaken to ensure the extinguishers are always in effective condition.

Mobile Equipment

Large extinguishers are too heavy to be carried and therefore are often wheel-mounted. These units are advantageous where fairly large capacities are required and where the extinguishers must be moved or wheeled to areas to or through which trucks cannot pass. Mobile equipment has capabilities of projecting streams much farther than portable equipment. Operators can therefore reach fires at greater and safer distances. In some instances fires inaccessible to persons with portable extinguishers can be reached.

Mobile equipment is sometimes made up of two extinguishers, one containing Purple-K dry chemical and the other light water. These are applied simultaneously through twin nozzles.

(text continues on p. 419)

SOME PORTABLE FIRE EXTINGUISHERS AND HOW THEY WORK [a]

Water Type

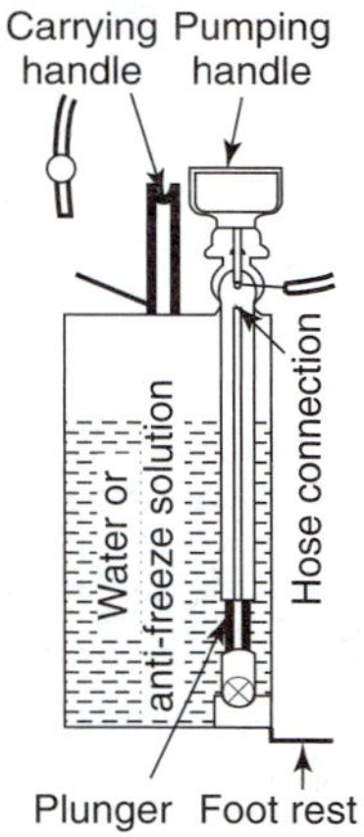

For Class A fires

Operation. May be operated intermittently (therefore useful on such operations as cutting and welding), but cannot be carried and operated at the same time. The operator should place his foot on the foot bracket and operate the pump handle with 6 to 8-in. strokes. The pump discharges water on both the up and down strokes. Direct the stream at the base of the fire, then follow up after the flames while working from side to side or around the fire if possible.

Maintenance. Periodically check the water level and test the pump by stroking several times while discharging the liquid back into the tank. Inspect the condition of operating parts annually and oil the piston-rod packing. Check the tank and foot bracket for corrosion. After use, simply refill with water or antifreeze solution.

Caution. Protect from freezing.

Range. 30—40 feet

Pump-tank fire extinguisher.

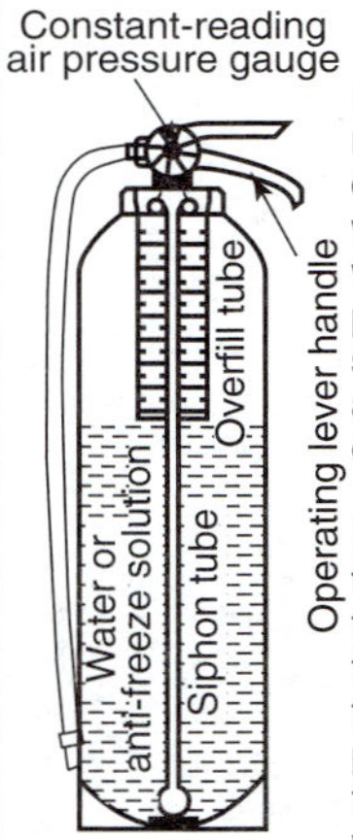

For Class A fires

Operation. Can be operated intermittently by removing locking ring and squeezing combination operating lever-handle. Extinguisher is pressurized with air or inert gas through a hose line equipped with an automobile-type air chuck. Charging pressures vary from 90 to 150 psi. Direct the stream at the base of the fire, then follow up after the flames while working from side to side or around the fire if possible.

Maintenance. Check pressure, hose, and nozzle frequently. When charging extinguisher, fill only to water-level mark so there will be sufficient room for pressurized air. Some models have an over fill tube that ensures filling to the proper level. When pressurizing, the gauge should indicate just above the full mark; then, when the pressure equalizes, it will not read less than full. After charging, let it stand for 24 hr to check for leaks.

Caution. Protect from freezing. Be sure shell is the corrosive resistant type before adding antifreeze charge. Do not refill with loaded stream charge unless extinguisher is designated as loaded stream type.

Range. 30—40 feet

Stored-pressure water fire extinguisher. May be labeled "loaded stream" if charged with a special alkali-metal salt solution instead of water.

Note: The above two types are seldom used, although they are legal.

FIGURE 22–12 Some Portable Fire Extinguishers and How They Work Continued on page 414

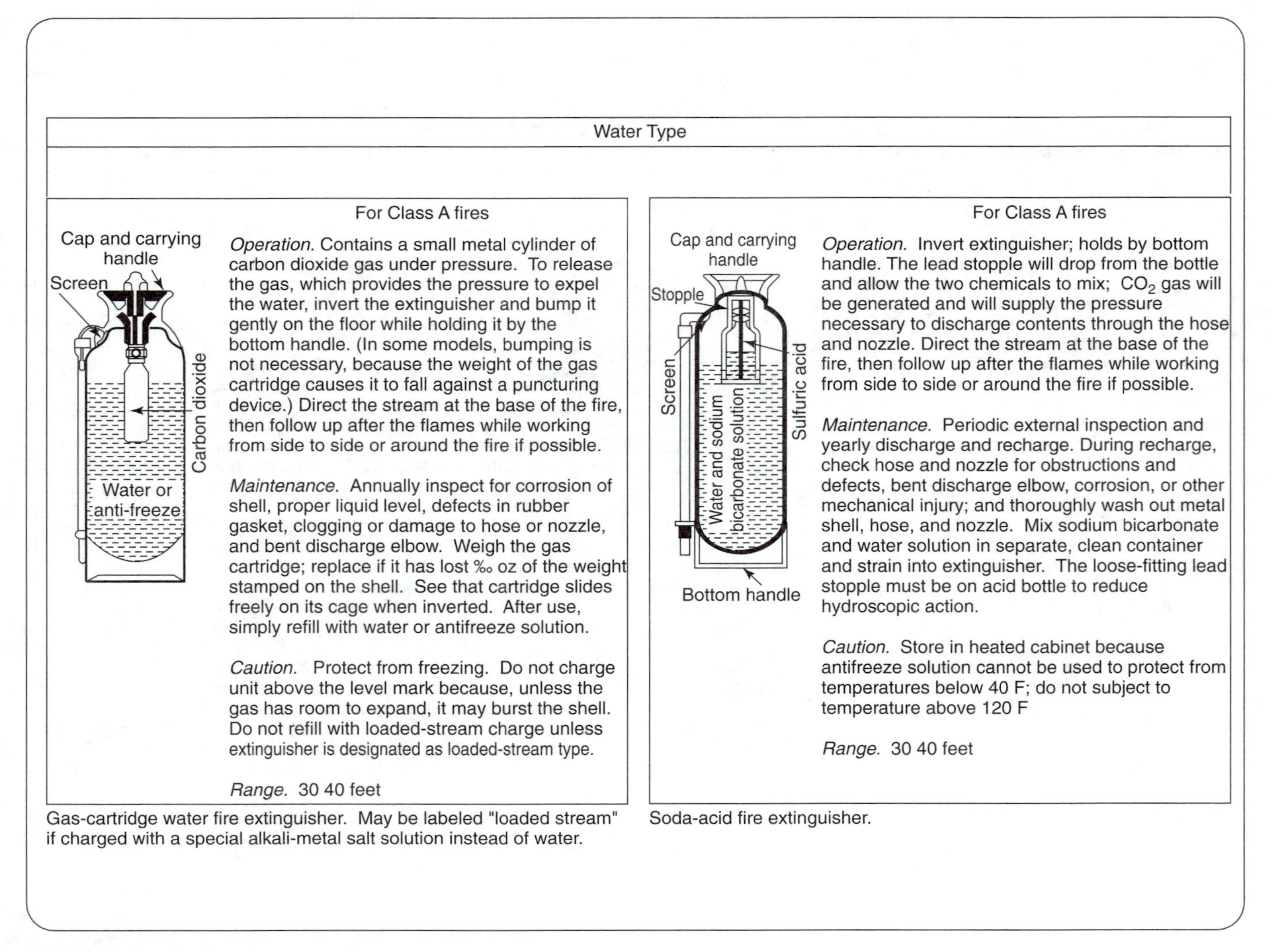

FIGURE 22–12 Continued

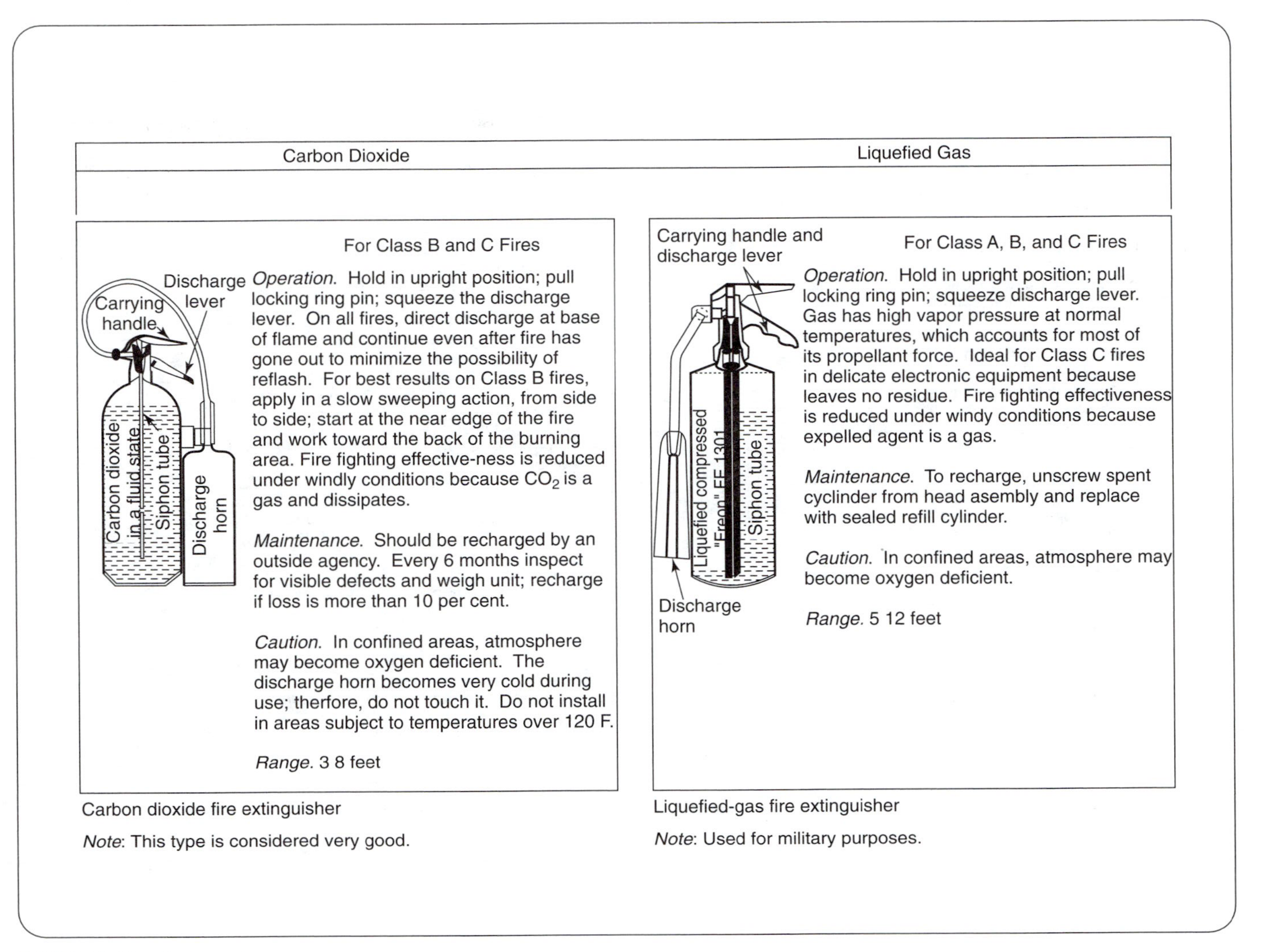

FIGURE 22–12 Continued

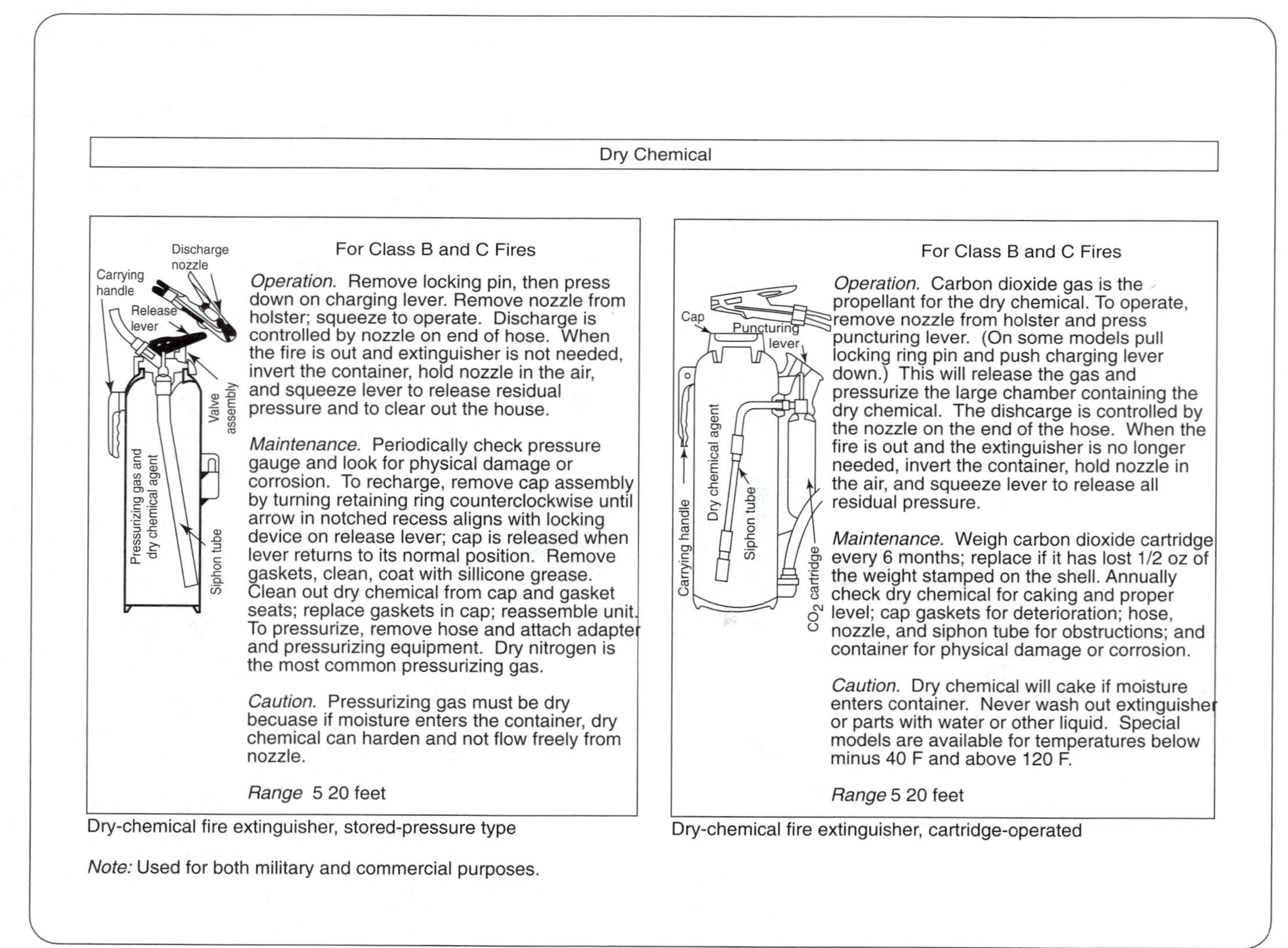

FIGURE 22–12 Continued

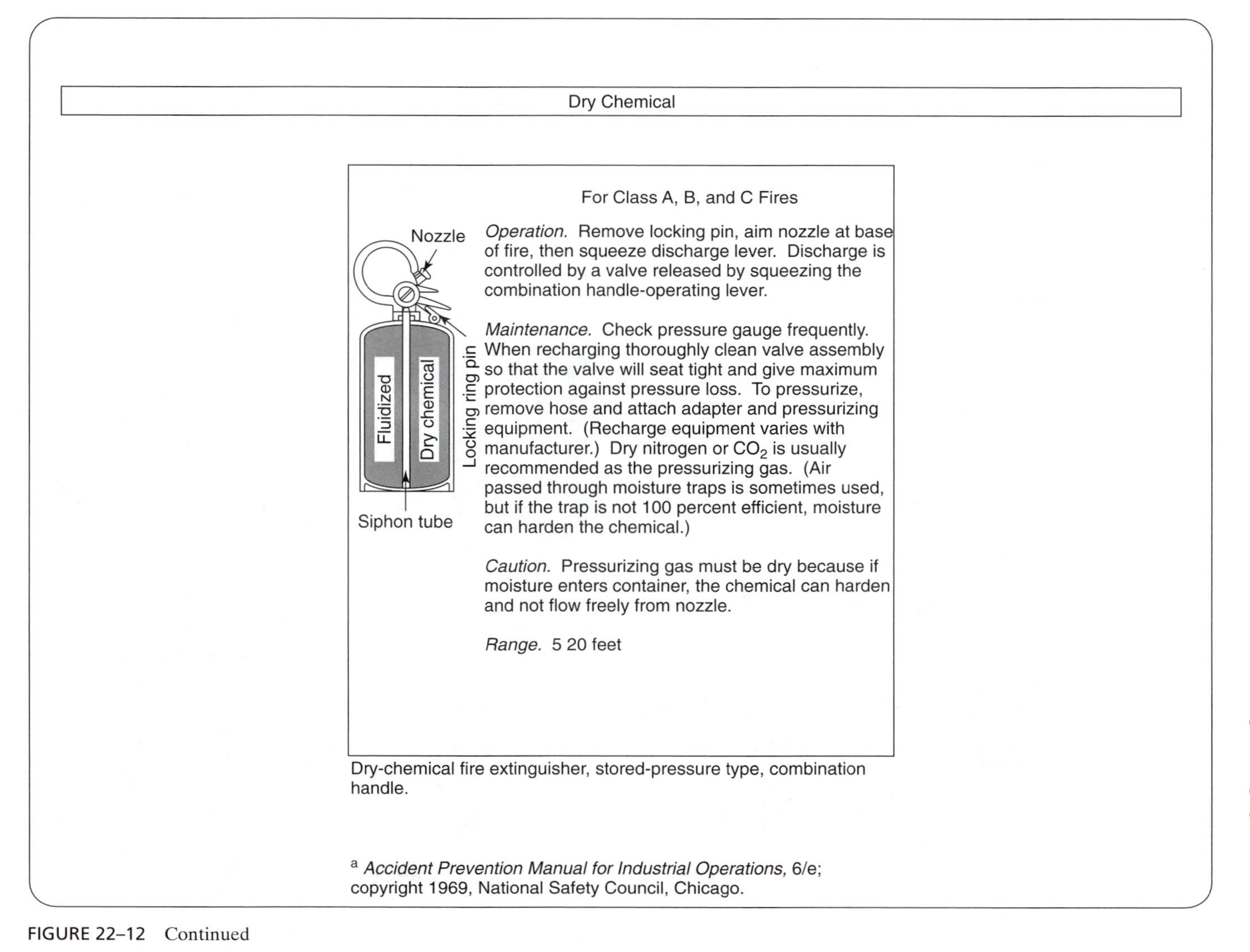

Dry Chemical

For Class A, B, and C Fires

Operation. Remove locking pin, aim nozzle at base of fire, then squeeze discharge lever. Discharge is controlled by a valve released by squeezing the combination handle-operating lever.

Maintenance. Check pressure gauge frequently. When recharging thoroughly clean valve assembly so that the valve will seat tight and give maximum protection against pressure loss. To pressurize, remove hose and attach adapter and pressurizing equipment. (Recharge equipment varies with manufacturer.) Dry nitrogen or CO_2 is usually recommended as the pressurizing gas. (Air passed through moisture traps is sometimes used, but if the trap is not 100 percent efficient, moisture can harden the chemical.)

Caution. Pressurizing gas must be dry because if moisture enters container, the chemical can harden and not flow freely from nozzle.

Range. 5 20 feet

Dry-chemical fire extinguisher, stored-pressure type, combination handle.

[a] *Accident Prevention Manual for Industrial Operations,* 6/e; copyright 1969, National Safety Council, Chicago.

FIGURE 22–12 Continued

FIRE PREVENTION MEASURES

1. Keep fuels separate from oxidizers.
   a. Locate fuel tanks, lines, and pressure bottles as far as possible from those for oxidizers and from the sources of ignition.
   b. Provide suitable containers for storage of fuel and oxidizers. Provide containers for fuel, oil, gasoline, cleaning compounds, and other flammables with covers to minimize contact with air.
   c. Containers should be suitably marked. Where container position is important, provide warnings and arrows to indicate proper attitude.
   d. Provide blankets of inert gas for extremely combustible or pyrophoric materials.
   e. Make certain that leaking fuel will not fall on an oxidizer (or vice versa), or on a hot surface. Provide containment devices to hold fuel or oxidizer leakage.
   f. Common drainage or vent lines should not be used for fuels and oxidizers.
   g. Provide means outside of a possible fire zone to shut off flow of fuel or oxidizer in case a fire should occur.
2. Use the least flammable and least hazardous material possible which will suit the prescribed operation requirement. Materials should be either nonflammable or flame resistant.
3. For equipment, lines, tanks, seals, and lubricants use only materials compatible with the fuel or oxidizer to be contained.
4. Avoid usage of those materials or their oxides which might act as catalysts in proximity to fuels.
5. Avoid having sources of ignition in the vicinity of a fuel or in an atmosphere that might be flammable.
6. Design or provide electrical equipment according to National Electrical Code provisions for hazardous atmospheres.
7. Provide adequate grounds for surfaces from which accumulations of static electricity might discharge to ignite flammables.
8. Make certain that vent or overflow lines from fuel tanks do not discharge near hot surfaces, open flames, or near oxidizers.
9. Insulate or protect hot surfaces, which might be sources of ignition, against spillage or leakage of fuel.
10. Avoid having exhaust gases or heat from combustion equipment pass near flammable materials.
11. Avoid use of quick-opening or quick-closing valves on oxidizer or monopropellant lines to prevent temperature increases by adiabatic compression. This might initiate combustion of a contaminant fuel in the oxidizer, or decomposition of the monopropellant.
12. Provide ullage space for fuels or oxidizers to permit them to expand at increased temperatures without overflowing.
13. Provide suitable insulation or cooling to reduce or eliminate evaporation of fuel or oxidizer.
14. Whenever possible, use welded instead of threaded, bolted, or riveted connections and seams to minimize leakage.
15. Connectors should be kept to a minimum, not carry any load, and maintain their integrity under all designed conditions.
16. Do not locate fuel or oxidizer tanks where they might be ruptured or damaged easily by impact of vehicles, handling equipment, or projectiles.
17. Limit the size of fuel or oxidizer tanks so that if one leaks, bursts, or catches fire, loss and damage will be constrained.
18. Fuel and oxidizer lines should be supported and secured to avoid bending loads, vibration, expansion, and hydraulic loads without failure.
19. Provide antisloshing devices in fuel tanks to minimize the formation of easily flammable mists and foams.
20. Make certain leaks or spills of fuels or oxidizers will not contaminate potable water supplies. Provide berms and dikes around large, above-ground fuel tanks to contain the liquid in case of a leak or tank rupture.
21. Paint fuel or oxidizer tanks white or coat them with reflective surfaces to minimize absorption of solar energy.

FIGURE 22–13 Fire Prevention Measures

22. In a zone where fire might take place minimize or avoid passage of critical lines whose loss might be catastrophic to the entire system.
23. Provide suitable fire suppression equipment in areas where flammable atmospheres may exist.
24. Provide flame suppressers, and fire barriers between spaces where fire might occur to prevent its spread or contain the damage it might create.
25. Provide automatic fire suppression equipment wherever possible. Where manual equipment is used, provide detection systems with warning signals that are easily understood.
26. Prohibit smoking in all areas where flammable liquids are present, unless designated as safe to do so.
27. Make certain that all personnel involved in handling or use of flammables are aware of their characteristics and hazards.
28. Provide clean-up and decontamination procedures and equipment so that leakage and spills of fuels or oxidizers can be cleaned up immediately.
29. Avoid welding on lines, tanks, or equipment filled or contaminated with combustible fluids. Weld only after the lines, tanks, or equipment have been cleaned.
30. Make certain that valve outlets, seats, and caps on pressure bottles containing oxidizers, such as compressed air or oxygen, are always clean and free of combustible materials.
31. Check atmospheres in enclosed spaces for the presence of flammable gases before permitting personnel to enter, and again before starting welding, maintenance, or repairs.

FIGURE 22–13 Continued

General fire prevention measures are listed in Fig. 22-12, while the following Figs. 22-13 and 22-14 are checklists to assist in reviewing for fire hazards. Checklists are provided in Figs. 22-15 and 22-16 for chemical reactions. These are provided principally because many of the reactions, but not all, lead to production of gases and fires. Figure 22-17 is a checklist for reviewing conditions on all types of chemical reactions.

BIBLIOGRAPHY

[86] *Fire in the United States*, U.S. Fire Administration, U.S. Dept. of Commerce, Washington, D.C., Dec. 1978.

[87] U.S. Air Force, *Handling and Storage of Liquid Propellants,* AFM 160-39.

[88] M. J. Miller, "Risk Management and Reliability," *Third International System Safety Conference,* Washington, D.C., Oct. 1977.

EXERCISES

22.1. What conditions are required for a fire to start?

22.2. What is a fuel? Give ten types of fuels.

22.3. What is an oxidizer? Which is the most common one? Name others.

22.4. Define and explain flash point, fire point, and autoignition temperatures.

22.5. What are the lower flammability limit and flammability range? What is a mixture called which has too little air present to burn? Too much air?

Hazards Checklist and Fires

Possible Effects

Injury to personnel:
- Burns
- Toxic gas and smoke inhalation
- Other heat and high-temperature effects
- Deprivation of oxygen for breathing

Destruction of material and resources:
- Carbonization and contamination of material
- Equipment rendered inoperative

Damage to the environment:
- Production of corrosive contaminants
- Destruction of wildlife and vegetation
- Production of airborne particulate matter

Possible Causes

Fuel/oxidizer mixture with ignition source:

Fuels:
- Heating fuels
- Engine fuels
- Paints and varnishes
- Wood and wood products
- Welding and process gases
- Lubricants
- Rubber and plastics
- Furnishings and upholstery
- Clothing
- Refuse and trash
- Vegetation
- Other organic materials
- Hydraulic and coolant fluids
- Normally low-combustible materials in the presence of strong oxidizers or high temperatures
- Normally nonflammable metals in finely powdered form
- Grain dust and other particulate matter
- Hydrogen from charging batteries
- Products of incomplete combustion of organic materials

Oxidizers:
- Oxygen in air
- Oxidizing compounds
- Oxidizing gases

Ignition source:
- Open flames
- Arcs and sparks
- Hot surfaces
- Lightning strikes
- Spontaneous ignition
- Adiabatic compression
- Hypergolic mixtures
- Pyrophoric mixtures
- Water-sensitive reactive materials

FIGURE 22–14 Hazards Checklist—Flammability and Fires

CHECKLIST—FIRES AND FIRE SUPPRESSION

1. Has the plant been surveyed to determine what combustible materials are present and how dangerous each is, whether special safeguards are required, and whether each is adequately controlled?
2. Is any condition present which will cause ignition of any of these combustibles?
3. If any of the combustible materials is a liquid, what is its flash point? Would its flash point put it in the category of a flammable material?
4. If any combustible material is a liquid, is it in a secure container from which it will not spilt or leak and in which it is protected from an ignition source?
5. Is each fuel container located where it is protected from damage which could cause it to leak?
6. Is each fuel container located where any leaking material or vapors will not come in contact with an ignition source?
7. Are combustion and fired heating equipment provided with controls to prevent uncontrolled burning?
8. Does any operation involve the evolution of a gas which might be flammable, such as hydrogen when a battery is overcharged?
9. Does the plant contain monitoring and warning devices which will indicate the presence of a flammable gas, or a fire, or of a condition which might precede the occurrence of a fire?
10. If the occurrence of a fire is possible, have means of fire extinguishment, manual or automatic, been provided for the types of fire which might occur? Are the number and locations of fire extinguishers in accordance with OSHA and local requirements?
11. Is the fire extinguishing equipment serviced, tested, and refilled periodically? If testing a sprinkler system requires it be shut off, is there a means by which the fact that the valve to do so when the test is over is open can be established positively?
12. Have warnings been provided to alert personnel to the presence of flammable materials and the means to avoid fires?
13. Are periodic inspections made by supervisory personnel, security personnel, safety personnel, and fire prevention personnel to ensure that the presence of combustible material is adequately controlled?
14. Have all plant personnel been indoctrinated on what actions they are to take in the event of a fire? Is prominent notice given to the meanings of signals which indicate the existence of a fire?
15. Are fire lanes and fire doors kept clear at all times? Are the passages to fire extinguishers clear?
16. Has a survey been made to determine whether any areas in the plant are more susceptible to fire than others and require more than average surveillance and protection?
17. Are containers available in which hazardous materials such as oil-soaked rags can safely be stored? Is there a procedure by which all such containers are emptied periodically?
18. Is any material present, such as an oxidizing material, which can cause a fire which is particularly difficult to extinguish or can have especially adverse affects, such as the generation of toxic gases? Are special precautions given to such materials, and is information on the need for special controls and the effects which might result made common knowledge?
19. Is housekeeping stressed to eliminate the presence of combustible refuse and trash?
20. Are paints fuels, solvents and other such flammables stored in segregated areas, where they are protected and in nonflammable storage compartments?

FIGURE 22–15 Checklist—Fires and Fire Suppression

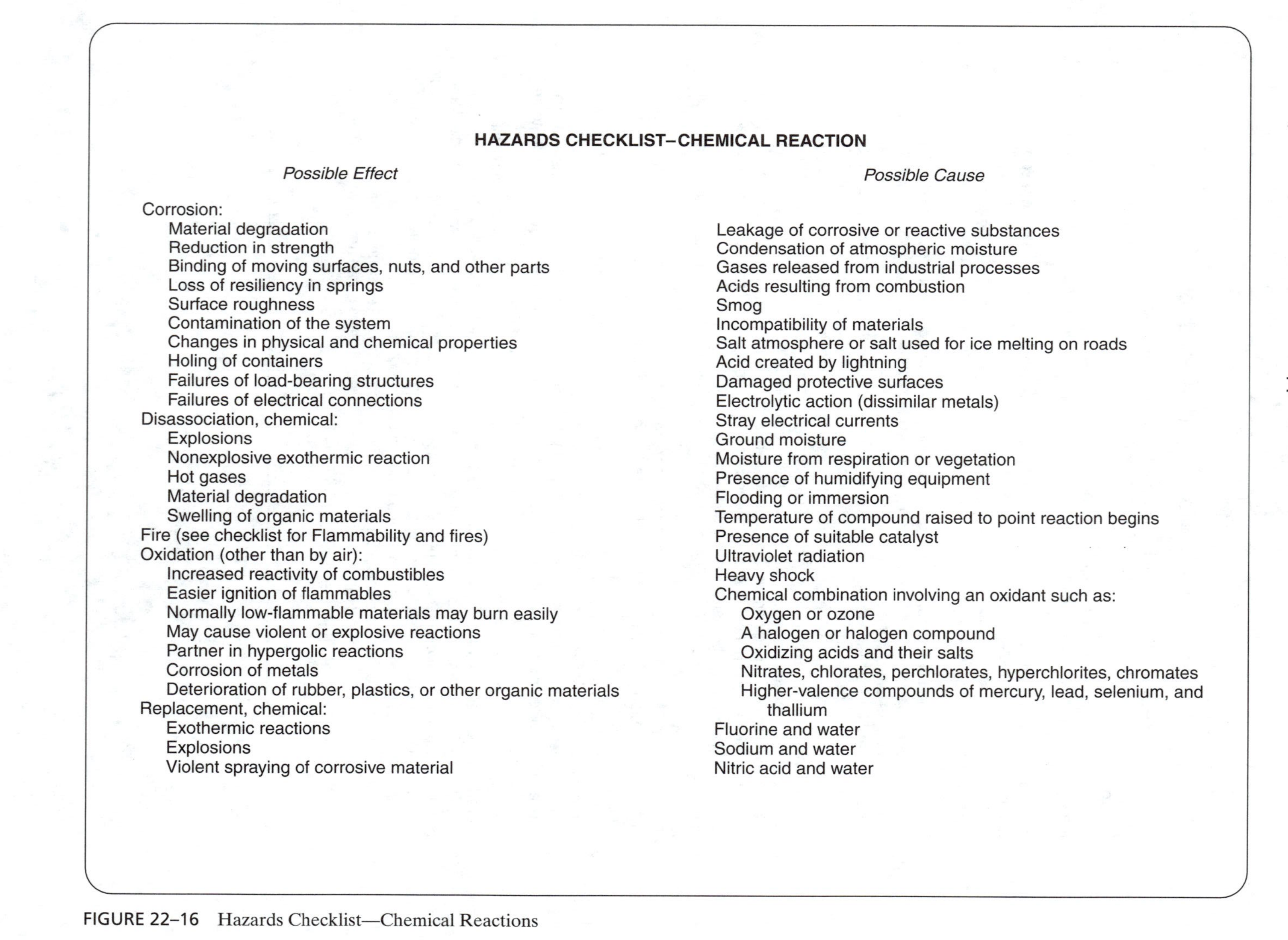

HAZARDS CHECKLIST–CHEMICAL REACTION

Possible Effect

Corrosion:
- Material degradation
- Reduction in strength
- Binding of moving surfaces, nuts, and other parts
- Loss of resiliency in springs
- Surface roughness
- Contamination of the system
- Changes in physical and chemical properties
- Holing of containers
- Failures of load-bearing structures
- Failures of electrical connections

Disassociation, chemical:
- Explosions
- Nonexplosive exothermic reaction
- Hot gases
- Material degradation
- Swelling of organic materials

Fire (see checklist for Flammability and fires)

Oxidation (other than by air):
- Increased reactivity of combustibles
- Easier ignition of flammables
- Normally low-flammable materials may burn easily
- May cause violent or explosive reactions
- Partner in hypergolic reactions
- Corrosion of metals
- Deterioration of rubber, plastics, or other organic materials

Replacement, chemical:
- Exothermic reactions
- Explosions
- Violent spraying of corrosive material

Possible Cause

- Leakage of corrosive or reactive substances
- Condensation of atmospheric moisture
- Gases released from industrial processes
- Acids resulting from combustion
- Smog
- Incompatibility of materials
- Salt atmosphere or salt used for ice melting on roads
- Acid created by lightning
- Damaged protective surfaces
- Electrolytic action (dissimilar metals)
- Stray electrical currents
- Ground moisture
- Moisture from respiration or vegetation
- Presence of humidifying equipment
- Flooding or immersion
- Temperature of compound raised to point reaction begins
- Presence of suitable catalyst
- Ultraviolet radiation
- Heavy shock
- Chemical combination involving an oxidant such as:
  - Oxygen or ozone
  - A halogen or halogen compound
  - Oxidizing acids and their salts
  - Nitrates, chlorates, perchlorates, hyperchlorites, chromates
  - Higher-valence compounds of mercury, lead, selenium, and thallium
- Fluorine and water
- Sodium and water
- Nitric acid and water

FIGURE 22–16 Hazards Checklist—Chemical Reactions

CHECKLIST–CHEMICAL REACTIONS

1. Are corrodable materials protected from exposure to the environment or materials which will cause them to corrode?
2. Is any material present which will react with the moisture in air or any other source of water or moisture?
3. Will even a small amount of any product in use in the plant cause injury to the eye by chemical reaction? If so, are suitable safeguards provided to prevent such injuries?
4. Is any material in use in the plant which will injure skin which it might contact?
5. Is the container of any material which could cause chemical injury, either alone or in combination with another material, provided with suitable identification and warning?
6. Have there been any reports of persons who might be allergic to a material used in the production process? Has action been taken to isolate those persons from that material?
7. If a material could cause injury if taken internally, is a practical antidote available?
8. Is any material present which will react with the oxygen in air to produce a toxic, corrosive, or flammable material which will ignite spontaneously?
9. Is any material present whose molecules will break down in the presence of heat or ultraviolet radiation to produce dangerous products when it is not intended to do so?
10. Is any reference document or referral agency available from which information on the reactive properties of chemicals to be used in the plant can be obtained?
11. Is any electrical equipment operating which can produce ozone by sparking?
12. Are two metals in contact, or which could come in contact, which would cause electrolytic corrosion?
13. Is each liquid or gas compatible with the material of the container in which it is held?
14. Is there any place where the leakage of liquid could result in a dangerous reaction with a nearby material?
15. Is any material present in the plant which is hypergolic (reacts to initiate fire) with any other material?
16. Is any material used which will react with a common substance to produce a toxic or flammable gas?
17. Is any material present which should not be used in an enclosed space or other location where it could be highly dangerous? Are janitorial personnel aware that two cleaning agents should not be mixed?
18. Is there any metal used in the plant which will act as a catalyst and cause chemical compounds to break down and form dangerous products?

FIGURE 22–17 Checklist—Chemical Reactions

22.6. What is the difference between a flammable and a combustible liquid?

22.7. Do liquids burn as such? What is the relationship between volatility of a liquid and its flammability?

22.8. Are sprays and mists easier to ignite than a larger mass of liquid? Why or why not?

22.9. How do most solids burn? How can a solid such as coal be volatile?

22.10. List six categories of ignition sources.

22.11. Explain a chain reaction in a fire.

22.12. Explain how spontaneous ignition occurs. Name some types of solids subject to spontaneous ignition. What is a hypergolic reaction and how does it occur? A pyrophoric reaction?

22.13. Give five principal effects of fire on personnel.

22.14. List seven harmful products of combustion which can be harmful.

22.15. Explain how some fire detection devices work. Is there any such device in your workplace, office, or home? What type is it?

22.16. How are fires classified?

22.17. Give some principal means by which a fire can be suppressed.

22.18. What are some common fire extinguishants?

22.19. What are halogenated hydrocarbons? What does each digit in a Halon number mean? What could be the Halon numbers of the following: CBr_3, CBr_2F_2, and CCl_4?

22.20. Explain how or why a suppressant might be unsafe.

22.21. Explain the advantages of having sprinkler systems. Explain wet, dry, and deluge systems. What is a major reason they sometimes fail to operate?

CHAPTER 23

Explosions and Explosives

On June 2, 1974, a chemical plant in Flixborough, England, exploded, killing 29 people, injuring more than 100, and devastating the town. The results of the explosion were considered to be worse than that of a World War II bombing raid. Nearly 20,000 people had to be evacuated from nearby towns and villages because of toxic fumes. Glass was shattered miles away.

On April 16, 1947, the French ship *SS Grandchamp* blew up at its pier in Texas City, Texas, destroying the city and killing 561 persons. On September 21, 1921, a blast in Oppau, Germany, killed an estimated 600 persons. In each case, and in numerous others that could be cited, the exploding material was ammonium nitrate. Ammonium nitrate is not classified as an explosive; it is a fertilizer.

On March 19, 1937, an explosion destroyed a school in New London, Texas, killing 413 pupils and 14 teachers. The explosion of propane from a tank in the State Fair Coliseum in Indianapolis killed 73 persons in December 1963. In 1944, a tank holding 80 million cubic feet of natural gas as a liquid failed in Cleveland, Ohio, pouring the cold liquid over the surrounding area and into the storm sewers. Its vapors ignited almost immediately and caused 135 deaths and widespread and severe damage. Most of the liquid natural gas vaporized rapidly and burned in the open air. Explosions occurred in the storm sewers where the gas was confined. At such points, streets cracked and manhole covers were blown away.

More recent and notable examples have included those involving the space shuttle *Challenger* and the nuclear plant at Chornobyl. Explosions that have taken place where no explosives were present could be cited endlessly; those mentioned include only a few outstanding examples. See Fig. 23-1 for a description of terms.

INDUSTRIAL USAGE AND PROBLEMS

From the examples cited it can be seen that activities at industrial plants can involve materials, processes, and equipment not considered explosive that can explode with catastrophic effects. Explosives for nonmilitary purposes are used for mining, quarrying, construction, demolition, and seismograph work. Approximately 3.7 billion pounds of explosives (most of it fertilizer-grade ammonium nitrate mixed with oil) and blasting

EXPLOSIONS AND EXPLOSIVES

Explosion: sudden and violent release of large amounts of gas. Damage may result from the rupture and fragmentation of a container, shock wave, heat or fire, or release of toxic gas. A compressed-air tank or boiler rupture due to high pressure is often called an "explosion," and the effects may be the same as those of an explosion due to a chemical reaction. An explosion involving a chemical reaction can be either a "deflagration" or a "detonation."

Deflagration: consists of a rapid reaction during which heat is transferred progressively from a reacting material to another nearby whose temperature is then raised to a point at which it, too, reacts. The rate at which deflagration takes place is high but less than the speed of sound. Large amounts of hot gas are produced, but unless they are confined, no shock wave (blast) will be generated. If the gases are confined, the resultant pressure due to the hot, expanding gases may cause a sudden rupture of the container.

Detonation: if the velocity of reaction through the reacting material reaches sonic or supersonic speed, the explosion is a detonation. A shock wave (blast) will occur even where there is no container. Some detonation velocities are: hydrogen-oxygen, 9,200 feet per second; TNT, 22,800 feet per second; nitroglycerin, 26,200 feet per second.

High-order explosion: the entire mass of reacting material reacts almost simultaneously because of the high reaction velocity.

Low-order explosion: the reaction is slow enough that the first particles of material to react blow the entire mass apart, so that only the original portion of reaction material takes part in the explosion. The smaller pieces produced generally will not explode.

Explosive: substance that is specifically used to produce an explosion because of the high-energy reaction it undergoes. The reaction may be either combustion or dissociation, but in either case it is an exothermic process in which large amounts of gas and heat are produced.

Low explosive: one that normally will deflagrate, and whose reaction rate can be controlled. The first explosives, such as black powder, were used as propellants (to propel rockets, cannon balls, shells, and bullets). Low explosives are therefore frequently called "propellant explosives." They include black powder, smokeless powder, and most solid propellants used in missiles and rockets.

High explosive: will detonate, with the shock wave it generates producing a shattering effect. The shattering power of an explosive is termed its brisance. High explosives can be categorized into bursting explosives, initiating explosives, and boosters. Main or bursting charges are generally very stable chemically and require a lesser explosion to initiate a reaction. This is usually done by firing an initiating explosive, such as lead azide, lead styphnate, or mercury fulminate, which is extremely sensitive to heat or shock. The shock wave that occurs may be used to set off a booster, or in termediate, explosive, which is less sensitive than an initiating type. The booster amplifies the shock wave so that the stable bursting explosive will react.

Explosive reactions: a deflagration is generally a combustion reaction in which a fuel combines with an oxidizer. In some instances, as with black powder, there is also a dissociation reaction first. With black powder, which in its elementary form consists of carbon, sulfur, and sodium nitrate, the sodium nitrate decomposes to provide the oxidizer necessary for combustion. Most high-explosive reactions involve highly energetic exothermic dissociations in which a complex molecule breaks down into simpler molecules, principally gases that expand rapidly because of the large amounts of heat generated.

TNT equivalence: weight of trinitrotoluene (TNT) which will produce the same effect as that generated by the explosion of another material. The common measure used for comparison is the peak overpressure (maximum instantaneous pressure produced in a shock wave). Nitroglycerin has a TNT equivalency of 1.42; ammonium nitrate, 0.57.

Permissible explosive: mining regulations stipulate that explosives used must not be capable of igniting methane, thus limiting the temperatures of the explosion products. Explosives that meet this requirement are permissibles.

Hazard classification: the categorization of explosives by characteristics so that those which are similar can be stored or transported together. The Department of Transportation classifies explosives into four categories: nonacceptable, and Classes A, B, and C. Nonacceptable explosives are those which require such special handling because of their unstable characteristics that they may not be shipped by

FIGURE 23-1 Explosions and Explosives

a common carrier. Class A explosives are acceptable types which detonate or otherwise present a maximum hazard. Class B includes explosives which are flammable hazards. Class C includes those rated as minimum explosive hazards. Compliance with current D.O.T. regulations, especially packaging and marking, is required before any explosive may be shipped. The Department of Defense classifies explosives differently, in eight classes.
Magazine: any building or structure, except an operating building, designed and used for the storage of explosives.
Barricade: an artificial or natural screen which will deflect any shock wave so that it does not reach any inhabited building, passenger railway, public highway, or magazine where other explosives are located.
Quantity-distance tables: the minimum distance that must separate the location of explosives and inhabited buildings, public traffic routes, railroads, and other explosives is dependent on the quantity and class of explosive, and whether any barricade separates them. Quantity-distance tables show these minimum distances.

FIGURE 23-1 Continued

agents are used annually. Greatest use of high explosives has been in coal mining (about one-third of the total), then in quarrying and nonmetal mining, metal mining, and seismographic work. (Recent reductions in mining and oil exploration have reduced the numbers of such activities.) A small percentage is still used for other purposes.

In industrial plants, the consumption of explosives is relatively minor, small amounts being used for explosive metal forming and for explosive tools. For the last two purposes low explosives are used, in which deflagration, and not detonation, is the explosive process. However, under certain conditions detonations will occur with materials that would usually burn at controllable rates.

Deflagrations

A large mass which burns without confinement will generally create a fireball whose size depends on the volume of flammable mixture present. A low-magnitude shock wave, somewhat like a loud "POOF," may be emitted.

Confinement will restrict expansion of the gases so pressure will build up. If the container in which this occurs is not strong enough to hold the pressurized gas, it may fail violently in a rupture that could produce a shock wave and other blast effects similar to those of a detonation.

Detonations

It was pointed in Chapter 22 that the term "explosive limits" is often erroneously used instead of "flammable limits." The table in Fig. 23-2 indicates the upper and lower flammability limits and explosive limits for some common gases. In small and extremely heavy concentrations hydrogen will burn; within its explosive limits, hydrogen will detonate violently. Factors other than fuel-gas concentration will cause reactive substances to detonate. A highly heated flammable gas mixture may detonate

Mixture	Lower Limit, Mole %		Upper Limit, Mole %	
	Flammability	Detonability	Detonability	Flammability
$H_2 - O_2$	4.65	15	90	94
H_2 – air	4.0	18.3	59	74
$CO - O_2$ (moist)	15.5	38	90	93.9
$(CO + H_2)$ – air	12.5	19	58.7	74.2
C_2H_2 – air	2.5	4.2	50	80
$C_4H_{10}O$ (ether – O_2)	2.1	2.6	40	82
$C_4H_{10}O$ – air	1.85	2.8	4.5	36.5

FIGURE 23-2 Limits of Flammability and Explosibility

after ignition. Normally, a mixture at room temperature will begin to burn at the site where it is ignited as the gas reaches a reaction temperature. The surrounding gas is heated progressively until each small volume also reaches the ignition point. On the other hand, in a highly heated gas mixture an ignition source may provide enough additional energy to bring the entire mass to a reaction temperature at almost the same instant throughout. The result is a detonation. The commonest example of this type of reaction is the "knocking" of a car's engine. (Overheated cylinders reflect the heat back into the middle of the mass of mixture, causing it to ignite all at once.)

The presence of a strong oxidizer, such as pure oxygen or chlorine instead of air, can cause a reaction so fast that it actually is a detonation. Numerous examples can be cited. Methane (the principal ingredient of natural gas) will burn in air with a flame velocity of 1 foot per second (fps) or less; a mixture of 35 percent methane and 65 percent oxygen will burn at a velocity of 10.8 fps; 33.3 percent methane mixed with 66.7 percent oxygen will burn at a rate of 7,040 fps. Similarly, hydrogen in air will burn at a comparatively low velocity, but a mixture of 66.7 percent hydrogen and 33.3 percent oxygen will burn at a rate of 9,246 fps. The oft-cited need for separating oxygen supplies, such as from gas cylinders, from other flammable gases, when not in immediate use, becomes readily apparent. Accelerated and explosive reactions can take place because of oxidizers such as chlorates, perchlorates, and others.

Liquid oxygen in contact with an organic material such as asphalt, grease, or oil will create a gel so sensitive it will detonate when subjected even to a slight vibration.

An increase in pressure may cause the rate of a reaction to increase. On April 1, 1971, an accident near Charleston, West Virginia, caused a fire that overheated a 3-inch line containing acetylene. The increase in temperature caused the acetylene to decompose. This reaction increased the gas pressure so that the rate of decomposition increased to the detonation velocity. The line had been built strong enough to contain the explosion without rupturing. The detonation progressed through 7 miles of pipe in approximately 6 seconds and was stopped only by flash arresters which had been installed to prevent damage to plant equipment at each end of the line.

A "donor" charge with high output energy can cause even certain stable compounds to detonate. In 1963 a tank of trichloroethylene detonated at an aerospace company, killing two technicians. Trichloroethylene was normally so chemically stable it had, until a few years earlier, been considered nonflammable. In this case, it had been used to clean missile engines and tanks but had been contaminated with hydrazine. Inadvertently, some nitrogen tetroxide was introduced into the tank. Hydrazine and nitrogen tetroxide are hypergolic and reacted. The energy produced was great enough to cause the trichloroethylene to disassociate in a violent reaction that blew the tank apart. In 1980 in Arkansas a similar accident destroyed the silo of a Titan II missile in an explosion so great that the concrete halves of the silo covers, weighing hundreds of tons, were thrown hundreds of feet away.

A catalyst will cause a reaction to proceed faster, generally by reducing the energy required for a reaction to begin. The presence of a catalyst can therefore permit more of a mixture of substances to begin reacting from the introduction of an initiating source and will cause the reaction to reach detonation speeds.

MATERIALS THAT WILL EXPLODE

Some substances that will explode have already been mentioned. However, a few more examples can be cited. High-energy substances react exothermally. The reaction rate increases with temperature; as the reaction proceeds, the temperature keeps increasing, so the reaction rate keeps increasing. Any substance that decomposes to form gases in an exothermal reaction which can accelerate in this way can create an explosion. Certain types of compounds which are potentially explosive for this reason include (but are not limited to) peroxides; ethers; chlorates; chlorites, and other radicals with chlorine; nitrates and nitrites; permanganates and chromates; and iodates and bromates.

Certain substances, such as sulfuric or other strong, concentrated acids, react violently with water to create minor explosions that have injured personnel. Personnel who must mix any of these with water should wear face shields, rubber aprons, and rubber gloves. Concentrated acids should always be poured slowly into the water while being mixed constantly. Water should never be poured into the acid.

A nonexplosive substance can decompose to give off a highly reactive gas which, when confined, might explode. Hydrogen is given off when secondary batteries are overcharged. If the gas accumulates in the battery or under the hood of a car or truck, a spark, lighted cigarette, or open flame can cause it to explode.

Combustible dusts can ignite easily in a violent reaction. During one winter (1977–1978), grain elevator explosions killed more than 100 persons in the United States; since 1900, 500 have been killed and 1,500 injured. The ease with which dusts ignite and the violence of the reaction are affected by numerous factors (see Fig. 23-3), including the nature and purity of the dust involved, particle size and concentration, heat of combustion emitted, and ignition sources.

DUST EXPLOSIONS

Any combustible solid material in finely divided form can create an explosion if in contact with air and ignited. A dust explosion is a deflagration that results in a sudden development of gas, heat, and pressure. A specific weight of dust occupies far less space than the gas that is produced in an explosion. The gas, in a confined space and in the presence of the heat produced, generates a pressure that may cause bursting of the container, equipment, or plant in which it is confined. The flame front by which the deflagration moves can cause flash burns to any personnel that it passes. When the dust is an organic substance, the carbon dioxide produced can asphyxiate anyone present, but the highly toxic carbon monoxide also created can cause fatalities so rapidly that the carbon dioxide is of little concern.

Factors affecting explosiveness:
The ease with which a cloud of dust can be ignited depend on numerous factors:

Combustible nature of the dust: numerous materials are of low flammability when in massive form. Finely divided, the same materials may be highly flammable. Metals, such as iron and aluminum, have been produced which are so fine that little energy is required for ignition; they will ignite at room temperature. They may, therefore be considered pyrophoric. Some solids, especially organics, contain volatile matter which is given off, ignites at a temperature lower than that of the solid matter, and produces heat that causes ignition of the solids.

Size of the dust particles: as dust becomes finer, the contact between the solid surfaces and air becomes intimate, and the dust becomes as easily combustible as a highly flammable gas. The ease of ignition is increased with the decrease in particle size and the roughness of the particles. In addition, finer dusts can be dispersed more easily and uniformly and remain in suspension in air longer, so that they constitute hazards longer. As particle fineness decreases, so does the concentration of dust required for ignition.

Concentration of the particle cloud: dust clouds will ignite only when the concentration of dust is between specific limits, similar to those for gases. A fine particle may ignite and generate heat; unless other particles are close enough to be ignited by that heat, however, the reaction cannot continue. Between the combustible limits there is enough oxygen for combustion of the dust present. Any excess dust will have no oxygen with which to burn and will therefore tend to stop the reaction. Somewhere between the two limits there is a particle size and concentration of dust at which the speed of reaction will be a maximum. As the dust concentration increases above the lower limit, the pressure that results after ignition increases until an optimum concentration is reached; after that, further increase in dust concentration reduces the pressure generated.

Temperature of the combustible mixture: as with any reaction, the outside energy required for initiation decreases with an increase in temperature. A dust cloud at an elevated temperature is far more susceptible to inadvertent ignition than one which is cool.

Presence or absence of moisture: any moisture present acts as a coolant. Heat generated by ignition of a particle is absorbed by the evaporation of the water. The vapor produced acts as an inert gas which dilutes the oxygen present. In addition, moisture makes air a better conductor of electricity. Accumulation of static electrical charges on the particles is inhibited, reducing possibilities of ignition by sparking. In addition, moisture causes particles to cling together, forming larger masses which have lesser tendencies to ignite.

Presence of inert solid particles: The presence of incombustible material reduces or inhibits tendencies of a cloud to ignite or to propagate. The incombustible material may be a part of each particle or may be separate particles.

FIGURE 23-3 Dust Explosions

EXPLOSIVE EFFECTS

There are three principal effects of explosions. The first of these relates to pressure.

Shock Waves

Detonation of a high explosive or the sudden rupture of a highly pressurized vessel, such as that produced by a compressed gas or a deflagration, will produce a shock wave, or blast. This can be damaging over comparatively long distances in air and even more devastating in a fluid. Most of the energy of the explosion is concentrated in the shock wave. This energy is transferred to succeeding volumes of air (or fluid) as the shock wave advances. However, because the volume of the shock wave increases with the cube of the distance from the source, the energy decreases similarly as it is spread over the increasing volume. The peak pressure produced by a detonation is therefore a function of distance and the weight, or equivalent weight, of the explosive activated.

Tables and charts have been prepared based on TNT equivalency, where the energy of the material involved in the explosion is compared with the weight of TNT. (Detonation of one pound of TNT contains 1.6 million ft-lb of thermal energy.) These indicate the overpressures that would occur at specific distances from a point where a known amount of TNT has been detonated. Quantity-distance tables are prepared from such calculations and from empirical information on the damage that can result from the various magnitudes of overpressures. Effects of overpressures (in psi) on external areas of structures, vehicles, and personnel are approximately as follows:

0.2	Limit for uncontrolled area; no specific damage to personnel or facilities.
0.4	Limit for unprotected personnel.
0.5 to 1	Breakage of window glass.
0.75	Limit for windowless, ordinary construction.
1 to 2	Light to moderate damage to transport-type aircraft.
3	Exposed person standing face-on will be picked up and thrown; very severe damage, near total destruction to light industrial buildings of rigid steel framing; corrugated steel structures less severely damaged.
3 to 4	Severe damage to wood frame or brick homes.
4 to 6	Complete destruction of aircraft or damage beyond economical repair.
5	Possible ear damage; exposed person standing side-on will be picked up and thrown; complete destruction of wooden frame or brick homes; severe battering of automobiles and trucks.
6	Moderate damage to ships.
6 to 7	Moderate damage to massive, wall-bearing, multistory buildings.
7	Possible internal injuries to human beings.
9	Complete destruction of railroad cars.
10 to 12	Severe damage and sinking of all ships.

12	Possible lung injuries to exposed personnel.
20 to 30	50 percent probability of eardrum rupture.
25	Probable limit of thermal injury.

Although a building will withstand an external overpressure of 3 to 4 psi, an ordinary building wall will not withstand an internal overpressure of 1 psi, and windows far less. A Navy manual[89] indicates that the explosion of one pound of TNT in an open-air surface explosion will generate the following overpressures (in psig):

Distance (ft):	2	4	6	8	10	20	40	100	200	400
Overpressure:	320	70	28	15	9.6	3.0	1.2	0.35	0.13	0.05

The ears would be the first part of the body to be affected (see Chapter 28). The ears have a certain small amount of inherent protection against a sudden overpressure; however, the level of self-protection is not great, and a blast can cause rupture of the eardrums. Next is the pressure effect on the lungs and circulatory system, which can also be damaged, especially if the person affected is in a confined location.

A shock wave approaching an interfering structure, such as a wall, compresses the air to a much higher pressure than the static (incident) pressure at or behind the front of a shock wave. The compressed-air pressure may be 5 to 6 times as great as the incident pressure. Personnel caught against a wall by the blast wave have been fatally injured while persons in the open have suffered only minor injuries.

Fragments

In any violent explosion, the second principal hazard is the fragmentation into missiles of the vessel in which the reacting material was contained. These missiles are projected outward at high speeds induced by the blast pressures and can cover considerable distances. Even an extremely small fragment can cause a penetrating and damaging wound (Fig. 16-6). Impacts by larger fragments can cause crushing of flesh and bone.

Figure 23-4 indicates where many fragments landed when a reactor vessel and high-pressure separator exploded at a chemical plant in Whiting, Indiana, on August 27, 1955. The reactor vessel had been alloy steel plates more than 2 inches thick; it was 127 feet in overall length and 23.5 feet in diameter. At the time of the explosion the vessel contained a mixture of 3 percent naphtha vapor, 19 percent oxygen, and the remainder inert gas, all at a pressure of 105 psig. Investigation and analysis indicated the explosion started in a pipe run and traveled to the reactor vessel in one direction and to the high-pressure separator in the other. The explosion threw one piece of the reactor, which weighed about 60 tons, almost 1,200 feet (piece no. 7 at the top of Fig. 23-4). It landed on and smashed a tank containing gasoline-based stock, igniting and scattering its contents. Other fragments of the reactor vessel were thrown as far away as 1,500 feet. Eventually, 63 tanks and 1,270,000 barrels of petroleum products were destroyed by the fire which followed.

The fragmentation that will result from any explosion will be unknown before the event takes place. This is because of the uncertainties of the strength of the container in which the explosion may occur and the nature and direction of the pressure

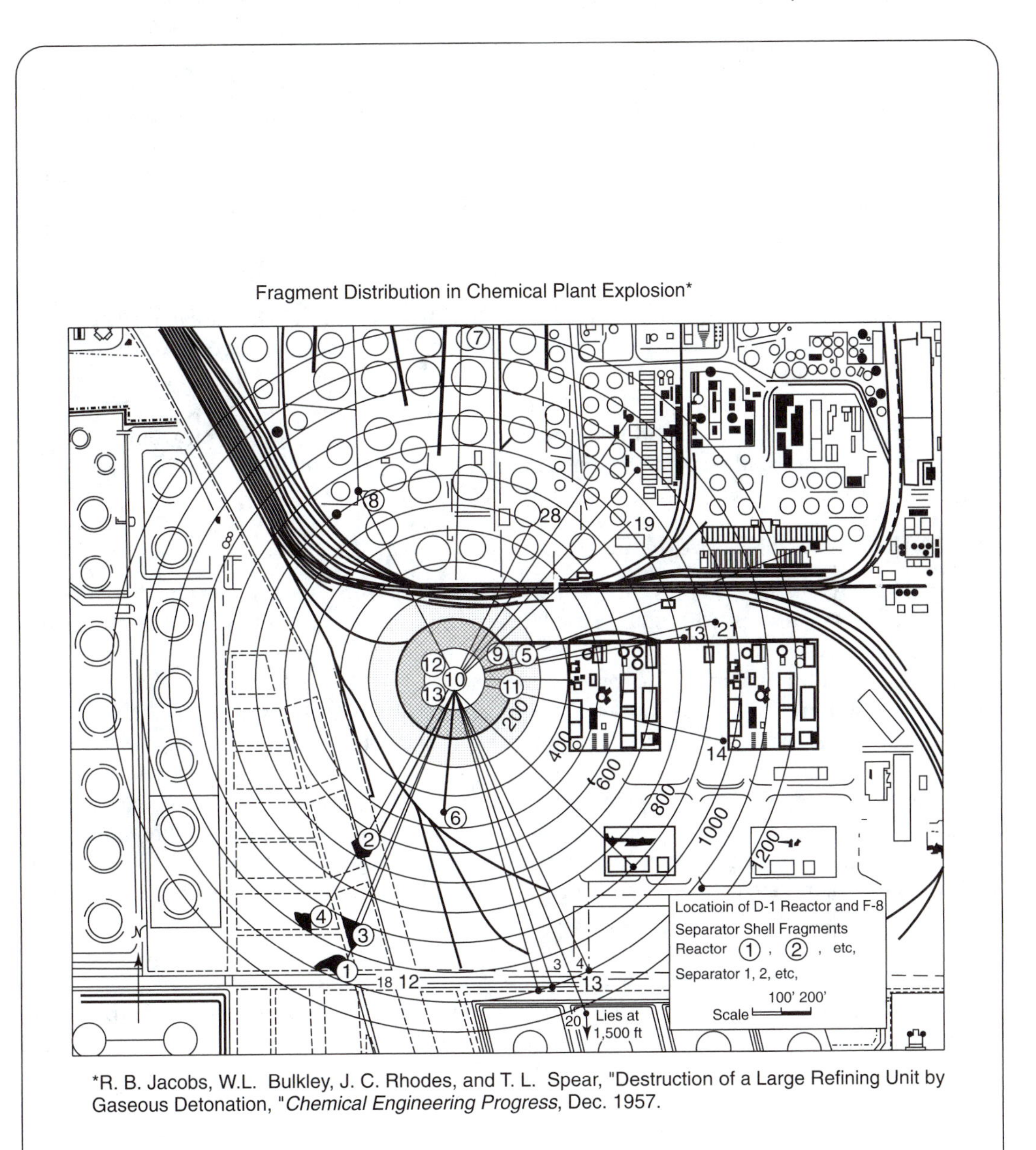

*R. B. Jacobs, W.L. Bulkley, J. C. Rhodes, and T. L. Spear, "Destruction of a Large Refining Unit by Gaseous Detonation, "*Chemical Engineering Progress*, Dec. 1957.

FIGURE 23-4 Fragment Distribution in Chemical Plant Explosion

rise. An air compressor may operate properly for a long time; then it may ingest air contaminated with acetylene, oil, or other material which is highly flammable in a hot, compressed-air environment. The resulting explosion is dependent on the amount and nature of the contaminant present. Explosions have occurred with compressors producing pressures as low as 100 psi. Combustion of the contaminant creates gaseous products whose partial pressure overpressurizes the compressor cylinder.

Body Movement

The third effect of a blast on personnel is that anyone hit by a shock wave may be picked up and thrown against a wall or other hard object. Here again, a person may suffer an impact injury. An exposed person facing (or with the back to) the source of the blast offers a larger frontal area to the pressure than for a "side-on" blast. A person would probably be picked up and thrown by a 3-psi overpressure, whereas it would require 5 psi to pick up and throw a person facing at right angles to the source of the blast.

PREVENTING EXPLOSION DAMAGE

The obvious way to prevent explosion damage is to prevent explosions. The precautionary measures for industrial plants, where explosions often involve flammable materials, are generally the same as those for preventing fires presented in Chapter 22. A few additional measures can be added.

Where explosives must be kept, the amounts present at any time should be as small as possible. Processes, equipment, and materials that might explode should be isolated from activities where people or vulnerable equipment are present. Where explosions might occur during a combustion process, "blowout" panels or plugs should be provided. Boiler furnaces using gas or oil are common examples. When a burner is first turned on in a cold furnace, the oil sprayed in may be too cold to ignite readily and may accumulate on the furnace floor or walls. As attempts are repeated to ignite fresh oil, the deposited oil may warm an excessive amount of oil until more vapor is present in a flammable mixture than would normally be present in the furnace. The mixture may ignite suddenly, creating an overpressure that could do serious damage to the furnace walls and to the boiler tubes or drum. Panels are therefore provided which blow out at a level at which equipment damage will not occur.

Explosion suppressors can sometimes be used to inhibit reaction (see Fig. 23-5). In this system a sensor detects when an explosion has occurred. A suppressant is released automatically which further inhibits reaction. Limitations are that a pressure rise of 1.5 psi may be required to activate the sensor, and that by the time suppression has taken place, a slightly higher pressure will have been reached. It is necessary that the vessel or structure being protected be strong enough to withstand this pressure. The gas used for suppression must be an effective inhibitor for the particular fuel that might be involved. Also, many detonations are the result of decomposition of complex explosions and not of combustion. Such explosive detonations are too rapid for suppressors to have any effect.

In Australia, grain elevators have been almost free of dust-explosion accidents. It is evident that methods do exist to prevent explosions or subsequent injuries or

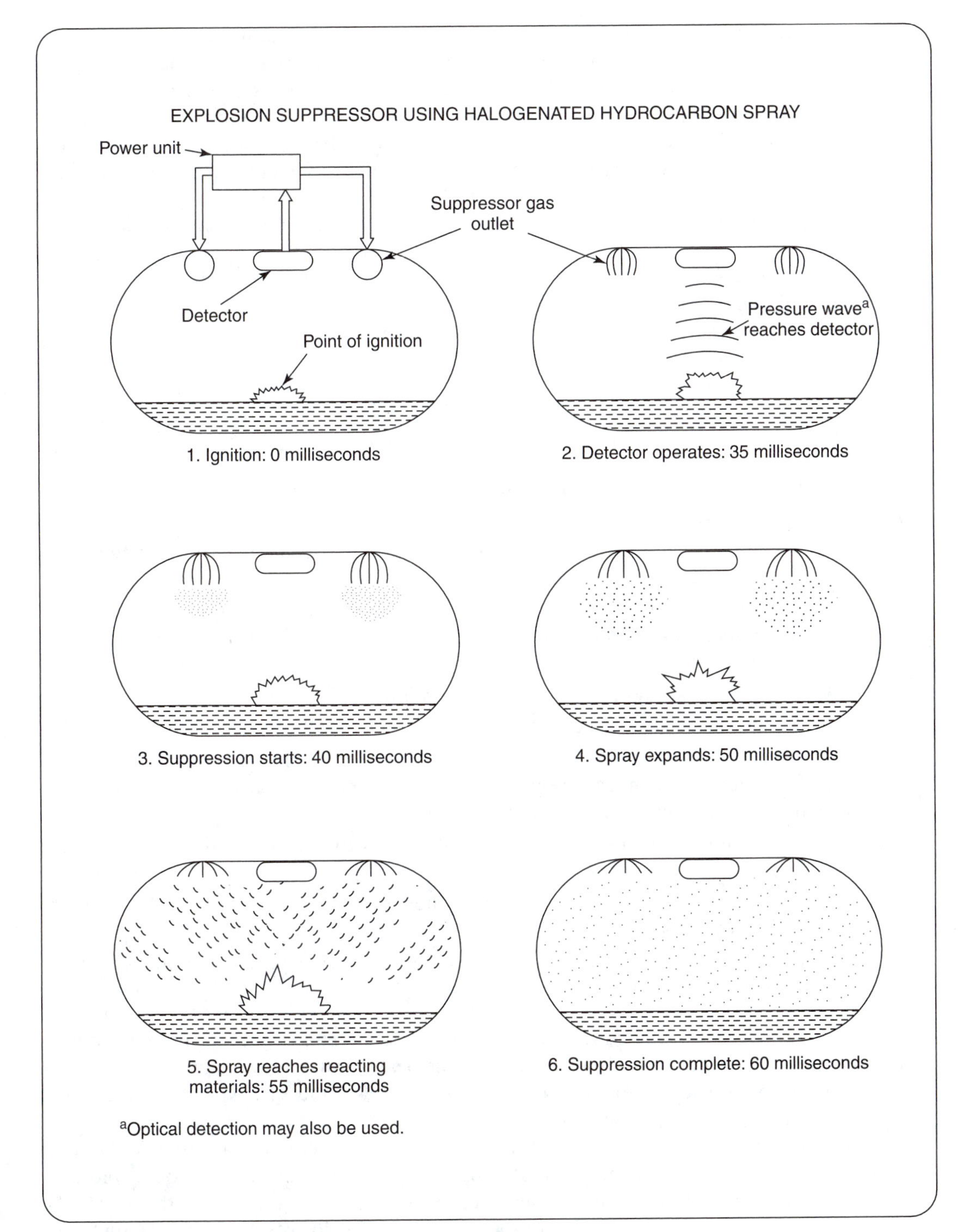

FIGURE 23-5 Explosion Suppressor Using Halogenated Hydrocarbon Spray

damage. Most of the prevention methods consist of eliminating the conditions that would lead to explosions. In Australia, investigating Americans found the most significant method is good housekeeping. The presence of fine dust is minimized by controlling the environment and preventing dust accumulations. Cleanup crews constantly monitor the plants to eliminate dust and use vacuum cleaners for its removal before it can reach hazardous concentrations. Instead of heavy concrete construction which would confine and intensify any explosion, light panels which will blow out fairly easily are provided for venting and relief of combustion products if an explosion should take place.

Good control and elimination of dust concentrations will reduce not only the possibility of explosions but also the adverse effect on personnel. Formerly, workers in dusty environments were instructed to wear filter masks which employers provided. Often the workers failed to wear them. OSHA now requires that to the greatest degree possible through engineering means the work environment be such that masks are not required. Actions which will reduce both the possibilities of dust explosions and the adverse effects on personnel can be summarized into:

1. Using processes which will minimize the generation of dust—for example, using wet instead of dry processes.
2. Isolating those processes which produce dust and then taking action to neutralize the possibility of an explosion.
3. Isolating dust under hoods or other collecting devices so that as soon as dust is generated, it is transported to a safe container in which it can be safely stored until it can be disposed of.
4. Using water sprays to maintain high humidity and to wash down particles.
5. Adding inert gas or inert solid particles which will interfere with the ignition process. This method cannot, of course, be used where personnel are located.
6. Reducing dust concentration by ventilation either generally (over an entire volume of space) or locally (where dust generators are).
7. Maintaining good housekeeping by eliminating dust accumulations. (Dust should be collected by a wet method.)
8. Keeping dust collections in either an inert or wet atmosphere or with very limited access to air.
9. Lessening the vibration which can keep dusts moving, preventing them from settling.

Some types of dusts are reusable, and collection and recycling can present good savings. Even when the initial cost of the material is low, the total amount saved may be substantial. The economic aspects of collection will increase with the value of the dust that would otherwise be lost.

Personnel employed in using explosives productively must be trained for such purposes, including safeguards that must be observed. Information on explosives, especially accident prevention measures which must be used with them, can be obtained on request form the Institute of Makers of Explosives, 1120 Nineteenth Street

Hazards Checklist— Explosives and Explosions

Possible Adverse Effects

Rupture of container

Blast effects:
- Overpressures
- Collapse of nearby containers
- Damage to structures, equipment, and vehicles
- Propagation of other explosions

Fragmentation effects:
- Holing of nearby containers, equipment, and vehicles
- Impact of pieces against personnel, equipment, vehicles, and structures
- Dispersion of burning, hot, combustible, or corrosive materials

Heat effects:
- Dispersion of toxic materials
- Injury to personnel

Possible Causes

Inadvertent activation by electric current, heat, electromagnetic radiation, lightning or other static electricity, impact or fire of:
- Explosive
- Combustible gases in containers or confined spaces
- Fine dusts or powders
- Combustible gases or liquids:
  - In high concentrations
  - In presence of strong oxidizers
  - At high temperatures

Afterburning of confined combustion products
Delayed combustion in a cold firing chamber
Ignition of hydrogen produced by battery charging
Warming a cryogenic liquid in a closed system
Warming a liquid with a high vapor pressure in a closed container
Ignition of sensitive gases, such as acetylene
Contact between water or moisture and a water-sensitive material such as molten sodium, potassium, or lithium; concentrated acids or alkalis; or similar substances

FIGURE 23-6 Hazards Checklist—Explosives and Explosions

CHECKLIST—EXPLOSIVES AND EXPLOSIONS

1. Does the plant use, or does the product being manufactured contain, any explosive or any material which could generate an explosion?
2. If any explosive is to be used, is it unduly susceptible to heat and height temperature, electromagnetic radiation, mechanical shock or other means of initiation? Have the characteristics of the explosive been obtained from the manufacturer or the formulator? Are safeguards provided against any characteristics which could result in an unintended initiation?
3. If the main charge of any explosive to be used needs an activating device to be initiated, is the activating device normally separated from the main charge?
4. Are the explosive and the initiating device suitably marked with warnings?
5. Are facilities provided where explosives can be stored in accordance with governing regulations?
6. If the product which contains an explosive has to be shipped, are the shipping containers, vehicles, and safeguards in accordance with Department of Transportation and local requirements?
7. Are cartridge-actuated devices used in the plant for such items as construction?
8. Can the cartridge-actuated devices take only cartridges of sizes intended for its use and not permit the use of oversize cartridges?
9. Does the cartridge-actuated device have a means of rendering it safe when loaded with a cartridge but not yet in use? Have safe operating instructions been provided and are users familiar with them?
10. Does the plant user or contain a compressed, liquefied, or piped gas, such as propane, butane, or natural gas, which can leak into a confined space, form an explosive mixture, and and explode when ignited?
11. Does any such gas used in the plant contain an odorant which will permit detection of leakage?
12. Do instructions for the lighting of gas or liquid fuel burners provide warnings on the need and means for avoiding excess gas or oil vapor accumulations when starting a fire in a furnace?
13. Is there any possibility a compressor could intake a gas such as acetylene which will explode when pressurized in the presence of air and heat?
14. Is there any large amount of material present, such as a nitrate or nitrite or similar compound which can disassociate violently in the presence of heat?
15. Is there any liquid present which is in a closed vessel and has a high vapor pressure which can be increased because of proximity to a hot surface, exposure to the sun, or other heat source?
16. Is any liquid present, such as trichloroethylene, which will explode when subjected to high-intensity heat such as a welding arc or cutting flame?
17. Is any reference source available from which the explosive characteristics of materials which might be used in the plant can be obtained?
18. Is any cryogenic system being used which could cause condensation of oxygen from the atmosphere where it will contact an organic material which will then become shock sensitive?
19. Have provisions been made to keep the presence of explosive dusts to a nonhazardous level by adequate ventilation or cleaning, or have other safeguards been provided to prevent a dust explosion?
20. Is any chemical present which could explode or form an explosive mixture when in contact with water or moisture?

FIGURE 23-7 Checklist—Explosives and Explosions

NW, Suite 310, Washington, D.C. 20036. Checklists to assist safety engineers in determining whether or not explosive hazards exist in their plants are included in Figs. 23-6 and 23-7.

BIBLIOGRAPHY

[89] Naval Air Systems Command Manual 06-30-501, *Technical Manual of Oxygen/Nitrogen Cryogenic Systems,* July 1, 1968.

EXERCISES

23.1. What is the difference between a deflagration and a detonation? How does a high explosive differ from other explosives?

23.2. How does the D.O.T. classify explosives?

23.3. In industrial activities, which constitutes the more common hazard: explosives or other substances that can explode?

23.4. What are some of the factors that can cause a reactive substance to detonate in a high-order explosion?

23.5. Describe some substances, other than explosives, which can explode.

23.6. What is the difference between the lower flammability limit of a combustible gas and the lower explosive limit?

23.7. What are quantity-distance tables? On what are they based?

23.8. How can an explosion cause injury or damage?

23.9. What is the best way to avoid explosion damage?

23.10. How does an explosion suppressor work? What types will not work?

23.11. Prepare a list of five questions on explosives or explosions which you believe could be added to the checklist in Fig. 23-7.

CHAPTER 24

Hazards of Toxic Marterials

Fear of toxic chemical releases have increased because of industrial accidents such as in Bhopal, India, tank car wrecks, and increased awareness of injuries to industrial plant workers, workers in the field, and the general public. It has been estimated that a new and potentially hazardous chemical is introduced into industry every 20 minutes. Highly reactive chemicals are being used more and more in industry, agriculture, research, space flights, and in the home. This has aroused great concern and apprehension about their effects on health. Many of these chemicals are belatedly found to be not only evidently and immediately toxic, but also insidious, carcinogenic, long lasting, and causes of previously unsuspected and long-lasting injuries. In addition to their toxicity, many substances are radioactive (a subject covered in Chapter 27).

It is therefore necessary that suitable safeguards be provided to prevent or lessen the injuries that can occur to workers in industrial plants and to the general public. To do this, every safety person must understand the physiological processes by which the human body can absorb a toxic material, the effects that can be produced, and the precautionary measures to be observed to prevent injury.

TOXIC MATERIALS

A material can be considered toxic, or poisonous, when a small quantity will cause injurious effects in the body of the average normal, adult human. Almost all materials are injurious to living organisms to some extent. Certain materials are toxic in rare cases but are not considered poisonous. Overdoses of common salt fed accidentally to infants have caused their deaths. Water, which we drink, and nitrogen and oxygen, which we breathe, can cause injury or death in overdoses.

Certain persons have unusual susceptibilities to substances that produce violent or fatal reactions when absorbed in small amounts harmless to most other persons. Such substances are called allergens, and a person is said to have an allergy or to be allergic to the substance. Allergies have been reported to almost every known common material.

Attempts have been made to determine the dosages of toxicants which will cause injury to the normal (nonallergic) adult. Only partial success has resulted because of the many factors involved. Because of personal susceptibility, other conditions affect the severity of the injury which may result: the size and duration of the dose, the route by which the material is taken into the body, the degree of toxicity, the rate of absorption, environmental temperature, and the physical condition of the affected person. Generally, the greater the toxicity, rate of absorption, and temperature, the more rapid the occurrence of the injury. The time during which a person is subject to the toxicant, whether it is cumulative or non-cumulative, or other factors may affect the size of the dose that could be injurious. Precise data on persons who have been injured accidentally by toxic substances are meager. It is impossible, therefore to predict exactly what an injurious dose would be for a specific substance. Researchers have used animals in attempts to determine effects of exposures, whether substances are toxic or not, and to what degree. Effects on persons may differ from those on animals, and comparisons because of their weights may also be uncertain or unreliable.

ROUTES TO INJURY SITES

Toxic injury can occur at the first point of contact between the toxicant and the body, or in later, systemic injuries to various organs deep in the body. Route of entry often has a profound effect on symptoms and injuries that can result. Systemic injuries will occur most rapidly by the route that permits easiest access of the chemical into the bloodstream. Routes by which this can occur are through the respiratory system, the skin, or the gastrointestinal tract. Of these, the route through the respiratory system (Fig. 24-1) is the most common and most dangerous, since the least amount of self-protection is afforded the body (Fig. 24-2).

The most rapid route is by direct injection into the tissues and bloodstream, but this in not common in industrial accidents. Absorption through an open cut or wound through the skin is more common. Absorption into the blood through the intestinal tract produces the same effects as entry into the bloodstream through the respiratory system, but the process is much slower, and food and body chemicals may themselves cause changes. A number of substances react quite differently if inhaled than if they must pass through the stomach. Food or the hydrochloric acid in the stomach may change and detoxify chemical compounds and render them harmless. Conversely, more damaging materials can be produced. Methyl alcohol (which is actually less toxic and narcotic than ethyl alcohol) is oxidized in this way to form formaldehyde and formic acid, which are highly irritating and cause inflammation and damage to the kidneys.

HYPOXIA

Hypoxia is the oxygen deficiency that results from any interference with oxygenation of the blood or the ability of tissues to absorb oxygen. The type and magnitude of tissue damage caused by hypoxia depends on the rapidity and duration of deprivation of oxygen.

One of the first and principal effects of hypoxia is a decrease in mental capabilities. The need for oxygen varies with different parts of the body, but the brain is the most sensitive to its lack. Brain cells must have oxygen constantly; consciousness may

THE RESPIRATORY PROCESS

Respiration: respiration is the exchange of gases between a cell and its environment during which oxygen is absorbed and carbon dioxide released. In a complex organism such as man, respiration is much more involved than in a simple organism such as a cell. Human respiration can be subdivided into external respiration, transportation of gases by the blood, and internal respiration. Respiration consists of *ventilation*, the movement of air into and out of the lungs; and *absorption*, or the interchange of oxygen for carbon dioxide in the lungs. Transportation consists of moving oxygen to the tissues and carbon dioxide back to lungs. Internal respiration involves tissue absorption and the interchange of carbon dioxide and other waste products for oxygen at the tissues.

Respiratory system: air enters the body through either the nose or mouth. Entering through the nose, it passes through the nostrils (external nares) into the nasal chamber. Air entering through the nostrils and nasal passages is filtered of particulate matter by hairs (cilia) moistened by the wet mucous lining, which also traps solid particles. The nasal chamber is a large cavity above the roof of the mouth which contains the sense organs of smell and the membranes by which air is warmed and cleaned. The air then passes through the internal nares to the pharynx (the cavity at the back of the mouth), where it meets additional air (or food) which may have entered by the mouth. The air then proceeds toward the lungs by way of the larynx and trachea (windpipe). Food is prevented from entering the larynx by a flap of tissue (the epiglottis) which automatically folds over the larynx opening whenever food is swallowed.

The lower respiratory tract begins at about the first rib, where the trachea divides into two main branches, called *bronchi*, each leading to a lung. Each bronchi then divides into *bronchiole*, which subdivide into smaller and smaller tubes, about five or six to each bronchiole. (The trachea, pharynx, bronchi, and bronchiole take no active part in the respiration process; their chief function is to act as air passages.) Each small tube ends in three to six air sacs. The walls of the small tubes and the air sacs contain minute cup-shaped cavities, *alveoli*. The walls of the alveoli are made up of a single, flat layer of cells. On the other side of the alveolar walls is a rich network of blood capillaries. Molecules of oxygen and carbon dioxide pass readily through the thin, moist alveolar walls into the capillaries. Oxygen from inhaled oxygen and carbon dioxide pass readily through the thin, moist alveolar walls into the capillaries. Oxygen from inhaled air is picked up by the blood while carbon dioxide brought from the tissues is released and passes into the alveoli.

The air sacs connected to each bronchiole constitute a *lobule*. The lobules make up wedge-shaped masses called *lobes*. Each lobe is separated from its neighbor, making it possible for only one lobe to be diseased or to be removed surgically without affecting the others. The left lung has two lobes; the right has three and is slightly larger. Both lungs are enclosed in membranes, called *pleura*, all within the bony structure formed by the ribs, spin, and breastbone.

Absorption: there are approximately 725 alveoli in the human lungs, with a total surface that has been estimated to be grater than 1,000 square feet. The extent of the alveolar surface and the thinness of the walls allow almost complete equilibrium to exist between each gas in the alveolar air and in the alveolar air and in the capillary blood. Normal oxygen concentration is higher in the alveoli than in the blood which the pulmonary artery brings to the lungs. A gas will pass from a region of higher to one of lower concentration. Oxygen therefore passes from the alveoli to the blood in the capillaries by diffusion. This easy and rapid absorption is a major reason small amount of toxic gas can be highly dangerous and injurious.

Ventilation: the normal breathing rate of an adult human at rest is 15-20 cycles per minute. Inspiration begins when the chest muscles contract and draw the ribs upward and outward. The muscle wall (diaphragm) between the chest and abdominal cavity distends at the same time, and the lungs expand. Pressure of the gas in the lungs (intrapulmonary pressure) is lowered 1-2 mm below atmospheric, causing the higher-pressure ambient air to enter the body. Between breaths, lung and atmospheric pressures are equal.

In the body cells, the oxygen is used in releasing energy in a process known as *metabolism*. Among the byproducts of this process is the carbon dioxide which is picked up by the blood; normally, it is held in solution as sodium bicarbonate. Alveolar air oxidizes the carbon-dioxide-laden blood which reaches the lungs, increasing its acidity and causing release of the carbon dioxide. The pressure of the carbon dioxide. The pressure of the carbon dioxide at this time is higher than that of the air in the alveoli, causing it to pass through the alveolar walls. It is then exhaled.

FIGURE 24–1 The Respiratory Process

Expiration: exhalation or expiration starts when the chest muscles and the diaphragm relax. The lungs contract because of their elasticity. The intrapulmonary pressure rises 2-3 torr above atmospheric, causing air to be discharged through the respiratory passages. By the end of expiration, pressure within the lungs is again equal to atmospheric. Expiration can also be induced by contracting the abdominal muscles to force the diaphragm up. This increases the pressure in the lungs so that air in an amount greater than normal is exhaled.

The amount of air inhaled and exhaled in a normal breath (*tidal volume*) is about 500 cubic centimeters. The amount of air that can be inhaled by forced breathing is about 1,500-1,700 more than the tidal volume. This excess is called the *inspiratory reserve volume or complemental air*. *Supplemental air* is the difference between tidal air and the maximum that can be exhaled. The total amount of air that can be breathed in and out during forced respiration is the *vital capacity*. Once the lungs are filled after birth, they are never emptied completely; approximately 1,500 cm^3 (*residual volume*) always remains that cannot be expelled, no matter how hard the effort.

Of the 500 cm^3 of air inhaled with each breath, only some 350 cm^3 actually reach the alveoli. The remaining 150 cm^3 remain in the upper respiratory passages, where no exchange of gases takes place. This air is the first to be forced out with the following expiration. The last 150 cm^3 expelled from the alveoli also remain in the air passages. This air, high in carbon dioxide, is the first to be drawn back into the alveoli on the succeeding breath. Air in the alveoli therefore contains more carbon dioxide (5.3 percent) and less oxygen (13.5 percent) than normal atmospheric air, while exhaled air contains 75 percent of the oxygen in normal atmospheric air and can be breather again.

Functions of the blood in respiration: blood is liquid tissue consisting of a yellowish, almost colorless, fluid portion called the *plasma*, and a solid portion composed principally of red corpuscles, white blood cells (leucocytes), and platelets (thrombocytes). The color of the red corpuscles and of the blood is due to the presence of hemoglobin. Hemoglobin is composed of *globin*, a protein, and *heme*, an organic compound containing iron ($C_{34}H_{33}FeN_4O_5$). Under normal conditions the rate of red cell production will equal the rate of destruction. Abnormal reduction in the number of red cells or in hemoglobin is called *anemia*. Anemia can be produced by bleeding, decreased cell production due to failure of the bone marrow (where it is made) lack of iron or vitamin B_{12}, or other causes. A person is generally considered anemic when the red cell count drops below 4.7 million per cubic millimeter, (male) and 4.13 million (female)

The blood plasma holds about 2 percent of the oxygen in solution. The remaining 98 percent is carried in combination with the hemoglobin. There are approximately 3×10^{12} (30 trillion) red corpuscles in an adult human body. Each corpuscle contains about 265×10^6 (265 million) molecules of hemoglobin, for a total of about 8×10^{20} molecules in the body. A hemoglobin molecule is attached to four atoms of iron, which gives it its ability to carry oxygen by loose chemical union. The compound created is *oxyhemoglobin*. The union takes place on a basis of one hemoglobin molecule to one of oxygen. Oxyhemoglobin is bright red, hemoglobin is purple. The difference is principally noticeable in the colors of the blood in the arteries and veins because of the presence or absence of oxygen.

The union that creates oxyhemoglobin takes place where there is an abundance of oxygen; disassociation and release, where oxygen is scarce. Oxyhemoglobin formation therefore takes place in the lungs, where there is a relatively high concentration of oxygen. Its disassociation occurs in the capillaries of the using tissues, where the partial pressure of oxygen is lower than that in the red cells. Oxygen transport direction of reaction is controlled by two factors-principally the *tension*, or partial pressure of oxygen; and, secondarily, alkalinity of the blood. Alkalinity, in turn, is dependent on the amount of carbon dioxide present. Increased blood alkalinity decreases the ability of hemoglobin to carry oxygen.

Carbon dioxide released from the tissues combines with water to form carbonic acid, a decrease in blood alkalinity. This, in turn, causes the release of oxygen. Hemoglobin is so sensitive to a change in alkalinity that the entire process of union with and release of oxygen is dependent on an extremely small range of alkalinity variations. Most of the carbon dioxide is carried as sodium carbonate in the cells and in solution in the plasma (8 percent).

Control of respiration: respiration is controlled by two chief means: neural, through the nervous system; and chemical, by changes in the blood alkalinity. Neural control operates through the respiratory center, located in the brain. The respiratory center is stimulated and controlled by impulses received through the sensory nerves, by the alkalinity and carbon dioxide content of the blood, or through activation from other parts of the brain. During inspiration, pressure sensitive nerve cells in the walls of the alveoli are

FIGURE 24–1 Continued

stimulated by expansion of the lungs. These nerve cells send impulses through the vagi nerves to the brain which inhibit the respiratory center and cause the succeeding expiration. This is called the *Hering-Breuer reflex.* Contraction of the lungs also stimulates the respiratory center, so that a regular cycle is maintained. In emergencies, breathing which has stopped can be reestablished by activating the nerve sells in the alveolar walls. The various means of artificial respiration are based on this principle.

Control of respiration by chemical change is effected principally by variations in concentrations of carbon dioxide in the blood. These changes are detected by the respiratory center, which then adjusts the rate and volume of breathing. Concentrations of carbon dioxide up to 3 percent in air produce little effect; amounts greater than this cause a greater depth of breathing. Further increases in carbon dioxide increase the rate of breathing.

Even small changes in the carbon dioxide content of the blood will cause marked changes in the respiratory rate and volume. Increasing the percentage of carbon dioxide in the alveoli and blood as little as 0.2 percent may double the respiratory rate. This fact is utilized in the gas supplied for pulmotor use: carbon dioxide concentration may be as high as 6 percent. This respiratory control is effective even at birth, where the gas mixture supplied to newborn babies may have as much as 10 percent carbon dioxide. In cases of necessity when other means are not available, carbon dioxide content can be increased by breathing into and out of a paper sack. Rebreathing this expired air would provide a mixture with about 4 percent carbon dioxide.

A lack of oxygen may also affect activity of the respiratory system. The carotid body, located at the carotid artery in the neck, responds to changes in oxygen content of the blood, affecting the breathing rate. Lack of oxygen, however, is less sensitive than an increase in carbon dioxide (or vice versa) as a regulatory mechanism. An increase in oxygen beyond normal limits will produce no significant change in breathing as long as the carbon dioxide content of the blood remains the same.

Nerve endings in the carotid body also respond to pressure. This pressure may be exerted by the adjacent carotid artery or by external force such as a blow. An increase in pressure causes immediate fall of the general blood pressure, and the heart slows. At one time the use of pressure at certain points was recommended to reduce loss of blood due to a wound. To stop raid bleeding from the head, pressure was to be applied against the carotid artery. This practice is now discouraged because it sometimes caused pressure against the carotid sinus and body. A sudden drop in blood circulation would occur at the time oxygen to the brain and body was most needed.

Respiration may also be affected by other involuntary reflex actions. A common example is the gasp and raid breathing that may follow being doused unexpectedly by a mass of icy water. Increase in respiration at this time may be a mechanism by which the body defends itself against the shock it has undergone.

Respiration may also be controlled voluntarily by an individual, to a limited extent. A person can hold his breath for a short time, increase or decrease the rate of breathing, and change the volume of each inhalation and exhalation. Xxxx breath, when held must be released and the body permitted to adjust itself to its needs indicates that voluntary control is limited.

FIGURE 24–1 Continued

be lost after only seconds of cerebral hypoxia. Irreversible changes occur if the supply is cut off for 5 to 7 minutes or more. Hypoxia causes an increase in reaction time and a reduction in the ability to concentrate. The condition is aggravated by the fact that the person affected generally does not realize the condition and its dangers. The person's moods may change rapidly, sometimes from anger and frustration to a sense of well-being. Failure of the heart function can occur when oxygen tension drops below 45 torr (mmHg), causing rapid decrease in blood pressure and pulse rate.

The affected person may grow pale, begin perspiring, have a headache, become dizzy, suffer from nausea, and faint. Ultimately, there might be complete collapse,

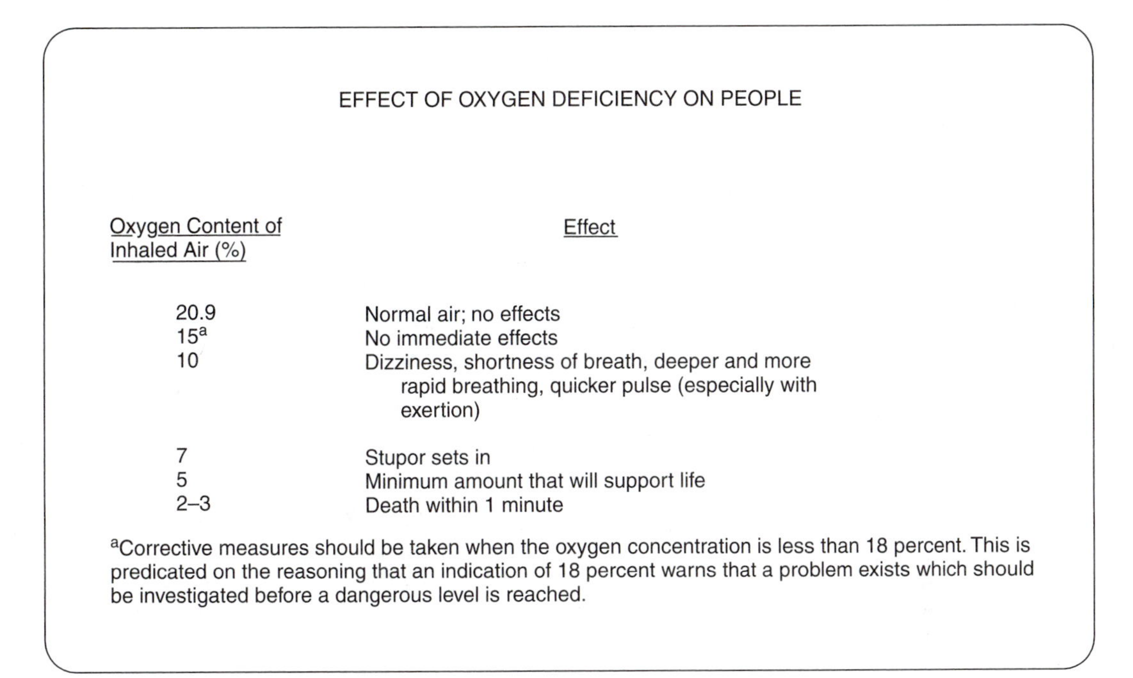

EFFECT OF OXYGEN DEFICIENCY ON PEOPLE

Oxygen Content of Inhaled Air (%)	Effect
20.9	Normal air; no effects
15[a]	No immediate effects
10	Dizziness, shortness of breath, deeper and more rapid breathing, quicker pulse (especially with exertion)
7	Stupor sets in
5	Minimum amount that will support life
2–3	Death within 1 minute

[a]Corrective measures should be taken when the oxygen concentration is less than 18 percent. This is predicated on the reasoning that an indication of 18 percent warns that a problem exists which should be investigated before a dangerous level is reached.

FIGURE 24–2 Effects of Oxygen Deficiency on People

convulsions, and death. There could be decreases in ability to control the eye muscles and in visual range and sensitivity. The senses of smell, taste, touch, and hearing also suffer, but to a lesser degree than does vision.

Four types of hypoxia may occur in an adult: hypoxic, hypemic, stagnant, and histotoxic.

HYPOXIC HYPOXIA

Hypoxic hypoxia is caused by low oxygen tension in the blood. There are a number of causes, primarily an inadequate source of oxygen. This can occur at high altitudes or where excessive amounts of other gases are present. Altitude due to high flights is generally not a problem that concerns an industrial safety engineer. However, dilution of the air so that the percentage of oxygen is reduced is a frequent cause of industrial fatalities. Figure 24-2 indicates the level at which various effects of oxygen deficiency occur.

Oxygen deficiencies can occur where there is leakage of an unbreathable gas, such as nitrogen, into a confined space. (NASA workers, who should have known better, were overcome when they entered a space filled with nitrogen to make a connection before a launch.) A fire will not only consume the oxygen present but will replace it with carbon dioxide, which is also inert. When one or more persons are in an enclosed space with inadequate replenishment of fresh air, the oxygen level may be reduced dangerously. Workers in blocked mines and children trapped in refrigerators are examples of persons suffocated this way. An excessive number of persons in a room can lower the available oxygen to the point where some occupants suffer from hypoxia. Figure 24-3 indicates the effects on and responses of persons exposed to various concentrations of carbon dioxide.

The blockage of respiratory passages can be complete or partial. Complete blockage can be caused by smothering, where an external barrier, such as sand from a cave-in, covers the head of a worker. The windpipe may be closed by external pressure such as a sharp blow or strangulation. Blockage can also result by a piece of food lodging in the throat or trachea; overwhelming amounts of water (as in drowning) or other fluids; accumulations of particulate matter, such as from silicosis, "black lung," or emphysema; or inflammation and swelling of the passage linings. In attacks of asthma, the muscular tissues of the bronchioles contract spasmodically, reducing the size of the passages and the amount of air that can pass.

The concentration of oxygen in the lungs may be sufficient for normal respiration, but passages from the alveoli to the blood capillaries may be blocked. One of the principal causes of such interference is pulmonary edema, the presence of plasma in the alveoli.

The plasma passes from the tissue spaces into the air sacs because of the increased permeability of the alveolar walls, sometimes caused by an irritant which produces inflammation of the tissues. The plasma fills the spaces usually occupied by air, diffusion of air into the bloodstream is stopped, and the person suffers from lack of oxygen. In an extreme case, the person literally drowns in his or her own body fluids.

Hypemic Hypoxia

Hypemic hypoxia is the disturbance of the ability for the blood to carry oxygen to the tissues, two of the commonest causes being hemorrhage and anemia. Both involve loss of red cells and reduction in the total quantity of hemoglobin. This in turn means a reduction in the ability of the blood to transport oxygen.

Interference with the process by which red blood cells absorb oxygen is another cause of hypemic hypoxia. Bonds similar to those between hemoglobin and oxygen can occur between hemoglobin and other gases. Hemoglobin has a much greater affinity for carbon monoxide, with which it forms carboxyhemoglobin (HbCO), than for oxygen. The worst effect is that the combination of carbon monoxide with its iron atoms results in the unavailability of the hemoglobin to combine with oxygen. This produces the same effect as though the red corpuscles themselves had been lost. The affinity of carbon monoxide is so great that its presence in a concentration of only one-half of 1 percent (5,000 ppm) in air breathed will reduce the ability of blood to transport oxygen by 50 percent. Because of this greater affinity, the release of carbon monoxide occurs very slowly, even when the victim is brought to uncontaminated air. However, the relief process with air alone may require several hours. Oxygen and carbon dioxide will both cause disassociation of carbohemoglobin, and the use of a mixture of both will accelerate relief from monoxide poisoning.

Other chemicals will also combine with hemoglobin to produce hypemic hypoxia (although this is infrequent). These affecting chemicals include nitrates, nitrites, and other oxidizing agents which react with hemoglobin to form methemoglobin (MetHb).

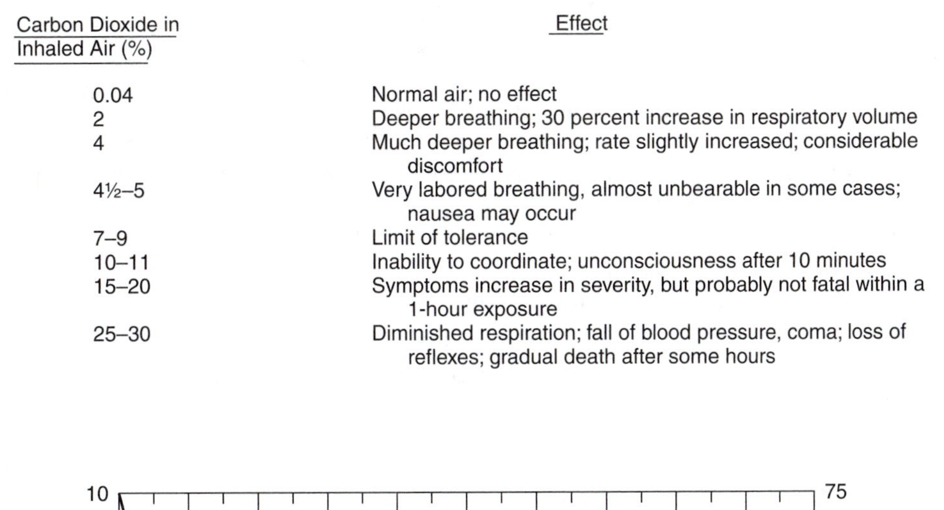

EFFECT OF CARBON DIOXIDE ON PEOPLE

Carbon Dioxide in Inhaled Air (%)	Effect
0.04	Normal air; no effect
2	Deeper breathing; 30 percent increase in respiratory volume
4	Much deeper breathing; rate slightly increased; considerable discomfort
4½–5	Very labored breathing, almost unbearable in some cases; nausea may occur
7–9	Limit of tolerance
10–11	Inability to coordinate; unconsciousness after 10 minutes
15–20	Symptoms increase in severity, but probably not fatal within a 1-hour exposure
25–30	Diminished respiration; fall of blood pressure, coma; loss of reflexes; gradual death after some hours

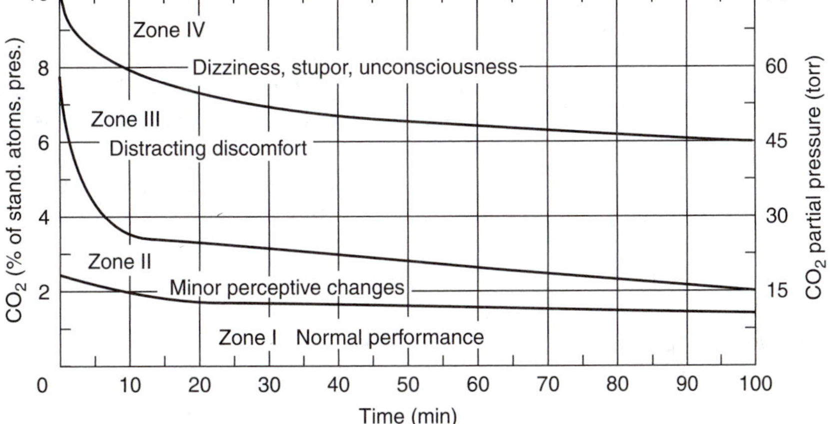

Toxicity of carbon dioxide.* Curves are adapted from the study of King and the review of Nevison. Zones fade into one another less clearly than shown and include symptoms common to most but not all subjects. In Zone I no psychophysiological performance degradation is noted. In Zone II small threshold hearing losses have been found and a perceptible doubling in depth of respiration. In Zone III are found mental depression, headache, dizziness, nausea, "air hunger," and decrease in visual discrimination. Zone IV represents marked deterioration with in ability to take steps for self-preservation.

*E. M. Roth, *Space-Cabin Atmospheres, Part II: Fire and Blast Hazards,* Lovelace Medical Foundation for Medical Education and Research, for National Aeronautics and Space Administration, NASA SP-48, Washington, D.C., 1964.

FIGURE 24–3 Effect of Carbon Dioxide on People

Stagnant Hypoxia

Stagnant hypoxia is due to inadequate circulation of the blood, and consequently of oxygen, through the body. Dehydration by perspiring excessively without adequate liquid intake will reduce fluid volume, sodium, and potassium. Losses of sodium and potassium will cause interference with absorption and release of oxygen and carbon dioxide by the blood. The loss of fluid causes an increase in viscosity of the blood and a reduction in the velocity with which it moves through the circulatory system. This decreases the volume of blood moved, the volume of oxygen delivered to the tissues, and the amount of carbon dioxide removed. Dehydration can be caused by burns, injuries such as crushing of tissues, or excessive excretory losses. Burns, injuries, or bleeding can cause shock, with a reduction in arterial blood pressure and therefore the volume of blood circulated.

Interference with circulation can also be caused by constricted blood vessels. Constriction can be due to disease or to pressure such as that created by use of a tourniquet. For this reason, the tourniquet is dangerous and its use is restricted to extreme emergencies, when it may be employed for only a short time.

Histotoxic Hypoxia

Histotoxic hypoxia is a disturbance of the tissues' ability to absorb and use oxygen carried to them by the blood. This is generally due to the presence of a toxic material, commonest of which are alcohol, narcotics, or strong poisons such as cyanide. Cyanide forms cytochrome oxidase, an enzyme in the oxidation process in the tissues, a highly complex, stable, and inactive combination. The cytochrome oxidase causes loss of the body's ability to carry out its normal function, preventing use of oxygen for normal metabolic processes.

Combinations

Combinations of these types of anoxia can occur. A person suffering from hemorrhage and shock may suffer from both hypemic and stagnant hypoxia. If there is unconsciousness or paralysis of the respiratory muscles, hypoxic hypoxia can also be present.

MECHANISMS OF TOXIC AGENTS

The mechanisms by which toxic agents cause injury differ. Some cause injury by more than one means, the effects of some being immediately apparent while those of others are delayed. They have been categorized broadly into seven types: asphyxiants, irritants, systemic poisons, anesthetics, neurotics, corrosives, and carcinogens.

Asphyxiants

Although the term "asphyxia" is commonly used with the connotation of "suffocation," it actually means hypoxia with the presence of high carbon dioxide in the blood and alveoli. Asphyxiants can be either "simple" or "chemical." Simple asphyxiants are generally considered to be those gases which dilute air to such an extent that the blood receives an inadequate supply of oxygen.

Some of the simple asphyxiants are carbon dioxide, nitrogen, methane, hydrogen, and the noble gases. All of these are especially insidious because they are odorless and colorless. Natural gas, which consists mainly of methane, is also odorless, except when it contains impurities that have distinctive odors. An odorant, such as ethyl mercaptan, is used to warn of leakage of natural gas. Ethyl mercaptan is so odorous that one part in one million can be detected easily.

Hydrogen and helium diffuse rapidly in the open, since they are both lighter than air. Nitrogen at room temperatures is slightly lighter than air, of which it constitutes the greatest part. At low temperatures, such as during leakage from tanks containing the cryogenic liquid or where rapid expansion cools the gas, nitrogen becomes heavier than the surrounding air. Cold nitrogen, carbon dioxide, and methane are all heavier than air and will settle in natural depressions such as swamps or excavations, sumps, pits, the bottoms of tanks, or other low spots where unwary persons could be asphyxiated.

Chemical asphyxiants are those toxic agents which enter into reactions to cause hypemic or histotoxic hypoxia. Some of these are carbon monoxide, nitrites, hydrogen sulfide, and aniline.

Irritants

Irritants injure the body by inflaming the tissues at the point of contact. The inflammation is evidenced by heat, redness, swelling, and pain. Mild irritants produce hyperemia, in which the capillaries dilate and fill with blood, causing the redness and increased heat. Blood vessel linings become sticky and white blood cells become attached loosely. The permeability of the capillary walls changes, and fluid passes from the blood into the spaces between the tissues. This increase in fluids induces swelling, which, when applied to nerve endings, causes pain. Strong irritants can produce blisters, which are localized collections of fluid between the layers of skin causing elevation of the outer layers.

Even a small amount of irritant can cause physiological injury to an extensive area of tissue. In the respiratory tract, the effects may differ depending on the exact location and type of the irritant. The irritation can be chemical or mechanical. Irritation of the upper tract can cause inflammation and pain in the larynx, pharynx, and nose. There might be increased flow of mucus and congestion of the throat. In the lower respiratory tract, irritation can cause pulmonary edema, the exudation of fluid which accumulates in and between the alveoli. The effect is not painful, but the injured person begins to suffer from hypoxic hypoxia. Another effect of some types of irritants is reflex action, caused by the stimulation of the brain, the holding of breath, and then gasping.

An irritant can be a gas, liquid, or fine particulate matter. Ammonia, acrolein, hydrazine, and hydrofluoric acid cause injury to the upper respiratory tract; chlorine, fluorine, and ozone along the entire length; and nitrogen tetroxide and nitric acid to the lower portion and the alveoli. Certain gases can be innocuous in the pure and dry state, but injurious when wet. Nitrogen tetroxide breaks down into nitric oxide, which reacts with moisture to form nitric acid. Moist nitric acid is more harmful to the tissues than is dry nitric acid. Silver and potassium nitrates, zinc and ferrous sulfates, and zinc chloride will irritate and damage the skin and other tissues.

Particulate matter may be inert and harmless in itself, not being toxic if ingested by mouth. However, its very presence in the respiratory passages and lungs can be mechanically irritating. Asbestos, for example, is a nonreactive substance whose fibers can be irritating merely by their presence in the respiratory system.

Asbestos is a prime example of the toxic effects of particulate matter (see Fig. 24-5). *Time*, January 28, 1974, observed that out of 869 persons employed for 17 years in an asbestos plant in Texas (which was closed in 1971), 300 workers "will die of asbestosis (a permanent and often progressive scarring of lung tissue and from inhaled fibers), lung cancer or cancer of the colon, rectum or stomach." After its inception, OSHA standards therefore stipulated that until July 1, 1976, no employee would be exposed to any environment containing more than five fibers longer than 5 micrometers per cubic centimeter of air for longer than 8 hours (time-weighted average), and in no case to an environment containing more than 10 such fibers. After July 1, 1976, the 8-hour time-weighted average exposure might not exceed two fibers per cubic centimeter. However, after asbestos was found to be a carcinogen, this limit was reduced to zero exposure. OSHA and NIOSH have established similar very low or zero levels for other substances.

The nasal passages contain the first defenses against harmful particulate matter entering the lungs. The hairs at the entrance act as a filter against larger particles, and the ciliated cells of the mucous membranes create a fluid flow that traps and washes out smaller particles. These ciliated cells are present over the nasal passages and accessory sinuses. Irritants in the nasal passages cause sneezing, a reflex action involving a deep inhalation, closing the oral cavity by the tongue, and then a sudden expiration through the nose and mouth.

Passages of an irritant solid, liquid, or gas in the pulmonary passages may then be blocked by an instantaneous muscular contraction which closes the glottis. Smaller particles may pass the glottis into the air passages, where again they might be caught in mucous fluid. Those caught are then moved back up the passages by the cilia lining of the larger tubes, or by rhythmic muscular motion of the smaller passages. Extremely small particles might pass into the circulating blood, where they are absorbed by white cells or into the lymphatic system of the lung.

Particulate matter 5 to 6 micrometers in diameter and larger might be removed by the trachea, bronchi, bronchioles, and alveolar ducts. Ninety-seven percent of the particles of 1-micrometer radius or more are taken out by the lungs, with only 3 percent exhaled. As particle diameter decreases, so does the percentage filtered out from further passage down the respiratory system. Findeisen has indicated what generally happens to particulate matter of various sizes:

Extremely fine material can sometimes be removed from the lungs or air passages by a cough. This involves an inhalation, closure of the glottis, then forceful exhalation of air from the lungs. This generally carries the irritating matter out of the air passages. Occasionally, however, the inhalation may carry matter deeper into the alveoli. The cough may attempt to remove not only particulate matter but any sputum created by the lungs.

Unfortunately, the defense mechanisms of the body are often overwhelmed by massive and continued invasion of the respiratory system by fine solids. Except for those minute amounts removed by coughs, the particulate matter will remain. Long-time exposure to coal dust has resulted in "black lung," to fine rock dust in silicosis, and to asbestos in asbestosis and, even worse, cancer. Cigarette smoke has caused emphysema, or coating of the respiratory passages with fine particulate matter which blocks normal respiratory function. Cigarette smoke has also been blamed as the irritant which causes cancer of the lungs.

Receptors in the linings of the larynx and pharynx provide other protective mechanisms. When stimulated, these receptors send impulses to the respiratory center to inhibit breathing. An irritating gas such as ammonia or acid fumes will provide the stimulation. The respiratory center inhibits breathing, causing an involuntary gasp (catch of the breath) which prevents the irritant from proceeding further. In the same way, food passing accidentally into the larynx stimulates other receptors. Breathing is temporarily inhibited, so food cannot enter and injure the lungs.

Eyes are sometimes affected by extremely small amounts of irritant, causing them to redden and water. An irritant of this type is called a "lachrymator." Eye sensitivity is becoming more and more a problem as the atmosphere increasingly is polluted. In addition, the necessity for personnel to spend long periods of time in totally enclosed atmospheres has focused interest on this problem. In such cases, extremely small amounts of contaminants can provide deleterious effects far more damaging than in outdoor atmospheres.

Systemic Poisons

Systemic poisons cause injury after they have been carried to the tissues of the body, especially to specific organs. Many of these systems cause histotoxic hypoxia by interfering with the use of oxygen in the metabolic process. Others interfere with other reactions necessary for the organs to carry on their normal functions; for example, some toxic agents affect the tubes of the kidney so they become nonfunctional. As a result, the filtrate normally produced and excreted by the kidneys seeps back into the circulating blood. Poisons may be excreted through the kidneys, lungs, liver, gastrointestinal tract, skin, and salivary glands. Most damaging effects of some poisons are found at these points.

Certain chemical agents are especially harmful to specific organs, less harmful to others, and harmless to some. Systemic poisons can be divided into four categories—those that:

1. Cause injury to one or more of the visceral organs such as the kidneys or liver. The majority of the halogenated hydrocarbons, besides having narcotic and anesthetic properties, belong in this group.
2. Injure the bone marrow, spleen, and the blood-forming system, causing anemia and reduced count of white blood corpuscles. These toxicants include naphthalene, benzene, phenol, and toluene.
3. Affect the nervous system and cause inflammation of the nerves and neuritis. The result is tenderness of the nerves, pain, interference with transmission of nerve

impulses, and even paralysis. Atrophy of the optic nerve and blindness can result from absorption of materials such as methyl alcohol or carbon disulfide.

4. Some toxic metals and nonmetals not only cause respiratory system damage as irritants, but can injure the body by being swallowed or entering the bloodstream through skin lacerations. These metals and nonmetals can be deposited in and interfere with functions of the body organs, bones, and blood. Generally, the effects of these substances are chronic in nature and take place only after continued and massive exposure. Arsenic compounds have acute effects when swallowed. A few, such as lead, are cumulative. Lead, mercury, cadmium, antimony, phosphorus, and sulfur are toxic nonmetals.

A toxic substance is called "reactive" when it changes through metabolism so its toxicity is eliminated. The elimination of reactive system poisons is influenced by the alteration undergone within the body.

Sometimes, metabolic changes create toxic products more injurious than the substance originally absorbed. Formic acid and formaldehyde are produced in the body from methyl alcohol in this way. The acidosis that results can cause death. Because time is required for some reactive substances to change within the body to other compounds, serious symptoms may not occur for several days after initial ingestion.

A nonreactive substance is one unchanged to any great extent within the body. The effects of a nonreactive agent depend on its solubility in body fluids, the rate of absorption, the frequency and depth of respiration, and the rate of circulation of the blood. Simple hydrocarbons, such as methane and ethylene, are examples of nonreactive gases.

Anesthetics. Anesthetics cause the loss of sensation in the body. They can be general or local. They cause respiratory failure by depressing the nervous system and interfering with involuntary muscular action. This interference may interrupt the reflex process controlling inspiration and expiration. A number of halogenated hydrocarbons generally used as cleaning or degreasing agents can produce this effect. Examples of narcosis-producing agents are trichloroethylene, ethyl ether, chloroform (trichloromethane), ethylene, and nitrous oxide.

Neurotics. Neurotics affect the nervous system, brain, or spinal cord. They may be either depressants or stimulants. Depressants exhilarate for a short period of time and then cause the person to become drowsy and lethargic. Breathing may be labored; there may be cyanosis, a clammy skin, and loss of consciousness. Alcohol and drugs can not only cause exhilaration and then deep depression but also affect the senses adversely. Drivers and other equipment operators under the influence of either alcohol or drugs are potential candidates for accidents. Concerns regarding such persons have led to laws regarding levels of such substances in the blood.

Depressants. A depressant can be selective and act on specific organs, such as the brain, while others are affected little. However, increasing the amount of the toxicant can affective sensitive organs little by little. Ethyl alcohol is a depressant (but not a stimulant), especially in the higher concentrations used for medical purposes or fuel. The body always maintains a residual level of 0.004 percent alcohol from the decomposition of glucose. Alcohol taken internally is distributed by the blood to the tissues, where 90 percent is oxidized to acetaldehyde, then to carbon dioxide and water. The remainder is excreted in urine, perspiration, and expired air. The carbon dioxide and water form carbonic acid, which reduces the alkalinity of the blood, depressing the inhibitory and respiratory centers of the brain. The rate of oxidation is fairly constant, a normal adult being able to metabolize about two-thirds of an ounce of liquor containing 50 percent (100 proof) alcohol. One and one-half to 2 pints of whiskey taken in a few minutes may prove fatal, because the violent drop in alkalinity causes an extreme reduction in depth and rate of respiration.

The effects of alcohol on a drinker are shown in Fig. 24-4. Effects may vary from person to person. In many states a person is considered under the influence of liquor when tests indicate the presence of 0.10 percent blood alcohol concentration (BAC) in the blood. Many states have dropped the BAC limit to 0.08. Police frequently look for and go by other indications (balance, speech, odor on the breath), but these indications are arguable. Actual test results generally are not.

Drugs such as marijuana, cocaine, hashish, LSD, and others also have adverse effects on the body (and safety). The effect of each depends on the type of drug, dosage, time since taken, susceptibility, and other factors. Although a blood test can indicate the fact that a drug was taken, there is no legal limit to its presence to consider a person under its influence as there is for liquor. Some organizations support the need for punitive action shown simply by an illegal drug's presence to any degree.

Stimulants. Stimulants accelerate the function of a body organ or system. They may excite the affected person, sometimes to a very slight extent, sometimes more radically. In a more extreme case, the person may have a rapid pulse, jerking of the muscles, disorganized vision, and sometimes delirium. Caffeine, found in small amounts in coffee, stimulates the brain and nervous system. Other drugs, taken to help overcome fatigue or workers, wear off, operators fall asleep, and accidents result.

Hypnotics. Hypnotics are sleep-inducing drugs. They include paraldehyde, chloral hydrate, and the barbiturates.

Carcinogens

Certain chemicals have been found to cause, or are suspected of causing, cancer to internal organs and systems of the body (see Fig. 24-5). It had previously been known that cancers of the skin could be caused by tar, bitumen, mineral oil, anthracene, and their compounds, products, and residues. Some of these chemicals are suspect because in their presence in some areas, environments, or conditions populations of persons have suffered an extraordinary amount from cancer. Cancers in laboratory animals, such as mice, have been attributed to other substances, but there is quite a degree of uncertainty as to whether they actually are so caused. Where doubt exists, and their unguarded use considered to be injurious, such substances are usually avoided or used while safeguarded against. A few of the widely used chemicals considered or suspected of being carcinogens include vinyl chloride, benzidene (and its salts), ethyleneimine,

SCALE OF TOXIC SYMPTOMS OF ALCOHOL*

Alcohol in the blood		Subjective States and Observable Changes in the Behavior Under Conditions of the Heavy Social Drinking
mg/ml	%	
0.10	0.01	Clearing of the head. Freer breathing through nasal passages. Mild tingling of the mucous membrane of the mouth and the throat.
0.20	0.02	Slight fullness and throbbing at the back of the head. Slight dizziness. Sense of warmth and general physical well-being. Minor body aches and fatigue are relieved. At this point, subject is not worried about personal appearance. He is willing to talk and is generally quite pleasant.
0.30	0.03	Mild euphoria. Time seems to pass very quickly. There is no sense of worry. Subject feels and acts superior. He does not want to go home.
0.40	0.04	The subject has lots of energy. He talks freely and in a loud voice. Memories are vivid and rich. His hands may tremble slightly and movements become a bit clumsy. He usually laughs loudly at minor jokes and is not embarrassed by mishaps he may cause.
0.50	0.05	At this point, normal inhibitions are absent. The subject acts upon impulse and takes personal and social liberties not usual to his character. He is long-winded and enlarges on past experiences. He feels as if he is sitting on top of the world, and it is at this point that he can "lick anybody in the joint."
0.70	0.07	Subject experiences feeling of remoteness. There may be odd sensations when he rubs his hands together or touches his face. The pulse is rapid and breathing is heavy. He becomes amused at his own clumsiness and asks other people to do things for him.
1.00	0.10	There is a very noticeable stagger. May talk to himself or sing loudly. Fumbles and has difficulty in controlling accurate motion. Has difficulty remembering and finding things. Feel drowsy.
2.00	0.20	Needs help to walk or undress. Easily angered. May shout, groan, and weep in turns. Is nauseated and has poor control of urination. Cannot recall details of the evening.
3.00	0.30	In a stuporous condition, very heavy breathing, sleeping and vomiting by turns. No comprehension of language. Strikes wildly at the person who tries to help him.
4.00	0.40	Deep anesthesia, which may be fatal.

*From *Drug Abuse, Chemical Recognition and Treatment*, Air Force Manual 160-33, April 25, 1966, with percentages of alcohol in the blood added.

FIGURE 24–4 Scale of Toxic Symptoms of Alcohol

CARCINOGENS

Inhalation of solids through the respiratory system products emphysema from cigarette smoking, silicosis (black lung) from coal dust, and byssinosis (brown lung) from textile fibers. But the effect that is feared most through inhalation is cancer generated via a carcinogen-causing agent, which may not otherwise be considers toxic.

The case most widely publicized concerns the results of the inhalation of asbestos fibers in even small amounts. It has been found that persons who have been subjected to otherwise comparatively minor exposures can be affected. Workers employed for only a few days at plants where asbestos is used or even handled, although they did not process it, have been afflicted. Workers have inadvertently carried asbestos on their clothing to their families at home where at members have sickened and died. Persons in automotive repair industries who replace or work with brake linings, clutch facings, packings, and other products that contain small amounts of asbestos have become cancerous and died.

The effects of asbestos are latent and may not appear immediately, but may take take themselves evident a month to 45 years after exposure. There is no recovery from asbestosis, nor from the noncancerous but debilitating emphysema, silicosis, or byssynosis, and asbestosis may be more painful. The problem was first recognized in 1898, but the effects were not immediately apparent. Workers stayed on the job because the serve effects of asbestosis were not known until it was too late.

The effects of the disease, in addition to cancer, are similar to the effects of other diseases which are due to the inhalation of solids, such as emphysema. There is coating of the respiratory system and blockage which diminishes the capacity to breathe. There is thickening and hardening of the lungs and scarring of the tissue, and other effects such as clubbing of the fingertips and enlargement of the heart and heart failure. Some of these effects, as with similar diseases, take place after extended exposure, but as pointed out, asbestosis can cause cancer even in case of relatively mild and short exposures.

Although there are several types of asbestos (chrysolite, crocodolite, riebekite, and others), they all produce cancer, but with different effects. Lung cancer cancer caused by asbestos generally takes place in the small respiratory passages that lead to the lungs, the bronchiole. Mesothelioma occurs in the linings of the lung or the tissues of the abdomen. Pleural plaques are shown on x-rays as localized, thickened, calcified patches that are not fatal in themselves, but may indicate the existence of diseases that will be fatal.

The insidiousness of asbestos is such that bags of asbestos, entire plants and products utilizing the substance, and clothing and materials in which it might be carried or embedded are disposed of by burial. Workers who must remove asbestos from buildings and other facilities must be protected with special equipment and procedures. Previously, some industries required that respirators be used where asbestos was handled. This was doubly dangerous: not only was the asbestos cancerous, but the workers frequently failed to wear the respirators because they were uncomfortable. Many workers considered the seemingly innocuous material safe but were affected later.

It has been estimated that over 50,000 lawsuits have been initiated against one company alone, The Johns-Manville Company, asking for compensation because of injurious and deaths due to asbestos inhalation. These suits have totaled more than $2 billion, a sum beyond the amount the company claims it can pay, and the company has declared bankruptcy. As a result, most of the claimants may receive no compensation from John-Manville. The magnitude of the claims brought forth the idea that the federal government, which has sometimes been named as codefendant, pay for the injuries and deaths, especially for federal workers who worked with asbestos. The government has, however, denied that it has any legal responsibility.

EDB (ethylene dibromide), a widely employed insecticide and anti-knock additive for gasoline, has been found to be deadly and to create an extraordinarily high risk of cancer. Statistics indicate that 990 or more of every 1,000 workers exposed to EDB regularly will develop cancer. EDB has no detectable odor until the amount present is hundreds of times greater than the limit that was formerly indicated to safeguard workers. The original list was predicted only on possible chemical toxicity damage to workers' lungs, livers, and kidneys, but studies have shown that EDB is also a greater cancer threat.

FIGURE 24–5 Carcinogens Continued on page 456

Other substances with widespread uses are becoming known or suspected of being carcinogens. Many, such as dioxin, affect entire families, especially workers. For women and some men, chemicals may harm or prevent future families because of their genetic and reproductive effects. The only way to mitigate injuries from the use of chemicals is either to eliminate or reduce public's exposure to them, or to institute close and strict control of all workers involved in their processing or use.

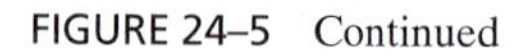

FIGURE 24–5 Continued

methyl chloromethyl ether, trichloroethylene, and a long list of others found at one time in the workplace. Many of these were found only after long complaints and tests. Gradually, use of such toxicants has been eliminated, but other chemicals also have been found to be carcinogens. The Department of Labor has stipulated that no persons be exposed to any chemical that has been suspected of carcinogenic activity. Each suspected agent may be employed only in controlled, safe areas.

Skin Absorption

The intact skin (see Fig. 19-2) is an excellent barrier to passage of most chemicals, especially against water solutions. However, many chemicals will be absorbed through the skin, while others, which will not be absorbed through the intact skin, may enter the body through cuts, blisters, or wounds. Absorption this way can be more dangerous than through the respiratory or digestive systems, which may provide defensive mechanisms, since chemicals are taken directly into the bloodstream. Those that will be absorbed even through an intact skin include tetraethyl lead, used as a knock suppressant in high-octane gasoline, aniline, hydrazine, the boranes, and nitroglycerine.

Corrosives. Corrosives damage by chemical reaction with the skin they contact. In addition to injuring the skin and underlying tissues, they provide a point of entry for the toxicant to reach the bloodstream, producing an effect worse than skin damage.

Corrosive burns can be caused by strong acids or alkalis. Alkalis can cause progressive burns, the injury increasing as the alkali moves through the damaged tissue. This is especially critical in injuries to the eye, where delicate tissues can be damaged little by little until vision is destroyed. If a corrosive chemical is swallowed, it will cause pain in the mouth, throat, and stomach. There usually will be vomiting, difficulty in swallowing and breathing, and distention and gas causing pain in the stomach.

Some corrosives are concentrated acids, such as nitric, hydrochloric, sulfuric, and oxalic; strong alkalis such as sodium or potassium hydroxide; and reactive elements, such as iodine, chlorine, or fluorine.

The severity of a corrosive burn depends on the concentration and type of corrosive chemical, whether the contact was covered or uncovered, and the length of time of contact. For this reason, a harmful agent should be washed away as soon as possible and neutralized with a mild antidote if one is available.

A covered skin contact usually creates a more severe skin reaction than one which is uncovered. Tricresyl phosphate evaporates rapidly through an uncovered skin, with a sensation of coldness and a brief mild redness. Clothing wet with tricresyl phosphate produces a burning sensation and blisters like those of a second-degree burn. Even less harmful liquids, such as gasoline, will produce similar reactions. The importance of prompt removal of contaminated clothing and of damaging chemicals is apparent.

Dermatitis. Dermatitis is an inflammation caused by defatting of the skin or by contact with an irritating or sensitizing substance. Exposure to solvents often causes removal of oils that keep the skin soft and pliable, making it dry, scaly, and somewhat thickened, and with a tendency to crack easily. Some redness may result from the irritating effects created by the absence of fats. Such skins often have poor resistance to bacterial infections and heal slowly when injured. Replacement of oil with creams and lotions to control the condition is only partly effective. Generally the only hope of recovery is removal from further direct contact with the damaging chemical.

Many irritants that will damage the respiratory passages will also affect the outer skin. The outer skin is tougher, but certain portions are extremely sensitive, especially when they are wet. Continued exposures, such as from cloths saturated with solvents, dusts, or other contaminants, can cause skin irritations and dermatoses. A primary skin irritant causes dermatitis almost immediately by direct action on the skin. A sensitizer may not cause injury immediately but will produce a susceptibility to a second attack or to other substances.

Ordinary dermatitis due to irritations can be cured even while the affected person is at work by reducing or eliminating contact by protective clothing for the skin. Where a sensitivity or allergy dermatitis exists and the skin is affected even by small amounts of the chemical, the affected worker should be removed from the vicinity of the offending chemical. A flare-up of the condition may sometimes result from only small traces of the substance. In any case, the best practice is avoidance of direct skin contact with chemicals by use of protective equipment. A few of the common chemicals that cause dermatitis are gasoline, formic acid, carbon disulfide, cresol, barium compounds, butyl alcohol, and naphtha.

Eye Damage. Of all skin contacts, those with eyes are the most damaging, because of their sensitivity. Most materials have the ability to injure the eyes to some degree. Solids can harm either by abrasion or by chemical action. The mildest injury is probably irritation, which causes redness, watering, and stinging. More severe irritation can damage the corneas, the transparent covering of the eye, causing a dry scratching feeling and various levels of pain.

A corneal burn is the commonest chemical eye injury. Corrosive vapors or fine spray can cause tiny burn spots. Contacts with a strong mineral acid or alkali can destroy vision. The tendency of alkali burns to spread, even though emergency treatment has been given, makes them particularly troublesome.

A hazard that has become common over the years is the possibility that a fine spray or splash of corrosive material might become caught between a contact lens and the eye. This keeps the chemical in close contact with the eye for an abnormally long time, aggravating the burn.

Gastrointestinal Absorption

The other important route by which toxic agents can enter the bloodstream is through the gastrointestinal system. Entry of the poison into the digestive system might occur when a toxic substance is ingested unknowingly or when contaminated fingers, food, or other objects are placed in the mouth. Washing the hands prior to eating is an important means of preventing accidental poisoning. Facilities for washing in industrial plants where a toxic substance might be present is a must to prevent this type of injury.

It has already been mentioned that the respiratory system may reject particulate matter, which might then enter the gastrointestinal tract. Solid particles filtered out by the cilia are moved up slowly to the mouth. Swallowing these accumulations may then permit absorption of the toxicant through the digestive system. It is estimated that 50 percent of the particulate matter deposited in the upper respiratory passages and $12\frac{1}{2}$ percent from the lower passages are subsequently swallowed.

Gastrointestinal absorption sometimes is affected by three processes that are generally not present if toxicants enter the body by other routes.

- Potentiation is the increased effect a toxic substance may have if it is mixed with a second substance. The second substance in this case causes an increased effect by the first substance, such as by sensitizing a location at which damage can take place.
- Synergism is a combined effect two toxic agents can generate which is more injurious than the total the two effects could generate by acting separately.
- Antagonism is the lessening effect one substance can have on another.

MEASUREMENT OF TOXICITY

Although some persons may be allergic and abnormally sensitive to specific substances, even persons who may be considered "normal" vary in their sensitivities to toxic materials. Attempts have therefore been made to formulate means of measuring the effects of various concentrations of toxicants. Unfortunately, the numerous effects that can be produced by a great number of toxic chemicals at various concentrations and over different lengths of exposure make a simple procedure impracticable. There are therefore a considerable number of systems in use, a few of which are discussed here. For any specific chemical, it is possible to obtain much toxicity data from the manufacturer or from a suitable reference.

LCtD, ID, and so on. One method expresses toxicity hazard as the product of the vapor concentration and exposure time that will produce a given response in a specific percentage of normal subjects. The concentration (C) is given in milligrams per cubic meter, and time is in minutes. The same response might be obtained by a short exposure to a heavy concentration as by a long exposure to a mild concentration. The time-concentration factor (C_t) is used for materials that may enter the body through the respiratory system. Toxicity entering through the skin is expressed by the dose (D) in milligrams of solid or liquid in contact with the body, which produces a similar response. Either factor, C_t or D, is generally qualified by an indicator (L) or (I) to show whether the amount expressed is lethal or incapacitating.

Minimum lethal dosage (*MLD*) is the smallest amount that will kill the most susceptible of a group of test animals. LD_{50} is the median dosage, the amount that will be lethal to 50 percent of the group. Limits other than the median, such as 25, 90, or 100 percent, may also be used. Thus, an IC_{t25} indicates the time-concentration factor of toxicant vapor that will incapacitate 25 percent of the subjects. Incapacitating dosages may be much less than lethal dosages. For example, LC_{t50} for chlorine is 19,000 mg-min/m^3, but IC_{t50} is less than one-tenth that amount—1800 mg-min/m^3.

The toxicity effects of various concentrations and dosages are usually obtained by tests on animals. Dosage is expressed in milligrams per toxicant per kilogram of body weight of the animal used in the laboratory. From that and extrapolation, a rough approximation of dosage that will affect humans can then be calculated for trial use. However, data obtained by experimentation with animals must be applied selectively and carefully when used this way, since a human may react differently.

Threshold Limit Values (TLVs). TLVs are primarily for toxic agents that enter the body through the respiratory system. A threshold limit value indicates the average concentration of toxic agent that can be tolerated during exposure for a 40-hour week continuously in a normal working lifetime. (Older literature used the term "maximum acceptable concentration (MAC)," which indicated the maximum concentration not to be exceeded during a 40-hour week. TLVs and MACs were actually the same in most cases.)

Threshold limit values established by the American Conference of Government Industrial Hygienists (ACGIH) for dangerous chemicals have been cited in the OSHA standards. OSHA now sets permissible exposure limits (PELs), which are regulatory, enforceable limits. OSHA PELs are based on an 8-hour time weighted average (TWA) exposure. Currently, approximately 500 PELs have been established (see 29 CFR 1910.1000). Values have been determined not only from experiments on animals but also from inadvertent human exposure and industrial experience. Information on worker injuries took place under accidental and uncontrolled conditions where the concentration of a toxic agent was not known accurately. For this reason, values do not constitute specific limits between safe and dangerous concentrations, especially since a toxic agent may vary in its effects on different persons.

Limit values are generally used only as guides and for comparative purposes. As added information and experience are gained on a substance's properties, the value listed for it may be revised. (The TLV for vinyl chloride was originally 500 ppm until it was found the chemical caused cancer. OSHA then reduced the permissible level to 0.) TLVs are not indicators of relative hazard unless other considerations, such as volatility, are taken into effect. A highly volatile material can be more dangerous than one less volatile but more toxic. At any temperature, the more volatile liquid will provide more gas, which will outweigh the toxicity of the other. The TLV for methyl alcohol is four times higher than that of furfuryl alcohol. However, its vapor pressure at room temperature is 18 times as great. Four and one-half times as much furfuryl would have to be spilled to provide a toxic hazard equivalent to that of methyl alcohol. Further, furfuryl alcohol has the same TLV as anhydrous ammonia. However, ammonia's much higher vapor pressure makes it 1,100 times more hazardous in case of a spill or leak.

Threshold limit values also have been established by many states, such as Pennsylvania, for substances not listed by the American Conference of Government Industrial Hygienists or OSHA.

Emergency Exposure Limit (EEL). The EEL is a measure of toxicity introduced by the National Academy of Science-National Research Council Committee on Toxicology. It was formulated to indicate the approximate length of time a person might remain without ill effect in an atmosphere contaminated with specific concentrations of toxic gas. As the concentrations increase, the time is reduced. Extremely short exposures even under heavy concentrations may not result in harm; conversely, a heavy exposure, even for a short time, might be damaging. Exposures, even short ones, should be avoided or limited, since they might lead to low-level discomfort, irritation, or intoxication even when lasting for less time than those indicated. Exposures within the EELs generally will not injure the average person, but an exposed individual should undergo a medical examination as soon as possible afterward to insure there have been no adverse effects. Each EEL should be regarded as an absolute ceiling: further exposure can be deleterious to health. EELs have been prepared for only a few chemical agents which were created for use as liquid propellants for missiles.

Comparison of Ratings. Industrial plants are obligated to observe criteria stipulated in OSHA standards. However, it has long been recognized that the threshold limit values, based on 40-hour-per-week exposures, are inadequate for determining the effects of, and limitations that should be set for, short-time massive exposures by toxicants. Smith[90] compared ratings of five chemicals by different means. The results indicated only one thing at the time: that no general relationship existed between the various means of rating toxicity.

Other Ratings. The method of rating toxicants should generally be interpreted only by a qualified industrial hygienist. However, often it is desirable to have other descriptive indicators by which personnel who are less technically knowledgeable can have an easily accessible measure of the degree of toxicity of a substance. Unfortunately a multiplicity of such methods is in use, each with a different numerical scale with a special definition for each rating. (N. I. Sax[91] uses U, 0, 1, 2, and 3, while Underwriters Laboratories uses six groups.) It can therefore be appreciated that when any rating is mentioned, it is necessary to know to what system it relates.

Sources of Chemical Agents

Chief sources of toxicants are industrial process chemicals and emissions, fuels or heating and power, products of combustion, and agricultural chemicals. Safety engineers should be intimately knowledgeable of any chemicals used in their plants, those developed during processing and those which result as end products, usable or waste. It was while an industrial chemical was being processed as an agricultural fertilizer that an accident created the disaster at Bhopal, India. In spite of such events and of the new and exotic industrial chemicals being developed, the most frequent offenders are generally those widely used or generated substances whose dangers have long been known. (More people were killed during maintenance of Air Force intercontinental missiles systems by asphyxiation with nitrogen and only mildly toxic solvents than by hydrazine, nitrogen tetroxide, aniline, and all other highly toxic missile propellants combined.)

Figure 22-9 indicates some of the hazards resulting from combustion. In addition, hydrocarbons, besides being combustible, provide a group of substances that vary from simple asphyxiants to systemic poisons. Commonest of these hydrocarbons is natural gas, which consists principally of methane and of lesser amounts of ethane, butane, and propane, the exact amount depending on the field from which it has been derived. None of these gases has a detectable odor in the pure state; an easily noticeable odorant is therefore added to allow detection of leaks. In spite of this, persons have been overcome and fatalities have ensued due to their presence. Toxicity of hydrocarbon liquids, such as gasoline and jet fuels, is generally higher than those of the gases. Both those and combustion gases are systemic poisons. Vapors from residues have been sufficient to overcome inadequately protected workers who have entered tanks for repairs, maintenance, or cleaning.

Cases of asphyxiation during excavations of deep pits or ditches, in marshes, or other low areas have occurred which have been attributed to hydrogen sulfide, methane, nitrogen, or other gases.

Fatal accidents also have occurred in industry, homes, and elsewhere with chlorine or mixtures that produced chlorine. Chlorine is highly toxic and dangerous, even though it can be recognized easily by its highly irritant, distinctive odor. It is especially deadly when it is produced or leaks into a small enclosed space. Accidents have taken place when bleach (sodium hypochlorite) was mixed with an acid-producing substance such as vinegar. A cleaning man or housewife, believing that if one cleaning agent is good two are better, would mix bleach with another chemical such as ammonia or lye, liberating chlorine, often with deadly results. Bleaching agents should not be mixed with ammonia, lye, vinegar, rust remover, toilet bowl cleaner, or oven cleaner.

DETECTION OF TOXIC AGENTS

In spite of those already mentioned, such as pure hydrocarbon, many toxic gases have distinctive odors or are irritating to the nasal passages. Others can be noted because of their color or taste. (Fuming nitric acid is distinctively colored red.) Many toxic materials have a bitter, sour, or otherwise objectionable taste. Information on characteristics that can be distinguished without the use of chemical or electronic aids can be helpful and should be provided workers involved with such substances.

Where a material does not have a distinctive odor, color, or irritating property, or the acceptable level might not be readily distinguishable by a human, devices have been created to warn of its presence. The level of a toxic (or flammable) gas is set to become apparent long before it reaches a harmfully dangerous point. A few examples of the principles on which these devices are based are presented in Fig. 12-5. Other principles and variations are so numerous that a complete description cannot be given here.

RESPIRATORY PROTECTIVE EQUIPMENT

Needs for this type of equipment can be separated into three broad categories: normal hazardous operations, investigations and corrections, and emergencies. Some equipment can be used for all three.

1. *Normal hazardous operations:* industrial operations are often undertaken which require the use of respiratory protective equipment (and protective clothing). These operations involve spray painting, cleaning with toxic solvents, use of fumigants, and sand blasting. In such situations, the type and expected concentrations of contaminant are known. Further comments regarding personal protective equipment are provided in Chapter 12.
2. *Investigations and corrections:* a situation might arise in which a contaminated atmosphere is suspected or has been detected. Personnel wearing protective clothing might enter an area because of a detected odor, leak, irritating effect, or remote-reading sensing device to determine or correct the problem. Corrective action would be taken as soon as possible, but time is not as critical as it is in an emergency.
3. *Emergencies:* fires, explosions, or ruptures of tanks containing toxic fluids are in this category. Reaction time of the emergency responders is important. Personnel might have been injured or rendered incapable of leaving an area where an imminent danger still existed. Equipment must be provided for them to escape through their own efforts or to be assisted and rescued by others. With a fire, entirely new, more dangerous, and more concentrated toxicants might be present. Equipment in this category must therefore meet more strigent criteria. The equipment must be suitable for high concentrations of extremely toxic contaminants and for varied uses in atmospheres deficient in oxygen, easy to don, ready for immediate use, highly reliable, and have the least possible weight and bulk.

Respiratory protective equipment is of three major types:

1. *Air purifiers:* this equipment purifies contaminated ambient air by mechanical or chemical means. It should be used only in atmospheres containing sufficient oxygen to support life.
2. *Self-contained breathing apparatus:* this equipment contains independent, although limited, supplies of air or oxygen. No outside supply is required during a unit's life. It can provide air or oxygen from a cylinder, or it itself can generate oxygen.
3. *Supplied air equipment:* this equipment provides air or oxygen from a central source through a connected pipeline.

Air Purifiers

Mechanical filter respirators offer protection only against dusts, fumes, mists, or smokes. Filters are fabricated of a fibrous material through which air to be breathed must pass. The filter will trap contaminant particulate but will provide no protection against toxic gases or atmospheres devoid of or deficiently low in oxygen. Mechanical air purifiers are used in industrial operations such as chipping metal, spray painting, or sandblasting.

Chemical cartridge respirators provide limited protection against organic gases. Manufacturers indicate they also have cartridges which will provide protection against some acid gases in concentrations up to 500 ppm and against ammonia up to 1,000 ppm. These chemical types can also be fitted with mechanical filters to remove dust, fumes, mists, and smokes. Chemical cartridge respirators will provide no protection in oxygen-deficient atmospheres and should not be used where organic gases in concentrations greater than 1,000 ppm might be present.

Canister-type respirator equipment is more commonly known by the term "gas mask." This is a misnomer, since masks are used with mechanical and chemical respirators, self-contained breathing apparatus, and airline equipment. To avoid ambiguity, the word "canister" will be used here to indicate the air-purifying container with its contents. The word "facepiece" will be used for the mask itself.

A canister serves functions similar to a chemical cartridge respirator, mechanical filter respirator, or both. Its size is due to the fact that it may serve multiple functions over a long period of time for protection against a number of contaminants. Canisters differ from each other in their abilities because of the chemicals or filter they contain.

Figure 24-6 provides cutaway views of two canisters. The accompanying table lists the types of canisters and the contaminants against which they provide protection. (To save weight and to lessen snagging problems because of long rubber hoses on a worker's body, many plants now provide types which resemble snouts. Their purposes are the same as those of the canisters described here, but their capacity is less.) Air enters through the bottom of the canister and passes through the various materials. Each bed is provided for a specific reason and its presence depends on the intended uses of the canister.

1. A mechanical filter may be provided to eliminate particulate matter. Not all canisters have filters because of their resistance to the passage of air. A filter increases the breathing effort, causing the user to tire rapidly. It is a major reason workers fail to wear such equipment even when they should.
2. Activated charcoal adsorbs contaminated gases. Charcoal adsorbs organic vapors well, and acid gases, such as chlorine and phosgene, to a lesser extent. Adsorption is a complex property of solids, the process of which is not entirely understood because of several affecting factors such as intermolecular attraction and molecular weight. Charcoal's ability to adsorb carbon monoxide, methane, or hydrogen sulfide is very limited. Carbon monoxide can be adsorbed to a slight degree, but a heavier contaminant can then replace it. Ammonia is adsorbed by charcoal to a greater degree than is carbon monoxide, but not to the point where an ordinary canister is adequate for its complete removal.
3. Silica gel will adsorb ammonia better than will charcoal. It is therefore used primarily for that purpose, although it will adsorb other contaminants, such as hydrocarbon gases. Its second function is to adsorb water vapor from the air passing into the layers of Hopcalite, where carbon monoxide is removed.
4. Hopcalite is a mixture of catalysts (60 percent manganese dioxide and 40 percent copper oxide) which induces combustion of carbon monoxide to form carbon dioxide and water. The water is picked up by the Hopcalite, reducing its capacity to provide protection against the carbon monoxide. For this reason, some canisters containing Hopcalite are provided with indicators which change color when a specific amount of water has been adsorbed, warning that the canister will no longer fulfill its intended functions. The reaction of carbon monoxide and oxygen in the presence of Hopcalite produces a considerable amount of heat, varying with the amount of monoxide present. The heat generated in an atmosphere

containing 1/2 percent or more of carbon monoxide may warm the air that is leaving the canister to an unbreathable temperature. In addition, any fuel or oxidizer penetrating the other layers may react exothermically because of the Hopcalite.

5. Molecular sieves of synthetic zeolite are also used as adsorbents. At relatively low humidity they are better adsorbents than silica gel. They will adsorb only small molecules because of their pore sizes; their effectiveness toward organic vapors therefore depends on the specific gas. They adsorb ammonia about as well as silica gel.
6. Soda lime is used to remove carbon dioxide. It consists mainly of calcium hydroxide, with a small amount of sodium hydroxide. Other ingredients include a small

COLOR ASSIGNED TO CANISTER OR CARTRIDGE*

Atmospheric contaminant(s) to be protected against	Color assigned
Acid gases	White
Organic vapors	Black
Ammonia gas	Green
Carbon monoxide gas	Blue
Acid gases and organic vapors	Yellow
Acid gases, ammonia, and organic vapors	Brown
Acid gases, ammonia, carbon monoxide and organic vapors	Red
Other vapors and gases not listed above	Olive
Radioactive materials (except tritium and noble gases)	Purple
Dust, fumes, and mists (other than radioactive materials)	Orange

Notes:

(1) A purple stripe shall be used to identify radioactive materials in combination with any vapor or gas.

(2) An orange stripe shall be used to identify dusts, fumes, and mists in combination with any vapor or gas.

(3) Where labels only are colored to conform with this table, the canister or cartridge body shall be gray, or a metal canister or cartridge body may be left its natural metallic color.

(4) The user shall refer to the wording of the label to determine the type and degree of protection the canister or cartridge will afford.

* This material is reproduced with permission from American National standard, *Identification of Air-purifying Canisters and Cartridges,* Standard K13.1-1973, copyright 1973 by the American National Standards Institute at 1430 Broadway, New York, N.Y. 10018.

FIGURE 24–6 Color Assigned to Canister Cartridges

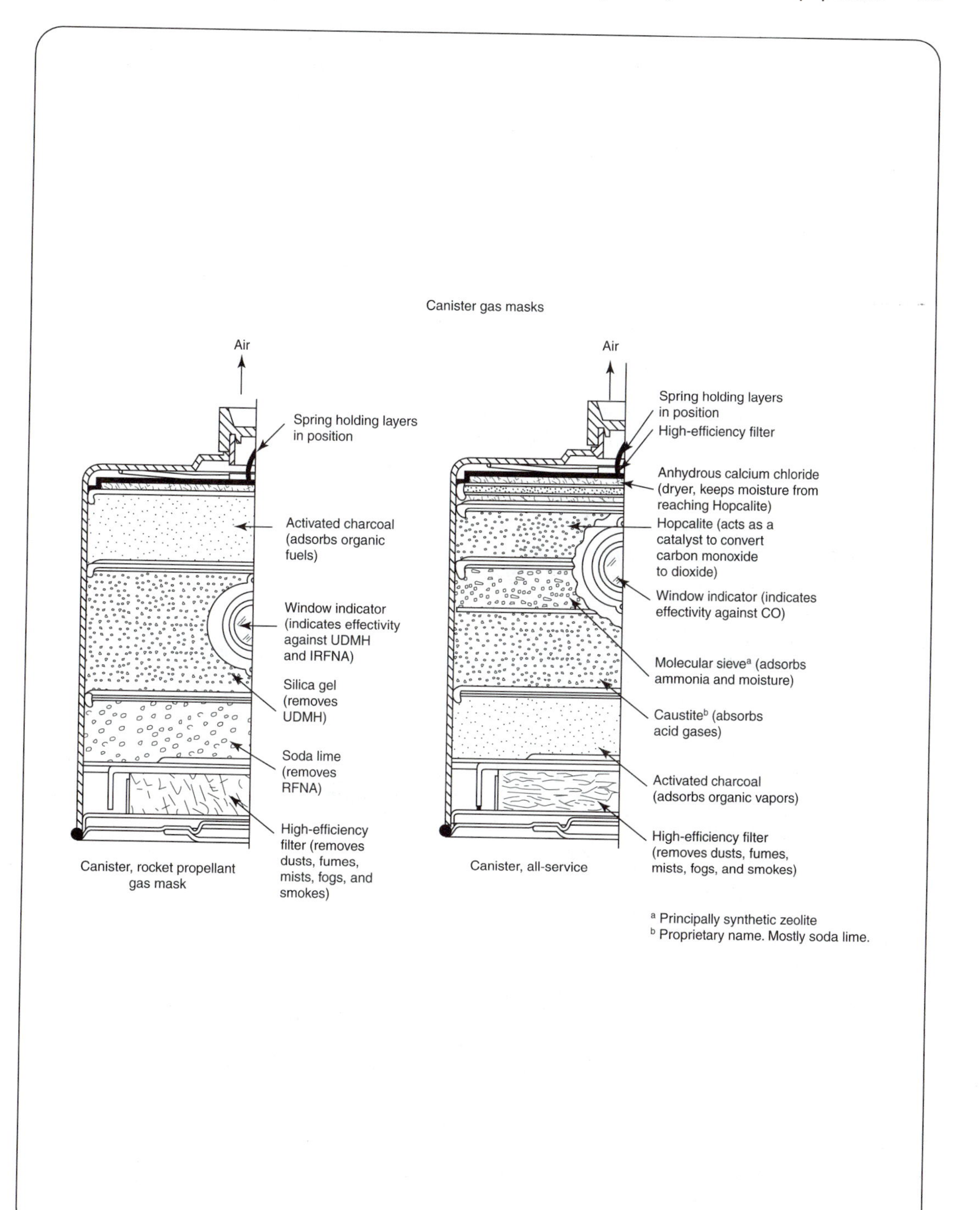

FIGURE 24–6 Continued

amount of cementing material, which permits the soda lime to be processed into small grains. The hydroxide combines with the carbon dioxide to produce carbonates and water. The effectiveness of the soda lime increases with humidity; therefore, it is placed in canisters so that the contaminated air passes through it before it gets to the silica gel and Hopcalite.

Respiratory protective equipment was first used in large amounts against poisonous gases in World War I. Although it was gradually and slightly improved, canister equipment has a number of serious disadvantages. Even where suitable, it impedes breathing effort. Its adsorptive capabilities are limited, and it provides protection only in atmospheres that are not deficient in oxygen. It will remove only those contaminants specifically intended by the chemicals provided, and even those can be overwhelmed. Even the "universal" or "all-service" canister will not protect against all gases. Contaminants for which the canister is designed will be taken out first; the remainder by other layers. When used for contaminants other than those for which the equipment is designed, the adsorbents may react unfavorably.

To pass Bureau of Mines tests for mining operations and the National Institute of Occupational Safety and Health (NIOSH) tests for other industries, gas mask canisters, except the universal type, must provide protection for at least 30 minutes. The universal type must give complete protection against carbon monoxide for 20 minutes, organic gases for 25 minutes, and acid gases and ammonia for 15. Tests are generally made at 77°F and 55 percent relative humidity. Use under conditions that exceed these values will decrease protection times. The service life for a canister designed for various combinations of gases and vapors is much less than that of one of similar size designed for one specific toxicant.

The first indication a user might have that a canister's capacity has been depleted is an odor or irritation caused by the contaminant passing through. Some canisters have indicators to show when adsorptive capacity no longer exists. However, this indicator is generally good for only one or two contaminants. It will not indicate loss of capacity against others that might be present.

Generally, canisters protect against most gases in concentrations of less than 2 percent. A mine canister will protect against 3 percent concentration of ammonia, but against only 0.5 percent hydrazine. Canister equipment should not be used in atmospheres where (1) the concentration of a toxicant is unknown and might be greater than these amounts; (2) toxicants lack odor or irritant effects to warn the user when the purifying capacity of the canister is expended; or (3) the toxicity of the contaminant is so great that the wearer will be endangered even by breathing an extremely small amount. For these reasons, canister equipment is being replaced increasing by self-contained equipment for more stringent requirements, such as for fire fighters.

Self-Contained Air- or Oxygen-Breathing Apparatus

This equipment uses a portable supply or source of air or oxygen to satisfy the respiratory requirements of the user. Within its own limitations, it makes the user independent of conditions of the ambient atmosphere. There are several types, each with its advantages and disadvantages, but they all have certain common characteristics. Each assembly consists of the air or oxygen supply, facepiece or helmet, flexible supply tube, gas regulator, and necessary valves, fitting, and gauges (see Fig. 24-7).

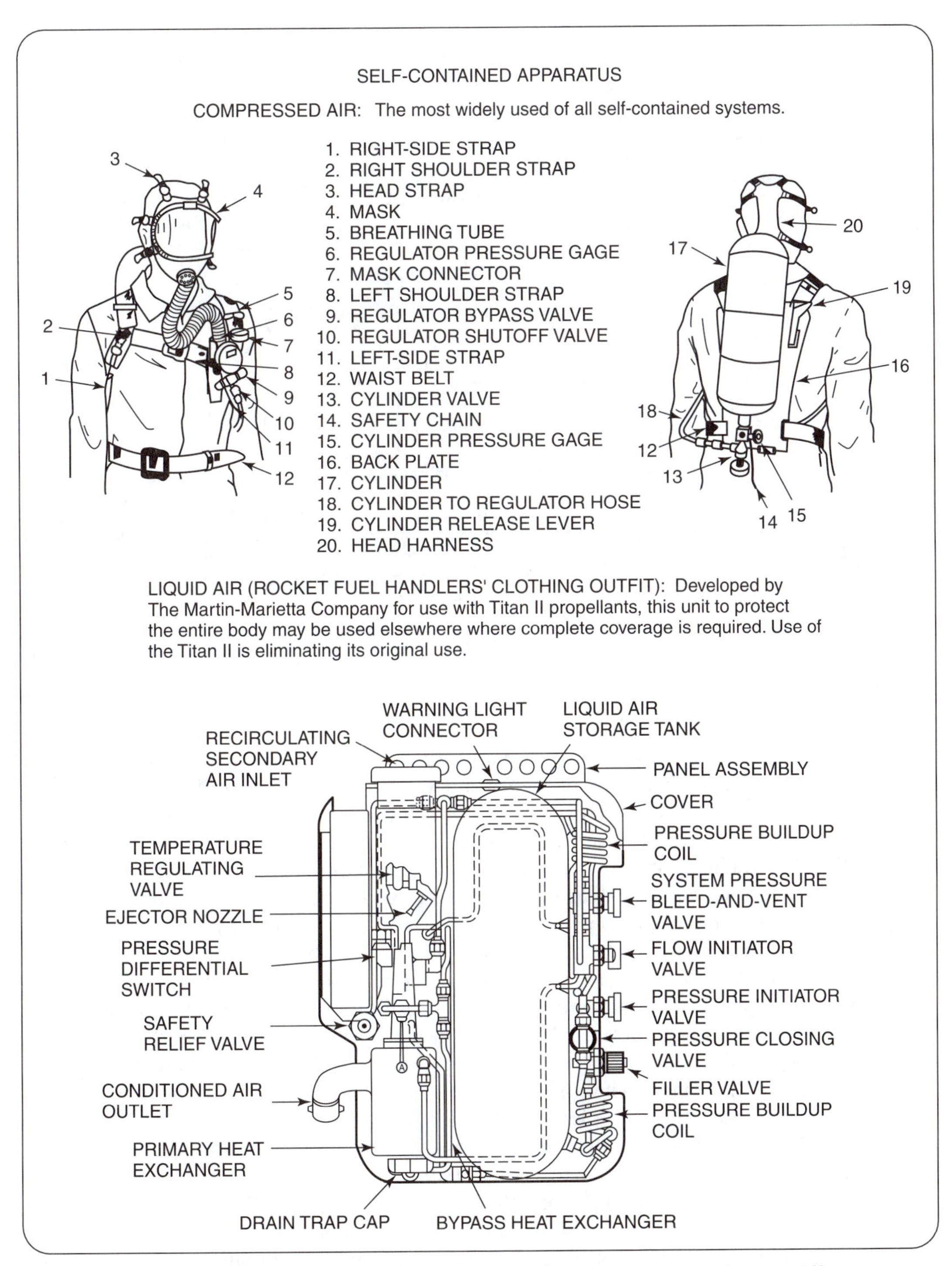

FIGURE 24–7 Self-Contained Apparatus, Respiratory Protective Equipment Continued on page 468

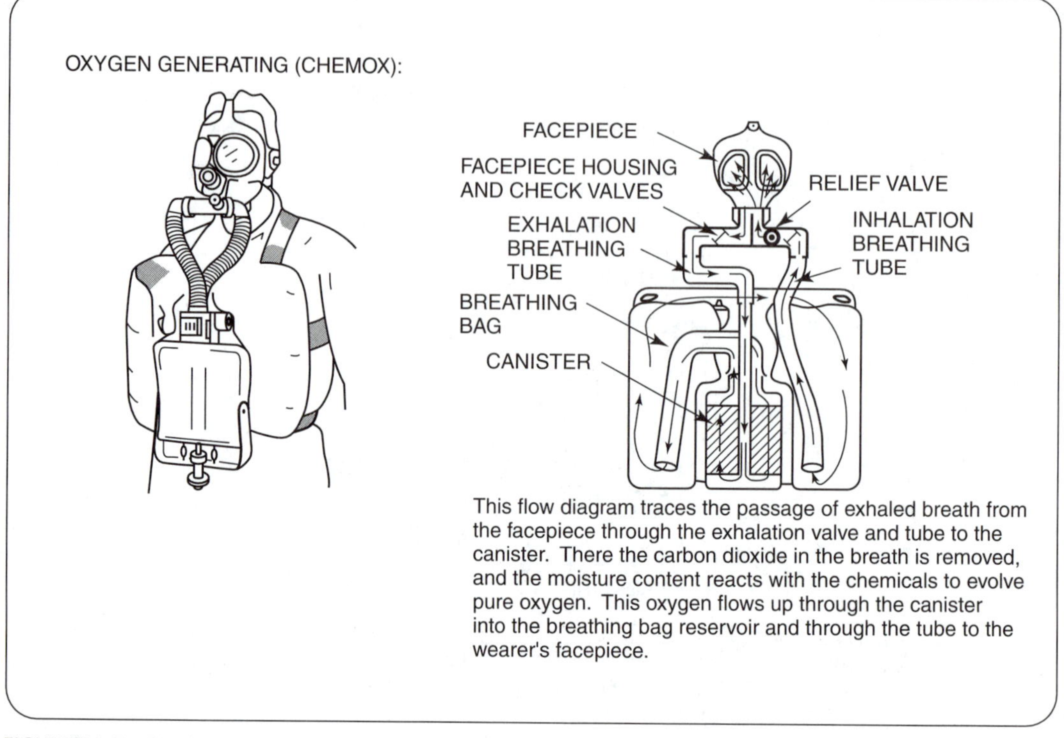

FIGURE 24–7 Continued

Regulators govern the pressure and amount of air supplied to the user. The simplest, but least efficient, type of regulator is the constant-flow type. A continuous supply of air is furnished the facepiece, maintaining a positive pressure at all times. The wearer uses some of this air; the remainder is exhausted. This method is wasteful, reducing the length of time a user can stay in a contaminated atmosphere. Also, when pure oxygen is used, the exhaust can create an oxygen-rich, flammable atmosphere.

Demand regulators open to admit air when the user inhales and creates a suction. On exhalation, flow from the tank is cut off, and the expired air is exhausted to the surrounding atmosphere. Since flow is not continuous, the amount of air drawn from the supply is only that amount used for respiration. The demand varies with lung capacity, so the flow is metered automatically with each inhalation. Demand regulators have one disadvantage in very toxic atmospheres: during inhalation there is a negative pressure in the mask, with the possibility that contaminated air might be drawn between a loose-fitting facepiece and the skin.

Where even the smallest inward leaks cannot be tolerated, a pressure demand type of regulator provides air to the breather on demand while it continuously maintains a positive pressure in the breathing tube and facepiece. The exhalation valve on the facepiece is spring loaded, so that exhalation takes place only when the pressure on the facepiece is above a preset value. Pressure demand equipment creates a lesser impedance to breathing than does the demand type.

Both the demand and pressure demand types of regulators have bypasses incorporated by which users can circumvent these mechanisms in the case of a failure or inadequacy.

Air and oxygen equipment can be classified as open or closed circuits. With an open circuit, the air or oxygen is used for breathing or pressurization and then is exhausted. With the closed circuit, exhaled air is recirculated through a regenerating unit which removes carbon dioxide and moisture. The regenerated air is added to a makeup supply to satisfy the needs of the user.

With the open-circuit compressed-air self-contained breathing unit, the user obtains air from the supply the user carries compressed in a cylindrical tank. Cylinders can be carried either vertically on the back (backpacks) or across the chest or back (slings). Backpacks will provide sufficient air for 30 to 45 minutes. Sling types are generally for emergencies to permit evacuations and are good for up to 15 minutes.

In another type of self-contained equipment, liquid air instead of compressed air is used. This permits a greater supply of air to be carried but is much bulkier and more complicated than compressed-air types. Liquid air involves low-temperature hazards, and vaporization of the air must be controlled to ensure it does not occur too rapidly, lest the valves freeze up. Such equipment is highly beneficial for use in mines.

Self-contained breathing apparatus, either open or closed circuit, also has defects: it is bulky, sometimes quite heavy, and can operate independently for a limited time only. To increase operating time, it is necessary to increase bulk and weight. The backpacks are extremely awkward to use in narrow, confined spaces or to carry through a restricted opening. Chest packs sometimes interfere when the wearer must perform a necessary task. The weight increases the worker's exertion, is tiring, and reduces efficiency. Fairly frequent inspections are necessary to ensure that valves will not stick or freeze when they are to be opened.

Another type of self-contained unit in use for emergencies is the oxygen-generating breathing apparatus with the proprietary name of Chemox (manufactured by Mine Safety Appliance of Pittsburgh, PA). It is a closed-circuit system, light in weight ($13\frac{1}{2}$ Pounds), and of good operational duration. A canister contains chemicals which provide oxygen for respiration and for removal of exhaled carbon dioxide. There are two types of canisters, standard and quick-start.

The standard canister contains potassium superoxide (KO_2) and sodium peroxide (Na_2O_2). Moisture exhaled by the user reacts with the oxides, releasing oxygen. The potassium superoxide and water form potassium hydroxide and oxygen; the oxygen is inhaled by the wearer. The hydroxide reacts with exhaled carbon dioxide to form potassium carbonate and water, which also reacts with the superoxide to release more oxygen. The sodium peroxide undergoes similar processes.

To start the reaction in the standard canister, the wearer's exhaled breath, with its moisture and carbon dioxide, must pass through the chemical beds. A number of breathing cycles are required, lasting 3 to 4 minutes. During this time the user must obtain air from the ambient atmosphere. The actual number of cycles required depends on the proficiency and lung capacity of the wearer and on the outside temperature and wind conditions.

A capability of initiating use of this type of equipment in a toxic atmosphere is provided with the quick-start canister. This canister contains potassium superoxide and a chlorate candle, which consists primarily of sodium chlorate ($NaClO_3$) and powdered iron (Fe). The candle also contains an ignition source, which is activated by the wearer's pulling a lanyard. When the candle is fired, the sodium chlorate reacts with the iron to form iron oxide, salt, and oxygen. This is adequate to supply the wearer until the moisture and carbon dioxide in the exhaled breath react with the potassium superoxide. In the past there was the possibility that the activating spark starting the reaction would be emitted from the canister. Because this is dangerous in a flammable atmosphere, canisters are now made from which no spark can be emitted.

Like all other equipment mentioned, the oxygen-generating apparatus has disadvantages. The rates of the reactions depend on temperature. The standard canister must be started at temperatures of at least 50°F, the quick-start above –20°F (in still air). In the presence of winds that increase the rate at which heat is conducted away, the ambient temperature required to start these oxygen-producing canisters is much higher. Canisters should be stored where temperatures are high enough to permit easy starting but not so high as to initiate reactions. If they must be started outdoors, they should be shielded from the wind.

Chemox equipment, worn as a chest pack, may interfere with the activity of the wearer. The oxygen generated is stored in a flexible sac, which collapses if the wearer

RESPIRATORY PROTECTIVE EQUIPMENT
(Rules for Proper Use)

1. Personnel required to use respiratory protective equipment should be familiar with its capabilities, limitations, and care.
2. Only respiratory protective equipment approved and prescribed for specific purposes should be used.
3. Air-purifying-type canister gas masks should not be used in oxygen-deficient atmospheres, or for gases whose concentration is unknown or is great enough to overwhelm the canister's capacity.
4. Personnel should be familiar with proper methods of fitting, testing, and maintaining protective equipment.
5. Personnel should make certain their equipment is fitted and working properly before entering the hazard area.
6. Protective equipment should not be removed while the user is in the hazard area, nor removed to enter the area.
7. The user should leave a hazard area as soon as a warming device indicates that the protective equipment is near exhaustion. No attempt should be made to exceed time limitations.
8. A person wearing protective equipment in a contaminated atmosphere should leave immediately if he feels sleepy, detects any unusual odor, or feels any irritation of the eyes, nose, or throat.
9. No attempt should be made to use equipment for a purpose for which it was not designed or for which it is unsuitable.
10. Equipment that is poorly fitted or in poor condition should not be utilized. If protective clothing is damaged while the person is in the hazardous area, he should leave immediately.
11. Personal protective items which seriously reduce vision, unduly reduce mobility or dexterity, or create other difficulties should be reported to the supervisor or safety engineer.

FIGURE 24–8 Respiratory Protective Equipment

SAFETY MEASURES FOR TOXIC HAZARDS

1. Design. Facilities and equipment designers should:
    a. Identify where toxic hazards exist with warning labels or signs to alert personnel who might be affected.
    b. Employ chemicals for a specific purpose which are the least toxic possible commensurate with design and operational requirements.
    c. Ensure that personnel protective equipment is not used as a substitute for proper engineering design, environmental control, or material selection.
    d. Endeavor to minimize the amount, weight, complexity, and duration of use of personal protective equipment.
    e. Designate only personal protective equipment and detection and warning devices which are suitable for the specific toxic hazards and for the operational conditions that can be expected.
    f. Prescribe suitable markings for identification of equipment, tanks, and lines containing toxic materials.
    g. Evaluate any location being considered for operations in which toxic gas might be released accidentally. The site should be determined safe under prevailing wind conditions before it is selected for use.
    h. Ensure that locations from which emergency and rescue personal protective equipment must be obtained are easily accessible, safe, close to the area where they may be required, and suitably identified.
    i. Prescribe procedures indicating actions and routes to be followed by personnel when a warning signal indicates a toxic hazard exists.
    j. Provide detecting and warning devices to indicate and evaluate atmospheres that might be toxic.
    k. Ensure that manholes of tanks or other enclosed spaces where a toxic hazard might exist and where maintenance, repair, or cleaning operations might be required are adequate in size to permit personnel to enter while wearing protective equipment such as self-contained breathing apparatus.
    l. Provide emergency showers and eyewash fountains to wash toxic or corrosive materials that could accidentally hit the body, legs, hands, or eyes of an operator.
2. Operations.
    a. Each safety engineer should be aware of any chemicals being used or being considered for use in his plant. He should know whether they are toxic alone, in combination with other chemicals, when undergoing any process, or when exposed to heat. If a hazard exists, he must ensure that suitable containment measures are taken, protective equipment provided, and emergency and first aid measures designated.
    b. Operations involving toxic chemicals should be isolated to the greatest degree practicable.
    c. Only trained personnel should be permitted to take part in operations involving toxic materials.
    d. The buddy system should be used for tank entry or other operations involving a toxic hazard.
    e. Only trained personnel should be authorized to operate, maintain, repair or handle equipment, containers, vehicles, or lines containing toxic materials.
    f. Operations involving toxic materials should be conducted only in accordance with procedures developed and approved for those particular operations and materials.
    g. Only respiratory protective equipment approved and prescribed for the specific hazard and purpose should be used.
    h. Personnel responsible for or taking part in operations involving toxic substances should be familiar with their characteristics, hazards, effects, and ,means by which their presence can be determined. They should be familiar with mishap prevention measures and emergency and first aid procedures.

FIGURE 24–9 Safety Measures for Toxic Hazards Continued on page 472

i. Immediate corrective measures should be taken when the atmospheric concentration of:
   - Toxic gas is in excess of the threshold limits value.
   - Oxygen is less than 18 percent.
   - Carbon dioxide is greater than 3.0 percent.
   - Carbon monoxide is greater than 0.1 percent.

j. Personnel required to use protective or emergency equipment should be familiar with its capabilities, limitations, and care. No attempt should be made to use equipment for a purpose for which it was not designed or authorized.

k. No person should enter a hazardous atmosphere without prescribed protective equipment, or remove it while in that atmosphere.

l. Personnel should check that their equipment is fitted and working properly before entering a hazardous area. The supervisor or safety engineer should ensure that the checks are made prior to entry.

m. Periodic exercise should be conducted to ensure that proficiency is maintained at a satisfactory level at all times in use of protective and emergency equipment.

n. Poorly fitted, damaged, or defective equipment should not be used. If protective clothing is torn, damaged, or made otherwise ineffective while a person is in an area with a toxic hazard, he should leave as soon as possible.

o. Safety showers and eyewash fountains should be inspected periodically and immediately before any operation involving a toxic or corrosive material.

p. Detecting and warning devices should be maintained in a state in which operations and readings will be dependable and accurate. They should be tested periodically and repaired immediately if defective.

q. Persons with physical conditions which make them particularly susceptible to specific substances should not be permitted to take part in operations during which they might contact or be affected by those substances.

r. Medical examinations should be required of personnel before they are permitted to handle or be otherwise involved with toxic chemicals, and at regular intervals during their employment. This is to ensure that they are physically qualified for such duties and that they are not being affected by the toxic material.

s. Check that workers handling irritating or corrosive materials wear impermeable gloves, aprons, and other clothing prescribed for that substance.

t. Ensure that personnel involved with toxic or corrosive substances have clean work habits and maintain their personal cleanliness so that skin problems will not develop. Clothing contaminated with solvents of other irritating or corrosive materials should be replaced as soon as possible. Showers, lavatories, and other cleaning facilities should be provided.

u. Ensure that thinners, solvents, or similar fluids are not used to clean the skin of paints and similar materials.

v. All containers of hazardous materials should be suitably marked. Where the attitude of a portable container is critical to avoid leakage or spillage, it should be marked with an arrow or other visual indicator to show the position in which it is to be maintained. Containers should be kept closed when not in use.

w. Spills of toxic liquids should be cleaned up immediately.

x. Outdoor operations should not be conducted if meteorological conditions are such that any inhabited downwind area might be endangered by a toxic hazard if an accident should occur.

y. No repairs should be made on equipment, tanks, or piping containing toxic liquids or gases, except with due care. Except in an emergency, the equipment, tanks, and piping should first be emptied before repairs are initiated. Where repairs must be made without emptying, provisions should be made for any contingency that might arise due to accidental release of the toxic substance.

FIGURE 24–9 Continued

Hazards Checklist–Toxic

Possible Effects

Injury to:
    Respiratory system
    Blood system
    Body organs
    Skin
    Nervous system
Irritation of eyes, nose, throat, or respiratory passages
Asplyxiation
Reduction in personal efficiency or capabilities
Cancer
Destruction of vegetation and animal life

Possible Causes

Any substance whose presence in relatively small amounts will produce physiological damage or disturbance
Gas which can be inhaled:
    From leak or release from pressurized system
    Evaporation of spilled liquid or from open container
    Product of reaction between two or more chemicals
    Product of combustion
    Outgassing of gasses in confined spaces
Liquid or solid which can be ingested or absorbed:
    Fine metal or other particle matter
    Food or material taken in by mouth
    Lack of skin protection
    Inadequate personal cleanliness
    Injected by high-pressure spray
    Through an open wound
Inadequate oxygen for respiration due to:
    High altitudes
    Dilution by inert gases
    Combustion that consumes all available oxygen
    Insufficient ventilation in occupied, enclosed space
    Atmospheric pollution by industrial, automobile, or other exhausts
    Blockage of respiratory organs by particulate matter in air
Use of food, cosmetic, or drug that is a carcinogen

FIGURE 24–10 Hazards Checklist—Toxic

CHECKLIST–TOXIC MATERIALS

1. Is any material used which in small amounts could be harmful to a person if it is inhaled or swallowed, or is absorbed through or chemically reacts with the skin?

2. Can the material affect the nervous system, act as an anesthetic, or cause cancer? Have all materials been checked to determine it they are carcinogenic or have any other adverse characteristics? Are the sources of the information reliable and up-to-date?

3. Can any two or more materials available in the plant react to create a toxic material? Are such materials separated? Is dry container of such materials marked with warnings against mixing it with any other material which could generate the dangerous substance?

4. If workers who might be affected or be involved in such a situation cannot read English, are the warnings in a language they can read, or are easily understandable logos used?

5. Can deterioration or combustion of any material result in a product (or products) which could he toxic?

6. Will any process being used in the plant generate carbon monoxide? Is it possible the monoxide can leak into an enclosed, occupied space?

7. If a gas which is toxic is being used in the plant, does it have a warning odor? Are persons in the plant knowledgeable as to what that odor is and what it signifies?

8. Has reliable information been obtained to determine if any new substance to be used is toxic, how toxic it is, what the effects might be, the precautionary measures to be taken, and any antidotes or treatments?

9. If a toxic material is a gas, has a Threshold Limit Value (TLV) or other rating been established? Have controls been instituted to ensure that no person is exposed to excessive amounts of the toxicant?

10. If a material is hazardous in an enclosed space either by itself or in combination with any other common material, have the potential users been warned and instructed in the conditions under which it is safe to use? Have janitorial personnel been warned against mixing cleaners?

11. Are persons who must or may have to wear respiratory protective equipment familiar with its use and with its limitations?

12. When a person must work in a closed space where there might be a respiratory hazard, such as in a tank, is the buddy system used? Does the man outside the tank know his duties and has he been informed he is to perform no other task than to safeguard the person in the tank?

13. Are the plant medical staff familiar with the types of materials used in the plant which might be injurious? Are they familiar with the characteristics of each such material, the symptoms which would indicate exposure to the material, and measures to be used for treatment?

FIGURE 24–11 Checklist—Toxic Materials

attempts to hold a bulky object against the chest, such as an inert person being rescued. In addition, the heats of reaction taking place within the canister may raise its temperature until it is too hot to touch.

The advantages and disadvantages of all types of self-contained apparatuses have to be evaluated for proposed use.

Air-line Breathing Equipment

A partial solution to the disadvantages listed for other devices is the air-line breathing equipment. As its name implies, air is supplied to the wearer through a connecting line from a fixed source. Air flow can be produced in a cylinder of air, an air blower, or a compressor. It finds its best use in situations where the wearer must stay in a very limited space for a long time (such as when cleaning a fuel tank). The simplest arrangement is via a salamander—a round flexible duct through which air is blown through manholes into sewers. A compressed-air line from cylinders will supply one person with air for 8 hours, or two people for 4 hours. Some organizations, such as the Air Force, require that personnel using air-line equipment in a confined space also carry a small cylinder of compressed air to be supplied automatically in the event of a failure through the line.

The chief disadvantage of air-line equipment is the restricted range imposed by the length of the line, which might be kinked, cut, or stepped on. It cannot be used between compartmented areas where doors must be kept closed. In addition, the hose may catch on protruding objects, impeding movement.

Use of Equipment

Advantages and disadvantages of the various types of respiratory protective equipment have been pointed out. Use of any type under conditions for which it was not designed can be more dangerous than use of no equipment at all, since the wearer may be falsely overconfident in its abilities. In addition, it should be pointed out that this equipment will protect the respiratory system, and a facepiece will protect the face. However, no protection is provided against harmful chemicals, for which suitable protective clothing may be required. Personnel who must wear respiratory protective equipment should be aware of the measures for their general maintenance. General measures are indicated in Fig. 24-8, but specific instructions of the manufacturer should be used wherever possible.

Some safety aspects and methods to be observed for toxic hazards are presented in Fig. 24-9. Checklists for toxic hazards are presented in Figs. 24-8 through 24-11.

BIBLIOGRAPHY

[90] F. S. Smith, *Air Quality Criteria for Propellants and Other Chemical Substances,* U.S. Air Force, Dec. 1965.

[91] N. I. Sax, *Dangerous Properties of Industrial Materials.*

EXERCISES

24.1. What is considered a toxic material? What is an allergy?

24.2. Give five major examples of toxic agents.

24.3. How can a toxic material cause injuries, and what sites in the body can be affected? Which part of the body is most easily affected by a toxicant?

24.4. Explain hypoxia, hypoxic hypoxia, stagnant hypoxia, and hystotoxic hypoxia, and give examples of each.

24.5. What is an asphyxiant, a simple asphyxiant, and a chemical asphyxiant?

24.6. How does an irritant injure the body? What are apparent symptoms of a skin irritant?

24.7. What is a systemic poison? Give examples of systemic poisons and how they injure the body.

24.8. What is an anesthetic, a neurotic, a stimulant, and a depressant?

24.9. How is dermatitis caused? How can it be avoided or eliminated?

24.10. How can an eye be damaged easily by a toxicant? How can a corneal burn occur?

24.11. Describe three principal means by which toxicity is either measured or indicated.

24.12. What are TLVs and what do they indicate? What is the difference between a TLV and an EEL? Which is most commonly used and why?

24.13. What are the three principal types of respiratory protective equipment? Which type was the earliest to be used? Which is the principal one in use nowadays?

24.14. What are the disadvantages of canister-type equipment? Describe some of the materials in canisters and the purposes for which they are used. Why are fire fighters getting away from the use of canister equipment?

24.15. What is a demand-type regulator?

24.16. How does oxygen-generating equipment work? What are its advantages and disadvantages?

24.17. What are the advantages and disadvantages of air-line breathing equipment?

24.18. List seven rules for use of respiratory protective equipment.

24.19. List seven safety measures that designers should observe for toxic hazards. List ten safety measures to be observed during operations where toxic hazards might be present.

24.20. Prepare a list of six questions which might be added to the checklists in Figs. 24-10 and 24-11.

C H A P T E R 2 5

Environments

Since the first days of industry, changes have come about in the priorities given to accident prevention. Today, the safety of persons is considered paramount to any other effects. Within the last few decades, judicial decisions and laws of the United States have given the next priority to care of the environment. After these come those for safety of equipment and then of operations.

The term "environment" has been used variously to connote the atmosphere or the place in which people work or live. An environment can range from a worker's protective suit, to a space such as a small shop, to a larger space where a micrometeorological environment exists (Fig. 25-1). Some enclosed structures have been so large, as in zeppelin hangars, they have had their own micrometeorological environments where it has rained. Thousands of people were killed and injured in Bhopal, India, because the environment became contaminated. The environment was of concern throughout the Northern Hemisphere after the nuclear accident in Chornobyl, Ukraine. Not only did the resulting fallout affect the environment in nearby countries, but it created apprehension regarding effects in other locations around the world.

OSHA AND EPA

In the United States, the OSHA is responsible for ensuring the safety of the environment in any industrial plant. Environmental problems beyond the limits of any industrial plant, such as occurred in Bhopal, would be the responsibility of the Environmental Protection Agency (EPA). A loss in a small area might be a micrometeorological problem (Fig. 25-1), limited to a small area such as a dip in the terrain so that fog or emissions from industrial smokestacks accumulate in depressions. This is what happened in 1948 in Donora, Pennsylvania, when 20 people were killed. This precursor of the tragedy at Bhopal took place when a meteorological temperature inversion blanketed the area for five days. A steel mill, sulfuric acid plant, and zinc smelter poured sulfur dioxide into the atmosphere until inhabitants sickened and died.

An even wider environment problem has existed because of sulfur dioxide and because fuel-burning industrial plants in the north central and eastern states dumped massive amounts of chimney exhaust into the atmosphere. The result was the creation

MICROMETEOROLOGY

The study of climate and weather over a small area is known as micrometeorolgy. Injuries and damage caused by external environmental pollution have gotten to be very serious health and safety problems as effluents from industrial plant smokestacks have increased. Most meteorological data presented in newspapers, over radios, or on television mention synoptic weather collected at widely scattered stations or from satellites. These general conditions cover thousands of square miles and many types of terrain.

Most people, however, are subjected to the weather that is in their immediate vicinity. Localized effects may result from offshore or land breezes, deserts, bodies of water or areas covered with vegetation; valleys, hills, or flat and open areas; high-rise buildings or other large structures that interfere with the wind; or even large expanses of concrete or reflective roofs. None of the micrometeorological weather from such features is visible from satellites or synoptic weather mapping.

Persons who must work or live in air polluted with gas, smoke, dust, or smoke in the environment can suffer from emphysema or other respiratory ailments engendered by industry. Animal life and vegetation can be blighted, or equipment corroded. Persons can be especially badly affected when atmospheric pollutants are concentrated in streets between high buildings or when there is little turbulence, such as during a temperature inversion.

For years, industrial plants have burned large amounts of coal in plants with high smokestacks to create high drafts of air that are needed for combustion and also to get rid of combustion products. The effluents have blown away in plumes of hot gases and vapors. With changes in wind direction, plant workers and nearby residents and other persons may be hit by discharges. Effects have been so deleterious that not only have states and municipalities restricted emissions of gas, smoke, dust, and vapors, but the federal government has created the Environmental Protection Agency to control generation and distribution of pollutants. This has created controversy between those industrialists who cite the economic need for plants and oppose the imposition of costs to reduce the pollution, and advocates of the health and safety of workers and the public.

Pollution sometimes has effects other than those on health and personal safety. In one case, very localized conditions caused explosions in an oxygen-generating plant. The polluted air entering the plant contained minute amounts of acetylene (which is highly sensitive) exhausted from a nearby plant, which accumulated. An explosion resulted in the compressed and concentrated oxygen being produced in the plant. Air cleaning had to be instituted before the oxygen plant could remain in use.

Micrometeorology has been studied intensively because of the development of highly toxic liquid rocket propellants to determine how chemical spills, accidental releases of gas, or test firings could affect workers. Based on intentional and highly instrumented test releases of detectable gases, theoretical equations of plume concentrations that would exist after various elapses of time, distance, wind direction, and velocity were proposed. Uncertainties because of changes in these factors and numerous others, such as differences in terrain, temperature, and turbulence, made predictable results undependable except for extremely local areas that had been highly instrumented and closely studied.

Local micrometeorological pollution problems from coal-burning plants got to be so bad in 1940 in Pittsburgh and the air was so blackened that street, bus, and car lights had to be turned on at noon. Because of intensive cooperative local government and industrial rectification action, the problem has been largely overcome in that area.

Although the original concept of micrometeorology involved areas of a few square feet, and then of a few square miles, the need for local weather evaluation and control has grown considerably. Accumulations of small amounts of pollution discharged into the air can produce major consequences. Numerous small emissions from industrial plants, including combustion products from coal and oil, all added to exhausts from automobiles in the eastern and northern tier of the midwest, have combined to create acid rain. This micrometeorological result has badly affected the northeastern states and adjacent areas of Canada.

When sulfur in coal or oil is burned, sulfur dioxide (SO_2) is produced. The SO_2 combines with moisture in air to form H_2SO_3 (sulfurous acid), which, in the presence of lightning, ozone (O_3), or other normal oxygen, combines further to form sulfuric acid (H_2SO_4). Precipitation of even small amounts of the highly corrosive acid damages and kills fish and vegetation and adversely affects human beings. Similar reactions occur to form nitric acid, which is also highly corrosive.

FIGURE 25–1 Micrometeorology

It had also been hoped that micrometeorological problems in the northeast would be eliminated with the development of nuclear power, but use of nuclear power has generated other local environmental problems.

FIGURE 25–1 Continued

of "acid rain" in the northeastern United States and southeastern Canada. Thus, environmental pollution control may be the simultaneous concern and responsibility of both the OSHA and EPA. To a degree, there must be and is cooperation between the two organizations.

Although there may be cooperation in the United States between the OSHA and EPA, cases have arisen elsewhere, as they did in Bhopal, India, and Industry, West Virginia, where environmental problems originating in a plant have affected surrounding areas elsewhere. Industrial accidents affecting the environment caused public anxiety about nuclear, chemical, and other plants thought to pose dangerous threats to the environment. One consequence was that many states required development of programs for possible evacuation of wide areas if there would be adverse effects on the environment following accidents. Lesser but more frequently occuring examples have been accidents with railroads and trucks carrying tanks of toxic chemicals. The environment was affected by the dispersal of chemicals over wide areas, leading to evacuations of persons, fatalities to persons and animals, and the loss of contaminating liquids into wells and other sources of potable water.

After midnight on March 24, 1989, the largest tanker spill in United States history occurred. The supertanker *Exxon Valdez* ran aground on Bligh Reef in Prince William Sound, Alaska. Almost eleven million gallons of crude oil was released. Portions of the shoreline of Prince William Sound, the Kenai Peninsula, lower Cook Inlet, the Kodiak Archipelago, and the Alaska Peninsula, including shorelines nearly 600 miles southwest from Bligh Reef, were contaminated. The oil affected a National Forest, four National Wildlife Refuges, three National Parks, five State Parks, four State Critical Habitat Areas, and a State Game Sanctuary. The natural resources affected by the spill included Sea Otters, Harbor Seals, Killer Whales, River Otters, Bald Eagles, Harlequin Ducks, Marbled Murrelets, Pigeon Guillemots, Pink Salmon, Sockeye Salmon, Pacific Herring, Cutthroat Trout, and intertidal and subtidal flora and fauna along 1,500 miles of coastline. Archeological sites were lost. Subsistence harvest in most of the spill-area villages declined substantially after the spill, and commercial fishing was harmed.

Both criminal and civil actions resulted. In 1991, the U.S. District Court approved a plea agreement in which Exxon received a fine of $150 million. It was the largest fine ever imposed for an environmental crime. However, $125 million was remitted in response to Exxon's cooperation in cleaning up the spill and paying private claims. In addition Exxon agreed to pay restitution of $50 million to the United States and $50 million to the State of Alaska. Additionally, Exxon agreed to pay $900 million in civil settlements.

TYPES OF ENVIRONMENTS

Environments can be divided into either natural or induced, and then into those that might be controlled or artificial, or free or closed. The effects of environment are listed in Fig. 25-2.

NATURAL AND INDUCED ENVIRONMENTS

A natural environment is one generally considered to be a climatological or meteorological condition without any man-made change or effect. Each type involves hazards or combinations of hazards that can reach out and cause fatalities and injuries. The magnitude and forces of some natural environments can be so great that little can be done to prevent them. On these occasions, it is possible to provide alerts of their impending occurrence and give warning to seek safe shelter—for example, for storms, hurricanes, or tornadoes.

The extremes of weather around the world have been compiled by the Earth Sciences Division of the U.S. Army's Natick Laboratories in Massachusetts and may be used for testing purposes for military equipment. Usual conditions in industrial plant areas are far less severe. Not only are activities usually undertaken in protected structures, but the probability is very low that any extreme cited in the data will be take place.

Finch[92] wrote about the environmental hazards: "Mortality is the rule in nature, and we tend to accept it, perhaps unconsciously, with indifference or philosophical acquiescence." Some types of environmental disasters against which measures can be and have been taken include floods, such as those that formerly ravaged the Mississippi Basin and its neighboring tributary rivers. The effects of some rain storms and floods have been reduced by suitable control systems of dams and levees. Many airports are located in areas which should not have been selected because of adverse environments that could interfere with safe airport operations.

Induced environments include those that have been affected in some way by human action: local temperatures rise because of heat emitted by a chemical processing plant, radiation from a concrete surface during a hot summer, or the heat generated by the friction between the metal and air of a high-speed aircraft; dust blows from mining or combustion of coal; the ground vibrates due to trains or other ground vehicles; or liquid particles gather from spray equipment or blown moisture.

Finch also pointed out the different regard in which the effects of natural and induced environmental disasters are held. After the tanker *Argo Merchant* sank on the Nantucket Shoals, releasing a large amount of oil, the Governor of Massachusetts asked the President of the United States to declare the coast a natural disaster area. The request was turned down because the spill was "man made." "On the other hand," wrote Finch, "the city of Buffalo, buried in snow quickly received the 'natural disaster' designation. But it seems to me that this 'disaster' was as much the result of human error (of underestimating the chances, and failing to prepare for it) as any delinquency on the part of the *Argo Merchant*'s captain." Finch appeared to imply that failures to prepare safeguards against adverse environments are no less acceptable than are failures to protect against man-made disasters.

CAUSES	POSSIBLE EFFECTS
High-Humidity Conditions Rain, clouds, fog, dew, snow Tides and floods Lakes, rivers, and other natural water sources Vegetation and animal respiration Temperature decrease without removal of moisture Condensation on cold surfaces Flooding and immersion in water Naturally high atmospheric humidity Personnel perspiring in inadequately ventilated enclosure, equipment, or impermeable covering Presence of humidifying equipment	Loss of visibility due to fog, clouds, or condensation Possibility or acceleration of corrosion Short circuits, inadvertent activations, or disruptions of electrical systems by moisture condensation in electrical devices Surface friction for traction reduced by wet surfaces Skidding and loss of control of vehicle caused by wet surfaces Flooding of facilities, shops, vehicles, and equipment Loss of bouyancy of boats and other vessels Washing away of foundations and equipment Drowning of personnel Swelling of water-absorbent materials Warping and sticking of wood doors, drawers, and similar items
Low Relative Humidity Hot weather with little moisture Heat in a closed room in winter Moisture removed by air conditioning	Drying out and cracking of organic materials Generating dusty conditions Increased tendency for creation of static electricity Easier ignition of accidental fires Increase in airborne salts, sand, dirt, and fungi
Sunlight	Ultraviolet radiation effects of sunlight Infrared radiation effects Snow blindness Difficulty in guiding a vehicle or in reading dials and meters caused by strong sunlight
High-Temperature Conditions Summer heat Tropical heat Heat from engines Heat from chemical processes and reactions Body heat Welding Friction	Melting of metals and sealants Fires or conditions which permit easy vaporization Skin burns Heat exhaustion, heat prostration Loss of ability of persons to function Rapid evaporation of liquids Reduced reliability of electronic equipment Loss of lubricating effects Increased gas pressure
Low-Temperature Conditions Winter cold Arctic and Antarctic Conditions High altitudes Refrigerated surfaces Cryogenic lines and equipment	Cold "burns" Frostbite, chilblains Plastics and metals become brittle Lubricants congeal Freezing of liquids Failure of engines and thermal processes

FIGURE 25–2 Environment and Weather Causes and Effects Continued on page 482

CAUSES	POSSIBLE EFFECTS
Airborne Salts, Dusts, Sand, Dirt... Desert and beach areas Dry areas subject to much traffic Sandstorms Windblown dust, dirt, and other solid matter Salt used to melt ice on roads Marine environment	Contamination by salt, sand, dirt, moisture, fungi Concentration of toxic gases, smog, or particulate matter caused by inversions Electrical conductivity of water increased by salt, thus reducing insulation value and permitting galvanic coupling and deterioration of adjacent dissimilar metals
Meteorological and Micrometeorological Conditions Wind Hail	Wind chill Structural overloads, movement, or toppling caused by pressure effects of wind Sudden accelerations due to turbulence and gusts of wind Energized power lines blown down by wind Impact damage caused by hail
Lightning	Shock to personnel Overloading of electrical circuits and equipment Ignition of combustible materials Other electrical effects
High and Low Pressure Water pressure, atmospheric pressure Reduced atmospheric pressure Changes in pressure	Implosions and crushing of closed vessels Pressure vessel ruptures Dysbarism and bends in humans
Radiation (see also Sunlight) Ultraviolet Visible light Infrared Microwave Ionizing	Eye damage: temporary and permanent blindness, snow blindness Skin burns, cancer, thickened skin (see sunlight) Overheating and high-temperature effects Excessive perspiration Internal heating Heating of metals Tissue sensitivity, cancer, anemia, cataracts, sterility
Vibration and Sound Impact equipment Unbalanced rotating parts Vibrating tools Loose, rattling parts Water hammer High velocity in air ducts Loud music	Breakage of metals or plastic parts, glass Metal fatigue Raynaud's Phenomenon (hand injury) Annoyance Degradation of hearing ability Interference with communications

FIGURE 25–2 Continued

The difference between a natural and an induced environment is not always easily determined with scientific certainty. Global warming is an example with far-reaching consequences to the economy and public safety and health. It is hard to know for sure when climate change is happening. Global average temperatures vary greatly naturally from year to year and decade to decade. These variations may have natural trends that extend over very long periods of time, waxing and waning. The problem is to know with scientific certainty that global average temperatures have actually changed. From the late 1980s, continuing through the "El Nino" year of 1998, a series of warm years continued. Over the last 100 years, the global mean surface air temperature has risen by 0.3° to 0.6° Celsius (C).[93] This warming has caused concern that human-induced global warming is happening. The heat from the sun would radiate back into space if not for the natural "greenhouse" effect of the atmosphere. This effect relies upon a balance of heat-trapping gases, such as water vapor, carbon dioxide, nitrous oxide, and methane. The argument is that the concentrations of these gases in the atmosphere have been increasing due to the burning of fossil fuels, the use of gasoline for transportation, and to deforestation, cattle ranching, and rice farming. These increases are said to upset the equilibrium of the natural greenhouse effect and result in global warming.

Industry requires energy. Much of this energy comes from burning carbon-based fuels, accompanied by a release of carbon dioxide to the atmosphere, which some predict to double in concentration by as early as 2050. This concern about the homeostasis of the atmosphere has raised fears that human health and commerce will be severely impacted by a human-induced heat-entrapment greenhouse phenomenon, if such is occurring. This concern is manifest regardless of scientific uncertainty, because, some argue, that by the time science can be certain of its existence it will be too late to recover. If allowed to go uncontrolled, some predict that this phenomenon could result in catastrophic, severe flooding of coastal lands from the melting of polar ice caps and the rise in the sea bed, extreme weather events, expansion of the earth's deserts, and severe stress on forests, wetlands, and other natural habitats.

Another human-induced environmental concern is that of ozone depletion in the atmosphere. The atmospheric layer endangered is the stratosphere, a layer of the atmosphere above the troposphere, where 90 percent of the Earth's ozone actually is. The ozone in the stratosphere protects us against ultraviolet radiation from the Sun. When common commercial chemicals with ozone-depletion potential are released in the atmosphere and travel to the stratosphere, they are struck by higher radiation which causes them to release chlorine. Chlorine acts as a catalyst, breaking apart ozone molecules. Chemicals such as chlorofluorocarbons (CFCs), methyl chloroform, carbon tetrachloride, and halons act similarly. These chemicals are used for air conditioning, industrial solvents, degreasers, foaming agents, and fire protection and control. They are part of the industrial scene. Each 1 percent depletion in stratospheric ozone is said to increase exposure to damaging ultraviolet radiation by 1.5 to 2 percent. It is believed that this can result in premature human skin aging, increased incidences of skin cancer, suppression of the human immune response system, damage to the human eye, damage to crops and various terrestrial ecosystems, damage to certain marine organisms, and harm to animals.

Concerns about global warming and ozone depletion have led to a series of international agreements to reduce the impact of industrial society and human activities on the atmosphere. International agreements from meetings in Montreal, Canada, in 1987 and London, England, in 1990 to regulate CFCs are designed to gradually reduce ozone-depleting chemicals in the atmosphere, until chlorine returns to natural levels in about 100 years. The 1990 Amendment to the Clean Air Act regulates CFCs to levels which some contend support these international agreements. In December, 1997, the United States and 37 other industrialized countries met in Kyoto, Japan, and gave commitments to reduce the global output of carbon dioxide and five other greenhouse gases as part of a global warming treaty. This treaty, unratified by the U.S. Senate, is an agreement to reduce emissions by 2012 to an average of 5.2 percent below what they were in 1990. This is a reduction in emissions up to 40 percent of what they would have been otherwise in 2012.

Ironically, this treaty conference occurred during the very month in which space-based measurements of the temperature of the Earth's lower stratosphere indicated it was the coldest month on record since measurements of this type began in 1979. Despite significant warming of the stratosphere following the eruptions of El Chichon in Mexico in 1982 and Mount Pinatubo in the Phillipines in 1991, the data over the past 15 years show a stratospheric cooling trend consistent with the depletion of ozone. National Aeronautic and Space Administration data indicate that the temperature of earth's surface is warming, the stratosphere is cooling, and there is no trend in the lower atmospheric layer over the past 19 years.[94] There is continuing uncertainty that global warming is a human-induced phenomenon, but these agreements, designed to control the atmospheric environment, can have considerable impact on the economy, industry, safety, and health.

CONTROLLED ENVIRONMENTS

A controlled environment is one that has been modified in some respect to mitigate or avoid an undesirable or adverse condition of the natural or induced environment. A space either fully or partially air-conditioned is a controlled environment. A controlled environment may change the temperature or humidity or the pressure in compressed-air operations in a caisson. Controlled environments have typically been small spaces; attempts to control the global environment are relatively new to the human experience.

An "artificial" environment will differ entirely or substantially from a controlled one. It may be one of nitrogen to prevent fires or material degradation of rubber or metal if oxygen was present, or one of low-pressure pure oxygen in a pressure suit. An environment might be artificially created for humans because without it the existing environment (for example, the moon) might be unbearable. The air in submarines and the oxygen in space capsules create artificial environments in that they protect humans against the environments of ocean water and airlessness of very high altitudes or interplanetary vacuum.

CLOSED OR FREE ENVIRONMENTS

A controlled or artificial environment can exist only in a restricted or confining space. Such a space might constitute one that is "closed" or almost so. A closed environment can even be one that protects against the natural environment without other modifications such as a building or other structure which protects against rain or wind. A "free" environment imposes no interference to the movement of air.

Although Fig. 25-3[95] presents the zones of luxury, comfort, and discomfort for persons in aircraft, they are similar to those in an industrial plant. However, these zones may vary by each person's metabolism, the activity being undertaken, and numerous other factors. Other than large disasters that have taken place, such as hurricanes, the most hazardous environments are those in small, limited spaces. This is because a very small amount of containment may be highly toxic and injurious. Persons have died in closed vehicles or rooms because of comparatively small amounts of carbon monoxide, absence of oxygen in breathable amounts, or presence of chlorine generated by the mixture of two chemical cleaners.

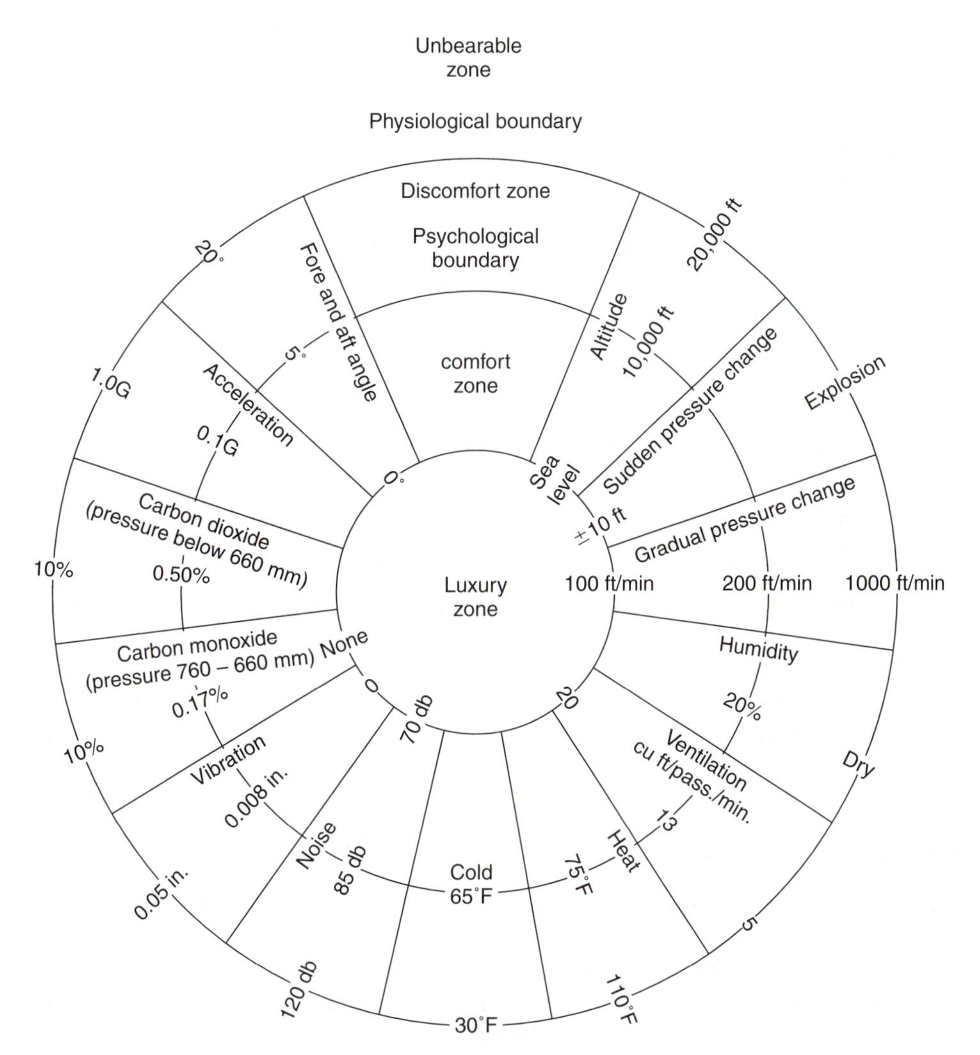

FIGURE 25–3 Zones of Luxury, Comfort, and Discomfort for Humans

HAZARDS OF THE ENVIRONMENT

Each environmental factor listed in Fig. 25-2 can be expressed individually but will never exist so. Thus, an environment can produce effects because it is hot or cold (or between), and at the same time it could be humid or dry, windy or calm, or involve to some degree any or all of these factors. Each factor at some times can constitute a hazard whereas at other times it can be acceptable.

For example, even moderate temperatures, when combined with high humidity, as in Hawaii, Florida, or other warm coastal areas, can create serious environmental problems. In such an environment, steel products must be protected against rust or other corrosion, or failures of mechanical equipment or structures. Moisture absorption can cause nonmetallic materials to swell and bind. High relative humidity and accumulations of moisture can lead to electrical short-circuits and malfunctions, or to deterioration of electronic equipment.

On the other hand, dry weather can also lead to environmental problems equally as destructive as humid weather, although in a different way. It can cause cracking of wood products, dry skin, losses of body fluid and other water, and lubrication inadequacies. It is a perennial problem for forested or grassy areas, creating conditions where fires can be generated easily. Low humidity may tend to increase generation of static electricity.

Still another common hazard in the natural environment is the presence of flammable or toxic gases, sometimes released from excavations of the earth, such as methane, sometimes called "swamp or marsh gas," or hydrogen sulphide. Many flammable gases are odorless and invisible, making it necessary to provide instruments to assist human detection.

The sun gives heat and light, but too much solar radiation can cause skin burning or excessive heat to pressurized unshaded tanks in the open. Few normal structures or vehicles are designed to withstand the wind speed of 225 miles an hour recorded on Mt. Washington in 1934, but even moderate winds have caused severe damage.

Some investigators, for example as indicated by Arnold,[96] described a technique to examine two or more simultaneous environmental conditions. Unfortunately, often so many factors are involved that only two or three of the most important can be studied to determine the combined effect. Tests under natural conditions are often very difficult to make because they are difficult to control. Even when they are done, they might not encompass the extremes that cause problems and disasters. Tests are sometimes made in large structures where environments can be simulated. An entire large piece of equipment or operation can be tested in an enclosed space where important environmental factors can be controlled and the effects evaluated.

DETECTING ADVERSE ENVIRONMENTS

Many contaminants in the environment, but not all, can be detected by human senses, and in this regard one of the most important is that of odor (Fig. 25-4). Odors can range from pleasant to neutral to highly offensive and disagreeable. Similarly, detectability will range from undetectable to highly apparent. It was because of the lack of detectability in dangerous underground environments that miners formerly carried

ODORS AND THE SENSE OF SMELL

The ability of humans to detect odors depends on stimulation of receptors of approximately 1 square inch in the mucous membrane in the upper part of the nasal cavity. Some substances stimulate both the nerve endings of the olfactory receptors and other nerve endings scattered through these areas. This makes difficult for investigators to determine whether a person's ability to discriminate the presence of a foreign substance is due to its odor or its irritant property.

Amoore and co-workers[a] indicate that there are actually only seven primary odors. His theory indicates that chemicals which exhibit five odors have molecules with shapes that fit into specific sites in the olfactory nerve endings in the nose.

Gases of widely different chemical composition will have the same odor if they have the some one of the five shapes. On the other hand, molecular isomers with the same chemical compositions but different atomic arrangements and molecular shapes will have different odors. A molecule that will fit into more than one site may have a complex odor made up of the corresponding primary odors.

The last two odor types are characteristic of chemicals that will not fit into the other sites but will have either positive or negative electrical charges. Molecules in these groups come in various shapes and sizes. A molecule that lacks and strongly attracts electrons will have a pungent odor; those with an excess of electrons will have a putrid odor.

A substance will have no odor unless it is a gas, so that it can reach the olfactory receptors, and is soluble in water and fat, so that it can penetrate the thin layers that coat the nerve endings of each receptor cell.

Other researchers differ from Amoore et al.'s findings and from each other. The numerous ideas on odor differ on the exact process of detection and recognition, but there appears to be general agreement that odorants with similar molecular shapes affect the same receptors, and that those with dissimilar shapes affect different receptors.

Ability to detect odors: as with other senses, the abilities of individuals to detect odors vary. A very small percentage of all people (approximately 1 in 1,000) may have *anosmia*, the inability to detect any odor at any time. Another 2 percent suffer from *paranosmia*, the inability to detect certain odors or odor groups. Anosmia and paranosmia may be permanent (an inherited tendency) or temporary. The ability to detect odors can be affected by illness such as colds or sinus infections. Continued or overwhelming exposure to a strong odor can cause desensitization of the olfactory sense so that temporary anosmia or paranosmia result. Smokers and indoor workers generally have less acute abilities to distinguish odors than nonsmokers and outdoor workers.

Odor levels: Attempts have been made to quantify odor levels. Suitable criteria can then be prescribed for maximum permissible odor levels, if desired. Amounts of contaminants present can be established, and data understandable by others can be recorded and transmitted. Odor intensity is important in determining or realizing what the effects of a specific level will be. Since apparent intensity varies with the sensitivity of a person, it is also necessary that comparisons and impressions be made under prescribed conditions. One scale now in use has six degrees of intensity which range 0—no odor to 5—very strong. No numerical relationship exists between intensity and measured amount of contaminant. Another method has therefore been recommended by which intensities are reported as multiples of the minimum detectable (threshold) concentrations. This minimum threshold value is sometimes called an olfactory. The amount of contaminant present, determined by an analyzer, would be divided by the threshold concentration to give a numerical intensity level.

In some instances nominal values have been established for gases that can be detected in an

FIGURE 25–4 Odors and the Sense of Smell [a]J. E. Amoore, J. W. Johnston, and M. Rubin, "The Sterochemical Theory of Odor," *Scientific American*, Vol. 210 (Feb. 1964), pp. 42–49. Continued on page 488

otherwise normal atmosphere. Nitrogen tetroxide and nitric acid are initially detectable at levels far below that at which injury will occur. However, because they are strong irritants, they reduce the ability to detect odors. In some cases such desensitization applies only to the gas which has caused the loss of ability; in others, it may affect the ability to detect any odor. Certain substances are detectable by their odors in such small concentrations that they are nontoxic, nonflammable, and noncorrosive.

Compound odors: Odors and odorants are rarely found individually in air. The possible ways in which two or more may act are:

1. One odor may overwhelm the second, making it imperceptible.
2. One odorant may cause desensitization of the capacity to distinguish any odor.

[a]J. E. Amoore, J. W. Johnston, and M. Rubin, "The Sterochemical Theory of Odor," *Scientific American,* Vol. 210 (Feb. 1964), pp. 42–49.

3. Odorants may act independently so that each can be perceived when it is above the threshold level.
4. Odors can have an additive effect, so that if two compounds each have one-half the threshold concentration together, they will produce a perceptible odor.
5. Odorants can produce an odor that is in proportion to and characteristic of the original ingredients.
6. An odor can be produced which differs from those of the original ingredients.

Odor avoidance: problems with odors can be avoided by two broad methods: preventing their occurrence, and abating any which are offensive. The presence of an odorant can be due to:

1. Industrial operations that release odorous gases or volatile liquids.
2. Leakage of fluids (gas or volatile liquid) from containers.
3. Burning of materials, which creates combustion products with odors.
4. Cooking and other vaporization of liquids or solids.
5. Body emissions of occupants of enclosed spaces.
6. Decomposition of food, vegetation, or other organic material.
7. Outgassing of volatile materials from solids, such as rubber, paint, linoleum, and waxes.
8. Microbial reactions.

Admission of odors: entrance of odorants into an occupied area can be prevented in a variety of ways, with each preventive measure having specific limited application.

1. Eliminating dissemination into the atmosphere of chemical wastes or combustion products which have odors or can react to produce odors.
2. Locating air intakes for ventilating systems at sources of odor-free air.
3. Keeping ventilating systems and occupied areas clean so that decomposition or oxidation will not create odorous products.
4. Controlling temperatures and humidity so that occupants perspire little, thereby reducing body odors, and so that food decomposition and microbial growth are minimized.

Elimination of odors: where the entrance of odorants cannot be eliminated, odors can be removed or their effects abated. Odors are most objectionable when first detected; olfactory fatigue reduces the effect. Other means by which odors can be abated are:

1. *Dilution:* odor-free air is used to reduce the concentration of objectionable contaminant to an acceptable level. This is the commonest, and usually easiest, method, exemplified by use of large amounts of ventilating air.
2. *Absorption:* the gas that produces the odor is passed into a liquid, where it goes into solution or reacts chemically and is absorbed.
3. *Adsorption:* a material such as activated charcoal adsorbs the molecules of offensive gas. Adsorbents can be made reusable simply by heating. However, their capabilities are limited to low concentrations of contaminants.

FIGURE 25–4 Continued

4. *Catalytic combustion:* the gas that produces the odor is burned in the presence of a catalyst. With organic materials, the result is carbon dioxide and water, and sometimes other end products. Care must be taken that these end products or products of partial combustion do not themselves produce odors. This method can be used only where the odorous substance is combustible.
5. *Oxidation or other reaction:* another chemical, such as chlorine, ozone, or formaldehyde, reacts with the odorous substance to create an undetectable or pleasant odor. This method is practicable only where the contaminant is reactive, the resultant products are not objectionable, and the reagents can be closely controlled.
6. *Masking:* a stronger, pleasant odor is used to hide the offensive one.
7. *Counteraction:* a second odorant is added, which greatly reduces the ability to distinguish either one. This is another form of desensitization.

FIGURE 25–4 Continued

with them caged canaries to detect the presence of harmful amounts of carbon monoxide or dioxide. (Canaries have been replaced only recently by sensing meters.) Some odoriferous contaminants such as hydrogen sulphide can be readily detected to provide warnings; some, like the oxides of carbon, have no odor. Hot emissions of small amounts of vapor can provide warnings through the sense of smell. The odor of "hot boxes" of locomotive bearings, burning rubber of tires because of skidding, overheated water of hot automobile engines, and fires can be detected as evidence of problems.

Although odors are detectable in any environment, free or closed, their benefit as a warning is often greatest in confined spaces. Because small amounts of contamination are often highly toxic, for safety purposes the ability to detect even smaller amounts of contaminants is necessary. Thus, many state, federal, and industrial organizations have prepared and published lists of environmental contaminants that should not be exceeded if harm is to be avoided. Some state health and safety agencies and OSHA publish lists (which may differ) of Threshold Limit Values (TLVs), which are changed as experience or testing indicates a substance to be more or less harmful than originally rated. Explanations regarding TLVs and other measures of atmospheric toxicity are given in Chapter 24.

Instruments are available to measure deleterious substances in amounts far smaller than are detectable by human senses. These instruments can indicate environmental hazards that may be flammable, toxic, or radioactive (Fig. 25-5) or merely such conditions as temperature, humidity, pressure, or wind speed. The chief problem with such environmental detectors is that they may give no readings or inaccurate readings if there is poor location of the intakes to the sensing device. The range of detection might be less than that necessary to activate the indicator or provide a warning of the contaminant's existence.

PROTECTION AGAINST THE ENVIRONMENT

Since time immemorial, humans have found needs and ways to protect themselves against the environment, using furs, hats, and similar protection. As technology and environmental hazards grew more complex, so did the types of protective devices. It is

EXAMPLES OF GAS DETECTOR PRINCIPLES

Type	Use	Operation	Limitations
Catalytic combustion	Combustibles	A heated resistor filament covered with a catalyst constitutes one leg of a balanced electrical bridge. The catalyst causes combustibles which are present in a sample of air even in very small amounts of ignite, heating the filament to a high temperature. This increases the filament's electrical resistance and unbalances the bridge, causing a change in output voltage which can be read in terms of combustible present.	Good only for combustibles which will be affected by the catalyst.
Thermal conductivity	Combustibles and noncombustibles	Similar to the device using catalytic combustion except that the filament resistance is changed by the rate at which a contaminant will conduct heat away. Changes in amount of contaminant will change the heat transfer rate and the filament's resistance.	May not be properly selected, since almost any contaminant will have an effect.
Infrared absorption	Any gas that will absorb infrared radiation	Gases that absorb infrared radiation will do so at specific wavelengths. Infrared radiation generated by heating two wound nichrome wires passes through a comparator cell and a test cell. Infrared radiation is absorbed by the contaminant which may be present in the test cell, reducing the infrared energy which reaches a detector at the far end of the cell. The detector contains gas which expands when hit by the infrared radiation. The other end of the gas detector is a thin diaphragm, which also forms one element of a capacitor. The amount of contaminant causes changes in the gas detector heating and pressure, and therefore in movement of the diaphragm, Varying the capacitance changes the signal output.	Cannot be used for gases such as hydrogen, fluorine, chlorine, or other halogens, which do not absorb infrared radiation.
Ultraviolet absorption	Any gas that will absorb ultraviolet radiation	Similar to infrared. However, gases for which it is used must be absorbers of ultraviolet radiation.	
Ionization	Fuels or oxidizers	A gas sample is burned or mixed with a reagent to form an aerosol and then passed into an ionization chamber. A radioactive source there generates ion pairs. When gas is ionized between two oppositely charged electrodes, current flows. Any aerosol present due to contamination increases the ion recombination rate and reduces ion mobility. This in effect, increases resistance within the chamber and current flow out. Changes in current will depend on the contaminant concentration. The ionization chamber is one arm of an electrical bridge circuit, with a reference chamber as the other arm.	This type of device has ultrahigh sensitivity (in some cases can read in parts per billion) and may therefore be too expensive, too heavy, and too complex for requirements that do not call for this sensitivity. Use is limited to those contaminants for which a reagent can be found which will form an aerosol with the contaminant.

FIGURE 25–5 Examples of Gas-Detector Principles

Type	Use	Operation	Limitations
Coulometric titration	Fuel or oxidizer detection	A fuel contaminant which impinges on the surface of the suitable sensing electrolyte is immediately oxidized by ions in the electrolyte. The reaction causes a change in ion flow and that of current through the electrolyte which is proportional to the concentration of the contaminant. A reducing electrolyte is used for detection of oxidizers.	
Colorimetric systems	Gases, liquids, or solids	A variety of methods is used. Example: Reagents are maintained in glass ampoules. When a test is to be made for a specific substance, the material or gas is drawn through. To detect carbon monoxide, yellow silica gel mixed with other compounds is used. If carbon monoxide is present, the color of the silica gel will turn green. The depth of the color can indicate the approximate concentration of the gas when a rubber aspirator is used to draw a measured amount of air through the reactant. The depth of color that results is checked against a series of standard shades called a comparator. The closest matching shade indicates the approximate amount of contaminant. Another type of colormetric detector uses a spray to react with the suspended contaminant so that it becomes visible. Sulfur dioxide has hydrochloride acid vapor will react with ammonia in air to produce a white smoke, which is really seen.	Not highly sensitive.
Electro-conductivity	Nitrogen oxides, ammonia, fluorine, hydrazine, and other gases that ionize in water	This detector operates on changes of conductivity caused by a contaminant introduced into pure deionized water. The increase in conductivity is proportional to the ionizable gas in a sample of air. Many gases that do not ionize are burned so that they will ionize in water.	Not specific, since it indicates all ionizable gases.
	Hydrogen, fluorine, hydrazine, UDMH, and nitrogen tetroxide	A radioactive isotope, such as krypton-85, is incorporated into a solid material which reacts with the gas being monitored. The isotope is released in proportion to the concentration of contaminant. The radioactivity of krypton-85 released is then measured by a Geiger counter.	Care required when the atmosphere has high relative humidity, since moisture will cause release of krypton from the material in which it is held.
Paramagnetic	Oxygen	A small glass of dumbbell is suspended on a quartz fiber in a nonuniform magnetic field. Oxygen is paramagnetic and attracted into a magnetic field, causing the field to change. In the detector, oxygen drawn into the chamber surrounding the dumbbell will alter the magnetic field and causing the dumbbell to rotate in proportion to the amount of oxygen present.	Limited to oxygen and a few other gases that are paramagnetic.

Too many detectors of too many kinds are in use or proposed for use to allow us to list them all.

FIGURE 25–5 Continued

a far cry from protection against body cold to protection against the localized environments in some industrial plants. To avoid injury or damage from the environment, engineers must consider environmental characteristics and their changes, the hazards involved, the problems and effects that could be generated, and the safeguards and protective measures that can be used.

Over the years means have been devised that have mitigated little by little the hazards of the natural environment. The damaging effects of lightning and other hazards can often be avoided through the use of protective devices. Lawrence[97] pointed out that in the fifty years following the start of the British-French War in 1793, more than 250 ships were struck and damaged by lightning. During one 16-year period there were 150 cases in which more than 70 seamen were killed and 133 injured. In 1853 an effective system of lightning protection for ships was patented and adopted that eliminated almost entirely the high incidence of deaths and damage due to strikes on ships.

Because of the vast number of causes, effects, and safeguards possible regarding environmental factors, a complete list is impossible. Much of the information in this book is on safety factors that relate in some way to the environment. Lists of possible effects and causes provided herein could be used to develop safeguards or precautionary measures. The information in checklists can also be used to review their needs, whether they had been safeguarded, and to check whether detection devices or protective measures or equipment are advisable. With them are listed protective measures, including descriptions of protective suits for workers in compatible environments. Environmental safeguards and accident prevention measures are presented later in this book. One such list of safety measures to be observed by workers against toxic hazards is given in Fig. 24-11. The needs for environmental protection against some electromechanical drive systems, together with some suggested remedies, are included by Lemmler[98] in Fig. 25-6.

INDUSTRIAL ECOLOGY

Environmental contaminants and human health are related. Some contaminants are present naturally; others are artifacts. Industry contributes to the contamination of the environment. As Graedel and Allenby[99] point out, our industry-related environmental concerns are on global, regional and local scales. Globally, there are concerns about climate change, ozone depletion, and loss of habitat and reductions in biodiversity. Regionally, there are concerns about surface-water chemistry changes, soil degradation, precipitation acidity, visibility, and herbicides/pesticides. Local concerns include photochemical smog, groundwater pollution, radionuclides, toxics in sludge, oil spills, toxics in sediments, and hazardous waste sites. These authors define industrial ecology, thereby conveying their conviction of the responsibility of industry toward our environment:

> Industrial ecology as applied in manufacturing involves the design of industrial processes and products from the dual perspectives of product competitiveness and environmental interactions.... The concept requires that an industrial system be viewed not in isolation from its surrounding systems, but in concert with them. It is a systems view in which one seeks to optimize the total materials cycle from virgin material, to finished material, to

ENVIRONMENTAL CHECKLIST FOR ELECTROMECHANICAL DRIVE SYSTEMS

CONDITION	POTENTIAL PROBLEM	SUGGESTED REMEDIES
Electrical Power Line		
"Soft" ac power lines droop under steady-state load or when heavy loads–such as large induction motors or electric arc welders–are applied.	Can affect unregulated power supplies; can drop ac line voltage out of limits specified for drive equipment (+ 10%, –5%); can cause fuse blowing on static regenerative drives.	Stiffen up source (substation, etc.); apply constant-voltage transformers where practical; shift load to different load center
Noisy ac line, full of voltage "spikes" or "holes"; high-frequency noise generated by SCR drives on ac feeder; transient high voltage on line.	Can cause SCR misfiring, erratic system performance/operation; overvoltages can destroy semiconductor devices, other components.	Shift to less noisy load center; suppress noise source or drive system if practical; use isolation transformer on offending system.
"Stiff" ac line (high short-circuit capability).	Can be hazardous to older equipment due to high short-circuit current levels.	Provide adequate short-circuit protection between drive and substation.
Line voltage dips for several cycles or disappears entirely, single or polyphase, due to lightning strikes on transmission lines; ac supplied through collector rings shoes.	Equipment may shut down, blow fuses, or behave intermittently.	Provide drives with automatic restart features; MG set can provide some ride-through due to system inertia (flywheel effect).
Electrical Noise		
"Noisy" power conductors induce noise into low-level signal wire by coupling; operating relays produce noise; improper shielding or no shielding on low-level signal leads can affect drive operation.	Any noise inserted into regulator or firing circuitry can cause misfiring, fuse blowing, and erratic drive operation.	Isolate signal leads from power leads, place in separate conduits; if leads must cross, cross at right angles; provide shielded leads for all low-level signals; ground shield at one end or cable only, preferably at power unit; ground power unit cabinet; suppress relay coils.
Mechanical Vibration		
Equipment subjected to vibrations external to system such as coupling misalignment, belt or chain drive whip, sprocket run-out, misaligned or eccentric rolls; drive vibrates due to vibrations from nearby equipment such as slow-speed compressor.	Can generate signals each revolution, causing system instability; can cause general deterioration of connections and wire breakage; can shorten life of bearings in drive motor.	Realign rotating members; shorten belt lengths; true up driven rolls; isolate system from external vibrating masses.

FIGURE 25–6 Environmental Checklist for Electromechanical Drive Systems Continued

ENVIRONMENTAL CHECKLIST FOR ELECTROMECHANICAL DRIVE SYSTEMS

CONDITION	POTENTIAL PROBLEM	SUGGESTED REMEDIES
Equipment subjected to vibrations generated within system such as large starters or contractors being operated.	Can cause general deterioration of terminations, wiring connections, wire breakage; parts can be shaken loose and dropped onto devices, reducing electrical clearances; contact tips of low-level signal relays can bounce, producing erratic drive operation.	Isolate vibration-sensitive devices; relocate offending devices; dampen vibrations by using some form of rubber cushioning.
More bearings are noisy indicating wear due to vibration	Can cause short brush life and poor commutation.	Replace bearings and remove any motor load vibration that may lead to further bearing failure.
Atmospheric Effects		
Temperature is too high due to adjacent heat sources such as ovens, process heat, or exhaust from other equipment; air flow blocked or restricted.	Can cause drifting of regulated parameters such as speed, higher than normal card and/or component failure, reduced motor bearing life, or shortened insulation life.	Remove heat source, relocate equipment; supply cooler air; remove any obstructions; keep filters clean; flood with conditioned air during extended downtime.
Temperature is too low in operating area; prior to installation, drive is stored in unheated warehouse.	Can cause accumulation of frost or moisture inside power unit/motor, poor regulation, ground hazards, or fuse blowing. Some semiconductors and capacitors can deteriorate. SCRs become more difficult to fire reliably. Rust can form on devices and equipment.	Supply space heaters for sustained storage or operation at low temperatures. Minimum storage temperature is –20°C, minimum operating temperature is 10°C usually.
Accumulation of dust or dirt in power unit, motor, or filters.	Particulate matter can reduce contact clearances on terminals and relays; contacts can become contaminated; leakage paths can be formed; contacts can close but not make circuits; motor brush life can be reduced; ground hazards can exist; motor temperature can be elevated due to improper air flow.	Maintain filters on a planned basis; inspect power unit and motor for dirt accumulation; clean equipment, but avoid use of air hoses that may be contaminated with oil particles; remove equipment from dirty air paths or use air ducts to bring in clean air.
Excess humidity in area caused by water vapor or steam from process; condensation from water and steam pipes; area humidifying equipment set too high.	Can cause collection of moisture in power unit or motor, rusting, reduction in creepage paths, or ground hazards.	Locate equipment in less humid area; remove source of moisture or shield drive equipment.

FIGURE 25–6 Continued

ENVIRONMENTAL CHECKLIST FOR ELECTROMECHANICAL DRIVE SYSTEMS

CONDITION	POTENTIAL PROBLEM	SUGGESTED REMEDIES
Humidity is very low and must be to meet process requirements.	Can cause insulation to dry out and crack, extremely short brush life, poor commutation, and static electricity.	Avoid dehumidified air by supplying moist air to power unit and motor through ducts.
Process vapors combine with moisture in air to form corrosive chemicals such as hydrogen sulfide, chlorine compounds, sulfur dioxide mixtures; silicone vapors in air; oil vapors in air.	Can cause corrosion of contacts, electrical components, and printed circuit card etchings; can cause deterioration of motor insulation and commutator films; can cause ground hazards.	Supply clean air to equipment; add special motor enclosures; place equipment in clean environment—such as in a motor control center.

FIGURE 25–6 Continued

component, to product, to obsolete product, and to ultimate disposal. Factors to be optimized include resources, energy, and capital.

The ISO 14000 series, mentioned briefly in Chapter 7, addresses environmental management and environmental auditing. Using that series, management sets environmental protection goals based on consideration of the demands of their customers, regulators, communities, and other interested parties. A company sets its own objectives and targets for performance. Like it or not, industry today is being held accountable for its contributions to the environment.

BIBLIOGRAPHY

[92] Robert Finch, "Technological versus Natural Destruction," *Technology Review*, Oct./Nov. 1977, pp. 16 *ff.*

[93] National Safety Council, *Reporting on Climate Change: Understanding the Science,* Chapter 1, Environmental Health Center: A Division of the National Safety Council, 1025 Connecticut Avenue, NW, Suite 1200, Washington, D.C., July 1997.

[94] National Aeronautics and Space Administration, "December 1997 is Coldest Month on Record in the Stratosphere," http://science.msfc.nasa.gov, Jan. 20, 1998.

[95] R. A. McFarland, *Human Factors in Air Transportation* (New York: McGraw-Hill Book Company, 1953).

[96] James S. Arnold, "Computer-Calculated Environments," *Machine Design*, April 25, 1963.

[97] Derek Lawrence, "Nature's Artillery," *Engineering*, Aug. 1972, p. 764.

[98] R. K. Lemmler, "Protect Your Drive Systems from Hostile Environments," *Automation,* Aug. 1972, p. 42.

[99] T. E. Graedel and B. R. Allenby, *Industrial Ecology* (Englewood Cliffs, NJ: Prentice Hall, 1995), 412 pages.

EXERCISES

25.1. What federal organizations are responsible for overseeing that there are no violations of laws regarding the environment?

25.2. Give examples of how environmental safety in the workplace can be violated.

25.3. If an accident occurs in an area under the control of the OSHA and generates an environmental problem there, is the problem restricted by law from spreading to the jurisdiction of the EPA? Provide an example of what might happen.

25.4. What is the OSHA's attitude regarding design and procedures for control on environmental problems?

25.5. What is micrometeorology and how could it affect you?

25.6. How does an induced environment differ from a natural environment? A free from a closed or restricted environment?

25.7. Which can be the most hazardous environment in the workplace? Why? Give examples of how some of the hazards in the workplace can be detected without the use of any sensing devices. Which cannot be detected without such means? With such devices?

25.8. How can protection be provided against the environment, both using and not using equipment?

25.9. Everyone talks about the weather and the environment. What can be done about it?

CHAPTER 26

Confined-Space Entry

The danger of work in confined spaces is well known and not new. Suruda et al. report that "the danger of work in confined spaces has been written about since Roman times, when the Emperor Trajan was noted to have sentenced criminals to clean sewers, an occupation considered one of the worst."[100] In 1556, Agricola wrote that stagnant air in mines impaired breathing, and mine fires soon killed those who came to work there. In 1906, The Pittsburgh Survey shocked America. This Russell Sage Foundation study of safety in the workplace in the Greater Pittsburgh area counted 526 work-related deaths and 500 seriously and permanently disabled workers in one year's time. Asphyxiation at work in industries such as steel making was cited in this study as one of the causes of injury and death. In 1925, Alice Hamilton wrote that decomposing organic matter in vats, tank, and manholes emitted hydrogen sulfide, resulting in death by asphyxiation.

NIOSH published a Criteria Document in 1979 containing recommendations for working in confined spaces.[101] In this document, 276 confined-space incidents were analyzed. Of the 193 fatalities, 78 (40 percent) resulted from atmospheric hazards. An Alert: Request for Assistance in Preventing Occupational Fatalities in Confined Spaces was published by NIOSH in 1986.[102] This was followed by a worker's guide in 1987.[103]

The National Institute for Occupational Safety and Health (NIOSH) studied fatalities related to confined-space entry during the period from December 1983 to December 1989.[104] NIOSH used data from a surveillance project called "Fatal Accident Circumstances and Epidemiology" (FACE), which was designed to collect descriptive data on selected fatalities to identify potential risk factors for work-related death, to develop recommended intervention strategies, to disseminate important findings, and to reduce the risk of fatal injury in the workplace. They reviewed 55 confined-space events which resulted in 88 deaths. Only three of the workers who died had received any training in confined-space safety, and only 27 percent of the employers had any type of written confined-space entry procedures. The mean age of the fatality was 35. All were men. Sixty-six percent were workers who were engaged in activities relating to water, wastewater, and sewerage repair, cleaning, inspection, etc. Thirty-five percent of the victims had supervisory duties. The cause of death was: asphyxiation (47 percent), drowning (21 percent), toxicity from chemical exposure (19 percent), blunt-force trauma (10 percent), electrocution (2 percent), and burns (1 percent). Three underlying factors were common to most of these deaths: (1) Failure to recognize the hazards associated with confined spaces, (2) failure to follow existing, known procedures for safe confined-space entry, and (3) incorrect emergency response. Before, during, and after this data-gathering period, attempts were made to advise industry of the dangers involved in confined spaces.

A study published in 1989 of OSHA investigations of asphyxiation and poisoning conducted in 1984 through 1986 showed 188 deaths in confined spaces.[105] Of these, 42 were from mechanical hazards such as engulfment by loose materials, and 146 were from oxygen deficiency or poison gases or chemicals. These confined-space fatalities constituted 4 percent of the fatalities investigated by OSHA during that period of time. The confined-space classification did not include deaths from trench cave-ins (190), electrocutions, or explosions, which occurred in confined spaces.

Finally, on April 15, 1993, OSHA issued a regulation entitled "Permit-Required Confined Spaces" as part of the Code of Federal Regulations, 29 CFR 1910.146. This regulation, because of its broad definition of a confined space, is applicable to an estimated 300,000 facilities with over 12 million employees.

In 1994, NIOSH published a summary of NIOSH findings about worker deaths in confined spaces, using the National Traumatic Occupational Fatalities (NTOF) surveillance system database.[106] NIOSH reviewed death certificates for 1980–1989 in which "injury at work" was indicated. There were 585 separate fatal incidents in confined spaces, with 670 fatalities. Seventy-two (12 percent) involved multiple victims, 61 involving 2 persons; 9, 3 persons; and 2, 4 persons. The confined-space death rate was 0.08 per 100,000 workers per year. In the mining/oil/gas industry that rate was 0.7, in agriculture it was 0.3, and in construction 0.2. Forty-five percent of the deaths were accounted for by asphyxiation, 41 percent by poisoning, and 14 percent by drowning. Clearly, this is a persistent problem.

WHAT IS A CONFINED SPACE?

There are two important definitions regarding confined spaces in the United States. First, in 29 CFR 1910.146 a *confined space* is an enclosed area which (1) is large enough to enable an employee to enter and perform assigned work, (2) has limited or restricted means for entry or exit, and (3) is not designed for continuous employee occupancy.[107] The second definition, as follows, is a subset of the first. In 1910.146, a *permit-required confined space* is a confined space with one or more of the following characteristics: contains or may contain a hazardous atmosphere, contains a material with the potential for "engulfment" of an entrant (such as sawdust or grain), has an internal configuration or shape such that an entrant could be trapped or asphyxiated, or contains any other recognized serious safety or health hazard.

Confined Space

The regulation's definition of a "confined space" is broadly inclusive. The first criterion indicates that a space that cannot be completely entered by an employee is not a confined space. The second criterion is less precise. It includes tanks, vessels, silos, storage bins, hoppers, vaults, and pits, because these have limited means of ingress and egress. Open-topped spaces can be difficult to ingress or egress. ANSI Z 117.1-1977 listed as one criterion for a confined space an open-topped space more than 4 feet in depth.[108] NIOSH says that "Confined spaces include but are not limited to storage tanks, compartments of ships, process vessels, pits, silos, vats, wells, sewers, digesters, degreasers, reaction vessels, boilers, ventilation and exhaust ducts, tunnels, underground utility vaults, and pipelines."[105] It is evident that confined spaces can be found in many, if not most, industries. The third criterion does not effectively limit the domain of confined spaces because it does not spell out the elements of design that must be included for continuous employee occupancy.

Permit-Required Confined Space

The criterion for permit-required confined space that it contains or has a known potential to contain a hazardous atmosphere includes chemicals, sludge, or sewage. "Hazardous atmosphere" means an atmosphere that may expose employees to the risk of death, incapacitation, impairment of ability to self-rescue (escape unaided), injury, or acute illness from one or more of the following causes:

1. Flammable gas, vapor, or mist in excess of 10 percent of its lower flammable limit (LFL);
2. Airborne combustible dust at a concentration that meets or exceeds its LFL;
3. Atmospheric oxygen concentration below 19.5 percent or above 23.5 percent;
4. Atmospheric concentration of any substance for which a dose or a permissible exposure limit is published in Subpart G, Occupational Health and Environmental Control, or in subpart Z, Toxic and Hazardous Substances, of 29 CFR part 1910, and which could result in employee exposure in excess of its dose or permissible exposure limit; and
5. Any atmospheric condition recognized as immediately dangerous to life or health (IDLH).

If a confined space contains a material that has the potential for engulfing an entrant, it is a permit-required confined space. "Engulfment" means the surrounding and effective capture of a person by a liquid or finely divided (flowable) solid substance that can be aspirated to cause death by filling or plugging the respiratory system or that can exert enough force on the body to cause death by strangulation, constriction, or crushing.

If a confined space has an internal configuration or shape that could trap an entrant or contribute to the asphyxiation of an entrant, it is a permit-required confined space. A confined space with inwardly converging walls or a floor which slopes downward and tapers to a smaller cross section is an example of such a potentially dangerous configuration for an entrant.

Finally, a confined space that contains any other recognized serious safety or health hazard is a permit-required confined space. The suggestion to this author is that if an employer has a serious accident or injury in a confined space, unless that accident or injury is unforeseeable, that space should have been a permit-required confined space. The key is prevention of that undesired event, which can be costly and embarrassing to the employer. The employer is under an obligation to recognize serious safety or health hazards before accidents or illnesses occur, and when these are recognized in a confined space, that becomes a permit-required confined space. These hazards can include physical, electrical, mechanical, chemical, biological, radiation, temperature extremes, and structural hazards.

ATMOSPHERIC HAZARDS

NIOSH has developed a classification scheme for atmospheric hazards in confined spaces. It is based on the oxygen content of the air, the flammability characteristics of gases or vapors, and the concentration of toxic substances present. These are shown in Figure 26-1. Classification is determined by the most hazardous condition present.

This classification scheme is useful in considering entry procedures and work practices in confined spaces. It is used in this way in NIOSH's "Criteria for a Recommended Standard: Working in Confined Spaces," 1979.[101]

CHARACTERISTICS

CLASS A
Immediately dangerous to life
Oxygen: 16 percent or less* (122 mm Hg) or greater than 25 percent (190 mm Hg)
Flammability: 20 percent or greater of lower flammable limit (LFL)
Toxicity: Immediately dangerous to life or health (IDLH)

CLASS B
Dangerous, but not immediately life threatening
Oxygen: 16.1percent to 19.4 percent* (122–147 mm Hg), or 21.5 percent to 25 percent (163–190 mm Hg)
Flammability: 10–19 percent LFL
Toxicity: Greater than contamination level, re: 29 CFR Part 1910 Subpart Z (IDLH)

CLASS C
Potential hazard
Oxygen: 19.5 percent–21.4 percent* (148–163 mm Hg)
Flammability: 10 percent LFL or less
Toxicity: Less than contamination level, re: 29 CFR Part 1910 Subpart Z

* Based upon a total atmospheric pressure of 760 mm Hg (sea level)
Table adapted from *Worker Deaths in Confined Spaces*, NIOSH, 1994[106]

FIGURE 26–1 Confined-Space Classification Table adapted from *Worker Deaths in Confined Spaces*, NIOSH, 1994.[106]

PHYSICAL HAZARDS

Physical hazards include those associated with the unwanted flow of energy in confined space. These include mechanical, electrical, and hydraulic energies; engulfment; communication problems; noise; and the size of ingress and egress openings.

Engulfment hazards often are associated with bins, silos, and hoppers used to store and transfer grain, sand, gravel, or other loose materials. These materials can behave unpredictably and entrap and bury a person in a matter of seconds. The flow path of a bottom-emptying bin forms a funnel shape, with the material at the center of the bin moving faster than elsewhere. The flow rate can become too great for a worker who is caught in the flow path to be able to escape. Material can cling to the sides of a container or vessel that is being emptied from the bottom, forming a "bridge." The bridge may collapse suddenly and without warning upon workers below. Engulfment can happen in seconds.

Activation of mechanical and electrical equipment (agitators, blenders, stirrers, fans, augurs, pumps, machinery with moving parts, etc.) can cause injury to workers in confined spaces. Release of material into or out of confined space can be life threatening to those who are within. Fluids under pressure pose a hazard. Objects falling into confined space present the entrant with the hazard not only of the object but also of limited evasive potential if the object is detected. Slippery surfaces (sometimes made slick by material residue), hot and cold extreme temperatures, inadequate lighting, limited work space, excessive noise, and sources of electrocution all are physical hazards that exacerbate confined-space entry and work.

CHEMICAL, BIOLOGICAL, RADIATION

Chemical hazards of industrial processes are varied and myriad. Hazardous wastes and useful materials that are threatening to life and health abound. There are threats not only of acids, corrosives, and toxics, but also of inert materials that displace oxygen and of materials that reduce oxygen content in confined spaces.

Biological hazards from infectious microorganisms and biological wastes are found in waste streams, pools and ponds, sludge pits, and sewers. Some industrial processes and research efforts involve this hazard intensely and are especially susceptible to its problems. The control of bacteria is a growing area of public health concern. Where these threats are potential within a confined space, entry to that space for work must be made with care.

Radiation potential in both nonionizing and ionizing forms where confined-space entry might occur requires special attention.

MANAGEMENT RESPONSIBILITIES FOR CONFINED SPACES

It is essential that employers have a well-defined and implemented confined-space program. In the United States this program must comply with the requirements of federal and state regulations. It is not the purpose of this text to restate regulations.

General program guidelines are provided by NIOSH (1994), without the distinction between "confined space" and "permit-required confined space" given in the federal regulations. These guidelines for elements in a comprehensive, written confined-space entry program are given below.[106]

The program should include, but not be limited to, the following:

- identification of all confined spaces at the facility/operation
- posting a warning sign at the entrance of all confined spaces
- evaluation of hazards associated with each type of confined space
- a job safety analysis for each task to be performed in the confined space
- confined-space-entry procedures
  - initial plan for entry
  - assigned standby person(s)
  - communications between workers inside and standby
  - rescue procedures
  - specified work procedures within the confined space
- evaluation to determine if entry is necessary—can the work be performed from the outside of the confined space?
- issuance of a confined-space-entry permit—this is an authorization and approval in writing that specifies the location and type of work to be done, and certifies that the space has been evaluated and tested by a qualified person and that all necessary protective measures have been taken to ensure the safety of the worker
- testing and monitoring the air quality in the confined space to ensure that
  - oxygen level is at least 19.5 pecent by volume
  - flammable range is less than 10 percent of the LFL (lower flammable limit)
  - all toxic air contaminants are absent
- confined-space preparation
  - isolation/lockout/tagout
  - purging and ventilation
  - cleaning processes
  - requirements for special equipment and tools
- safety equipment and protective clothing to be used for confined-space entry
  - head protection
  - hearing protection
  - hand protection
  - foot protection
  - body protection
  - respiratory protection
  - safety belts
  - lifelines, harness
  - mechanical-lift device—tripod
- training of workers and supervisors in the selection and use of
  - safe entry procedures
  - respiratory protection

  - lifelines and retrieval systems
  - protective clothing

- training of employees in confined-space-rescue procedures
- conducting safety meetings to discuss confined-space safety
- availability and use of proper ventilation equipment
- monitoring the air quality while workers are in the space.

Safe work in confined spaces is teamwork. It requires a trained confined-space team which includes:

- the entrant(s), the person(s) who does (do) the work
- the attendant, who stays outside while the entrant is inside the confined space
- the person who authorizes permits, and
- the rescue team.

A person may perform more than one of these tasks. For example, the person who authorizes the permits may also be trained to do one of the other tasks. The National Safety Council has prepared educational materials which outline the duties of this team.[109]

The entrant does the work assigned to be done in the confined space. The entrant:

- reviews the permit before entry; wears/uses appropriate personal protective clothing and equipment;
- uses, attends and appropriately responds to monitoring equipment;
- is sensitive to and appropriately responds to personal physical reactions that may indicate unsafe conditions;
- maintains communications with the attendant and obeys evacuation orders;
- signals the attendant for help as appropriate and leaves the confined space promptly.

The attendant is responsible to make sure the entrant remains safe. This responsibility is discharged by:

- reviewing the permit before any entry,
- constantly keeping track of the personnel in the confined space,
- keeping unauthorized persons out of the area,
- maintaining continuous communication by sight or voice with all persons in the confined space,
- making sure the ventilation equipment, if any, is working,
- monitoring the atmospheric testing equipment,
- tending the lifeline of the entrant,
- tending the air line, if any, to prevent tangles and kinks,
- remaining alert for signs of trouble,
- watching for hazards outside and inside the space,

- maintaining clear ingress and egress to the space,
- notifying the entrant and ordering evacuation if conditions warrant, or the permit limits expire,
- being prepared to call for emergency help,
- remaining at the entry point unless relieved by another trained attendant.

The person in charge of issuing the permits:

- plans each entry
    - describes the work to be done
    - identifies the workers involved
    - evaluates the hazards of the space
    - performs or oversees atmospheric testing and monitoring
    - develops rescue plans
- ensures the permit is properly completed
- determines equipment needs
- ensures appropriate atmospheric testing
- verifies that all necessary procedures, practices, and equipment for safe entry are in effect
- cancels the permit and terminates the work if the conditions are not acceptable
- trains (or provides training) for all workers on the confined-space entry team
- Keeps records on training, safety drills, test results, equipment inspections, and equipment maintenance
- cancels the permit and secures the space when the work is done
- determines if a written rescue plan is necessary for a particular confined-space entry
- verifies that emergency help is available and that the method of summoning assistance is operable.

Members of the rescue team may be from within the plant or from an arrangement with an outside rescue team. The team members, if from within the plant, must be trained in appropriate personal protective equipment, be trained as authorized entrants, drill at least once every twelve months (by rescuing mannequins from simulated confined spaces), and at least one member must have a current CPR and first aid certification. Regardless of the rescue arrangements, the team should be able to respond quickly to the site with the necessary equipment in proper working order. They should be cross-trained, and preferably all should have current CPR and first aid certification. Several should be trained in advanced first aid and first responder.

BIBLIOGRAPHY

[100] A. J. Suruda, D. N. Castillo, J. C. Helmkamp, and T. A. Pettit, "Epidemiology of Confined-Space-Related Fatalities," In *Worker Deaths in Confined Spaces*, National Institute for Occupational Safety and Health, Jan. 1994.

[101] National Institute for Occupational Safety and Health, *Criteria for a Recommended Standard: Working in Confined Spaces,* DHHS (NIOSH) 80–106, 1979.

[102] National Institute for Occupational Safety and Health, *Request for Assistance in Preventing Occupational Fatalities in Confined Spaces,* DHHS (NIOSH), 86–110, 1986.

[103] T. A. Pettit and H. E. Linn, eds., *A Guide to Safety in Confined Spaces,* National Institute for Occupational Safety and Health, DHHS(NIOSH), 87–113, 1987.

[104] J. C. Manwaring and C. Conroy, "Occupational Confined-Space-Related Fatalities: Surveillance and Prevention, *Journal of Safety Research,* Vol. 21, pp. 157–165, 1990.

[105] A. Suruda and J. Agnew, "Deaths from Asphyxiation and Poisoning at Work in the United States 1984–6," *British Journal of Industrial Medicine,* Vol. 46, 1989, pp. 541–546.

[106] NIOSH, *Worker Deaths in Confined Spaces: A Summary of NIOSH Surveillance and Investigative Findings,* National Institute for Occupational Safety and Health, 1994, 273 pp.

[107] Code of Federal Regulations, 29 CFR 1910.146 [1993], *Permit-Required Confined Spaces,* U.S. Government Printing Office, Washington, D.C.

[108] American National Standards Institute, *Safety Requirements for Confined Spaces,* American National Standard, Z-117.1, New York, 1977.

[109] National Safety Council, *Confined Spaces: Training the Team,* Chicago, 1991.

EXERCISES

26.1. How many facilities are covered by the 1993 OSHA regulation, "Permit-Required Confined Spaces"? How many employees are in these affected facilities?

26.2. 1994 NIOSH reviewed death certificates of occupational fatalities for the 80's decade. How many fatal incidents were there? With how many fatalities?

26.3. Name some of the spaces included by NIOSH as confined spaces.

26.4. Name the criteria for permit-required confined space.

26.5. Name each class and describe the various characteristics used in the classification of confined spaces according to atmospheric hazards.

26.6. Name the elements in a comprehensive, written confined-space entry program for the entrant, attendant, and permit issuer.

26.7. Can the rescue team be from within the plant?

CHAPTER 27

Radiation

This chapter includes information on all types of radiation, but ionizing radiation is the one which has become highly feared. This apprehension was of minor concern until the intended explosions of the atomic bombs; these fears escalated with the construction of nuclear power plants throughout the world and peaked after accidents at Three Mile Island in Pennsylvania and then at Chornobyl in the Ukraine.

One writer stated his belief that nuclear power plants were accidents waiting to happen. Protesters have prevented new power plants from being built; some whose construction had been completed were prevented or delayed from starting operations; and there were calls for shutdown of those already in operation. A resulting problem is very high construction costs even before plants open. After the start of operations, problems have been: (1) inadequate incorporation of safe designs by engineers; (2) lack of proper engineering to negate errors or confusing ambiguities resulting from (1); and (3) lack of adequate operator education, training, conduct, and supervisory control. In any case, knowledge by safety engineers has become vital in all aspects of ionizing radiation for nuclear workers.

Mr. Polly Story told the United States Senate:[110] "Radiation is radiation. It cannot be seen; it cannot be felt; it cannot be smelled; it cannot be touched. Yet it exists, and though its proper use has been of immeasurable benefit to mankind, its abuse or improper use presents great hazards."

The speaker was concerned with X-radiation, but his statement is applicable with minor exceptions (certain radiations can be seen) to all types of electromagnetic radiation. These are indicated in Fig. 27-1; fundamentally they are similar in that they all travel at the speed of light. The range of each type of radiation is not clear, as shown on the drawing, and may overlap other types. For example, hard X-rays radiate at wavelengths that are the same as those for gamma rays. They actually have identical properties, the sole difference being their origins: X-rays are produced in a vacuum tube and gamma rays are produced by radioactive materials, natural or manufactured. X-rays can be turned off by shutting off the equipment; radioactive materials cannot.

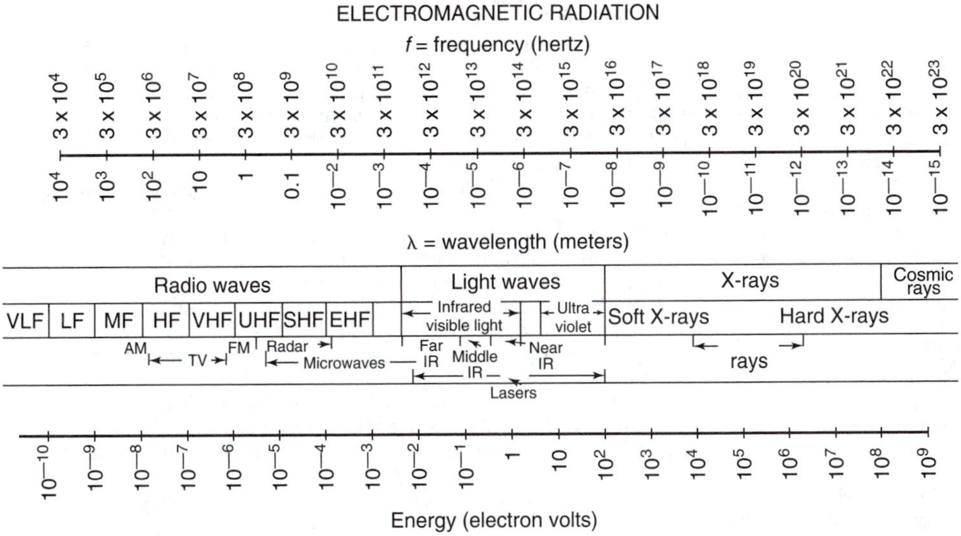

ELECTROMAGNETIC RADIATION

f = frequency (hertz)

3×10^{4}, 3×10^{5}, 3×10^{6}, 3×10^{7}, 3×10^{8}, 3×10^{9}, 3×10^{10}, 3×10^{11}, 3×10^{12}, 3×10^{13}, 3×10^{14}, 3×10^{15}, 3×10^{16}, 3×10^{17}, 3×10^{18}, 3×10^{19}, 3×10^{20}, 3×10^{21}, 3×10^{22}, 3×10^{23}

10^{4}, 10^{3}, 10^{2}, 10, 1, 0.1, 10^{-2}, 10^{-3}, 10^{-4}, 10^{-5}, 10^{-6}, 10^{-7}, 10^{-8}, 10^{-9}, 10^{-10}, 10^{-11}, 10^{-12}, 10^{-13}, 10^{-14}, 10^{-15}

λ = wavelength (meters)

Radio waves | Light waves | X-rays | Cosmic rays

VLF | LF | MF | HF | VHF | UHF | SHF | EHF | Infrared visible light | Ultra violet | Soft X-rays | Hard X-rays

AM | TV | FM | Radar | Microwaves | Far IR | Middle IR | Near IR | IR | Lasers | rays

10^{-10}, 10^{-9}, 10^{-8}, 10^{-7}, 10^{-6}, 10^{-5}, 10^{-4}, 10^{-3}, 10^{-2}, 10^{-1}, 1, 10, 10^{2}, 10^{3}, 10^{4}, 10^{5}, 10^{6}, 10^{7}, 10^{8}, 10^{9}

Energy (electron volts)

Electromagnetic radiation: energy emitted or absorbed as small discrete pulses called *quanta* or *photons.* The quanta differ in magnitude in proportion to the frequency. X-rays, gamma rays, and cosmic rays are much shorter, have higher frequencies than radio waves, and move with greater energies.

Wavelength: λ = velocity/frequency = $3 \times 10^{8}/f$, where 3×10^{8} meters per second is the speed of light, the velocity of all electromagnetic radiation; f is the frequency in hertz, and λ is in meters. The various types of radiation are generally expressed in various types of units.

Wavelengths are sometimes indicated in *angstrom units* ( ), one angstrom unit being equal to 10^{-10}m. To read the wavelength scale in angstrom units, add 10 to the exponent in meters. Thus a wavelength of 10^{-8} meter is 10^{-2} .

Electron volt (eV): kinetic energy an electron acquires by being accelerated between two points having a potential difference of 1 volt. In Systeme Internationale (metric), 1 eV is equal to 1.602×10^{-19} joule. This is so small that radiation energies are usually expressed as thousands (keV), millions (meV), or billions (beV) of electron volts.

Ionizing radiation: ionization is the process by which atoms are made into ions by the removal or addition of one or more electrons. Ionizing radiations (x-rays, gamma rays, and cosmic rays) produce this effect by the high kinetic energies of the quanta they emit.

Isotopes: atoms which are chemically identical but which differ in the number of neutrons in their nuclei. For example, hydrogen has three isotopes: hydrogen the most abundant which has one proton and no neutron; deuterium, which has one proton and one neutron; and tritium, which has one proton and two neutrons.

Radioactivity: a radioactive substance is one whose atomic nucleus is unstable. The nuclei are continually disintegrating spontaneously into nuclei of lower-atomic-weight elements, with the emission of nuclear particles and energy. This process is called *radioactivity*. Radioactivity may be natural or it may be produced by artificially induced nuclear reaction.

Radioactive isotope: radioactive form of an element.

Half-life: disintegration of radioactive nuclei cannot be altered by changes in physical conditions. The process goes on continually, but each radioactive isotope differs from all others in the number of atoms that disintegrate per second.

Half-life: is the length of time necessary for the disintegration process to consume exactly one-half of the original radioactive mass. Half-life is a constant for that material regardless of the quantity involved.

FIGURE 27–1 Electromagnetic Radiation Continued

Curie: a curie is the amount of any radioactive isotope that will produce 3.7 x 10^{10} nuclear disintegrations per second. A *millicurie* is a thousandth and a *microcurie* a millionth of a curie.
Alpha particles: essentially nuclei of helium with two positive charges. When emitted by the integration of the nucleus of a radioactive isotope, an alpha particle will travel at 2000–20,000 miles per second. They are comparatively large, and since they attract electrons because of their two positive charges, they have little power to penetrate other substances. By collision, alpha particles cause other atoms to release electrons, which they then absorb to form a stabilized helium atom. The atoms that have lost electrons remain ionized until they can replace their losses.
Beta particles: fast-moving electrons. They are lighter than alpha particles and have much greater penetrating power, but are less ionizing.
Gamma radiation: result of excess energy leaving a disintegrating nucleus immediately after losing an alpha or beta particle. Gamma radiation travels at the speed of light and has great penetrating power. Gamma radiation passing through a normal atom will sometimes cause the loss of an electron, leaving the remaining atom as a positive ion. This ion and the expelled electron are called an *ion pair*.
X-rays: radiation energy similar to gamma radiation. They are produced by sudden acceleration or deceleration of a charged particle, such as when high-energy electrons hit metal targets or the walls of vacuum tubes.
Roentgen (r): unit of measure of a material's absorption of gamma or X-radiation. One explanation is that a roentgen is the amount of gamma or X-radiation whose absorption by a cubic centimeter of air at standard conditions will create 83 ergs per gram of dry air. A *milliroentgen* (mr) is one-thousandth of a roentgen.
Rad, or roentgen absorbed does: amount of ionizing radiation which will result in the absorption of 100 ergs per gram of tissue.
Rem, or roentgen equivalent man: quantity of radiation which produces a physiological effect equivalent to that produced by the absorption of one roentgen of gamma or x-radiation. Each type of radioactive particle has its own *relative biological effectiveness* (RBE) to destroy tissue by ionization.
The rem is equal to the number of rads times the RBE.

Radiation	*Energy Level (Me V)*	*Absorption Ratio*	
		Roentgen	*Rad*
X-ray	8 keV to 2	1	1
Gamma	0–3	1	1
Beta	0–3.5	–	1
Alpha	4–8	–	20

Biological half-life: if a radioactive material has been taken into the body, normal life processes will tend to reduce or eliminate the amount. The time required to eliminate one-half the total quantity of a radioactive material absorbed internally is called its *biological half-life.*

FIGURE 27–1 Continued

IONIZING RADIATION

Alpha, beta, neutral particles, X-rays, and gamma rays (cosmic rays are not of concern in this book) are ionizing radiations. Each of these may cause injury by producing ionization of cellular components, leading to functional changes in the tissues of the body. The energy of that radiation is great enough to discharge electrons from the atoms that make up the cells, producing ion pairs, chemical free radicals, and oxidation products. The body's tissues differ in form, composition, and function. Their sensitivities and biological responses to ionizing radiation are therefore different. In some instances, radiation damage to the cell is replaced by natural processes. (Sensitivity to radiation is directly proportional to the reproductive capability of the cell.) The radiation effects in such cases are said to be "reversible," whereas permanent or nonrepairable injury is "irreversible."

Sensitivities of various tissues and organs of the body are shown in Fig. 27-2. The effects that can be produced by various levels of acute exposure are given in Fig. 27-3, with probable symptoms that will be shown by affected persons. The figures indicate that the dosage is received by the "whole body." Much higher dosages have been applied to small areas for therapeutic purposes, but they are very local and not of the whole body. Low-level radiation will cause cell damage, not death. At from 100 to 400 rems there will be radiation sickness and some deaths. Above 400 rems, delivered at one time to the whole body, 50 percent of all victims will die. Dosage results can range from radiation burns and cancer to genetic mutations in future generations.

The effects generated during a massive single exposure or over a short period are said to be "acute" and can produce both immediate and delayed effects on the body. Low but repeated radiation exposures are said to be "chronic" and generally have delayed effects. Acute exposures usually result from mishaps, whereas chronic exposures are due principally to sustained and unrecognized conditions.

Radioactivity does not lose its potency by such processes as absorption or ingestion by living tissues. Thus, the radioactive material from airborne fallout may land on grass, where it may be eaten by cattle or sheep and subsequently by humans with continued radioactivity.

Alpha Particles

Although alpha particles are emitted from atomic nuclei at high speed and with high energy, they have short ranges in dense materials. Alpha radiation is the most energetic but least penetrating. Even those with maximum energy will just penetrate the epidermis. Since the epidermis is composed of dead cells, there will be little effect. It can therefore be considered that alpha particles from an external source constitute no danger. They can be most harmful when they enter the body through a wound, inhalation, or the digestive system. If ingested or inhaled into the body, they will be close to living tissue that can be damaged. The injury will be local but intense, so that essential organs containing a source of alpha particles may be completely destroyed.

Beta Particles

Some fission products in spent fuel and wastes are beta particles, such as strontium-90, iodine-131, and cesium-137. A beta particle emitted from the nucleus of an atom is far smaller in mass than is an alpha particle (about 1/7300). Because beta particles are more penetrating and have longer linear-energy-transfer (LET) paths than do alpha particles, the radiation absorbed dose (rad) is generally lower. However, occasionally a beta particle will ionize an atom, but far fewer than do alphas. Because of chemical similarities with calcium, strontium-90 might be concentrated in the bones. A beta source maybe an external hazard if the energy of the betas is great enough, since the particles may penetrate the dead cells of the skin. A thin sheet of metal, such as aluminum, can act as a shield, but the metal must be selected with care to ensure it is not

EFFECTS OF IONIZING RADIATION ON BODY TISSUES AND ORGANS

Tissue or Organ Affected	Effect
Blood	White cells (leucoeytes) are the most sensitive to radiation, which will reduce their number and leave the body open to infection. In severe cases the number of platelets will drop after 1 week so that the blood's clotting capability is reduced. Weeks later, the number of red cells will decrease to the point that anemia results.
Bone marrow	Damage to the blood can be overcome if there is replacement of the injured blood cells by new ones from the bone marrow. However, radiation damage can injure the bone marrow so that cell replacement cannot take place. Damage to the body will be permanent in such cases.
Eyes	The eyes are the parts of the body most sensitive to radiation, and the lens cells are the most easily damaged by ionizing radiation. The lenses gradually become opaque with "cataracts," since the cells are not replaced as are blood cells. Other parts of the eye, such as retina, are less sensitive, but will be affected by high exposures as would any other body cell.
Skin	Skin is easily damaged. However, since the outer layer is composed of dead cells, the skin has great recuperative powers. The dead layer prevents lower-level damage by alpha particles, and attenuates the effects of beta particles.
Reproductive organs	Immediate effects would be the same as those on other cells. To produce sterility in a person would require almost a fatal dose. Genetic effects produced by radiation damage to reproductive cells can only be surmised and would not be known for several generations after exposure.
Lymphatic system	The lymph nodes (which filter out foreign matter from the lymph) would be the first affected in this system by a heavy dose of radiation. The spleen (which normally filters out dead cells from the blood) would be next.
Digestive system	The various portions of the digestive system vary in sensitivity and types of damage. The small intestine is probably the most sensitive. When walls of the digestive tract are damaged, the dead cells are released into the passages, obstructing their normal processes so that nausea and vomiting occur. Breaking away of surface cells on the lining may lead to ulcers and inability to absorb food. Infection may set in.
Nervous system	Damage to the brain may occur if blood vessels and the blood to it are damaged. The spinal cord and the nerves are highly resistant to radiation.
Hair	Radiation will lead to loss of hair. The effect is generally temporary after exposure stops; however, the new hair may be of a new color or have other characteristics different from the original.
Other organs	Other major organs, such as the kidneys, circulatory system, respiratory system, and liver are generally highly resistant to ionizing radiation and will be injured only by very high dosages, such as those which result from the presence of an internal source.

FIGURE 27–2 Effects of Ionizing Radiation on Body Tissues and Organs

SYMPTOMS AND EFFECTS RESULTING FROM ACUTE WHOLE-BODY EXPOSURE TO RADIATION*

Exposure (Roentgens)	Symptoms and Effects
0–25	No observable reactions. Delayed effects may occur.
25–100	Changes in blood detectable by clinical tests. Disabling sickness not common; individual should be able to continue usual duties. Delayed effects possible, but serious effects on average individual unlikely.
100–200	Produces nausea and fatigue, with possible vomiting above 125 roentgens. Changes in blood detectable by clinical tests. Delayed effects may shorten life expectancy of exposed individual on the order of 1 percent.
200–300	Produces nausea and vomiting on first day following exposure. Latent period up to 2 weeks or longer, then other symptoms appear, but are not severe. Symptoms are loss of appetite, general illness or discomfort, sore throat, pallor, petechiae (crimson spots in skin or mucous membrane), diarrhea, and moderate emaciation. Recovery is expected in about 3 months unless complicated by poor previous health, additional injuries, or infections.
300–600	Produces nausea, vomiting and diarrhea in first few hours following exposure. Latent period, perhaps as long as 7 days, with no definite symptoms. Symptoms are epilation (loss of hair), loss of appetite, general illness or discomfort, and fever during second week, followed by hemorrhage, petechiae, inflammation of the mouth and throat, diarrhea, and emaciation in the third week. Some deaths in 2–6 weeks. Possible eventual death of up to 50 percent of the exposed individuals for exposures of 450 roentgens.
600 and over	Produces nausea, vomiting, and diarrhea in first few hours following exposure. Short latent period with no definite symptoms in some cases during first week. Diarrhea, hemorrhage, inflammation of mouth and throat, and fever toward end of first week. Rapid emaciation and death as early as the second week with possible eventual death of up to 100 percent of the exposed individuals.

*Air Force Communications Service, *Electromagnetic Radiation Hazards*, Air Force Technical Order 31Z-10-4, Aug. 1966.

FIGURE 27–3 Symptoms and Effects Resulting from Acute and Whole-Body Exposure to Radiation

one from which X-rays can be generated. Like alpha particles, betas are most dangerous if inhaled or ingested.

Gamma Rays and X-Rays

Both rays are similar, but the first is natural, from fission products; X-rays are produced by high-speed electrons striking a suitable target. Electrical potentials required to accelerate electrons to speeds great enough to generate X-rays must be at least 15,000 to 16,000 volts. Electrical equipment operating at voltages lower than this level will not produce X-radiation. At voltages higher than this, the possibility of this hazard's being present is substantial.

X-rays and gamma rays ionize matter in three principal ways. The photons which constitute the X-ray or gamma ray strike the atom so that:

1. An electron is emitted at such velocity that all photon energy not used in dislodging the electron increases its kinetic energy (photoelectric effect).
2. An electron is emitted without absorbing all the photon energy. The photon therefore continues on, but with less energy than before (Compton effect). The electron and photon may both continue to ionize other atoms, depending on their energies.
3. An electron and positron (positive electron) are formed by the photon if its energy is high enough near the nucleus of the atom and are emitted (pair production). The electron and positron both lose energy as they ionize other atoms. The positron will react with an electron it strikes to form two new photons, of lesser energy than the original, which also go on to ionize other atoms (pair production).

The energies of X-rays and gamma rays give them high penetrating power even through fairly dense materials. In air and other substances of low density, because they have long ranges, they will be hazardous at comparatively long distances. Because of their penetrating power, X-rays and gamma rays will injure tissue throughout the body in addition to damaging the skin.

Internally, the energies of X-rays and gamma rays are not as readily absorbed by tissue as those of alpha or beta radiation. The shorter ranges of alpha, beta, or neutron particles in tissues result in the absorption of radiation energy, which produces damage to a small volume of tissue, with a correspondingly higher absorbed dose.

FACTORS AFFECTING EXPOSURE AND RISK

The injury that can result from electromagnetic radiation is dependent on two factors: time and intensity of exposure. An increase in either increases the injury.

Intensity depends on a number of factors: strength of the source, its distance, and the existence and amount of any shielding. That intensity depends on the strength of the source is obvious in one respect: a larger or more powerful source will radiate more strongly than will one smaller or less powerful. Some radioactive isotopes are more hazardous than others, depending on the type and where the source is located. Certain types constitute hazards only at short ranges.

The intensity of radiation received from a point source will be inversely proportional to the square of its distance (inverse-square law). When a source is large compared to the distance between it and the object, the inverse-square law may not hold exactly; however, for each small portion the intensity will decrease inversely with the distance.

The type and magnitude of shielding required to attenuate radiation to safe levels depends on the radiation involved. The types of radiation have been noted earlier; precautionary measures and recommendations for shielding are shown in Figs. 27-4 and 27-5.

SOURCES OF IONIZING RADIATION

The commonest source of ionizing radiation in industrial plants are not only for X-rays, but for radioactive isotopes for nondestructive purposes. Chemical laboratories are frequent users of radioactive materials, generally in small amounts, but with high

PRECAUTIONARY MEASURES AGAINST IONIZING RADIATION

1. Only personnel familiar with the hazards involved should be permitted to use, operate, handle, transport, or store equipment or material which produces ionizing radiation. Where certification is required, ensure that operators of ionizing radiation equipment are currently certified.
2. Prior to activation or use of a facility where ionizing radiation equipment will be produced, a survey should be made by a qualified expert to determine the adequacy of the facility, its equipment, protective devices, and safety precautions.
3. Safety engineer should be aware of any equipment or material in his plant which could produce ionizing radiation, and ensure that suitable precautionary measures are stipulated and in use. Ensure that a license has been obtained for each ionizing radiation source prior to initiation of operations.
4. Operating and emergency procedures for radiation safety should be prepared and posted at the accesses to any facility where there may be ionizing radiation. Personnel working in such facilities should be familiar with these procedures.
5. Access to areas in which equipment or materials producing ionizing radiation are present should be restricted to personnel directly concerned with operation, maintenance, or other required activity.
6. All accesses to areas in which ionizing radiation may be present, either intentionally or accidentally, should be posted with suitable warnings. Accesses to locations where equipment generating ionizing radiation is in use should be provided with flashing or rotating purple warning lights which operate automatically when radiation is being produced.
7. All electronic devices capable of producing X-radiation should be operated only in shielded enclosures which attenuate the radiation to a permissible level on the outer surface.
8. Radiation levels outside of shielding around equipment or material emitting ionizing radiation should be monitored periodically to ensure that the effectivity of the shielding is maintained.
9. Records should be maintained of individual exposures to ionizing radiation to ensure that permissible levels are not exceeded.
10. Periods during which personnel remain in areas where ionizing radiation may be present should be kept to a minimum. This measure should be secondary to eliminating any possibility of exposure.
11. Any person who believes that he has been subjected to ionizing radiation should report to the medical facility for examination.
12. Personnel who enter areas where ionizing radiation may be present should be provided with personal dosimeters to warn them against excessive dosages. Protective clothing and equipment worn into such areas will depend on the level of radioactivity that might be present. Laboratory coats should be worn even where tracer amounts only will be employed. Coveralls, hoods, masks, gloves, shoe coverings, and other equipment will depend on the risk and level of possible contamination.
13. No edible material of any kind, cosmetics, or cigarettes should be brought into an area where there might be the possibility of radioactive contamination. Hands should be washed and outer protective clothing discarded after leaving the hazardous area and prior to eating, applying cosmetics, or smoking.
14. Cleanup techniques, equipment, and procedures should be developed and ready before any radioactive material is permitted in the plant. Procedures should require:

- Any person who spills a radioactive material to notify the proper office or person responsible for containing and cleaning up the spill.
- Rapid action to minimize spread of spilled radioactive material. Fans and blowers to and from the affected area should be turned off, and windows and doors closed and sealed. Dry spills of radioactive materials should be covered with wet absorbent paper, or cleaned up with a vacuum cleaner specifically designated for the purpose if one is immediately available. A wet spill should be covered immediately with absorbent paper.
- Accesses to affected areas should be marked with warning signs.
- Anyone who may have been in the area where the spill occurred should be checked for contamination. If contaminated, he should be directed to obtain medical assistance.

FIGURE 27–4 Precautionary Measures against Ionizing Radiation Continued

- Cleanup and decontamination should be undertaken only by personnel designated and trained for the purpose. Personnel decontaminating affected areas or equipment should do so only while wearing appropriate protective clothing and respiratory equipment.
- No other person should be permitted to enter the area until it has been monitored and certain that no contamination exists.

FIGURE 27–4 Continued

Shielding Recommendations for Ionization Radiation Protection[a]

Radiation type	Range of Rays in air	Shield Material	
		Type	Thickness
Alpha particles (4 million electron volts)	2.8 cm	Aluminum sheet	$\frac{1}{64}$ inch
		Paper	$\frac{1}{64}$ inch
		Ordinary clothing	$\frac{1}{64}$ inch
Beta particles (3 million electron volts)	13.0 meters	Lead	1.4 mm
		Aluminum	5.3 mm
		Pyrex	6.6 mm
		Lucite	12.4 mm
		Water	14.8 mm
Gamma rays (4 million electron volts)		Shielding is accomplished by reducing intensity of incident gamma radiation by scattering interactions within a shield (the probability of completely absorbing the nuclei of atoms in a shield is slight). Thickness of material required to reduce radiation to one-half is called the half-value layer. Half-value layers for typical materials are:	
		Lead	0.3 inch
		Iron	0.5 inch
		Aluminum	2.7 inches
		Concrete	2.7 inches
		Water	8.3 inches

[a]Based on R.E. Barbiere et al., *A Radiobiology Guide*, Wright Aeronautical Development Center (WADC) Technical Report 57-118 (1), 1958.

FIGURE 27–5 Shielding Recommendations for Ionizating Radiation Protection

dangers because of the possibility of mishandling or of accidents that could cause release of the material. Persons present might then inhale or absorb the radioactive material through the skin, especially where there has been a cut or other point of entry.

BENEFICIAL USES OF IONIZING RADIATION

X-rays are produced intentionally for a number of purposes:

1. Medical diagnoses by which fractured bones can be detected and treated and the presence of foreign bodies can be determined; where passages in the body are constricted or blocked; and where other internal conditions are suspected.
2. Treatment of cancers which can be reached by a source of ionizing radiation.
3. Examination of welds, other fastenings, and internal structures for the existence of cracks, voids, or contaminants; determining conditions in the interior of sealed containers and the presence of foreign materials.
4. Examinations of packages and baggage for illegal articles.

The greatest projected industrial use of radioactive material has probably been for generation of heat in nuclear power plants. Enough fissionable material is brought together to constitute a critical mass in a nuclear chain reaction. The heat generated must be controlled, and for this, rods of neutron-absorbing material such as graphite or boron are used. When the rods are inserted completely, fission ceases, and the reactor shuts down. (Because of residual heat in the system, the effect is not immediate.)

At one time, it was believed that through atomic fission the cost of electricity would be so low, power could almost be given away free. Such circumstances never came to pass, mostly because of high construction costs, poor and unsafe engineering design, and the need to rectify both.

Beta particles are sometimes used to neutralize the static electricity generated during the manufacture of paper or cloth.

FEARS OF NUCLEAR RADIATION

Injuries by radioactive materials even in minute amounts were found soon after their discovery. Atomic scientists suffered from radiation effects, such as cancer, as radioactive elements were isolated and separated. Radiologists who used X-rays for beneficial purposes suffered in a similar way. Years ago, many serious cases of injury occurred during painting of dials with radioactive material to make them self-luminous. To properly shape the end of the brush, the worker would put the tip in her mouth and wet it after each application. The accumulated material caused horrible effects to the mouth, jaws, intestinal tract, and bones.

The public had accepted the great number of accidents and fatalities of railroads because of the benefits derived. They might have accepted the high construction costs of nuclear power plants. However, they were never willing to accept the risks of a nuclear accident. Apprehensions regarding use of any nuclear material grew tremendously after some of the effects of the bombings of Hiroshima and Nagasaki were seen. Since then, many persons have become fearful of the injuries and deaths that could take place. There were claims after tests that radioactive fallouts had caused cancers and deaths. Some effects would last for many years.

Nuclear Plant Accidents

Figure 27-6 lists a few of the accidents that have taken place in nuclear power plants. Although most involved minor problems such as piping leaks, there has been notable damage as the list points out. At Three Mile Island and Chornobyl, damage was so severe that there was melting of the metal-containing rods of the fissionable fuel. At Chornobyl, the great amount of heat created destroyed the reactor and the plant, but there was no "meltdown," considered the greatest possible accident. (A meltdown includes loss of control of a reactor, with such great overheating that the hot nuclear material burns through the concrete foundation and substance of the plant. Contact with the moisture below would generate such a massive amount of steam in a restricted space that the result would be a great explosion.) At Three Mile Island, it was estimated and reported the plant had been within 30 to 60 minutes of meltdown before the accident was controlled.

Because of the growing fears, all contracts for new nuclear power plants and those in the progress of construction (97 in all) in the United States were canceled. It was reported no new plant orders in this country were expected before 1990, but the accident at Chornobyl pushed back even that date. At the change of millennia, the U.S. Department of Energy maintains a no-new-orders policy. The reactor design at Chornobyl lacks a containment structure. U.S. nuclear power plants have a containment structure as a safeguard.

Although no one was physically hurt at Three Mile Island, the psychological trauma was great. Fundamental design and procedural changes had to be made at all nuclear power plants to lessen the possibilities and probabilities of any more nuclear accidents. Radioactive releases have been feared not only from highly energetic power plants but also from materials in mining and processing.

Uranium Mill Tailings

Mill tailings are the residue from uranium ore mining and processing. Mining may result in the release of slightly radioactive radon gas; milling results in a uranium oxide called "yellowcake." The next processes, conversion, enrichment, and fuel fabrication, result in another compound of uranium oxide which is formed into ceramic pellets, then placed in zirconium-clad fuel rods. Any gases and liquids remaining contain small quantities of uranium and thorium, both of which are radioactive.

Fission of the uranium-235 generates heat, but fragment products accumulate, lessening the sum of the reaction going on. After three to four years, the rods must be removed and replaced.

Nuclear Waste Disposal

High-level waste (HLW) is irradiated reactor fuel, liquid wastes from spent fuel reprocessing, or solids into which such liquid waste have been converted. Even with "no new orders" there will be an accumulation of about 80,000 metric tons of high-level nuclear

The public's apprehension regarding a nuclear disaster is such that any unusual occurrence at a nuclear power plant has become a subject for mention by the news media. A review of the *New York Times* in 1983 showed that the number of news items regarding nuclear accidents during the year far outnumbered mentions of any other type of accident. The problem involves the potentialities not only of massive explosions of nuclear power plants and the widespread distribution of airborne radioactive debris, parts, and equipment, but also of causing cancer in workers, neighbors, and others of the public.

Incidents at nuclear power plants have varied from mishaps involving minor nonnuclear events to some which have resulted in death and destruction, or their narrow avoidance. At Three Mile Island, it was estimated and reported that the plant was within 30 to 60 minutes of meltdown before the accident was controlled. Major reactor accidents have included melting of reactor fuel, release of radioactive materials, and explosions. Lesser results have been shutdowns due to minor malfunctions of operational or safety equipment, or release of small amounts of radioactive material. Common radioactives released have xxxx then xenon-133 (with a half-life of 5.25 days) and iodine-131 (with a half-life of 8.05 days) to strontium-90 and cesium-137 (with half-lives of 28.1 and 30.2 years, respectively.) Less common are carbon-14 and technetium-99 (used for medical purposes) with half-lives of 5,730 years and 210,000 years.

October 1957, Windscale Plant, England: Temperature of reactor was permitted to rise too rapidly so that fuel cartridges were burned releasing a cloud with large amounts of iodine-131. Use of milk produced over a 200 square mile area was banned for 25 days. *Finding:* operator error.

January 1961, AEC test area, Idaho: Three men were on top of a test reactor where they had to raise a control rod manually. It appeared that a rod ejected and pierced one of the men. The exposed radioactive material had badly contaminated all three men, two of whom died immediately, the third two hours later. The reactor was very badly damaged by the heat so that radioactive iodine-131 escaped. *Finding:* human error.

October 1965, Enrico Fermi Plant, Michigan: An emergency was declared when the plant began overheating so that nuclear fuel melted. A piece of zirconium sheet-metal flow liner had torn loose and blocked passage of coolant. *Finding:* poor design, because the liner was unnecessary. Repairs required five years.

September 1970, Hanford Forces, Washington: A clogged coolant strainer caused blockage in flow, and emergency shutdown was attempted. Attempts to drop safety rods containing neutron-absorbers failed until a backup system was used. *Finding:* poor design of the diodes permitting release of safety rods.

March 1975, Brown's Ferry, Alabama: When workers used a candle to check whether draft of air had been stopped, foam and vinyl caught on fire. Extinguishing methods failed and the fire spread so that operators lost control, use of instruments, and use of emergency, normal installed equipment. After ten hours the chance of meltdown was avoided by pumping cooling water into system. *Finding:* poor design and lack of redundancy of crucial systems.

March 1979, Three Mile Island, Pennsylvania: Failure to open water valve on secondary circuit caused nuclear fuel in primary circuit to overheat and melt. Other problems were failure of steam relief valve to open, confusion due to operators lack of knowledge of existing situation, and inadequate instrumentation. In addition to damage to reactors, and losses of radioactive gas and water, the greatest effect was a psychological one upon nearby residents. *Finding:* operator error and poor human factors design.

April 1986, Chornobyl, Russia: An explosion completely demolished one of two reactors. It was announced that 31 persons had died shortly after the accident and 237 had been hospitalized because of radiation sickness. Over a decade later the number of deaths and illnesses related to

FIGURE 27–6 A Few Nuclear Accidents Continued

the disaster is unknown and is estimated by some to be into the tens of thousands. Milk from cows in Russia, Poland, Sweden was not to be used, nor was meat from animals as far away as Lapland. There is still a potential for further serious complications from this disaster. Immediately afterward, a concrete "sarcophagus" was built over the remains of the damaged reactor. Some believe it is in danger of collapse. *Finding:* cause unknown, but reported to be operator error during a "safety" research project.

FIGURE 27–6 Continued

waste in the United States, mostly from spent fuel for nuclear power generation. This volume could be reduced considerably if the reprocessing of spent fuel rods were to be practiced. Reprocessing was rejected as an option by President Jimmy Carter for national security reasons. Plans are for HLW disposal at a site or sites that can host the waste in underground geological isolation. Presently the waste from nuclear power generation is held on-site at the power plants around the nation, in water pools, or in a few instances in casks placed above ground in dry storage. Only one site is presently under study to determine if its geological characteristics are scientifically adequate for a high-level waste repository. That site is Yucca Mountain, about 95 miles north of Las Vegas, Nevada. If the site is used for this purpose, the waste will be placed in volcanic rock, 1,500 feet below the surface and 500 feet above the water table. It should start receiving waste some time after 2010.

Several nations have plans to place their nuclear high-level radioactive waste underground in geologic isolation. Canada, Sweden, and Finland, for example, plan to place their high-level nuclear waste in host rock. Germany plans to place its high-level waste in domed salt mines 3,000 feet beneath the earth's surface. Belgium plans to use underground isolation in clay. Presently international laws forbid dumping nuclear waste in the ocean.

If geological isolation in underground mines is used for HLW as expected, present estimates are that the disposal containment canisters and engineered barriers should not leak radioactive nuclides for 10,000 years. The host rock should continue that isolation to some extent for between 100,000 and 300,000 years to ensure public safety.

Low-level waste includes contaminated filters, liquid filter resins, wiping rags, protective clothing, hand tools, vials, needles, test tubes, animal carcasses, and equipment parts resulting from the use of radioactive materials. These come from academic, government, industry, medical, and utility sources. The nature of this waste does not have the same long-term isolation requirements that high-level waste has. Various states have formed consortiums to provide disposal sites. Some sites have operated for some time, including Barnwell, South Carolina, Richland, Washington, and Beatty, Nevada.

IONIZING/NONIONIZING INTERFACE

There is no sharp juncture between ionizing and nonionizing radiation. The OSHA standards have few requirements regarding nonionizing radiation. Section 1910.133, on eye and face protection, states: "suitable eye protectors shall be provided where machines or operations present hazards of . . . injurious radiation . . ." Section 1910.96, on ionizing radiation, indicates that the word "radiation" does not include radio waves, visible light, or infrared or ultraviolet light. Section 1910.97, on nonionizing radiation, restricts that term to the radio-frequency region, including microwaves.

NONIONIZING RADIATIONS

Normally nonionizing radiations include those electromagnetic regions extending from ultraviolet to radio waves. Four types that can cause injury are discussed here: ultraviolet, light, infrared, and microwave. Persons who must work in the sun generally tan, but long exposures sometimes cause small skin cancers. Usually these can be prevented by shading or skin creams.

Ultraviolet Radiation

Other injuries due to ultraviolet radiations may be either thermal or photochemical. Experiments have shown there is little temperature rise in affected areas. Adverse effects of ultraviolet radiation on the eyes, therefore, appear to be photochemical rather than due to heat.

Much of the ultraviolet radiation the sun emits is filtered out by the atmosphere. However, the remainder is still adequate to cause painful sunburns and, under certain conditions, blindness. Ultraviolet radiation will not penetrate the intact skin, although certain wavelengths will penetrate as far as the dermis (see Fig. 19-2). Ultraviolet radiation on the skin will cause a rapid increase in cells of the lower layer of the epidermis, which tends to thicken. Greater-than-normal numbers die and become part of the corneum, which also thickens, reducing the ultraviolet that will be absorbed. The "suntan" that results is due to migration of melanin pigment toward the outer skin.

An acute exposure to solar ultraviolet radiation, especially at wavelengths of approximately 3×10^{-7}meters, will produce first- and second-degree burns. Reddening of skin which has not been tanned previously will occur when an acute dosage of 8 milliwatt-seconds per square centimeter is received. After the reddening of the initial exposure has disappeared, a higher dosage will be required to produce reddening again, since the skin takes defensive measures. Repeated exposures, controlled to prevent second-degree burns, will cause the skin to darken and become dry and wrinkled. Many persons consider this a healthy, outdoor look. However, certain fair-skinned persons who tan poorly may develop skin problems, including skin cancers.

No reports of cancer or excessive hardening and scaling of the skin from industrial operations have been noted. This is probably because workers find that overexposures due to artificial sources generally result in painful effects they prefer to avoid.

The principal industrial source of ultraviolet radiation is electric arc welding. Others include ultraviolet lamps, plasma torches, and, increasingly, lasers (Fig. 27-7).

The cornea and conjunctiva of the eye (Fig. 27-8) are the principal absorbers of ultraviolet energy, causing them to become bloodshot, irritated, and to tear (lacrimate) through photochemical, rather than thermal, injury. These effects, especially with mild exposures, may not be immediate, but may appear anywhere from 6 to 12 hours later.

(As a young plant engineer, I had to monitor construction and installation work for which I had prepared designs. I spent one day in an area where the cast-iron form of a machine was being cut by a welder using a carbon arc. All the millwrights, machinists, other workers, and I knew enough not to look directly at the arc, and an enclosure around the welders had been provided. However, the light was reflected off the white-painted walls. After I went to bed, and awoke about 4:00 the following morning, I felt

PRECAUTIONARY MEASURES FOR LASER OPERATIONS

1. Personnel should know whether the laser to be used is capable of injuring the eyes or the skin. The following measures should be used with any laser capable of causing eye injury.
2. Personnel supervising or conducting laser operations should be aware of the hazards involved with the equipment being used.
3. Only authorized and trained personnel should be permitted to operate lasers. *(Note:* Some states have considered and may even require that laser operators be licensed.)
4. In no case should a person look into a beam directly or when it is reflected from a shiny (specular) surface.
5. A laser beam should never be directed another personnel.
6. Whenever possible, laser operations should be conducted in brightly lighted areas to reduce pupil size.
7. Wherever possible, laser beams should be totally enclosed in a firing tube. Otherwise, laser operators and personnel in the vicinity of laser operations should wear eye protection chosen specifically for the wavelength and intensity of the laser being operated.
8. Lasers, laser work areas, and accesses to such areas should be suitably marked and entry controlled.
9. Enclosed areas in which lasers are used which are not totally enclosed should have walls, ceilings, and equipment with dull or matte finishes so that reflections are minimized. Eliminate or reduce as much as possible all objects with shiny (specular) surfaces. Doors should have suitable warnings, and a warning flashing red light should be provided. The light should be activated any time the laser is being operated. A sign should warn against opening the door when the light is flashing.
10. Targets used with lasers should have dull, nonreflective surfaces, must be noncombustible at the beam intensity of the laser, and must not emit any toxic gases, or must be ventilated to prevent emission of toxic gases into the workroom air.
11. Personnel working with lasers should have eye examinations periodically and wherever they may have been exposed to laser radiation. Any accidental eye exposure to a laser beam or reflected light should be reported immediately to the medical office.
12. Lasers or laser heads should be secured in position so that they may not be inadvertently misdirected, swung, or tilted during operation.
13. Laser operations should not be conducted in the vicinity of flammable liquids or in the presence of combustible vapors.
14. Operating lasers should not be left unattended.

FIGURE 27–7 Precautionary Measures for Laser Operations

THE EYE*

The eye is the part of the human body most susceptible to radiation and most easily damaged. The eyeball is surrounded by a tough protective membrane, the sclera, whose primary function is support and protection of the eye. The sclera is unpigmented and appears white. Inside the sclera and covering about two-thirds of the inside of the eyeball is a highly pigmented membrane called the *choroid.* This membrane supports the retina and provides for the needs of the retinal cells. The retina also covers about two-thirds of the rear portion of the inner eye and is also highly colored with melanin, the common dark pigment of the body in general. It is composed of light receptor calls. The optic nerve enters at the rear of the eye and branches out into the retina. The depression shown on the diagram in the center of the optic nerve entrance is a blind spot; there is no perception at this point in the visual field.

Filling the posterior cavity of the eye is the fluid called the *vitreous humor,* which provides support for the eye and maintains a path for nourishment and other processes necessary to the retinal cells. The posterior cavity is bounded in front by the lens and its supporting muscles. The lens focuses the incoming light on the retina by tensioning the muscles attached to it until the appropriate lens shape is reached. In front of the lens is the iris, generally highly pigmented but with a color much different from that of the choroid or the retina. Its functions are both to regulate the amount of light entering the eye to help reduce the spherical aberration of the eye by limiting the aperture of the lens as necessary.

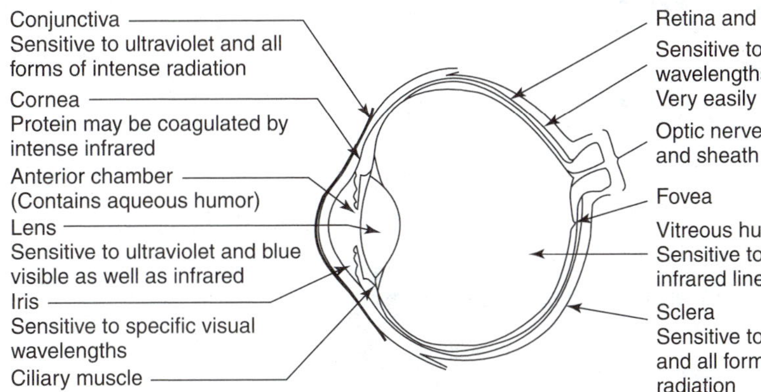

Construction and Radiation Sensitivity of the Human Eye

The anterior cavity of the eye is also filled with a liquid, known as the *aqueous humor* and much less viscous than the vitreous humor. Its purpose is also to maintain the shape of the eye and to bathe the cells of the iris and lens. It is believed that an exchange of fluid occurs between the anterior and posterior cavities, which renews the vitreous fluid and passes certain materials from one area to another.

Light enters the eye through the cornea, a clear tissue that is the front boundary to the aqueous cavity. The cornea has both optical and protective functions. It has a very high transmission in the visible range but is fairly thick and tough.

The special transmission of the eye is nearly unity for wavelengths from 0.3 to 0.9 micrometer and is very high at a number of infrared wavelengths. Therefore, most of the visible light incident on the eye is transmitted to the cornea. The pigmentation of the various parts of the eye has a great deal to do with their susceptibility to light-induced damage. Areas of the eye may be distinguished by their sensitivity to particular wavelengths of light. Thus the infrared wavelengths are primarily absorbed by the aqueous and vitreous humors because of the large number of infrared absorption bands in the absorption spectrum of water. Here the primary interaction of the eye with radiation is selective absorption of energy, with a resulting thermal effect. Wavelengths in the far infrared are also absorbed in those areas of the eye, but the absorption is much less selective. In this region, however, the energy of the light quanta is low, and at usual intensities these wavelengths do not interact significantly with biological materials (that is, they cause no major biological damage).

*T.0. Huston, *Human Biological Interactions with Laser Light,* AD-660-361, Naval Electronics Laboratory Center, San Diego, Aug. 2, 1967.

FIGURE 27–8 The Eye Continued

The visible wavelengths are transmitted efficiently through the eye and impinge upon the highly pigmented retina where they are strongly absorbed. Most of the light that penetrates the retina is absorbed by the highly pigmented choroid. Intense light anywhere in the visible range has significant biological interaction with the retina and choroid. These areas also strongly absorb such infrared energy as is transmitted through the "windows" in the absorption spectra of the aqueous and vitreous humors. The iris also interacts strongly with visible wavelengths. The ultraviolet wavelengths interact strongly with the proteins and nucleic acids of the cornea and sclera. As the eye becomes aged, the tens becomes increasingly yellowed and also begins to absorb the blue-violet portions of the spectrum.

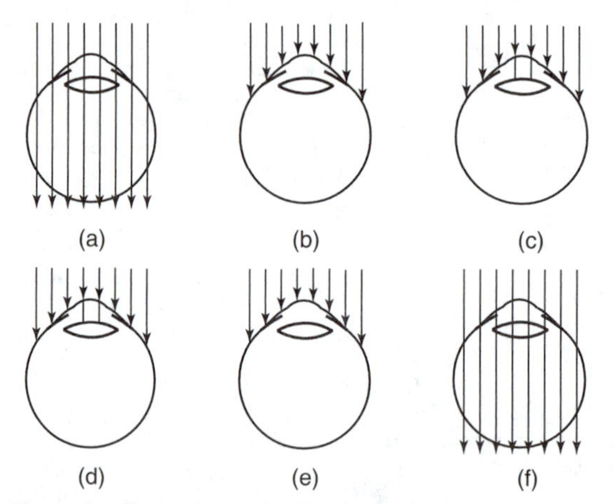

Absorption Properties of the Eye for Electromagnetic Radiation*

(a) Most higher-energy X-rays and gamma rays pass completely through the eye.
(b) For short ultravioiet, absorption occurs principally at the cornea.
(c) Long ultraviolet and visible light are refracted at the cornea and lens and absorbed at the retina.
(d) Near-infrared energy is absorbed in the ocular media and at the retina; near-infrared rays are refracted.
(e) Far-infrared absorption is localized at the cornea
(f) Microwave radiation is transmitted through the eye, although a large percentage may be absorbed.

*Department of the Army Technical Bulletin MED 279, *Control of Hazards to Health from Laser Radiation, Washington, D.C.* Feb. 24, 1969.

FIGURE 27–8 Continued

as though someone had thrown a handful of sand in my eyes. After reaching the bathroom I managed to wash out my eyes with cold water until I could open them. My eyeballs were bloodshot almost entirely red. I continued the eye washing and used cold compresses to provide relief until I got some sleep. When I reached the plant in the morning I found that half the force that had been working in the area on the previous day had had similar experiences. The only persons who had not suffered were the welders. They had had eye protection. —Willie Hammer)

Protection against injuries caused by ultraviolet radiation is easy to provide. As any sunbather knows, any opaque cover, no matter how thin, will block ultraviolet light. (The light may heat the shield and increase its temperature.) Certain glasses and plastics will also be opaque to ultraviolet radiation and may be used where transparency is required.

The principal emitter of ultraviolet radiation is, of course, the sun. Solar energy is used to heat water for cooking and, when properly focused, to ignite flammable materials. More dangerous when used without adequate care are lasers. [A laser (Light Amplification by Stimulated Emission of Radiation) is a device that generates and emits coherent light. Coherent light is light transmitted at one frequency, in a single direction, and in a single, regular pattern in which interference does not occur. The intensity of the energy in a laser beam can be far greater than that received from the incoherent light of the sun or a welding arc and can be extremely damaging, especially to the eye.] The power and directability of lasers has made them extremely useful for delicate incisions in human organs and for fine cutting of various types of materials. However, they are dangerous when not properly controlled in magnitude and direction.

Avoiding the Effects of Ultraviolet Radiation

Sources of ultraviolet radiation can be shielded so that workers will not be affected by welding or laser operations. Such operations can be conducted in enclosed rooms or partitioned areas, where passers-by or others will not be subjected to injurious radiation. Protective goggles or face masks with eyepieces of glass with high ferric oxide content which will absorb ultraviolet light are effective. Plastic materials differ in their absorptive capabilities; therefore, care must be taken to choose only those suitable for the radiation. Certain additives can be added to plastics to act as absorbers, but here, too, the capabilities of the additives differ. When selecting eyewear, the supplier should be provided information on the type of source and, if possible, the principal wavelength of radiation, so that the supplier may recommend the best suitable protective equipment.

Visible Light

The range of electromagnetic radiation a human can see is a narrow one, with wavelengths of 3.8 to 7.5×10^{-7} meters. Injury that can result will be only to the eyes, the organs by which radiation can be detected, and will result from "overloads."

The eye has two protective mechanisms, the iris and the eyelids. However, overloading may occur before either of these mechanisms can function. A bright source will cause restriction of the iris until the minimal-size pupil (3 mm) is reached. The eyelid requires approximately 150 milliseconds to react. These combined actions under normal conditions are generally adequate to prevent injury to the retina. They may be inadequate where the intensity of the energy hitting them is extremely high. The intensity of light radiation which may cause eye injury could be any source that permits the eye to absorb more than 50 calories per square centimeter per minute for any sustained period of time.

In spite of the natural protective means of the eye, additional safeguards are required for operations with high intensity light, of which there are many that can cause damage. Some such industrial sources include welding and industrial arcs and lasers. Intermittent sources, such as an area where a welder strikes an arc, welds a bead, and so on, are often more damaging than steady sources. The iris of an observer's eye will open in the lesser light before the arc is struck. Much light is absorbed before the iris closes after the arc is struck. Gupta and Singh[111] found that out of 520 welders, 398 (76.5 percent) suffered color-perception deficiencies. Almost 20 percent were completely unable to identify color on test slides. Safeguards for lasers (Fig. 27-7) are applicable to other light sources, even though their intensities and the protection required may not be as great.

Less injurious are spotlights, floodlights, high-intensity lamps, and so on. However, persons rarely stare at these sources for the length of time required to produce damage. More injurious are the sources that generate light at intensities far above the minimum damage level. The sun is the commonest of these. Injuries have resulted from persons looking into the sun through optical magnifying instruments, especially when they see it unexpectedly as an eclipse passes. The resultant burn injuries to the eye can produce temporary or permanent blindness.

Overexposures because of reflections by snow, sand, or large bodies of water are more gradual ways in which the eyes can be subjected to injurious levels of illumination over long periods of time. In industrial plants intense light sources of long duration are unusual. However, sustained glare under which some workers may have to work may cause eye fatigue, discomfort, or headaches. This has become an increasing problem for office workers who must continually stare at computer monitor screens.

Infrared Radiation

All bodies with temperatures above absolute zero radiate in the infrared region, and such radiation may be received by other bodies which are at lower temperatures. Because infrared radiation (IR) is electromagnetic in nature, it can be used for detection purposes. Radiant energy is easily converted to heat, especially when hitting another body. The problems that can be generated by IR are due to thermal effects. Infrared radiation will be produced in industrial plants from any high-temperature process, especially those involving molten metals or glass, chemical reactions, and paint or enamel drying.

The injurious effects that can be generated include all those possible from high temperatures or heat. Skin burns (see Chapter 19) are the principal effects. In addition, there will be dilation of the capillaries in the affected skin, so that it reddens and becomes hot. Long and continued exposure to high-intensity IR can result in permanent capillary dilation, so that the skin always appears red.

Continued exposures to intense infrared radiation and heat produced can cause excessive perspiration and loss of body salts. This, in turn, could result in heat exhaustion, heat stroke, or heat cramps (see Chapter 19).

The eyes, being sensitive, can be damaged by IR. Recent research indicates it may be a cause of cataracts, in that the lens of the eye gradually becomes opaque, so that vision is lost. This may be because the eye is remote from blood vessels that could dissipate heat received through IR. Surface irregularities in the cornea can also occur.

The retina can be injured because of its sensitivity to all forms of radiation. The lens and cornea focus the radiation on a small area, so its intensity is increased. If the rate at which radiation is received is great enough, a retinal burn could result. Most industrial processes do not produce IR with intensities adequate to cause retinal damage. However, the invention, development, and use of lasers has created a source of IR with damaging intensities.

Clothing, gloves, and face masks will protect the skin against infrared radiation. The heating of the skin as a person approaches a source of IR generally provides adequate warning. Glasses are available for protection of the eyes, even for the high intensities of lasers. These glasses differ from those used for ultraviolet radiation.

Microwave Radiation

Microwave absorption by the body increases the kinetic energy of the absorbing molecules, manifesting itself as heat, sometimes with an increase in temperature. The chief hazard therefore lies in the incapability of an organ to dissipate the heat of the absorbed microwave energy. Where this heat cannot be dissipated adequately, temperatures will rise, possibly with burn injury.

Studies have shown that power densities greater than 100 milliwatts per square centimeter (100 mW/sq. cm) for 1 hour or more at frequencies from 1.2 to 24.5×10^9 Hz can affect the body thermally. Effects at less than 100 mW/sq. cm appear to be nonexistent; however, opinions differ regarding this. With a safety factor of 10, the permissible level of exposure was set at 10 mW/sq. cm.

[Soviet investigators indicate the safe level is one-thousandth that used by U.S. personnel (1 micro W/sq. cm). The contention is that at higher levels, effects may be headaches, cataracts, falling hair, sleeplessness, irritability, loss of appetite, loss of sexual potency, and memory difficulties. American researchers consider that these claims are unjustified, that the Soviet standard is based on the no-effect level, while the U.S. limit indicates the no-damage threshold.]

As with infrared radiation, a body tends to cool any exposed part by blood circulation. The blood flow increases in the affected parts, carrying heat to cooler parts of the body. In addition, the body begins to sweat so that the heated skin is cooled by evaporation, convection, and radiation. Here again, organ susceptibility to damage is most where blood circulation is ineffective, such as in the lens of the eye. Exposures greater than 10 mW/sq. cm for one hour or longer with microwaves at frequencies of approximately 2.5×10^9 Hz can produce cataracts.

Since the effects of microwave radiation are thermal, it has been found that sensitive areas of the skin will feel pain even at fairly low intensities (20 to 60 mW/sq. cm) for 1 second even at certain frequencies. A fairly moderate acute dosage (energy dosage) could therefore cause pain. OSHA has therefore stipulated that for radiation of frequencies from 10^7 to 10^{11} hertz, the following maxima will apply:

- *Power density:* 10 mW/sq. cm for periods of 0.1 hour or more.
- *Energy density:* 1 mW • hr/sq. cm during any 0.1-hour period.

These values will apply whether the radiation is continuous or intermittent, will be determined by the average over any 0.1-hour period, and will apply to either whole- or partial-body radiation.

All microwave ovens sold in the United States must leak no more than 1 mW/sq. cm at the time of sale, and no more than 5 mW/sq. cm during the lifetime of the equipment. All measurements are to be made 2 inches from the oven door.

Sources of Microwave Radiation

Microwave ovens, dryers, and heaters are some of the commonest uses and sources of this type of radiation. High-powered radar systems used in military detection and warning operations, aircraft and aircraft control radars, and ship and boat radars are others. Microwave communication systems, alarm systems, diathermy equipment, and signal generators also produce radiation in this range. This type of heating is used in plywood plants to hasten glue drying and bonding.

Other Hazards

In high-intensity fields, microwave radiation can cause inductive heating of metals and induced currents that can produce sparks. Unintended heating of metals in fields of strong radars or high-powered communications transmitters has resulted in temperatures that caused burns when the metal was touched by personnel. Rings, watches, metal bands, keys, and similar objects worn or carried by persons in such fields can be heated until they burn the bearers. Metal containers can also become so hot in such fields they can ignite flammable or explosive materials.

The current that can be induced in a microwave field can also be high, the magnitudes depending on the intensity of the field and the characteristics and size of the metal structure in which the currents are generated. There are cases on aircraft carriers in which powerful sparks, generated by radars, were observed when a part of a body or a tool was brought near a charged aircraft. Sparking of even lesser magnitude would be adequate to ignite a flammable gas-air mixture.

Protection Against Microwave Radiation

Equipment that produces microwave radiation can often be provided with shielding to protect users. In other instances, such as with radars and microwave communications equipment, it may be impossible to provide such shielding. However, accessible areas that have a power density of 10 mW/sq. cm or more should be posted with warning signs. The microwave radiating device should be similarly marked.

- No microwave antenna or other emitter should be inspected, nor should work be done on one by a person in the radiation path, during periods it is energized.
- Dummy loads or water loads should be used when possible to absorb the energy output of transmitting microwave equipment while it is being operated or tested. If this cannot be done, absorbent screening should be used to isolate the radiation.
- Operating microwave equipment should not be pointed toward inhabited areas.
- Rings, watches, keys, or other metallic objects should not be worn or carried by persons working in areas where there might be microwave radiation, even of low intensity.

- Tools or other metallic objects that were in a microwave field should be grasped carefully because they may be very hot.
- Flammable or explosive materials in, or in contact with, metallic containers, should not be left in a metallic field generated by microwave equipment.
- A warning device should be provided on each piece of microwave equipment to indicate when it is radiating.

Potential radiation problem causes and effects are listed in Fig. 27-9. Figure 27-10 is a checklist with which safety engineers can review their plants and operations to determine whether or not some of these problems exist.

RADIO FREQUENCY RADIATION OF WIRELESS COMMUNICATION DEVICES

Radio-frequency (RF) and microwave (MW) electromagnetic radiation are in the frequency range of 3 KHz and 300 gigahertz (GHz). Microwave radiation, discussed above, is usually considered to be a subset of RF, but some treat them as two spectral regions. (MW frequencies are between 300 GHz and 300 megahertz (MHz). RF frequencies are in the range between 300 MHz to 3KHz.) Some define the RF spectrum as extending from 0 to 3000 GHz. For purposes of human RF exposure guidelines, the Federal Communications Commission's (FCC's) range of interest is from 300 KHz to 100 GHz.

At the start of the new millenium there were over 80 million Americans using wireless communications devices, with about 25 thousand new users daily. This surge comes from the popular use of mobile phones. However, exposure to this radiation energy is increasing at an increasing rate. Wireless devices are used for a wide variety of purposes such as the monitoring and control of irrigation systems, vending machines, security systems, gas detection systems, personal communication, and access to the Internet. Wireless communication is flourishing in this digital and broadband technological era. It allows rapid transmission of voice, video, or data. Wireless is involved in voice, video, and data convergence. Along with its success have come concerns about safety.

The kind of hand-held mobile phone that has a built-in antenna that is position close to the user's head during a normal telephone conversation has been questioned. Some users of mobile phones have been diagnosed with brain cancer. Brain cancer occurs in the U. S. Population at an annual rate of 6 new cases per 100,000 people. That means that out of 80 million mobile phone users, 4800 would get brain cancer whether or not they used their phones. A few animal studies suggest that cancer might be related to low levels of RF exposure. In one study, genetically altered mice were exposed to a digital phone signal. The exposed mice developed twice as many non-lymphoblastic lymphomas as the unexposed control group. Other studies found no evidence of cancer. At this time, long-term animal studies, with multi-dose exposure treatment paradigms, do not exist. There presently are not sufficient data to conclude that wireless communication technologies are safe or are not safe.

In 1997, the Federal Communications Commission established guidelines for Maximum Permissible Exposure (MPE) limits to RF. In general, applications to the

Hazards Checklist — Radiation

Possible Effects	*Possible Cause*
Ionizing: Tissue damage Degradation of electronic components and changes in their characteristics Degradation of material strength	Ionizing: Inadequate containment of radioactive materials Accidental exposure to ionizing source Inadvertent production of rays by radar, communications, or TV components operating at potentials over 15,000 V Use of X-ray equipment Nuclear reaction
Microwave: Heating of metals and tissue by induction Cataracts or other eye injury Interference with operation of other electronic equipment Activation of sensitive electroexplosives	Microwave: Radar equipment operation High-power and microwave equipment operation Other microwave generator operation (ovens)
Infrared radiation: Undesirable heat gain or temperature rise Increased temperature in enclosed space Overheating Skin burns Charring of organic materials Initiation of combustion of flammables	Infrared radiation: Flames Solar radiation Infrared heaters Highly heated surface Lasers
Visible light: Temporary blindness	Visible light: Strong sunlight High-intensity lights and flashlamps Electric arcs
Ultraviolet light: Vision damage and other eye injuries Deterioration of rubber, plastics, and other materials Ozone or nitrogen oxide generation Decomposition of chlorinated hydrocarbons Color fading of fabrics	Ultraviolet light: Sunshine Electric welding arcs Germicidal lamps Lasers Photocopying Machines

FIGURE 27–9 Hazards Checklist—Radiation

CHECKLIST – RADIATION

1. Does the plant or the product being manufactured use any radioactive isotope for any reason?
2. If so, what type of radioactive emissions does each generate?
3. Is the quantity available at any one time adequate to cause injury to a person close to it?
4. Are the containers of such materials suitably marked?
5. Is there a strict accounting of all radioactive materials so that none is lost or misappropriated?
6. Is there a procedure in effect to clean up any spill or loss which might accidentally occur?
7. Is there any machine in use which produces X-rays by design or could do so unintentionally? Is there any electrical device using a potential of 15,000 volts or more which could generate X-rays? Has such equipment been tested to determine whether it does generate X-rays?
8. Are periodic surveys conducted to determine whether there is any presence of ionizing radiation where such emissions might be present?
9. Are the facilities, equipment, and warnings where ionizing radiation is or might be present in accordance with the standards set by OSHA and by the Bureau of Radiological Health?
10. Does any operation, such as welding, generate high-intensity ultraviolet radiation against which protection must be provided? Is equipment which will produce high-intensity radiation protected to contain the radiation or to limit access to areas where such radiation will exist at a harmful level?
11. Is there a warning against wearing rings, watches, or other metallic jewelry in areas where microwave radiation might exist which will cause heating of the metal?
12. If a laser is being used for any purpose, has it been categorized in accordance with the requirements of the Bureau of Radiological Health?
13. Is the warning label on the laser in agreement with the category of the laser?
14. Does the laser have the special safeguards, such as a key switch, it must have because of the category in which it has been classified?
15. Have laser operators been qualified to operate the laser safely? Are they aware of the precautions necessary to avoid injury to others when operating it?

FIGURE 27–10 Checklist—Radiation

Commission for construction permits, licenses to transmit or renewals thereof, equipment authorizations or modifications in existing facilities must contain a statement or certification confirming compliance with the limits. These limits are shown in Figure 27-11.

In the figure, Occupational/Controlled limits apply where persons are exposed as a consequence of their employment and who are fully aware of and can exercise control over their exposure. General population/uncontrolled exposures apply where the general public may be exposed or where employed persons may not be fully aware of the potential for exposure or cannot exercise control over their exposure. MPE limits are defined in terms of "plane wave equivalent power density". In *far-field* plane wave conditions, power density (milliwatts per centimeter squared), electric field vector (volts per meter), and magnetic field vector (amperes per meter) and the direction of propagation can be considered to be mutually orthogonal. They are related by the following equation.

$$S = E^2/3770 = 37.7\ H^2$$

Where: S = power density (mW/cm^2)
E = electric field strength (V/m)
H = magnetic field strength (A/m)

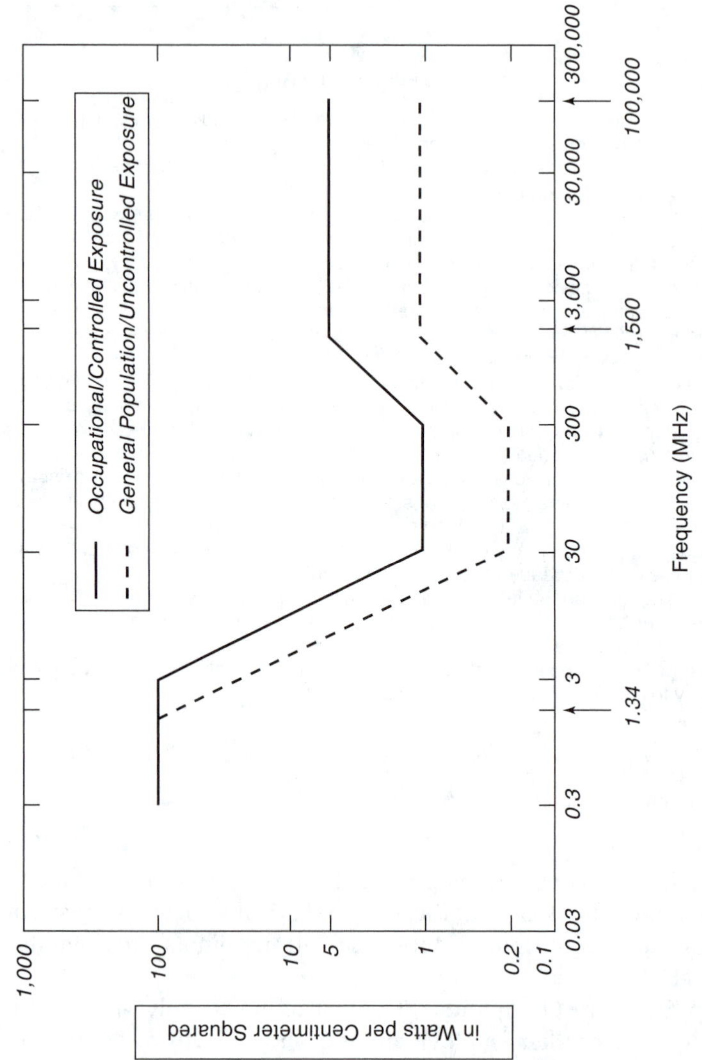

FIGURE 27–11 FCC Limits for Maximum Permissible Exposure (MPE) *Source:* Evaluating Compliance with Federal Communications Commission Guidelines for Human Exposure to Radiofrequency Electromagnetic Fields, Appendix A, OET Bulletin 65, Edition 97-01, August 1997, FCC, Washington, D.C.

"Plane wave equivalent power density" is a quantity calculated by using *near-field* values of electric field and magnetic field strengths in this equation. For plane waves, power density (S), electric field strength (E) and magnetic field strength (H) are related by the impedance of free space, i.e., 377 ohms. The most restrictive limits occur between 30 and 300 MHz where whole-body absorption of RF energy is most efficient.

BIBLIOGRAPHY

[110] Mr. Polly Story, President, the American Society of Radiologic Technologists, to the United States Senate Committee on Commerce, March 8, 1973.

[111] M. N. Gupta and H. Singh, "Ocular Effects and Visual Performance in Welders," Directorate General Factory Advice Service and Labour Institutes, General Labour Institute, Sion-Bombay, India, 1968.

EXERCISES

27.1. Of what type of radiation is the public most apprehensive? Explain why it is more so than for other types of hazards.

27.2. Describe some accidents that have occurred in nuclear power plants.

27.3. Plant workers might be asked to do the work of removing and disposing of radioactive materials, parts, equipment, or wastes from a nuclear power plant. What might be some of the effects on workers? What could be some of the symptoms of an affected person?

27.4. How does ionizing radiation generally injure the body? Describe the effects on skin, eyes, blood, hair, and bone marrow.

27.5. Describe the protective measures that can be used against ionizing radiation.

27.6. Describe the types of nonionizing radiation. How can each cause injury?

27.7. What part of the body can be most easily hurt by nonionizing radiation? Why? How can each injure the skin? How can injury be prevented?

27.8. Even though welders use protective lenses, how can their eyesight be damaged? What type of damage can occur? How can persons other than welders have their eyes injured when in the vicinity of welding operations?

27.9. List some types of hazards other than eye injury that can be caused by microwave radiation. Why are some persons advised not to wear rings?

27.10. Why are lasers hazardous? List some precautions to be observed during their use.

27.11. Why is Yucca Mountain being studied?

CHAPTER 28

Vibration and Noise

Every state makes loss of hearing compensable if the loss occurred because of a traumatic (sudden, violent) occurrence, such as a nearby explosion; and most make it compensable if the hearing loss was caused by long-term exposure to noise. It is expensive to society. In 1995, the U.S. Veterans Administration alone paid $252 million to military personnel for hearing-loss compensation. In 1990, an estimated 28 million people in the United States suffered from hearing loss, and about 10 million of these had an impairment at least partially due to exposure to loud sounds. Occupational exposure to loud sounds is the most common cause of what is often called today "noise-induced hearing loss" (NIHL). NIHL is preventable, but our increasingly noisy environment puts more and more people at risk.

In 1936 the federal government, as part of the Walsh-Healy Public Contracts Act, stipulated requirements for safeguarding against hazardous work conditions for employees of companies that received federal work contracts. The provisions of the Walsh-Healy Regulations were updated in 1960; however, a fundamental problem remained: an employer could be considered in violation only if the government could prove the employer maintained an unsafe, unhealthful, or unwholesome environment for employees working on the federal contract. Not only was this difficult to do, but the Walsh-Healy Regulations were limited to employers that had federal contracts.

Under the OSHAct, which included the provisions regarding noise of the Walsh-Healy Public Contracts Act, compliance with OSHA standards is mandatory. An employer who does not control noise so as to minimize fatigue and reduce the probability of accidents can be charged with a violation of the OSHAct.

Both employers and employees are therefore obligated to observe existing noise standards. An employee who does not comply with previously described procedures for his or her welfare and suffers a loss of hearing can be charged with "misconduct." In 1973, a North Carolina Appeals Court upheld a lower-level decision denying benefits, under the state's workers' compensation law, to a worker who was discharged for refusing to wear protective hearing devices in an area where the employer was responsible for seeing the employee did so. The court indicated that his discharge was for "misconduct" and was proper.

In 1948 the New York State Court of Appeals[112] ruled that a worker who had suffered a hearing loss through exposure to noise from a drop forge which he helped operate was entitled to damages. He was awarded $1,661, even though he could continue to work at his job and receive his usual wages. Hundreds of similar claims by other persons, all over the country, were filed against employers. Later rulings indicated that since much of the loss is temporary, the permanent loss cannot be determined until the affected employee has been away from injurious noise levels for at least six months. Workers rarely have the opportunity or desire to leave their jobs for this length of time to determine whether their hearing loss is permanent or temporary.

Although liabilities under workers' compensation laws are still of concern, violation of legal requirements is a more immediate problem. In addition, it will be pointed out that when noise is involved, even meeting existing legal requirements may permit conditions to exist in which injuries to personnel can occur. Also, excessive noise may exceed Environmental Protection Agency (EPA), state, and municipal legal levels.

EFFECTS OF VIBRATION, SOUND, AND NOISE

The commonest injury due to vibration is sound-induced hearing loss. In industrial activities the sounds which cause these losses constitute noise—unwanted sound. Figure 28-1 indicates some typical noise intensities. (In other activities, desired sound, such as amplified rock music, can have adverse effects similar to those of loud noise.) The adverse effects include (1) loss of hearing sensitivity, (2) immediate physical damage (ruptured eardrums), (3) interference (masking), (4) annoyance, (5) distraction, and (6) contribution to other disorders. Each will be discussed at length.

Hearing Loss

Hearing loss is an impairment that interferes with the receipt of sound and with the understanding of speech in sentence form (as opposed to use of test words). The most important frequencies for speech understanding are those between 200 and 5,000 hertz. It is generally losses in this frequency range which are compensable under workers' compensation acts.

A young person with normal hearing can detect sounds with a frequency range that extends from 20 to 20,000 hertz. Less than normal ability to hear speech indicates there has been degradation. Degradation of hearing can also result from aging, long-term exposure to sounds of even moderately high levels, or a sudden, very high intensity noise. (The effects of illness on hearing will not be considered here.)

Tests have shown that hearing loss occurs as persons age. Figure 28-2 presents hearing loss data at different frequencies for long exposure times. These losses have been plotted for various frequencies against ages of men and women. As these curves show, hearing losses are greater for the higher than for lower frequencies.

Research has indicated that much of this degradation with age may be due to continuous exposure to environmental noise of modern society rather than to simple aging (Fig. 28-2). Hearing losses occur even at noise levels lower than those permitted by OSHA standards (Fig. 28-3). It can therefore be assumed that even where a company complies with the standards, personnel will suffer hearing loss and be eligible for

VIBRATION, SOUND, AND NOISE*

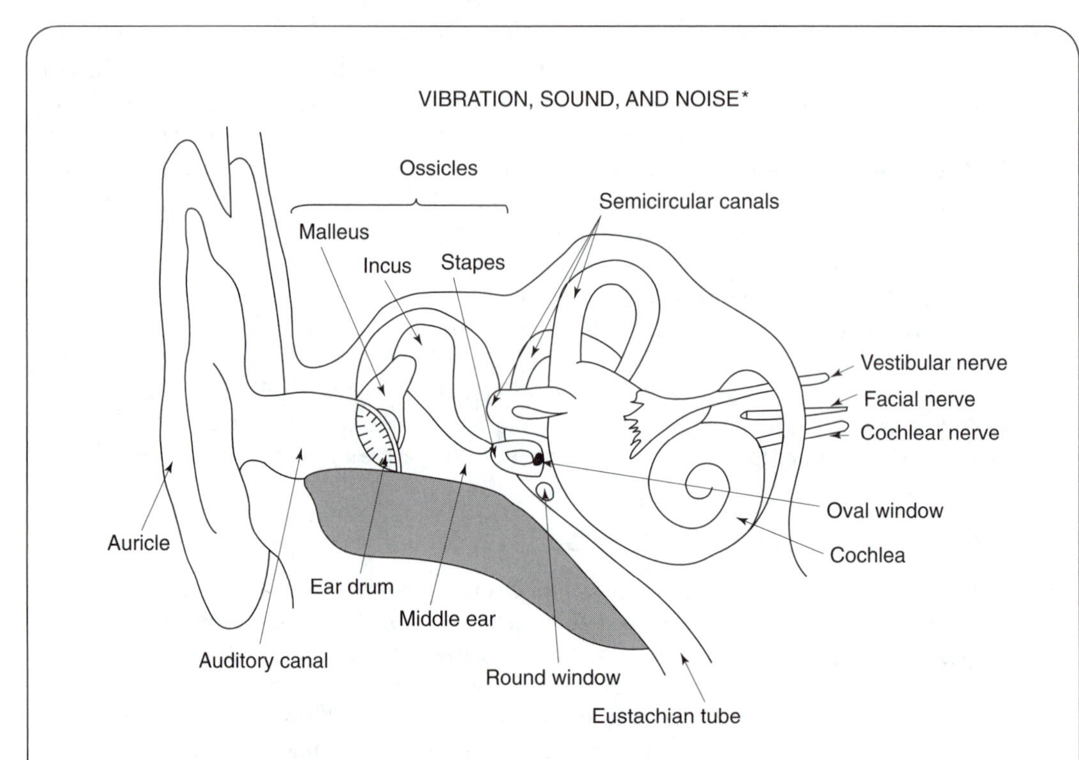

Cross Section of the Ear

Sound waves are collected by the auricle and pass along the auditory canal (about 2.5 centimeters) to the eardrum (*tympanic membrane*). Changes in sound pressure cause movements in the eardrum proportional to the sound's intensity. The eardrum is connected to a bone called the *malleus* (hammer), the first of the three bones which constitute the *ossicles*. The second bone of the ossicles is the *incus* (anvil) and the third is the *stapes* (stirrup). Sound is transmitted across the middle ear through the ossicles to the *oval window* of the *cochlea*, which forms part of the inner ear. The cochlea is a spirally wound tube, resembling a snail shell, filled with a fluid which moves back and forth with the vibrations of the stapes. The fluid's movement causes stimulation of very small but extremely sensitive hair cells, and the generation of nerve impulses. There are almost 25,000 of these hair cells located in the *organ of Corti*. These impulses are transmitted along the cochlear nerve to the brain. Cells located nearest to the middle ear in the cochlea are stimulated by the highest audible frequency sounds. Cells farthest from the middle ear at the tip of the cochlea are excited by sounds of low frequency.

The middle ear is connected with the pharynx of the mouth by the *Eustachian tube*, which serves as a pressure equalizer for air on both sides of the eardrum. Normally the tube is closed, opening only on changes of pressure or by action of muscles used in yawning or swallowing. The vestibular portion of the inner ear, which includes the three semicircular canals, is concerned with the equilibrium and the sense of motion. The three semicircular canals are located at right angles to each other, thereby permitting balance to be maintained regardless of the head's position in space.

Frequency: number of vibrations or hertz (Hz) (cycles per second) of a sound wave.

Tones: a *pure tone* is a continuous single frequency vibration within the audible range. A *musical chord* is a number of tones in which the frequencies are simply related, including a fundamental tone and its harmonics. A *fundamental tone* is the one which has the lowest frequency of all present. *Harmonics*

FIGURE 28–1 Vibration, Sound, and Noise

are tones which are whole number multiples of the fundamental tone. Differences in quality of a musical tone are due to variations in the distribution of energy among the harmonics of the fundamental frequency. *Speech* is a more complex frequency relationship between tones. *Noise* is discordant and unwanted sound in which the continuous tone characteristic has been lost, or it may be sound energy at other than the fundamental tone or its harmonics.

Octave: band of frequencies between one frequency and a second which is twice as great. Bandwidth is 70.7 percent of the center frequencies, which by international agreement have been established at 31.5, 63, 125, 250, 500, 1,000, 2,000, 4,000, 8,000, and 16,000 Hz.

One third octave: a band of frequencies in which the ratio of the higher end frequency is the 1/3 power of 2 times the lower frequency of the band ($f_H/f_L = 2^{1/3}$).

Narrow band: a band of frequency whose width is less than one third but at least 1 percent of the center frequency.

Resonance: when two objects have the same natural frequency or period of vibration they are said to be in resonance. A system is in resonance when any change in the frequency of forced oscillation causes a decrease in the response of the system.

Interference: two or more continuous sounds produced together may interfere with each other. When the energies reinforce each other, the interference is constructive; when they oppose each other, destructive. If tones of 2,500 and 3,000 Hz are sounded, an additional tone of 500 Hz will also be heard. Reinforcement and opposition of energy may occur to cause regular increases and decreases in the volume of sound. These increases and decreases are called beats. If tones of 600 and 605 Hz are sounded together, five beats per second will be heard.

Intensity, I, of a sound wave: average time rate of flow of energy transported per unit area. It is expressed in watts per square centimeter. Intensity is proportional to the square of the amplitude and the square of the frequency of the wave. An intensity of 10^{-12} watt per square centimeter is an arbitrary reference that was thought to correspond to the faintest sound which can be heard at a frequency of 1,000 Hz by a normal, healthy, young ear. (This is equivalent to a pressure of 0.0002 dyne per square centimeter in air at 20 degrees celsius.) This basic intensity is designated as I_0 and used as the reference point from which the intensity of the sounds may be computed.

Decibel: the relationship between other sound levels and I_0 may be expressed in bels by the logarithm to the base of 10 of the ratio between the two. A bel is large, so the decibel or one tenth of a bel is used:

$$\text{no. decibels} = \log (I/I_0)$$

Sound Intensity Level is 10 times the log (base 10) of the ratio of the measured intensity in watts per square meter to the reference intensity of 10^{-12} watt per square meter. It is sound power flow per unit area. Since the early 1990s, sound measurement instruments have been marketed which measure sound intensity level.[113]

Sound pressure level (SPL) is the ratio between the measured pressure, p, and the reference pressure, p_0, of 0.00002 pascals or 20 micropascals (The reference level of 20 micropascals is normally implied; other pressures must be in similar units.) In practice, a sound level meter is calibrated to read decibels relative to 20 micropascal. (Sound pressure level is a measured value of the compression and rarefaction of air caused by sound vibrations.) Sound pressure level meters have been used for many decades to measure noise levels in industry. Expressed as a relationship between pressures, the decibel rating of any sound would be:

$$\text{no. decibels} = 10 \log (p^2/p_0^2) = 20 \log (p/p_0)$$

FIGURE 28–1 Continued

If a sound exerted an acoustical pressure 10 times greater than p_0, the ratio p/p_0 would be 10/1 the logarithm to base 10 would be 1, and the sound would have a decibel rating of 20 x 1, or 20 decibels.

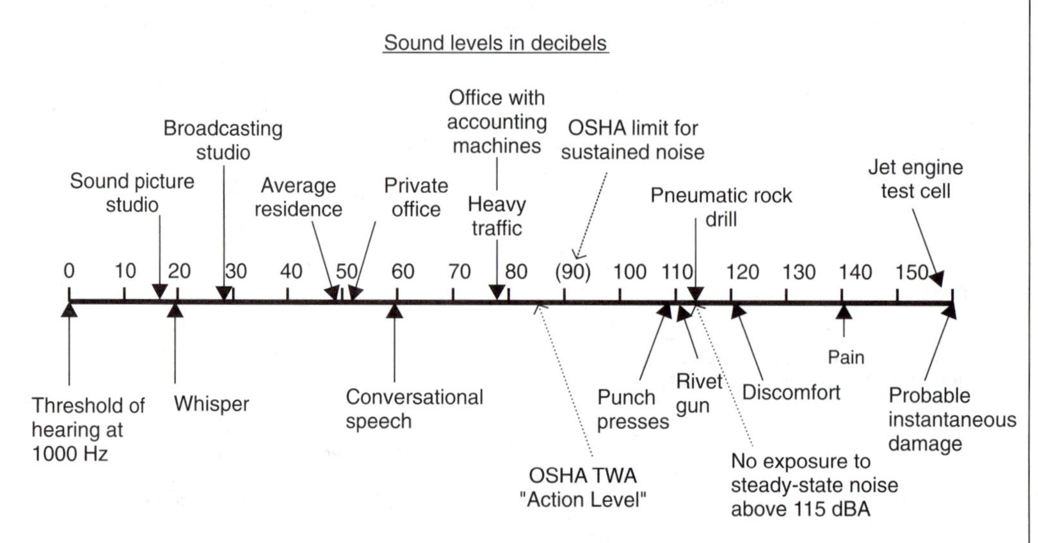

Audibility: the audible range for the average young person with unimpaired hearing is from 20 to 20,000 Hz. Below 20 Hz, sound is subaudible or infrasonic; above 20,000 it is ultrasonic. The most sensitive range is between 500 and 5,000 Hz generally about 4,000 Hz, the approximate resonant frequency of the ear canal. The ratio of the intensity which the ear can tolerate to that which it can detect is 10^{12}. The minimum sound intensity the ear can detect varies widely with the frequency. At 50 Hz, a sound to be perceived must be 10^6 times as intense as one at a frequency of 3,000 Hz. Loudness is a function of intensity and frequency. Loudness level is expressed in phons, which are expressed on a logarithmic scale, or in sones, which are expressed on a linear scale. Loudness level varies with different individuals (and often between the two ears of the same individual).

Phon: A means of evaluating the subjective equality of loudness of sounds at different frequencies. For such comparisons, sound pressure levels, expressed in decibels, are related to equivalent sounds at 1,000 Hz, and given in phons. Phon curves are equal-loudness contours relating frequency and sound pressure levels to various decibel levels of the standard tone of 1,000 Hz. For example, tones of various frequencies and sound pressure levels which fall on the 40-phon contour are subjectively equivalent to a tone of 40 dB at 1,000 Hz.

Sone: means for rating the relative loudness of a sound to a reference tone. One sone is considered to be equal to 40dB of a 1,000-Hz tone. A sound with a rating of 2 sones is judged to be twice as loud as one of 1 sone.

**Missile and Space Safety*, adapted from Air Force Manual 127—201, Washington, D.C., March 1967.

FIGURE 28–1 Continued

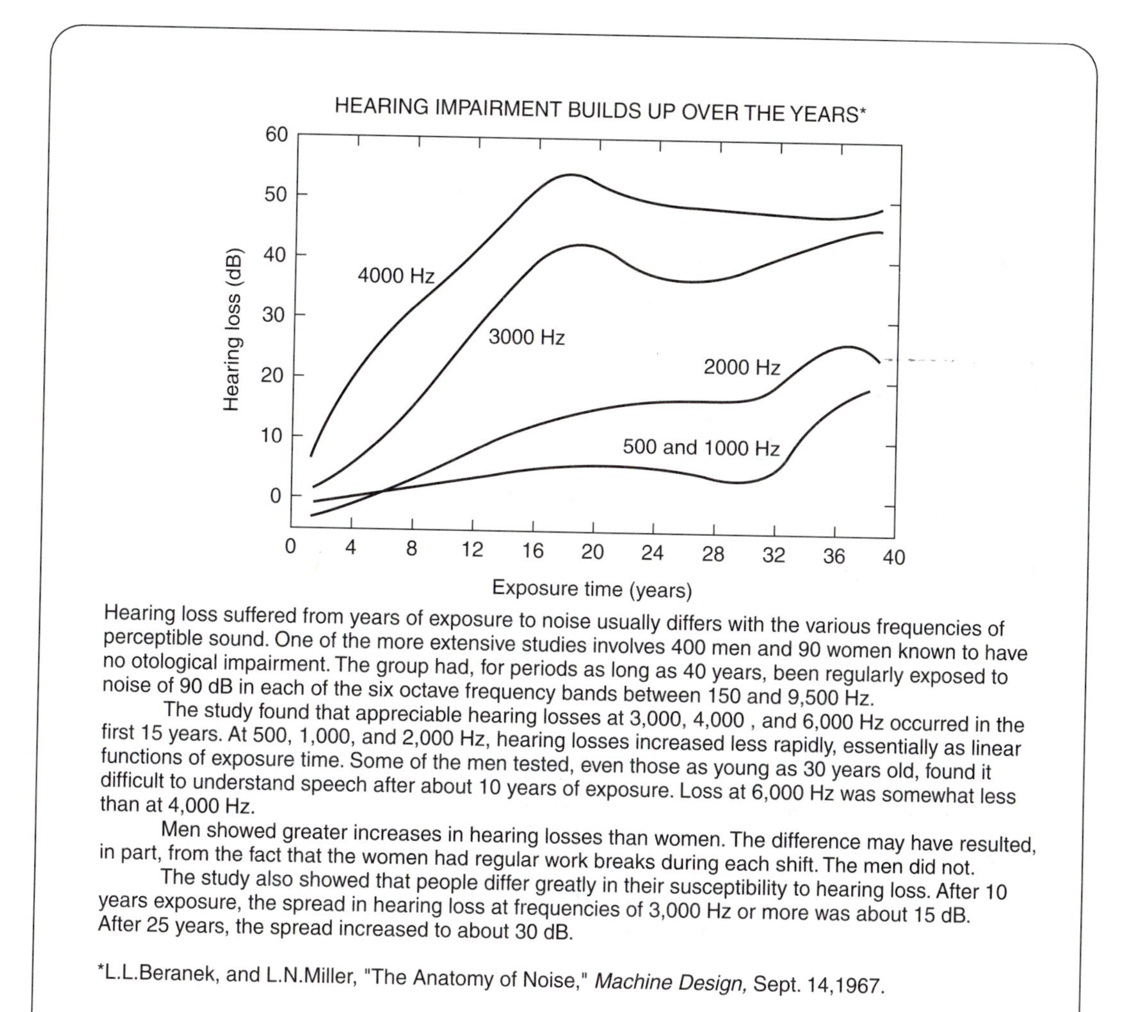

Hearing loss suffered from years of exposure to noise usually differs with the various frequencies of perceptible sound. One of the more extensive studies involves 400 men and 90 women known to have no otological impairment. The group had, for periods as long as 40 years, been regularly exposed to noise of 90 dB in each of the six octave frequency bands between 150 and 9,500 Hz.

The study found that appreciable hearing losses at 3,000, 4,000 , and 6,000 Hz occurred in the first 15 years. At 500, 1,000, and 2,000 Hz, hearing losses increased less rapidly, essentially as linear functions of exposure time. Some of the men tested, even those as young as 30 years old, found it difficult to understand speech after about 10 years of exposure. Loss at 6,000 Hz was somewhat less than at 4,000 Hz.

Men showed greater increases in hearing losses than women. The difference may have resulted, in part, from the fact that the women had regular work breaks during each shift. The men did not.

The study also showed that people differ greatly in their susceptibility to hearing loss. After 10 years exposure, the spread in hearing loss at frequencies of 3,000 Hz or more was about 15 dB. After 25 years, the spread increased to about 30 dB.

*L.L.Beranek, and L.N.Miller, "The Anatomy of Noise," *Machine Design,* Sept. 14,1967.

FIGURE 28–2 Hearing Impairment Builds Up Over the Years

benefits under workers' compensation laws. Nowikas[114] pointed out (Fig. 28-4) that if a group of persons is exposed to noise at the legal limit stipulated by the OSHA standard, one out of five employees will have suffered enough hearing loss to qualify for compensation after 10 years. (This was based on the fact that an 8-hour exposure was not to exceed 90-dBA time-weighted average (TWA). A TWA permits exposures above 90 dBA, provided that there are compensatory exposures below the 90 dBA during the 8-hour period. No values below 80 dBA can be included in calculating the TWA. The National Institute of Occupational Safety and Health (NIOSH) and other organizations have recommended the OSHA standard level be reduced to 85 dBA.

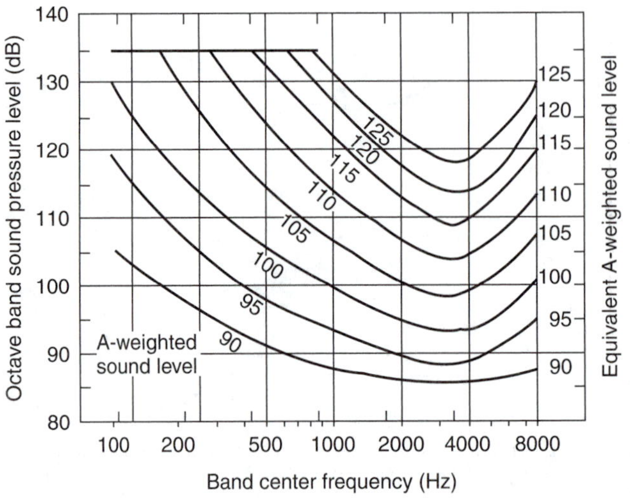

Table 1
PERMISSIBLE NOISE EXPOSURES

Duration per day (hours)	Sound level (dBA)
8	90
6	92
4	95
3	97
2	100
$1\frac{1}{2}$	102
1	105
$\frac{1}{2}$	110
1/4 or less	115

Equivalent sound-level contours. Octave band sound pressure levels may be converted to the equivalent A-weighted sound level by plotting them on this graph and noting the A-weighted sound level corresponding to the point of highest penetration into the sound-level contours. This equivalent A-weighted sound level, which may differ from the actual A-weighted sound level of the noise, is used to determine exposure limits from Table 1.

When daily noise exposure is composed of two or more periods of noise exposure of different levels, their combined effect should be considered, rather than the individual effect of each. If the sum of the following fractions: $C_1/T_1 + C_2/T_2 + \ldots + C_n/T_n$ exceeds unity, then the mixed exposure should be considered to exceed the limit value. C_n indicates the total time of exposure at a specified noise level, and T_n indicates the total time of exposure permitted at that level.

Exposure to impulsive or impact noise should not exceed 140-dBA peak sound pressure level.

OCCUPATIONAL NOISE EXPOSURE

a) Protection against the effects of noise exposure shall be provided when the sound levels exceed those shown in Table 1 when measured on the A scale of a standard sound-level meter at slow response. When noise levels are determined by octave-band analysis, the equivalent A-weighted sound level may be determined from the chart and instructions provided above.

(b) When employees are subjected to sound exceeding those listed in Table 1, feasible administrative or engineering controls shall be utilized. If such controls fail to reduce sound levels within the levels of the table, personal protective equipment shall be provided and used to reduce sound levels within the levels of the table.

(c) If the variations in noise level involve maxima at intervals of 1 second or less, it is considered continuous.

(d) In all cases where the sound levels exceed the values shown herein, a continuing, effective hearing conservation program shall be administered.

FIGURE 28–3 Permissible Noise Exposures

ADDING NOISE LEVELS IN DECIBELS

Sound is a pressure phenomenon. If sound pressure were doubled, the level in decibels would be increased by slightly more than 3 dB. This can be seen from the following:

$$\text{no. dB} = 20 \log \frac{p_1}{p}$$

and with p_1 double that of p, $\log \frac{p_1}{p_0} = \log \frac{2}{1} = \log 2 - \log 1 = 3.0103 - 0.0000$

When two individual sounds are present, the sound level of the combination of the two cannot be more than 3 dB greater than the higher level. Locating a piece of equipment that has a noise level of 85 dB in an area that already has a noise level of 85 dB will result in a combined noise level of 88 dB (85 + 3).

When one noise level of a combination is greater than the other, the resultant can be computed from the following:

Difference in Levels (dB)	Add to Higher Noise Level (dB)
0–1.5	3
1.5–3	2
3–5	1.5
5–9	1
More than 9	0

To compute the approximate noise level of a group of noise emitters, they are combined progressively. For example: six noise emitters have noise levels of 75,77,85,68,78, and 80 dB. Then

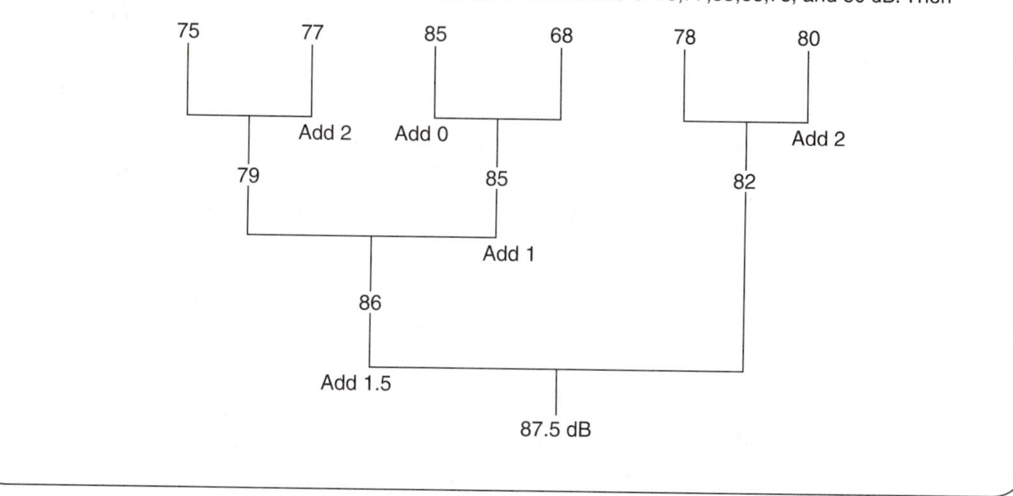

FIGURE 28–3 Continued

PERCENTAGE OF PEOPLE WITH IMPAIRED IN A NOISE EXPOSED GROUP*

Equivalent Continuous Sound Level (dBA)		Years of Exposure									
		0	5	10	15	20	25	30	35	40	45
≤ 80	Risk, %	0	0	0	0	0	0	0	0	0	0
	% with impaired hearing	1	2	3	4	7	10	14	21	33	50
85	Risk, %	0	1	3	5	6	7	8	9	10	7
	% with impaired hearing	1	3	6	9	13	17	22	30	43	57
90	Risk, %	0	4	10	14	16	16	18	20	21	15
	% with impaired hearing	1	6	13	18	22	26	32	41	54	65
95	Risk, %	0	7	17	24	28	29	31	32	29	23
	% with impaired hearing	1	9	20	28	34	39	45	53	62	73
100	Risk, %	0	12	29	37	42	43	44	44	41	33
	% with impaired hearing	1	14	32	42	48	53	58	65	74	83
105	Risk, %	0	18	42	53	58	60	62	61	54	41
	% with impaired hearing	1	20	45	57	64	70	76	82	87	91
110	Risk, %	0	26	55	71	78	78	77	72	62	45
	% with impaired hearing	1	28	58	75	84	88	91	93	95	95
115	Risk, %	0	36	71	83	87	84	81	75	64	47
	% with impaired hearing	1	38	74	87	93	94	95	96	97	97

Note: Percentage of people with impaired hearing in a non-noise-exposed group is equal to percentage in a group exposed to continuous sound levels below 89 dBA. Age = 18 years + years of exposure.

The line "% with impaired hearing" under "80 Risk %" indicates the degradation in hearing that will occur to persons after the years shown in the column headings, from aging alone. At the higher sound levels, the higher figure in any column indicates the percentage of persons exposed who will be suffering some hearing loss from the excessive noise. The lower figure in the same column indicates the total percentage of the population who will have some hearing loss from both aging and exposure to the sound level shown. For example: 14 percent of the people exposed to sound at the 90-dBA level 8 hours per day for 15 years will suffer hearing losses. In addition, 4 percent will have losses due to aging.

*W.M. Nowikas, "The Noise of Sound," American Machinist Special Report 652, *American Machinist*, Dec. 11, 1972

FIGURE 28–4 Percentage of People with Impaired Hearing in a Noise-Exposed Group

For most industries, if noise levels exceed 85 dBA TWA for an 8-hour day, OSHA requires the employer to institute a hearing conservation program. Therefore, 85 dBA is called the "action level" (Fig. 28-1). It is evident that if a company wants to avoid loss claims under workers' compensation laws, it must not only meet prescribed legal standards, but attempt to reduce noise to the lowest possible level and not let it be higher than 80 dB. In 1990, the National Institutes of Health held a Conference on Noise and Hearing Loss. They developed a consensus that sound levels of less than 75 dB(A) are unlikely to cause permanent hearing loss (Noise-Induced Permanent Threshold Shift, NIPTS), while sound levels about 85 dB(A) (for an 8-hour work day) exposure will produce permanent hearing loss after many years.

The ear's greatest sensitivity is in the frequency range from 3,000 to 5,000 hertz, and the loss of hearing almost always occurs first at about 4000 Hz. The earliest audiometric sign of NIHL is usually increased thresholds of hearing from 3,000 through 6,000 Hz. The most important frequencies for speech comprehension are those between 200 and 5,000 Hz. The "speech interference level" is the average sound-pressure level of 500, 1,000, and 2,000 octave bands. The percentage contribution to speech intelligibility at 200 Hz is about 2 percent. As the frequency increases from 200 Hz to about 2,000 Hz the percentage contribution to speech intelligibility rises monotonically to about 12 percent. From 2,000 to 5,000 Hz the contribution to speech intelligibility decreases monotonically from 12 percent to about 6 percent.[115] The fundamental tool used in industry to evaluate hearing loss is the pure-tone audiometer. It produces tones which vary in frequency between 250 and 8,000 Hz. Tests are performed by determining the "hearing thresholds" or lowest levels at which persons being tested can detect sounds.

MECHANISM OF HEARING INJURIES

Hearing losses in older persons (presbycusis) were considered to be due to changes in the small bones of the middle ear, which caused a reduction in their ability to transmit higher-frequency vibrations. However, it is now believed that much of the loss formerly considered due to aging is actually due to almost constant exposure to loud sounds in our noisy modern society. This type of hearing loss (sociocusis) involves deterioration of tiny ciliated cells in the inner ear. These cells convert the vibrations they receive to nervous impulses that are transmitted to the brain. When the ciliated cells in the inner ear are damaged, the accompanying hearing loss is irreversible. Presbycusis results principally in higher-frequency-range hearing loss, so that the persons affected have trouble distinguishing consonants and therefore in understanding what is said. Sociocusis and industrial noise injury involve reduced hearing capabilities at the frequencies of the noises that have caused the losses.

Occupational noise-induced hearing loss occurs over a period of several years' exposure to continuous or intermittent loud noise. This is distinguished from occupational acoustic trauma, which is a sudden change in hearing after a single exposure to a sudden burst of loud sound such as might come from an explosion. Occupational noise-induced hearing loss almost always affects the hair cells in both inner ears, but occasionally the effect can be asymmetric. The loss is usually not profound, and once the exposure is removed, the progress of the hearing loss is inhibited. Low-frequency losses are usually limited to around 40 dB and high-frequency losses to 75 dB.[116]

Impulsive Noise

The degradation of hearing capabilities just described will take place when workers are exposed to steady-state noises for hours each day. Similar effects and some even worse can be generated by loud, impulsive noises. An impulsive noise is one that occurs suddenly, such as that due to the impact of a heavy steam hammer, an explosion, or a rifle shot. (Noise is classified as "wide band," which can be found where machinery produces sound over a broad spectrum, or "narrow band," from one piece of equipment such as a power saw.)

When an impulsive noise occurs, most of the mechanisms of the ear (see section on ear protection) are incapable of providing self-protection. In a most extreme case, such as from the shock wave of an explosion, the air pressure may be so great as to cause rupture of the eardrums. Generally, however, this is a pressure effect rather than one of vibration.

A very loud impulsive noise can cause ringing in the ears (tinnitus) and immediate loss of hearing sensitivity. If there is no further exposure to high noise levels, the tinnitus will disappear, and hearing will return to the normal hearing level of the person exposed. Tinnitus also can be a result of the aging process. Tinnitus from aging might not disappear, and might occur almost continuously.

A loud, impulsive noise may cause tightening of the blood vessels, which reduces the flow of blood through the body. In addition, a sudden loud noise will cause adrenaline to be released into the blood, as the body instinctively prepares to defend itself. The least that will happen as a result is fatigue, and with it possibly some headaches.

The OSHA standards stipulate that personnel exposure to impulsive or impact noise should not exceed 140 dB peak-sound-pressure-level fast response. The prohibition is for an instantaneous noise. Although noise of lower levels can be time weighted (see the section on measuring sound levels), personnel at no time may be exposed to impulsive noise levels in excess of 140 dB.

ELEMENTS OF A HEARING CONSERVATION PROGRAM (HCP)

U. S. OSHA regulations require that a hearing conservation program be instituted if the noise dose exceeds 85 dBA TWA in a workplace. This program should include record keeping of these activities: audiometric testing, monitoring of noise exposure, using hearing protection devices (HPDs) appropriately, employee training, and noise-control engineering. These elements of a hearing conservation program are discussed briefly below. Records must be kept for at least the last two years for all noise-exposure measurements. Audiometric testing records must be maintained throughout the worker's employment. As Casali and Robinson (1998)[113] point out, these records should be used as "feedback for improving the [HCP] program."

Audiometric Test Programs

A worker's hearing may have been defective before he or she was employed. Unless the company is aware of this, after a few years the worker might institute a claim for workers' compensation. Therefore, for its own protection, a company must have an audiometric test program by a qualified and certified audiologist.

It is important that accurate records of the test results be maintained. Each record should contain the audiogram itself; the employee's name, pay, and social security number; date and time of test; name, title, and signature of the person conducting the test; type of work in which the employee is engaged and its location; available information on any exposure to high noise levels; and any other significant data obtainable from the employee or audiologist.

Exposure Monitoring: Measuring Sound Levels

Sound surveys to assess the degree of exposure to hazardous sound levels should be conducted regularly. In general, noise measurements should include measurements taken close to where a worker's ear might be when working. However, measurements should not be made right at the on-duty worker's ear, since the head can cause sound diffraction and alter the sound field. It is most important that sound levels measured are typical of those encountered by the worker. Proper survey techniques should be rigorously observed. Care in calibrating and transporting instrumentation is essential. Operating instructions should be observed, and exposing the instrumentation to prolonged thermal extremes and to moisture should be avoided. Spurious readings from electric and magnetic fields or from wind blowing across the microphone also should be avoided.

Workplace monitoring uses sound-pressure-level (SPL) meters. SPL meters measure the small pressure changes initiated by a vibrating source and transmitted through the air. These small pressures cause a diaphragm in a microphone to vibrate at the frequency (or frequencies) of the air waves. The amplitude of the waves, a function of the intensity of the sound, is also transmitted to the microphone. The inputs are then converted into readable outputs. The capability of humans to detect sound varies with the sound's frequency and intensity. It is therefore necessary that circuitry within the meter modify any inputs to compensate for their variations. The different types of outputs desired determine exactly how this is done. One meter may, however, have capabilities for providing any or all of the different types of outputs.

Weighted Sound-Level Meters. Three weighting circuits (A, B, and C) are incorporated into the standard sound-pressure-level meter. Readings are indicated by the letter designating the scale used, such as 80 dB(A) (or dBA). Because the ear is less sensitive to low frequencies, the A network attenuates very low frequencies to approximate the response of the human ear.

Octave-Band Analyzers. Octave-band analyzers are used to determine at which frequency bands sounds are being generated. If a piece of equipment has been found, through use of a sound-level meter, to be excessively noisy, an octave-band analyzer can be used to determine the noise source. This type of analyzer is designed to obtain readings by switching in suitable filters which block out noise of frequencies outside the octave to be measured. Octave-band sound-pressure levels can be converted to the equivalent A-weighted sound levels through use of the graph presented in the OSHA standards (Fig. 28-3).

One-third-octave-band analyzers provide even narrower responses than do octave-band analyzers, for each band is divided into three parts. This provides more accurate analysis through a greater capability for distinguishing between noises at frequencies that are fairly close.

Meters may also have capabilities for "fast" or "slow" responses. A fast response permits measurements to be made of transient, peak sound levels such as those generated by punch presses and other impact noises. A slow response averages out high-level noises rather than measuring peaks.

In addition to hearing loss, blood-vessel constriction, fatigue, and headaches, noise can have other adverse effects. Exposure to a noisy environment can cause nervousness, psychosomatic illnesses, and inability to relax. Noise can upset balance, disturb sleep, and alter electroencephalograms.

Hearing Protection: Minimizing Adverse Effects on Hearing

There are many ways to lessen effects of loud noises on the ears, as suggested in the following paragraphs.

Natural Ear Protection. The ear itself has protective mechanisms that help reduce possible effects of loud noises and their attendant pressures and vibrations. The ear canal (Fig. 28-1) is curved so that sound waves cannot impinge directly on the eardrum. Eardrum muscles contract in response to a loud noise, causing the ossicles to stiffen and thereby to dampen the vibrations transmitted. In addition, the means by which the stapes conducts vibrations is changed, also producing a dampening effect. The latter effect will occur even with sudden, intense noises. The actions of the muscles tied to the eardrum are less responsive and provide little or no protection against impulsive or sudden noises.

Hearing Protection Devices. When noise levels to which workers may be exposed exceed those indicated by the OSHA standards (Fig. 28-3), personnel protection must be provided. The means by which noise levels can be measured have been described. However, hearing protection devices (HPDs) should not be resorted to until steps to reduce noise levels through engineering and administrative controls are exhausted. There are two basic types of HPDs: passive and active. The passive HPDs are the most common in industry. Casali and Berger (1996)[117] identify the principal types of passive hearing protection devices as: earplugs that are inserted into the ear canal, ear canal caps that seal the canal at or near its rim, and earmuffs that encircle the outer ear.

Passive HPDs include:

1. *Plugs:* rubber or plastic devices fit snugly against the ear canals, blocking the passages against transmission of sound. They come in a variety of sizes and types. Each person who must wear them should be fitted initially by a qualified audiologist, who will ensure they fit snugly and effectively and without discomfort. They are effective for 8-hour exposures only up to 95-dB noise levels. They provide no protection through the bony areas around the ears. Another disadvantage is that it is sometimes difficult for a supervisor to determine from a distance whether a worker is wearing them.
2. *Foam plugs:* small foam rubber cylinders are compressed and twisted between the thumb and forefinger, then inserted into the ear canal. The foam "memory" then causes the plug to try to return to its original shape, expanding it into the ear canal. The worker's skill in inserting these plugs properly is essential and requires employee training. This is one of the most common types of hearing protection.
3. *Wools:* the earliest ear protection devices were probably masses of cotton or wool forced into the ear canals like ear plugs. New types, such as waxed cotton (ordinary cotton is not considered acceptable) or "Swedish wool" (extremely fine glass fibers), have proved only slightly less effective than plugs. They are usually more comfortable for workers who suffer from the use of plugs. However, since they are used once and then disposed of, they are more costly.
4. *Muffs:* these cover the entire ear and some of the bony areas around it through which sound might be conducted. They offer greater protection than do plugs or wools which are not fitted properly. They are easily fitted and adjusted and are put on and taken off easily. However, workers sometimes complain of headaches caused by the compression effect against the head. They are often uncomfortable

in high-temperature environments. Special types must be used if they are to be worn with hard hats or other headgear.

Active HPDs are earplugs, canal caps, ear muffs or even noise-attenuating helmets that incorporate electronic components and transducers. Active HPDs provide active noise cancellation, communications features, and attenuation which is level-dependent. They reduce noise by introducing destructive cancellation by applying opposite-phase sound waves at the ear. Some are designed for hearing protection and others for one- or two-way communication. These have become viable in the past decade because of advances in miniature semiconductor technology and high-speed signal processing. [117]

Research has shown that some manufacturer's claims for the effective reduction of noise by their hearing protection devices do not indicate the kind of hearing protection actually received by workers in the field. Safety engineers must exercise caution in selecting hearing protection devices and must take steps to ensure that the devices selected are properly selected and used by workers in the field. The noise-reduction rating on the HPD is a laboratory-derived value. The evaluation protocol may have relied upon an "experimenter fit," which has been shown not to represent always the kind of fit obtained by users in the field.[118] HPDs rely on a good fit. Protection is compromised without it. For employees who need to wear HPDs, a training program in their proper use is essential. Of course, failure to wear an HPD, when it is needed and provided, must be protected against by vigilance. HPDs offer no protection unless worn!

Employee Training

Employees need to be trained to:

- understand the danger to hearing that comes from noise exposure
- recognize noise exposures which are harmful
- evaluate noise levels of exposure in a practical way
- take action to protect themselves from harm from noise

Engineering Control: Eliminating Vibration Causes

Administrative procedural control to require personnel to use protective equipment is regarded by the OSHA as secondary to the use of good engineering. Measures that can be taken to reduce noise levels are:

- Select equipment for installation which has low vibration and noise characteristics. Stipulate permissible maximum noise levels in specifications for new equipment.
- Determine whether an operation, process, or piece of equipment that is noisy can be avoided or eliminated by use of a quieter one. Use presses instead of drop hammers wherever practicable.

- Select and operate rotating and reciprocating equipment, such as pumps, fans, motors, and presses to operate as slowly as feasible. Use large, slow machines instead of small, fast ones with the same capacity.
- Mount equipment that might vibrate on firm, solid foundations. Make certain foundation bolts are kept tight.
- If equipment vibrates, determine whether or not its characteristics can be changed by use of devices such as dynamic dampers, rubber or plastic bumpers, flexible mountings and couplings, or resilient flooring.
- Keep the velocity of fluids, such as air flowing through ducts and jets or liquids in piping, at the lowest speeds possible. Avoid sudden directional and velocity changes in pipelines and ducts.
- Do not use quick-opening valves in liquid systems which could cause water hammer. If water hammer does exist, replace valves with slow-acting ones or install accumulators.
- Make certain pipe is firmly secured to prevent rattling.
- Provide noise-absorbent lining on air ducts and mufflers on openings through which air must pass.
- Ensure that periodic maintenance minimizes vibration and noise which might be generated by keeping the equipment free of chatter, impact, and similar noise-generating motions. Replace, adjust, and tighten worn, loose, or unbalanced machine parts.
- Locate noisy activities and equipment far from other operations.
- Keep moving parts lubricated to minimize noise and vibration.

Isolating Sources

- Locate noisy activities, such as individual machines, in sound-absorbing enclosures, or provide barriers between such activities and other locations where personnel are present.
- Ensure that floors, walls, and other structural features do not vibrate and transmit vibrations and noise to other locations.
- Where vibrations of fixed equipment cannot be eliminated, mount the equipment on vibration isolators to prevent transmission of motion.

Isolating Personnel

- If only one or two workers must remain in a large, noisy area, determine whether they can be isolated in an acoustically quiet booth or other enclosure.
- If the noise level cannot be reduced to a maximum of 85 dB, provide workers with hearing protection devices.
- Attempt to schedule personnel so they remain in high-noise-level areas for periods which are as short as possible. Arrange schedules so that work at high noise levels or extended periods is divided among two or more persons. Ensure that

personnel who must work at noise levels greater than 90 dB are also scheduled to work at low noise levels for durations that will bring the time-weighted exposure to less than 90 dB.

- Ensure that areas where personal hearing protection devices should be worn are posted, warning workers of the hazard and the necessity to use the protective devices.
- Ensure that workers who must wear hearing protection devices such as ear plugs are initially fitted with them by approved audiologists and that each worker is aware of the proper type and size to use.
- All supervisory personnel should ensure that workers wear their protective equipment at all times while they are in high-noise areas.
- Noise levels should be checked as often as practicable to ensure that areas with noise levels which exceed the 8-hour limit are recognized, so that suitable precautions are taken and that any changes are recognized. Where surveys indicate that noise levels have been reduced, areas in which protection was formerly required may no longer need it.

There are three places where noise reduction can be attained: (1) at the source, (2) along the noise path, and (3) at the host (e.g., the worker that is in the noise path). The elimination and engineering control of vibrations can be applied to the source and path. Personal protective devices and administrative controls can be used to protect the host, if that site must be addressed.

ANNOYANCE

Noise annoys people; however, the types and levels that do so are difficult to determine. Rock music at any sound level may sound like noise and disturb people who like operatic music, which in turn may sound like noise and disturb persons who like rock. In addition, many persons who have been exposed to certain noises over long periods of time develop a tolerance so that they may not even hear those noises without conscious effort. On the other hand, the same noises may annoy other persons, who have not developed the tolerance, to such a degree that their efficiency is degraded. They may become more prone to make errors which lead to accidents.

Noises that annoy need not be loud noises. A slow water drip in the silence of the night is one example. In general, though, the louder the sound, the more annoying it can be. In addition, an unexpected impulsive noise can be most annoying. A sudden impact or a sonic boom which startles will not only annoy persons who hear it, but may keep them on the alert against being startled again.

DISTRACTION

The same types of sounds that annoy persons can also distract them from their normal activities. In addition, other sounds may not annoy but music over public address systems can sometimes distract a person's attention when the song played is one he or she likes. The talking of persons in the vicinity can distract the attention of other persons even if it does not annoy them. The ringing of a telephone, the signal for beginning or ending an activity, and an announcement over the public address system are additional examples. Accidents have occurred when persons engaged in hazardous activities were spoken to, thus distracting their attention momentarily so that they failed to respond during a critical instant. Buses are posted with signs telling passengers not to hold conversations with their drivers when the buses are in motion.

INTERFERENCE AND MASKING

Any operation that requires oral communication will suffer from a noisy environment. Interference with communications can create misunderstandings about information transmitted from one person to another. When such communications relate to hazardous activities, any misunderstandings can lead to accidents.

Technically, masking is the level, measured in decibels, by which a sound must be increased to be understood in the presence of another, interfering sound. Generally, this is determined experimentally by establishing the threshold level at which a message can be understood in a quiet environment. The threshold level is then determined in the presence of the masking sound. It has been found that the degree of interference depends on the frequency or frequencies affected, whether the interfering sound is of one frequency or many, and whether intensities vary. In an industrial environment, many of the masking sounds can be highly complex.

Tests were conducted of reception of pure-tone (single-frequency) communications where masking was also provided by pure tones. It was found the intensity of the message to be communicated had to be increased 15 to 30 decibels to be understandable.

Numerous methods have been proposed to measure the effects of noise levels on speech communications. These methods are indicated in Figs. 28-5 and 28-6, which show the relationships among noise levels, voice levels, and distance between speaker and listener.

The noise level in an industrial plant, even if high, is fairly constant. However, hearing and comprehension may vary, especially in an area of fluctuating sounds. Some persons are more powerful speakers than others; the frequencies of their voices differ; some articulate more clearly; the mouth-to-ear distance between speaker and listener will vary; and visual cues from the speaker's eye, face, or hands can contribute a great deal to speech intelligibility. The selection of active HPDs with built-in electronic communication enhancement might be required for safety in some situations, especially where communication of warnings or other information is critical to avoiding danger.

OTHER VIBRATION EFFECTS

Noise is the commonest vibration problem; however, vibration can have other adverse effects. High-intensity, low-frequency sounds can cause the skull, other bones, and internal organs to vibrate with injurious or annoying amplitudes. Resonances will occur at certain frequencies so that these effects become much more noticeable.

Figure 28-7 indicates vibration tolerance criteria. In actual practice, vibration tolerances of personnel may differ substantially. A person who must do fine work while

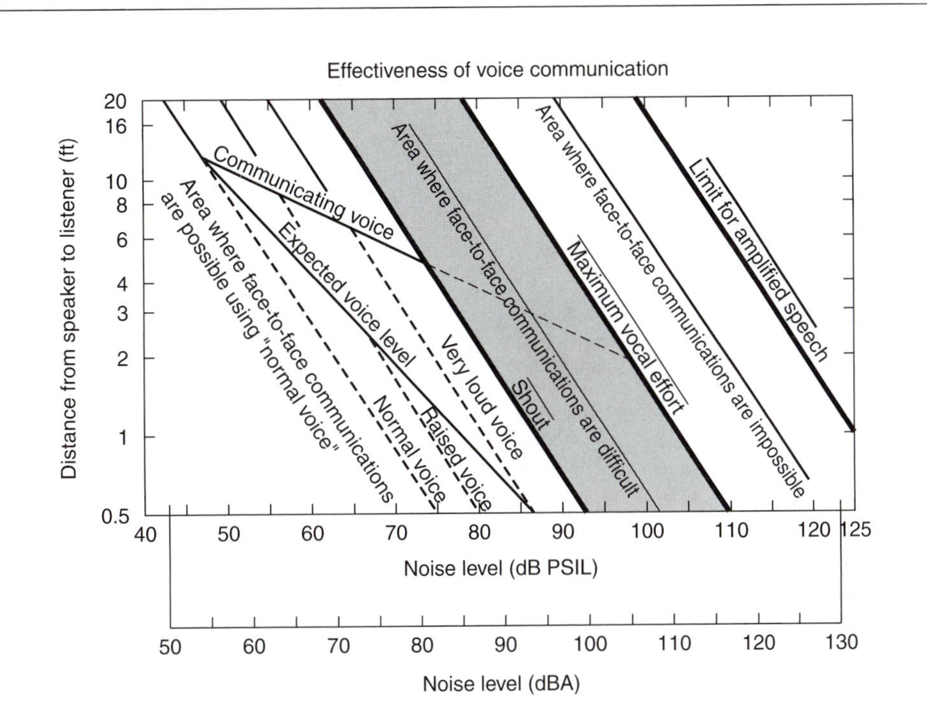

The effects of varying noise-exposure levels on satisfactory face-to-face speech communication are expressed here in terms of voice level and distance between talker and listener.

Preferred speech interference level (PSIL), in decibels, is the average of the three octave bands centered on preferred frequencies: 500, 1,000, and 2,000 Hz. Noise levels in dBA equal noise levels in dB PSIL plus 7 dB (63 dB PSIL equals 70 dBA):

Normal voice	= usual speaking level
Raised voice	= normal voice + 6 dB
Very loud voice	= raised voice + 6 dB
Shout	= very loud voice + 6 dB

Expected voice level represents the increase of voice level a speaker located in a noisy field normally adopts. The expected noise level increases from normal to very loud as the noise level increases from 55dBA (48 dB PSIL) to 95 dBA (88 dB PSIL). The communicating voice level is the voice level that a speaker can produce, over the range of sound levels shown, when forced to communicate (achieve a 95 percent word score; with positive, instantaneous feedback that shows success or failure).

**Hazardous Noise Exposure,* Air Force Regulation 161-35, July 27, 1973.

FIGURE 28–5 Effectiveness of Voice Communication

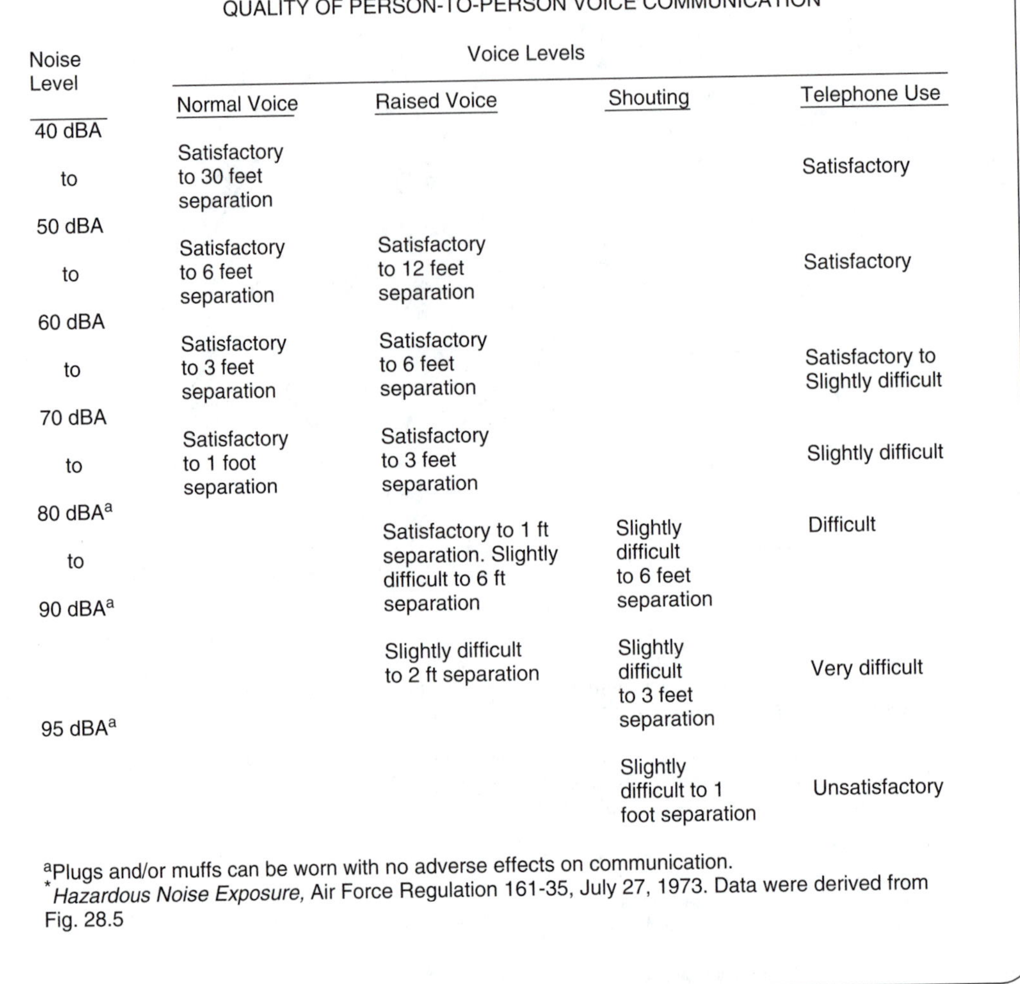

QUALITY OF PERSON-TO-PERSON VOICE COMMUNICATION*

Noise Level	Voice Levels			
	Normal Voice	Raised Voice	Shouting	Telephone Use
40 dBA to 50 dBA	Satisfactory to 30 feet separation			Satisfactory
50 dBA to 60 dBA	Satisfactory to 6 feet separation	Satisfactory to 12 feet separation		Satisfactory
60 dBA to 70 dBA	Satisfactory to 3 feet separation	Satisfactory to 6 feet separation		Satisfactory to Slightly difficult
70 dBA to 80 dBA[a]	Satisfactory to 1 foot separation	Satisfactory to 3 feet separation		Slightly difficult
80 dBA[a] to 90 dBA[a]		Satisfactory to 1 ft separation. Slightly difficult to 6 ft separation	Slightly difficult to 6 feet separation	Difficult
95 dBA[a]		Slightly difficult to 2 ft separation	Slightly difficult to 3 feet separation	Very difficult
			Slightly difficult to 1 foot separation	Unsatisfactory

[a]Plugs and/or muffs can be worn with no adverse effects on communication.
Hazardous Noise Exposure, Air Force Regulation 161-35, July 27, 1973. Data were derived from Fig. 28.5

FIGURE 28–6 Quality of Person-to-Person Communication

his or her arm rests on a work table that vibrates even slightly can become extremely annoyed and uncomfortable in trying to control vibrations. Coordination and visual capabilities can also be degraded substantially. A person operating a piece of mobile equipment in the field might not be affected by vibrations, but a person's tolerance depends on the type of equipment used, the type of terrain, the operating speed, the vehicle's suspension system, and the padding and springs in the vehicle's seat.

Vibrations can be produced within the equipment itself or induced. In addition to many of the causes of vibration and sound already pointed out, sources of vibration by the equipment include:

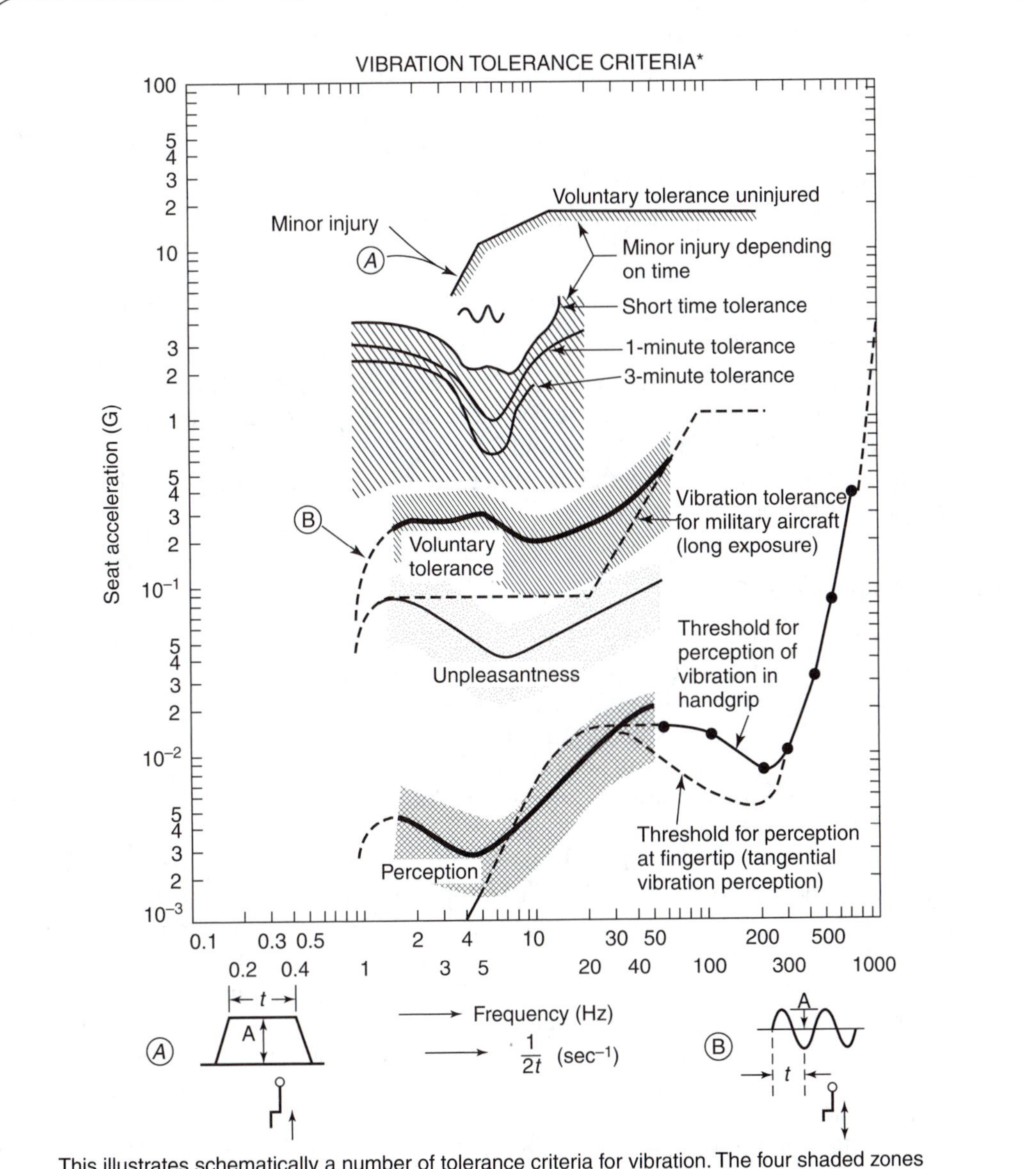

This illustrates schematically a number of tolerance criteria for vibration. The four shaded zones represent: threshold for perception; the unpleasant area of vibration; the limits of voluntary exposure, unprotected, for 5–20 minutes; and the voluntary tolerance limits for subjects with lap belt and shoulder harness for 3 minutes, 1 minute, and less than 1 minute. Above this, minor injuries occur, depending on time. At the top of the chart is plotted the large magnitude. The conventional way of measuring durations (*t*) and amplitudes (*A*) is shown in the small diagrams at the bottom.

*Air Force Systems Command, *Systems Safety,* Design Handbook 1-6.

FIGURE 28–7 Vibration Tolerance Criteria

- Rotation with unbalanced loads or bent shafts.
- Misalignment of driving and driven equipment.
- Impacts of a moving part against another part, moving or stationary.
- Water hammer in hydraulic systems.
- Lack of adequate snubbing devices in pneumatically operated equipment.
- High-velocity air in ducts.
- Looseness of equipment parts, which causes them to rattle.
- Flat spots on bearings or steel wheels.
- Worn or separated treads, or nails or other objects embedded in tires.
- Belt or gear slippage.

Vibration is transmitted more easily through solid materials than through air. It may happen, therefore, that a heavy piece of equipment can transmit vibrations through a structure, such as a frame and flooring of a building, to other equipment (Fig. 28-8). Operators in contact with this equipment may become aware of and be affected by the transmitted vibration.

Effects on Equipment

Even worse than the annoyance of vibration to personnel is the metal fatigue induced by vibration. Metal fatigue can cause failures of rotating parts and other stressed mechanical equipment. The result can be loss of or damage to the part or equipment, and possible injury to personnel. A prime effect is loss of control because of steering-mechanism failure, resulting in an accident to a vehicle and injury to the vehicle's operator and passengers. Vibrations can cause leakage of fluid lines, pressure vessels, and containers of hazardous liquids and gases.

On January 10 and April 8, 1954, two B.O.A.C. de Havilland Comets, the first jet airliners, exploded after leaving Rome, at approximately the times they reached their cruising altitude of 30,000 feet. Until the possibility was ruled out, it was thought the violent explosions were caused by bombs aboard the aircraft. Investigation and tests determined that in each case vibration of a direction-finder antenna had induced metal fatigue in a fuselage opening. Cracks started. The total expansive difference between the internal and atmospheric pressures was 8 psi, equivalent to that which would have been created by 500 pounds of explosives and sufficient to tear out a large section of each aircraft.

Similar ruptures, in which vibration was the initiating cause, have occurred in pressurized industrial equipment. In most cases the pressures involved have been far higher than 8 psi. The degradation in strength leading to a rupture would be far less.

Effects on Tools

Hunter[85] pointed out that use of vibrating tools can lead to arthritis, bursitis, injury to the soft tissues of the hands, and blockage of blood vessels. Of these, the last, known as Raynaud's phenomenon, is probably the most prevalent and most serious. Raynaud's phenomenon involves paleness of the skin from oxygen deficiency due to reduction of

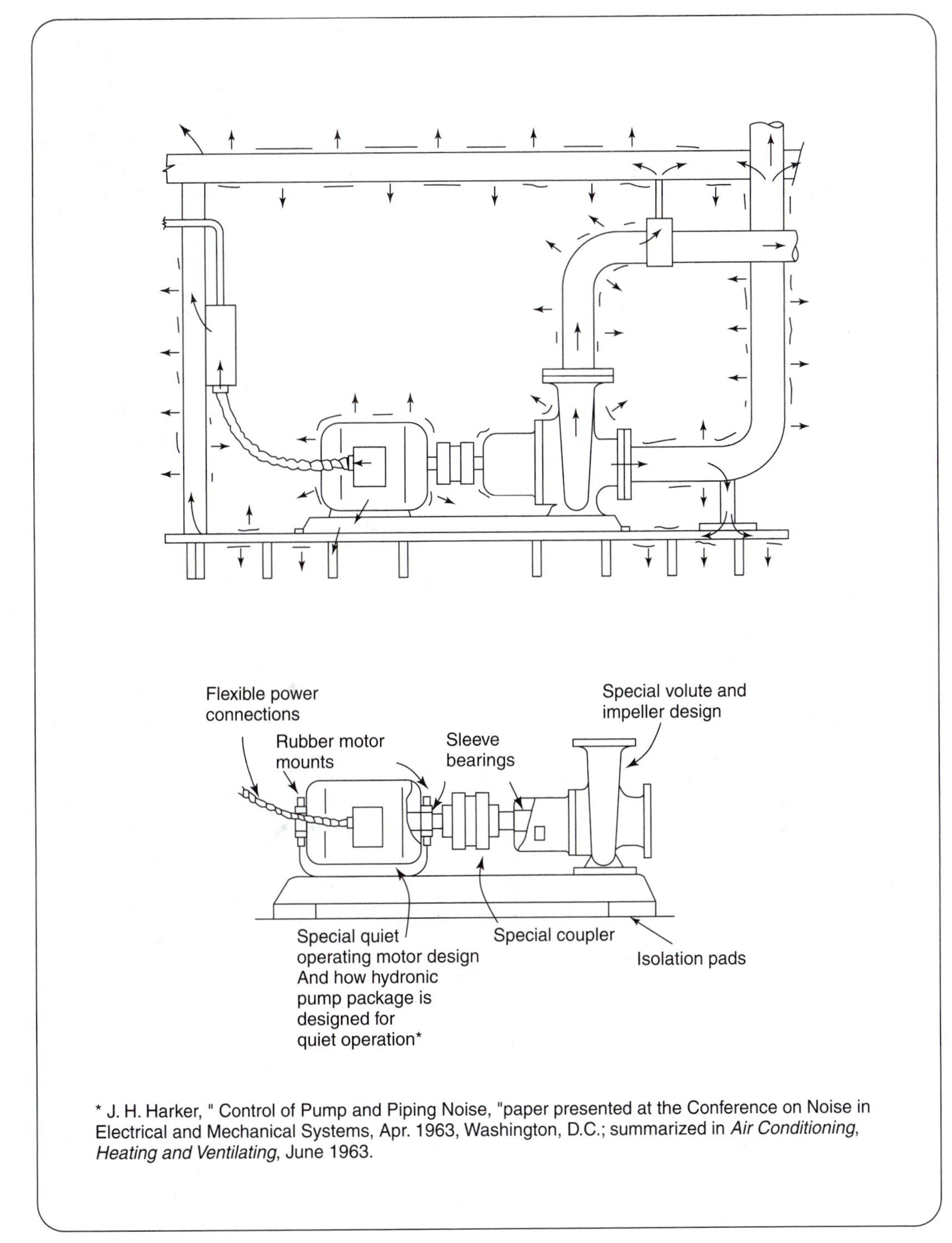

FIGURE 28–8 Vibration and Noise of a Pump and Its Surroundings

blood flow caused by blood-vessel and nerve spasms. Because of the deficiency in blood flow, the hands feel cold, often with decreased sensation. The disease is produced by vibration directly on the fingers or hands. It has been a problem with persons using pneumatic chisels and hammers and hand-held rotating grinding tools. Vibrations at 2,000 to 3,000 beats per minute appear to generate the worst effects with hammerlike tools. With rotating grinding tools the most dangerous frequencies were from 40 to 125 hertz, especially if the amplitudes of vibration were greater than 100 micrometers.

Hunter reported that a large group of workers suffered from this disease after working in jobs polishing duraluminum and steel castings with hand-held tools. Of 278 men and women, 6 percent were affected within two years. Eventually 77 to 90 percent of the group were affected, a few even after they had ceased to do such work. The symptoms were especially pronounced during cold weather; when the worker was warm, there was little problem.

Figure 28-9 summarizes safety-problem causes and effects due to vibration and noise. Figure 28-10 provides another checklist by which potential problems due to vibration and noise can be reviewed.

BIBLIOGRAPHY

[112] *Slawinski v J. H. Williams Company,* 208 N.Y. 546 (1948).

[113] J.G. Casali and G.S. Robinson, "Noise in Industry: Auditory Effects, Measurement, Regulations, and Management," in W. Karwowski and W. Marras, eds. *Handbook of Occupational Ergonomics* (Boca Raton, FL: CRC Press, 1998).

[114] W. M. Nowikas, "The Noise of Sound," Special Report 652, *American Machinist,* Dec. 11, 1972.

[115] J. J. Crocker and A. J. Price, *Noise and Noise Control,* Vol. I (Cleveland, OH: CRC Press, 1975).

[116] American College of Occupational and Environmental Medicine Noise and Hearing Conservation Committee, "Guidelines for the Conduct of an Occupational Hearing Conservation Program," *Journal of Occupational Medicine,* 1987, Vol. 29, pp. 981–989.

[117] J. G. Casali and E. H. Berger, "Technology Advancements in Hearing Protection circa 1995: Active Noise Reduction, Frequency/Amplitude-Sensitivity, and Uniform Attenuation," *American Industrial Hygiene Association Journal,* Vol. 57, pp. 175–185, 1996.

[118] J. G. Casali and M-Y. Park, "Laboratory versus Field Attenuation of Selected Hearing Protectors," *Sound and Vibration Magazine,* Vol. 25, No. 10, pp. 28–38.

EXERCISES

28.1. What is noise?

28.2. List five effects of vibration.

28.3. List five sources of industrial vibration and noise. What types of industrial equipment are great noise generators?

28.4. Can employees be discharged for failing to use personal protective devices against noise when required?

Hazards Checklist –Vibration and Noise

Possible Effects	*Possible Causes*
Effects on personnel: Fatigue Inability to read instruments or to activate controls Involuntary reaction to sudden loud noise Injury to hearing ability Raynaud's disease Interference with ommunications	Irregular motion of rotating parts Bearing deterioration and misalignment Irregular or cyclic motion Loose or undersized mountings Pump or blower cavitation Reciprocating motion Vibrating tools Misaligned equipment in motion Lack of vibration isolators Scraping of hard surfaces against each other Bottoming or failure of shock mounts or absorbers
Damage to equipment: Metal fatigue and other changes in crystalline structure Loosening of bolts or other fastened parts Breaking of lead wires, filaments, and supporting parts Crazing and flaking of finishes	Fluid dynamics: Excaping high-velocity gas High-velocity fluid hitting a surface or object that can vibrate Pneumatic or hydraulic shock (water hammer) Aerodynamic flutter or buzz Jet engine exhaust Sonic booms and other shock waves
Operational effects: Loss of calibration of monitoring and measuring devices and other equipment Chattering of spring-type contacts, valves, and pointers Possible false reading on pointertype devices. Static electricity generated between plastic surfaces	Highly amplified music or other sounds Explosions or other violent ruptures Lack or failure of sound isolation devices such as mufflers

FIGURE 28–9 Hazards Checklist—Vibration and Noise

CHECKLIST—VIBRATION AND NOISE

1. Is there any equipment in the plant which appears to be making an undue amount of noise? Have readings been taken to determine what the noise levels are? Have the levels been increasing over those which existed when the equipment was installed?
2. Is rotating or reciprocating equipment mounted securely to avoid or minimize vibration and noise? If vibration sources cannot be eliminated, is the equipment provided with vibration isolators or dampers?
3. Are periodic surveys made of the noise levels in the various parts of the plant? Is there a program to monitor the personnel in the plant to determine what their hearing levels were when they were hired, and whether deterioration has taken place since then?
4. If a piece of equipment or process is noisy, is it possible to reduce the noise level, or could the equipment or process be isolated?
5. Are any power tools used in the plant subject to vibration which might cause Raynaud's Phenomenon to persons who use the tools for long periods of time?
6. Is the piping in the plant noisy when a fluid flows through it? Has the piping been fastened securely to supports so it will not vibrate when in operation? Are the lines big enough for the amount of liquid which must flow? (A larger line means a lower velocity and less noise at fittings.) If there are quickly closing valves, do they create water-hammer? Have such lines been provided with accumulators to eliminate water-hammer?
7. Are thin surfaces avoided in air streams where they could "flutter" and cause noise?
8. Can the use of noisy high-speed rotating devices be minimized to avoid vibration and noise by use of slower operating devices?
9. Have bolts and other fastened parts for safety-critical assemblies been tightened securely to prevent motion between parts?
10. Are all sound-isolation devices and materials in good shape so that their effectiveness is not lost?

FIGURE 28–10 Checklist—Vibration and Noise

28.5. What constitutes a hearing loss, and to what degree could a worker be compensated? At what frequency does hearing loss first occur?

28.6. Other factors being equal, would younger or older workers suffer hearing losses? Why? What are presbycusis and sociocusis?

28.7. What is the maximum permissible impact noise level?

28.8. What is an octave-band analyzer?

28.9. Describe three types of passive personal ear protection devices, with advantages and disadvantages of each.

28.10. What is an active personal ear protection device?

28.11. List ten ways in which vibrations can be eliminated or controlled.

28.12. What is Raynaud's phenomenon and how is it affected by vibrations?

CHAPTER 29

Computers and Safety

It has been said the Industrial Revolution has been succeeded by the Age of Computers. The speed and delight with which the general public has accepted and begun using modern computers has been phenomenal. Computers have entered our daily, personal, commercial, and military lives. Of their many aspects, this chapter focuses on only one: their safety aspects.

Software safety is at the heart of the issue of computer safety. Software, i.e., a program that the computer processes, is inert. It possesses no hazard by itself. However, it is used in equipment and systems to control energy. An accident is as dependent upon the flow of energy as a fire is dependent upon oxygen. When software fails to control equipment and system energy properly, then accidents can occur. Improperly written software can result in tragedy. The great difficulty is that code can be thousands of lines long, with the bug that can deliver the dangerous release of energy embedded somewhere in a code fragment, perhaps waiting a long period of time for the cycle of events that trigger the undesired event. To attempt to determine that a code is safe, Charles L. Piechota indicates, is like entering the twilight zone for many safety engineering professionals.[119] He cites three events involving software failures that show why the safety engineer must venture into this zone.

The first occurred when an F-18 fell 20,000 feet after the pilot fired a missile attached to its wing. The computer that received the signal to fire was supposed to send ordered signals that opened the clamp that held the missile, fired the missile, and closed the clamp. The clamp opened and closed before the missile could fire. The missile then fired, while attached to the wing. The aircraft crew regained control just in time to avoid a crash, but not until they had experienced the ride of their lives.

The second event was a simulation exercise of the F-16. During the exercise it inverted when it crossed the equator. The inversion was due to a software failure. The third event occurred during maintenance on the weapons bay door of the B-1A bomber. The door was opened and held open by using a mechanical inhibit/override. During the period it was open a test in the cockpit was conducted which issued a software-based command to "close weapons bay door." At the end of the maintenance action, the inhibit/override was released, whereupon the bay door closed suddenly and unexpectedly.

During the flight of a civilian jet plane, a computer problem caused a loss of engine power and a long drop in altitude, engine ice began to form, and then the engines began to overheat. Computers caused the inadvertent firing of an automatic weapon; unprogrammed application of power to a fighter plane while on an aircraft carrier's deck (so it went overboard and into the ocean); and *unintended* destruction of a group of high-altitude weather balloons. In these last cases, the accidents were due to erroneously programmed actions.

The computer control of equipment and systems is now a part of almost every complex new design. Aircraft are conceived in which the pilot flies by a "virtual cockpit." Displays and controls are computer generated. Displays are virtual images in 3-D without hardware except what is necessary to generate images to the pilot. Scenes from outside the cockpit are detected and then generated for viewing by the pilot in her/his virtual environment. Medical diagnostic equipment that is capable of delivering fatal doses of radiation to the patient in scans relies on computer controls to prevent mishaps. Industry uses robots with dangerous strength and speed to increase productivity and quality in manufacturing. Even the amusement park controls its exciting and dangerous rides by dependence upon computers and software for safety. It is clear that the safety engineer must be concerned about software safety.

SAFETY USES OF COMPUTERS

When most persons think of computers and safety, they tend to think immediately of direct effects. Computers as such, however, present few hazards to persons. Electricity might shock an unwary person who ignorantly tries to probe the inner workings of a power unit (although damage would probably occur to the computer, rather than injury to the person). In very large units, overheating fires have taken place in the incoming power line. In 1988, the National Fire Protection Association reported there had been approximately 500 significant fires in computer rooms each year during 1980 through 1984. (A significant fire was one reported to the fire department.) About 15 percent began in computer hardware, while a major portion of the remainder began in the electrical equipment. For small units such as home computers, other than for initial input power, demands for power within the circuitry are small and most lines are contained in nonflammable material.

Safety of humans is a prime cause for the development and use of computerized equipment. Many actions that formerly were performed only at great personal risk are now being undertaken by machines. Examples of such equipment uses have been the replacement of workers in hazardous environments such as dangerous work in steel production. Computerized equipment has been used for the routine handling, disposal, and replacement of nuclear materials. In addition, it has been used for the removal and disposal of contaminated radioactive material resulting from accidents, such as those at Three Mile Island and Chornobyl. Such usage will increase as nuclear plants are shut down and the materials contained must be disposed of. Computers have been used in deep sea searches to locate sunken ships and to find and retrieve parts that remain after a disaster.

There are now a multitude of software products designed to assist the safety engineer. These products range from aids to the management of industrial safety programs to ensure OSHA compliance and cost effectiveness, to system safety programs for HAZOP, fault-tree analysis, consequence modeling for chemical spills, etc.

SAFETY PROBLEMS TO WORKERS

The chief problems have reportedly been eye strain and physical and psychological stresses because of the constant use of the video display terminal (VDT) and computer keyboard. Some claim that constant use of the eyes to view computer monitors may increase strain, especially where there have already been eye problems. Possible harmful effects of radiation from monitors and TVs have been found not to exist. It has also been claimed that the speedup of computer operators in offices has increased psychological stress problems and lessened birth rates of working mothers. Such claims need scientific verification before they are accepted. However, overuse disorders, discussed earlier in this text, have increased greatly because of the electronic office and sustained work at the computer keyboard. Carpal tunnel syndrome is a commonplace phenomenon of the computer age.

ACCIDENTS WITH COMPUTERIZED EQUIPMENT

Computerized systems are often in the forefront of technology, the realm of civilian aircraft, military devices, and space systems, where some accidents have occured.

In 1984, newspapers stated in a minor headline: "Robots Seen as Posing Threat to Worker Safety." It mentioned that the only confirmed instance in Japan was one in which a worker had been killed in 1981. The worker had been performing repairs on a piece of computerized equipment in which a fault had occurred. Occurance of a fault normally triggered a signal for the robot to return an operating part to its rest position. In this case, in cutting off all power, the maintenance man stopped all programmed operations. He made the repair and then switched on the equipment. The arm of the robot moved unexpectedly but as programmed, crushing the man against another machine.

Two other robotic problems were reported in 1987. In the first, robots making key welds had been skipping many of these required on an automobile production line. Another robot designed to lift filled boxes of soup cans dropped them, causing thousands of dollars in damage. In addition, software errors had caused burns by subjecting patients to excessively intense nuclear radiation during medical treatments.

COMPUTER INABILITIES

Figure 29-1 presents the advantages and disadvantages of abilities of computers as compared with those of humans. In addition, computers and computerized equipment must be extremely massive and complex to have only a subset of human capabilities. Within their designed capabilities computers can accomplish only programmed actions. They lack flexibility and have limited mobility, even when incorporated in very specialized equipment. (The adult brain contains more than ten billion nerve cells, or neurons, but weighs only about three pounds.) A computer-operated piece of

PERSON	MACHINE
Can reason and make decisions inductively.	Has no inductive capability.
Can follow a random and variable strategy.	Always follows the programmed strategy.
Can improvise and exercise judgment based on memory, experience, education, and reasoning.	Is better at routine functions.
Can make judgments and take action when preset procedures are impossible.	Programs for all conceivable situations, such as emergencies, and corrective or alternative actions are impracticable.
Can adapt his or her performance, since he or she learns by experience, education, and reasoning.	Cannot learn other than those facts and capabilities that it is programmed to learn.
Has high ability to reason out ambiguities and vague statements and information.	Is highly limited if input lacks clarity.
Can interpret an input signal accurately even in the presence of distraction, high noise level, jamming, or masking.	Can have performance degraded by interference so that it may fail entirely.
Can fill in lacking portions to supplement superficial training.	Pertinent facts and programming must be present and complete for accomplishment of function.
Can undertake new programs without extensive or precise programming.	Reprogramming must be as complete and precise as initial programming.
Can sometimes overcome effects of failure of one part of his or her nervous system through use of other parts.	Electronic systems will sometimes fail completely if only a single circuit element fails.
Can maintain himself or herself or requires comparatively little care.	Maintenance is always required and increases vvith system complexity.
Is small and iight in weight for all functions that can be performed, and requires little power.	Equivalent capabilities are generally heavier, bigger, and with high power and cooling requirements.
Is in good supply and inexpensive for most functions.	Complexity and supply are limited by cost and production time.
Can override preset procedures and plans if necessary or preferable.	Can accomplish only preprogrammed actions within their designed capabilities.
Can add reliability to system performance by ability to make repairs on associated equipment.	Generally has no repair capabilities.

FIGURE 29–1 Capabilities: Person vs. Machine

Can detect and sometimes correct own mistakes.	Machines make few mistakes once their programs have been checked out. Programs frequently have self-check routines.
Can sometimes tolerate overloads without complete failure; in other cases, performance deteriorates slowly.	Even small overloads can cause complete breakdown or disruption of operations.
Has high performance flexibility.	Performs only tasks for which it was built and programmed.
Performance can be degraded by fatigue. boredom, or diurnal cycling.	Performance will be degraded only by lack of calibration or maintenance.
Long repetitive tasks will impair performance.	Performs repetitive or precise tasks well.
Can refuse to perform even when capable of doing so.	Will always respond to proper instructions except when there is a malfunction.
Can detect low-probability events impracticable in machine systems.	Many unexpected events cannot be handled adequately because of the size and complexity of the equipment required.
Can exert comparatively small force. Generally cannot execute a large force smoothly or for extended periods.	Can exert large forces smoothly and precisely for almost any periods of time.
Is not adapted to high-speed search of voluminous information.	Searching of voluminous information is a basic function of computers.
Is interested in personal survival.	Lacks consciousness of personal existence.
Is emotional in relations with others and in stress situations.	Has no personal relations or emotions.
Performance may deteriorate with work-cycle duration.	Performance is impaired relatively little with long work cycles if maintenance has been adequate.
Great individual differences can take place in performances by different personnel.	Only very minor differences in performances will take place by similar types of machines.
Has certain sensing abilities machines do not have: smell and taste.	Range of abilities generally extends outside human limits for those abilities it does have: can see into infrared and ultraviolet
Quickly saturates capacity for accomplishing diversified functions.	Can be designed to accomplish a large number of functions at once. Ability to do each rapidly increases its ability to do many sequentially. Can frequently do many simultaneously.

FIGURE 29–1 Continued

equipment might be programmed to go around an obstacle such as a fence. A person could not only go around the fence, but climb over or burrow beneath it, selecting the optimal means and path. Conversely, when programmed erroneously, computers will continue to make mistakes repetitively.

The Los Angeles *Times* (Dec. 6, 1984) pointed out that lightning might follow wiring into a computer, and once inside, could burn up the fine wires inside microcircuits. At the very least, an electrical surge or "spike" of power might wipe out all data. Lightning and other strong electrical emissions, such as very strong radio waves or magnetic interference from industrial motors, might cause loss of information held in a computer's memory.

Because of adverse occurrences and considerations of what might happen, the article stated, "American scientists are becoming seriously worried about the fallibility of computer systems and the dangers of a fiasco . . . They are thinking big—at least a power plant meltdown; or at most an accidental nuclear war. . . ."

PROGRAMMING ERRORS

Because computers are machines that cannot make mistakes, erroneous computer actions are almost always the result of programming errors. And because software must be programmed by people, and people make errors, there may be programming errors in the software. To avoid errors or omissions, experienced programmers have developed certain basic features to incorporate, programming rules to observe to avoid faults, and tests to ensure the absence of faults.

AVOIDING HUMAN ERRORS

To minimize human errors in entering data into a computer, one technique is to repeat an entry and then have the computer compare the two entries to ensure they coincide. This is a common use of computers in that they can be programmed to detect human mistakes. Immediately after a possible input error, such as one by an operator, an error-detection capability can be included. Because operators might make errors, it would also be a programming error not to incorporate a detection capability at the interface between the operator and the equipment. Since input data are very likely to contain operator errors, it is advisable that the interface and computer have approval immediately after input data are entered and before any further human or computer operation is to be performed. Programmers frequently lack suitable knowledge of safety and fail to incorporate into software programs preventive measures to ensure accident-free operations.

SAFETY DATA PROCESSING

Computers lack emotion or consciousness of personal survival, health, safety, or heat or cold. They can perform indefinitely as long as proper maintenance and repair have been provided, and they are not subject to the boredom that often comes from long and repeated work with numbers.

Computers are far better than are humans for routine repetitive tasks, which they can do repeatedly, precisely, and well. Their performance does not deteriorate with operator fatigue or boredom. The operator provides data that are used as a base for future processing and operations. Subsequent data are updated, new data entered, stored, and used to provide outputs. The vast number of numerical calculations and uses of lists, comparisons, specific fields of interest, persons, or terms, charts, or other items is known as "number crunching." Information outputs can be provided as printed records or as inputs for use by other computerized equipment.

In the safety area, the first use of data processing with computers was probably for recording and analyzing great numbers of accident occurrences. Analysts could use them to determine the numbers that had taken place; names, sexes, ages; fatalities or degree of injury; general causes; type of employment of workers; time lost; monetary losses; and whatever other detail appeared interesting, pertinent, or otherwise desirable. The capabilities of computers permit them to search quickly through voluminous safety information, select and segregate applicable data, and present results as called for.

AVOIDING SAFETY PROBLEMS

When safety is involved, both hardware and software operations must be analyzed to ensure avoidance of computer outputs that could cause accidents. The initial step is the preliminary hazards analysis. First, all the units that could be affected by a controlled system should be listed, including all possible computer-initiated events and potential effects. The analysis might then review possible causes and effects of the adverse events and possible safeguards.

An adverse event may be the result of a hardware or software fault, or both. Both hardware and software contribute to control or prevent any action that could lead to a dangerous condition. Where the hardware might create a hazardous situation, it may be possible to abort, limit, or rectify the unsafe action through suitable software. If the safeguard provision has not been incorporated properly, the software program is deficient.

Computers can impose restrictions that permit, initiate, abort, lessen, or alter the flow of energy. An input signal generated by the exceeding of a limit, a photoelectric cell, an interlock being activated, or an interruption of some kind can also be tied through a software program to inhibit or permit further action. There have been cases where hardware faults (which may be in a computer chip) have provided inputs to the computer so that danger signals or aborts resulted. (Probably the most famous of these cases have been "false alarms" in missile air defense systems.) Computers should also be designed to indicate the presence of a fault.

COMPUTER CONTROLS AGAINST HAZARDS

Computers for control of safety can be highly beneficial. Properly programmed, computers might have avoided the severe damage that occurred to an Atlas F missile. Because of the rapid evaporation of the liquid oxygen (LOX) when first coming into contact with comparatively warm metal, an escape valve had to be opened before pumping began. The operator failed to do so, and as a result the pressure of the vaporized oxygen caused severe distortion damage to the thin-skinned missile. The error permitted (by poor design) could have been negated by proper computer programming to ensure pumping could not begin until after the valve opened.

A few devices besides interlocks that could be used to good effect include time delays, parameter sensing devices, sequential controls (as for the Atlas F), or those for machine activation, delay, or stoppage. In addition, computers can be used to provide vocal or visual warnings. Suitable programming can permit computers to control and take effective action in the event of a failure of a piece of equipment by shutting down any accident-causing operation or by activating a possible safeguard. Starts of operation of equipment can be prevented until all conditions called for in a program have been satisfied.

COMPUTERS AND HAZARD ANALYSES

Just as humans originated computers and methods, they have also developed computer programs to prepare or contribute to other safety analyses. Medical doctors have begun using computers to determine possible causes of illnesses. Potential causes of accidents could be established in the same way.

Computers are put to highly advantageous use not only for qualitative analyses but also for quantitative determinations, such as for fault-tree analyses. Computers have already been programmed to draw trees after inputs have been provided by analysts. But even more valuable are the rapid solutions of the complex Boolean arithmetical equations. Once the various interrelationships between events have been established, and the numerical values entered, the final probability of an accident can be determined. Any calculated designed end result not acceptable calls for a new design, changes can be made, and the results quickly recalculated until a desirable and acceptable design is achieved.

Another advantage found even with use of fault-tree analyses was the ability to trace and pinpoint potential or actual malfunction or accident causes. Computerized fault trees have enhanced the ability to search and find possible causes after malfunction symptoms or accidents have occurred. Such analyses have involved automobiles, aircraft, missiles, and other systems. Fault-tree analyses have been used to determine the probabilities of possible but unknown causes of accidents—for example, the wreck of an automobile in which everyone was killed. The probability of each possible contributing factor or event is entered (or eliminated if zero). The probabilities of all remaining final possible causes of the accident can then be listed, based on which investigations could be conducted.

SIMULATIONS

Simulations can be used to determine where hazard reductions would be most effective and what change in overall failure probability would result. Calculations by computers using fault trees for such efforts have been used often, chiefly for studies of safety systems in nuclear power plants. Major problems are that humans must still originate the designs (which may be erroneous), such plants are extremely complex, and the cost of computer programs may limit any effort. Complete fault-tree analyses and computer usages are extremely expensive.

SOFTWARE HAZARD CATEGORIES

A computer program may be incorrect so that a computer does not recognize a dangerous situation, incorrect inputs are not recognized, or a desirable operation may be inhibited. Software hazard categories include: unwanted event occurrence, out-of-sequence event occurrence, planned-event nonoccurrence, and wrong magnitude or direction of a planned event. These involve the failure to perform a required/requested function, performance of a nonrequired/nonrequested function, and performance of a wrong function to a requirement/request. It must be added that such errors could be caused by hardware. For example, a failure to respond to a request could be due to loss of communication with the computer due to hardware failure, and an incorrect response could occur because of incorrect signal interpretation due to background electrical noise or data loss.

SOFTWARE ANALYSIS

Software Life Cycle. System safety analysis is a discipline that is applied throughout the life cycle of a system—from cradle to grave. The phases of the life cycle—from system concept through system requirements determination, design, development, test, operation, change, further operation, and disposal—should include system safety analysis. It is an iterative process in which a phase, such as system operation may cause changes to system requirements, which result in design changes, etc. Software safety analysis is part of system safety and should also be considered throughout the system life cycle and its iterations. As user and hardware requirements are determined, software requirements should be defined. Hardware design and user layouts are accompanied by software design. Hardware fabrication and user documentation are accompanied by code writing. Software is integrated with hardware assembly and test and with user procedures development. These are subjected to system integration and test. As system safety analysis is applied throughout the system cycle and its iterations, so must system software analysis. The life cycle of software itself goes through the phases of concept, design, code, test, and maintain.

Software Safety Approach. The traditional approach to software safety has been to wait for a crash to occur, find the error that caused the crash, and fix it! Of special concern, of course, are not simply computer "crashes," but software failures that contribute to an accident or incident. The system safety approach is to investigate hazards, evaluate, and control software.

Elements of Software Analysis. MIL STD 882 B presents the basic elements of software hazard analysis program to be: software requirements hazard analysis, software design hazard analysis, code-level software hazard analysis, software safety testing, software/user interface analysis, and software change hazard analysis.

- Software requirements hazard analysis (SRHA):
  1. examines the system requirements, ensures that system safety requirements have been clearly defined,
  2. reviews the system's preliminary hazard list and preliminary hazard analysis,
  3. identifies software requirements documentation to determine that system safety requirements can be traced to the system specifications, software requirements specification, software design documents, software test plan, software configuration management plan, and the project management plan.

The report of this analysis provides recommendations and inputs to each of these.

- *Software design hazard analysis.* Software design hazard analysis (SDHA) identifies, defines, and analyzes safety-critical software components. The preliminary hazard analysis and SRHA are inputs to the SDHA. It should be completed before software coding is begun.
- *Code-level software hazard analysis.* Code-level software hazard analysis (CSHA) examines the actual source and object code, system interfaces, and software documentation.. The analysis results in recommendations to change the software design, code, and software testing. Software documentation recommendations to include safety requirements can also be made.
- *Software safety testing.* The possible existence of a hazardous condition can be identified to some degree by simulation and test. A hazard could be intentionally incorporated under controlled conditions. A proper response should be evinced by the computer even under the most severe circumstances. There should also be a means incorporated for identifying an erroneous output or a failure by a chip—for example, by a comparison routine.

  Both hard and software must be tested to ensure proper operations. Hardware, like all equipment and devices, varies in reliability. In most cases, computer problems have come about because of faulty "software" that acts as the brain of a system. With hardware that operates properly, software can direct reiterated operation so that any problems ("bugs") are gradually eliminated. Repeated uses of the accepted software program can then be made.

  However, computer specialists have said that testing a software program until it is completely correct to certify that it contains absolutely no errors is generally such a lengthy procedure that it is impracticable. One specialist described a very simple program for testing a small circuit involving three small "loops," some in series and some in parallel. To make the complete test of all possible paths would require 10^{20} different routes. Even if a person had started in A.D. 1 to check out all computer paths at the rate of one path per nanosecond (one-billionth or 10^{-9} second), the job would now be only half completed. Another writer pointed out that the major weakness of all software testing methods is their inability to guarantee a program has no errors. Apparently, then, more fruitful than testing is proper analysis to ensure that software faults are not built into any program.

  Software safety testing is designed to determine that all hazards have been eliminated or controlled during the system design phase. Safety-critical components are tested under normal and abnormal environment and input conditions to identify corrections needed to eliminate or mitigate hazards. The corrected software is then retested under the same conditions. Software testing is the controlled execution of a program with the intent to find errors. Boundary testing is testing on or near the boundaries of dimensions in the program. Equivalence testing is the division of software into equivalence classes and the testing of one member of the class, on the assumption that if it works, all will work. Positive tests are those in which the test input is valid; negative tests are those in which the test input is invalid. Software may be tested by testing a unit of the program, by testing selected integrated components of a system, or as part of the total system tests.

Certain units of the software require special attention. Safety inhibits, traps, interlock, etc. should be verified by test or simulation. Some advocate that safety-critical software functions be verified independently of the original programmer, to ensure objectivity.

- *Interface analysis.* Software/user interface analysis provides a means to control or manage hazards that were not eliminated or controlled in the system design phase. It recommends means to provide the user early detection, warning, and mitigating response or recovery from a hazardous condition or situation. Sometimes this is as simple as providing a way to shut down or terminate a process.
- *Software change hazard analysis.* Software change hazard analysis (SCHA) reviews all changes proposed or made to the software to determine their impact on safety. It is done to ensure that a change does not create a hazard or affect any existing hazards, as well as to make sure that it is properly incorporated into the code.

SOFTWARE HAZARD ANALYSIS TECHNIQUES

Table 29-1 identifies the analytic techniques cited by various sources.[120] Some of these techniques are briefly explained below.

Code Walk-Throughs. This is a team effort in which a programmer leads the team members through a code. The members ask questions and make comments to improve the code and identify possible errors. A code walk-through for software safety analysis should be a dedicated effort to examine, particularly, safety-critical code segments or modules. A safety professional should be part of the walk-through team.

Event Tree. This technique models the sequence of events that results from a single initiating event. When applied to software, that event(s) is taken from a segment(s) of the code that is safety critical, suspected of error or code inefficiencies. For example, Fig. 29-2 is a partial event tree related to the incident mentioned at the start of this chapter, in which the failed sequence of clamp release, missile ignition, clamp close caused the fighter plan to be thrown out of control.

This is a partial tree because it is unfinished and not a rigorous detailing of events. It is used for example. It should be taken to completion so that the events that

TABLE 29-1 Sources That Specify Software Hazard Analysis Techniques

Source	Technique(s)
AFISC SSH 1-1, "Software System Safety," Headquarters Air Force Inspection and Safety Center, Sept. 1985	Nuclear Safety Cross-Check Analysis Petri Nets Software Fault Tree (Soft Tree) Software Sneak Circuit Analysis Desk Checking Code Walk-Through Structural Analysis Proof of Correctness
FDA 89 "(DRAFT) Reviewer Guidance for Computer-Controlled Devices," Medial Device Industry Computer Software Committee, Jan. 1989	Code Walk-Throughs Failure Mode, Effects, and Criticality Analysis
IECWG9'91 "Software for Computers in the Application of Industrial Safety-Related Systems," British Standards Institution, Sept. 1991	Cause Consequence Diagrams Event-Tree Analysis Failure Mode, Effects, and Criticality Analysis Fault-Tree Analysis Hazard and Operability Studies Monte-Carlo Simulation
IEEEP1228-C, D, E, "Draft Standard for Software Safety Plans," IEEE Standards Dept., Nov. 1990, Mar. 1991, Jul. 1991	Event-Tree Analysis Failure Modes and Effects Analysis Fault-Tree Analysis Petri Net Analysis Sneak Circuit Analysis
MIL STD 882B, "System Safety Program Requirements," Dec. 1983	Code Walk-Throughs Cross Reference Listing Analysis Design Walk-Throughs Nuclear Safety Cross-Check Analysis Petri Net analysis Software Fault Tree Software/Hardware Integrated Critical Path Software Sneak Analysis

Adapted from: L.M. Ippolito and D.R. Wallace, "A Study on Hazard Analysis in High Integrity Software Standards and Guidelines," U. S. Department of Commerce, National Institute of Standards and Technology, NISTIR 5589, Jan. 1995.

follow after events such as "Pilot D/N initiate emergency procedures" are identified, and it should be in greater detail. However, in its partial state, it plainly shows that this sequence of events can be disastrous. If the clamp does not (D/N) release the missile before the missile ignites, the plane is thrown out of control. Even if the missile does not ignite, the pilot has a "hang-fire" emergency. In the event sequence described at the start of this chapter, the clamp opened, failed to release the missile, closed, then the missile fired. A simple event tree, such as the one shown above, would have alerted the software designer to the need to verify that the missile is away before initiating the "missile ignite" command.

The initiating event is found in the safety-critical code segment that contains the command to "Fire" the missile. The event tree developed from that single event will show the sequence system timing and that verification of "clamp open" and "missile

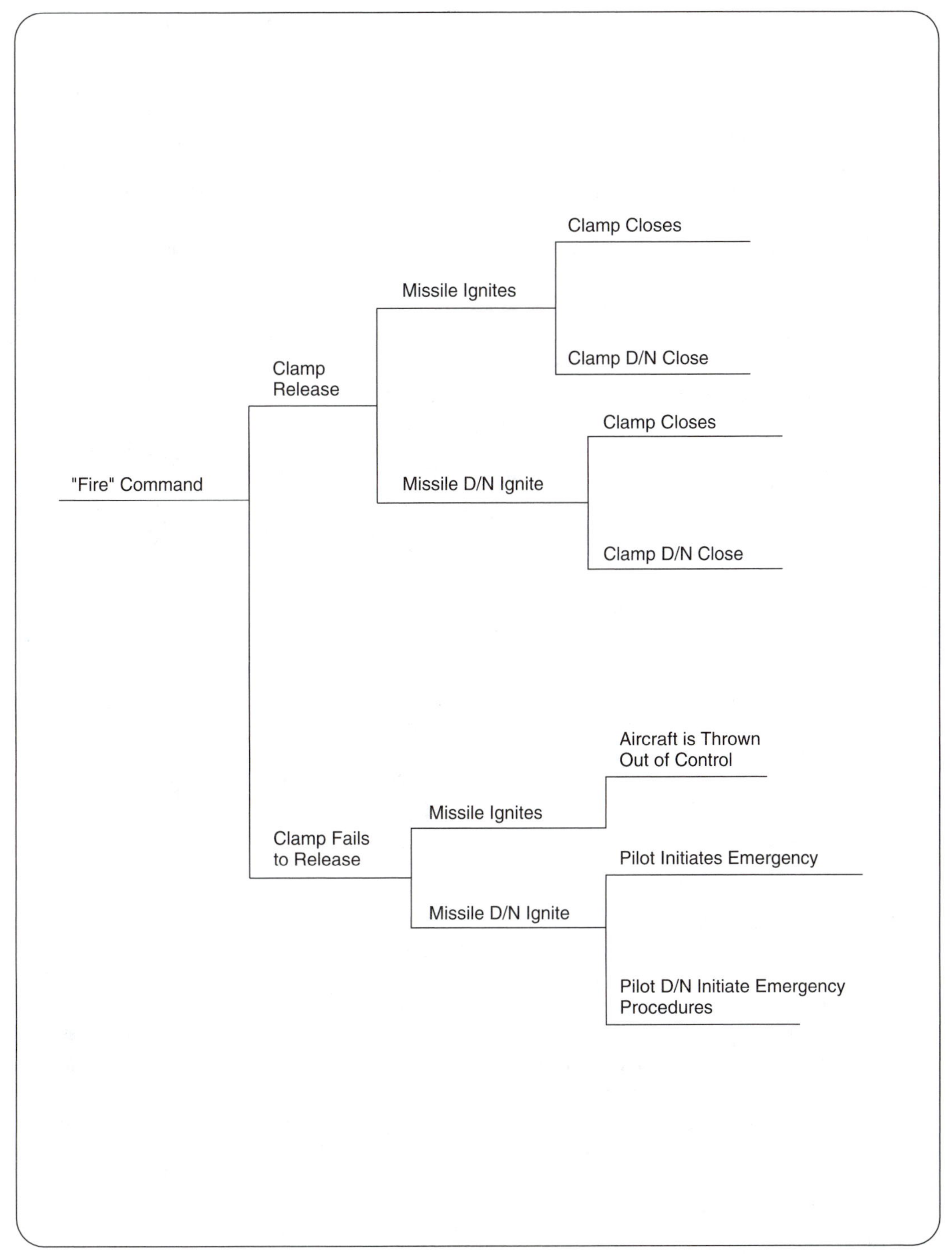

FIGURE 29–2 Event-Tree (Partial) Example

away" by this software are critical to assure that the missile is free of the aircraft before missile ignition occurs.

The event tree starts with the selected initiating event (either desired or undesired) and develops the consequence of each subsequent event through considering the system/component failure-and-success alternatives. The top sequence of events in the figure above follows the success alternatives of a "normal" sequence. All other routes in the figure follow "off normal" or failure-impaired routes and their consequences.

Fault-Tree Analysis (Soft Tree). The fault-tree analyst should include both software and hardware in the tree development. Elementary aspects of fault-tree analysis (FTA) have already been covered in this book. The idea of a "soft tree" might imply to some that analysis of software is separate from hardware in a system. Care must be exercised to ensure that software is considered as part of the overall system safety. The goal of the software FTA is to demonstrate that software logic will not produce system safety failures.

A review of a Preliminary Hazard Analysis (referred to previously in this book), Software Failure Modes and Effects Analysis (FMEA, discussed elsewhere in this book), and/or the hardware FTA should be conducted to determine the top events in the system which have a software interface. A top event is selected for analysis based upon a classification of its system impact, usually on a scale from "catastrophic to negligible." The application of this technique is expensive and must be used with stewardship of resources. The software program flow is followed in reverse to include only those conditions that would cause the top event selected for analysis. These might include potential: input errors, processing errors (such as incorrect signal generation, unexecuted/improperly executed code), and output errors. The fault-tree is constructed according to the rigor and rules of fault-tree analysis. The analyst can use the FTA evaluation techniques of cut-set analysis and common-cause analysis to understand the software contributions to paths to the top undesired event. The analyst can examine paths to the top event that involve software, hardware, and procedures to make recommendations designed to eliminate the paths or make them less likely to occur.

Nuclear Safety Cross-Check Analysis. This technique was developed originally to evaluate military nuclear systems; however, it is generally useful. It has two components: technical and procedural. The technical component is designed to ensure that system safety requirements are met. The procedural component is designed to provide protection and security for critical software components.

The technical component analyzes and tests the software. The extent to which each software function affects safety objectives is first determined. Then the software is decomposed to the lowest-level functions. These are examined, and those which do not affect critical events are not examined further. A criticality matrix is developed which plots software functions against safety objectives. Each cell in the matrix assigns influence ratings of "high, medium, or low." Each software function is given recommendations for evaluation techniques which should be applied to them.

The procedural component is dedicated to security and control measures. Background investigations for personnel clearances, configuration control and document security, facility security, and product control are instituted. Both components are conducted by an organization that is independent of the software developer. Thus it gets its name because it is an independent cross-check that the system software contains no improper design, programming, fabrication, or application that could contribute to loss of safety.

Software Sneak Circuit Analysis. Program source code is converted into topological network trees. Six basic patterns are used to model the code, as shown in Fig. 29-3.

Each node of the software is modeled using these patterns linked in a network tree that flows from top to bottom. The analyst then asks questions about the use and interrelationships of the instructions that are elements of the structure. The answers provide clues that identify typical sneak conditions that could produce undesirable outputs. The analyst searches for four basic types of software sneaks:

1. The occurrence of an undesired output
2. The undesired inhibit of an output
3. Incorrect timing
4. A program message that does not adequately describe the actual condition

The clue-producing questions are taken from the topograph that represents the code segment. For example, if the code segment is modeled by the "single line," some of the questions the analyst might ask are shown below.

1. Can the segment be executed?
2. Are the proper parameters available for entry?
3. Are the proper parameters available for exit?
4. Are the proper data available for operations in the segment?
5. Are data destroyed in the segment? Is this correct?
6. Does the segment implement the functions shown in supporting documentation?

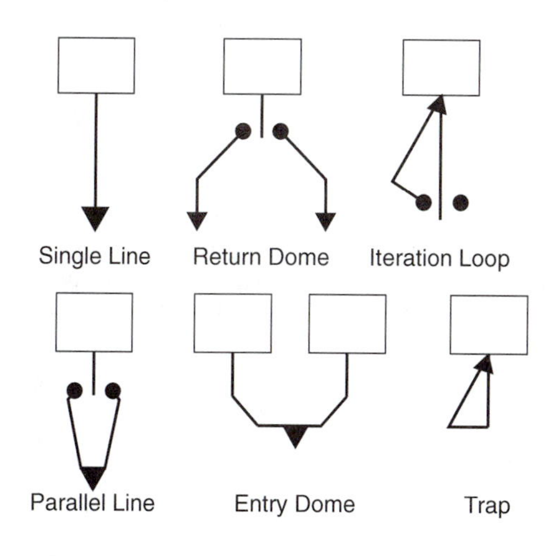

FIGURE 29–3 Sneak Analysis Code Patterns

When a sneak is found, the analyst verifies that the code does in fact produce the sneak. The impact of the sneak is then assessed and corrective action recommended. Code analysis uses the methods of desk checking, code walk-through, structural analysis, or proof of correctness. Code walk-through has been described above.

Desk Checking. This is a poorly defined method which looks at a program for faults and compliance to requirements, performs arithmetic calculations to verify output-value correctness, and manually simulates program execution in order to understand and verify the program logic.

Structural Analysis. This method applies an automated tool designed to analyze the program for errors in its structural makeup. Two types of software testing might be "black-box" testing and "white/glass-box" testing. The former tests the program for functionality without getting into the program itself. The latter looks at the structure of the program. Therefore, some refer to this analysis as "white-box" or "glass-box" analysis.

Proof of Correctness. The program is broken into logical segments. The input and output assertions for each segment are defined. The analyst then verifies that each input assertion and its related output assertion are true and that if all input assertions are true, then all output assertions are true. It uses mathematical theorem-proving concepts to verify that a program is consistent with its specification.

The above are samples of some of the techniques mentioned in various standards. Other techniques such as Failure Modes and Effects Analysis are simply the application of system safety analysis to the software arena. Others, such as Petri Net Analysis modeling using reachability graphs, are useful but are not included here because of space limitations.

Safety Critical. It should be evident that a basic aid to analyzing software safety is a well-written, well-documented, carefully modularized program. Documentation and modularization should be designed to clearly identify those parts which are safety critical. Safety-critical parts, according to MIL STD 882B, are those software operations that, if not performed, performed out of sequence, or performed incorrectly could result in improper control functions which could directly or indirectly cause or allow a hazardous condition.

TAILORING SOFTWARE ANALYSIS

The thoroughness and extent of the software analysis needed is determined by what the software controls, the consequence severity, and the program size and complexity, according to Michael Brown.[121] Categories of software energy control are given in Table 29-2.

Severity-of-consequence categories often used in system safety analysis are:

I Catastrophic

II Critical

III Marginal

IV Negligible

TABLE 29-2 Software Control Categories

I	Autonomous control over potentially hazardous energy, without likely human intervention that would preclude the occurrence of a hazard
IIa	Control over potentially hazardous energy, allowing time for human intervention or the intervention of safety systems to mitigate the hazard
IIb	Software displays information requiring immediate operator action to mitigate a hazard
IIIa	Software issues commands over potentially hazardous energy requiring human action to complete the control function
IIIb	Software generates information of a safety-critical nature used to make safety-critical decisions
IV	Software does not control safety-critical functions and does not provide safety-critical information

Adapted from: M.L. Brown, "Tailoring MIL-STD-882B 300 Series Tasks," *Proceedings of the Ninth International System Safety Conference,* July 1989.

These two ratings, Brown points out, can then be put into a matrix to arbitrarily assign criticality ratings on a scale of 1 to 5, with 1 being most critical, as shown here.

		Software Control			
		I	II	III	IV
Hazard	I	1	1	2	3
	II	1	2	3	4
	III	3	4	5	5
	IV	5	5	5	5

Next the program size and complexity is categorized: small (less than 2,500 lines), medium (2,500 to 10,000 lines), large and complex (10,000 to 100,000 lines with 100 to 1,000 modules), very large and complex (100,000 to 1 million lines with extensive documentation required and years to develop), and extremely large and complex.

Finally, the criticality ratings and program size and complexity are used to determine how extensive and detailed the software analysis can be, and what analytical strategies can be used, given the resources available. Judgments must be made on a case-by-case basis, but obviously criticality ratings of 1 and 2 will require more effort in analysis than 4 and 5, and small and medium programs can be analyzed with different strategies and (sometimes) levels of detail than can the larger and more complex programs.

ROBOTS AND ACCIDENTS

Robots have advanced from mythology and the theater to acceptance in industry and for action in hazardous situations. So-called first-generation robots had limited uses, generally in conjunction with operators who guided them. Those robots could not adapt to changing situations because they could not "see" or "feel."

Second-generation robots are controlled by minicomputers and can "sense" and adapt to changing conditions. The most useful "sensing" abilities that computers emulate have been found to be vision and touch. Computers that "see," for example, have inspected, sorted, and guided in such operations as following and welding irregular seams. Robots have been used not only in hazardous environments, such as for spray painting or near nuclear materials, but also for handling and disposing of items such as bombs and other explosives.

Another advantage of robots is to decrease the distances normally used for passage of humans and thereby to save space. In the case of the Japanese company where the maintenance man was killed, there had been little room for the worker. Since robots are mindless (and inexorable), they will only do what they are programmed to do. If a human is to be present at any time, a major advantage of use of computers is lost. The worst situation in any case is where there is an intermix of operations by humans and computers.

Possible problems resulting in accidental damage have been postulated. A simple example which might have used a first-generation computer can be used to illustrate problems in programming that have to be considered. (Later computers and growing experience have superseded this state of ability.) Suppose a computer had been designed and built to direct a robotic machine to make electric spot welds in an automobile fabrication plant. The following would be only a few of the many programmed actions that would be required: the robot would first have to position the welding electrode properly, then to provide suitable electrical power, and finally to shut it off at the proper instant when the operation was completed. The robot would first position the welding electrode in elevation and then rotate it in azimuth. After that, the electrode would be extended until it was close to the work to be welded. Electric power would be applied and then shut off at suitable times.

Assuming the software program had been adequately tested and all "bugs" eliminated, an analysis might be made of the problems that could occur. Any damage that might result would probably be due to interference or multiple failures. The types of possible problems include: loss of proper directional control, so that welding action occurs in the wrong location; firing when the electrode is too far from the location where the weld is to be made, with no weld or a poor one. If there is no firing after a proper signal, there will be no weld. If its duration is too short, the weld may be weak. On the other hand, if the power output is too great or too long, it may cause a metal burn. Such an analysis indicates the safety measures that should be incorporated into the programming.

Robot safety analysis requirements are the same as for any computer-controlled equipment or system. There are special concerns for the unique hazards they present. They require careful analysis to determine guarding, tagout/lockout, and emergency shutdown equipment and provisions. They may involve unexpected movements that demand special safety provisions. For example, the robot may spring into action suddenly after a dwell-time and surprise someone within its reach. The danger zone of a robot includes not only its reach in normal operations, but also its maximum reach limits and beyond, because it can throw things from its grip. Failure modes and effects must be examined carefully to avoid consequences such as the robot "throwing things." Maintenance requirements must be examined carefully with job safety analysis. "Teaching" and "programming" routines require careful safety analysis. Administrative controls for personnel exposure and training must be considered carefully.

FOR THE FUTURE

One use of the computerized equipment being looked into has been for paraplegics and other handicapped persons. Equipment has been voice controlled. Production facilities, unmanned, except for maintenance men and workers to receive and ship material, are already in use, especially for large chemical plants. As pointed out much earlier, production plants will continue to grow larger and be of greater complexity so that operations can be controlled only by computers.

New and larger vehicles of all types will be employed within the earth's atmosphere and under the seas, such as for offshore deepwater oil drilling. Orbital and outer space journeys, manned and unmanned, will be used increasingly. The Viking Lander permitted use of a robot on Mars, where the atmosphere was unbearable for unsuited men, digging, collecting, and analyzing soil samples and studying the weather.

The most important and remaining portions of all operations, those that cannot be accomplished by computers, will still be conceived and designed by humans. It is also there that accidents will be initiated: through engineering deficiencies, lack of proper procedures, or inadequate programming.

BIBLIOGRAPHY

[119] Charles L. Piechota, "The Twilight Zone," *Professional Safety,* Jan. 1992, pp. 32–35.

[120] L. M. Ippolito and D. R. Wallace, "A Study on Hazard Analysis in High Integrity Software Standards and Guidelines," U. S. Department of Commerce, National Institute of Standards and Technology, NISTIR 5589, Jan. 1995.

[121] Michael L. Brown. "Tailoring MIL-STD-882B 300 Series Tasks," *Proceedings of the Ninth International System Safety Conference,* July 1989.

EXERCISES

29.1. What is "code walk-through"? How is it useful in software safety analysis?

29.2. What hazards exist in the use of computers?

29.3. List three aspects of computers for which their advantages make them preferable to humans.

29.4. List three aspects where humans are better.

29.5. Describe some ways computers can be used in hazardous situations.

29.6. How can computers be used as safeguards to prevent accidents?

29.7. A home telephone bill for $8 million is received. The company's printout was evidently an error. Where do you believe the error might have been in the computer operations?

29.8. How can computers be used with fault-tree analyses to study new designs and probabilities of accidents? In attempts to find unknown causes of accidents?

29.9. What is the chief use of computers in accident recording, determinations, and studies?

29.10. A premilinary hazards analysis was made for computerized welding of automobiles. Describe some of the hazards. What failures might have been possible in the hardware? Software? What safeguards could have been incorporated, and where?

29.11. How can computer simulations be used in safety matters?

29.12. Explain the use of event-tree analysis applied to software safety.

29.13. Describe Software Requirements Hazard Analysis, Software Design Hazard Analysis, and Code-Level Software Hazard Analysis.

Bibliography

[1] University of Michigan, Survey Research Center, *Survey of Working Conditions,* prepared for the U.S. Dept. of Labor, Employment Standards Administration, Washington, D.C., Aug. 1971.

[2] H. Hollister and C. A. Traut, Jr., "On the Role of System Safety in Maintaining 'Affordable' Safety in the 1980s," SAND 79–1671 C (at the Fourth International System Safety Conference, San Francisco, CA, 1979).

[3] R. N. Anderson, K. Kochanek, and S. L. Murphy, "Report of Final Mortality Statistics, 1995," *Monthly Vital Statistics Report,* Vol. 45, No. 11, Supplement, 2, (PHS) 97–1120, 1997.

[4] E. E. Ludwig, "Designing Process Plants to Meet OSHA Standards," *Chemical Engineering,* Sept. 3, 1973.

[5] *Hazards Survey of the Chemical and Allied Industries* (Technical Survey No. 3), American Insurance Association, New York, 1979.

[6] CTD News, "News Briefs: Workers Win $10.6 Million in Lawsuit," *LRP Publications,* Oct. 1997, Vol. 6, No 10.

[7] D. L. Price, "Risky Business: Creating a Safe Environment," *Personnel,* Nov. 1986, American Management Association publication, pp. 62–67.

[8] E. L. Bowers, *Is It Safe to Work?* (Boston: Houghton Mifflin Company, 1930), p. 170.

[9] New York Employers' Liability Commission, *First Report,* 1910, Vol. 1, pp. 19 ff.

[10] H. M. Somers and A. R. Somers, *Workmen's Compensation* (New York: John Wiley & Sons, Inc., 1945).

[11] C. Eastman, *Work Accidents and the Law* (New York: New York Charities Publication Committee, 1910), p. 1783.

[12] D. Gagliardo, *American Social Science* (New York: Harper & Row, Inc., 1949).

[13] Texas Sunset Advisory Commission, "Texas Workers' Compensation System Report," 6 Dec. 1996 revision, 6 pp.

[14] American Express Company, "Keeping Worker's Compensation Costs Down," 1997, http://www.aexp.com/smallbusiness/resources/managing/workcomp.shtml.

[15] N. H. Seigel, "The Fellow-Employee Suit: A Study in Legal Futility and Misguided Legislation," *Risk Management Today* (New York: American Management Association, 1960).

[16] Monroe Berkowitz and J. F. Burton, Jr., "The Income Maintenance Objective in Workmen's Compensation," *Industrial and Labor Relations Review,* Oct. 1970, pp. 3 ff.

[17] The National Commission on State Workmen's Compensation Laws in *Compendium on Workmen's Compensation,* Washington, D.C., 173, p. 103.

[18] *Business Insurance,* Feb. 26, 1973, p. 1, col. 4.

[19] H. W. Heinrich, *Industrial Accident Prevention,* 3d ed. (New York: McGraw-Hill Book Company, 1950), p. 18.

[20] J. S. Lee and W. N. Rom, eds., *Legal and Ethical Dilemmas in Occupational Health* (Ann Arbor, MI: Ann Arbor Science Publishers, 1982).

[21] *Federal Register,* Vol. 37, No. 202, Oct. 18, 1972, Part II, is titled "Occupational Safety and Health Standards," whereas Vol. 30, No. 75, Apr. 17, 1971, is titled "Safety and Health Regulations for Construction," although the types of requirements are the same.

[22] Extracted from testimony before the Joint Committee on Atomic Energy, Congress of the United States, Washington, D.C., March 19–20, 1970.

[23] David Pittle, Commissioner, Consumer Product Safety Commission, remarks to 1977 Design Safety Conference, Chicago, Illinois, May 9–12, 1977.

[24] Ralph Manaker, "Standards—Effects on Liability," *Proceedings of the 1974 Product Liability Prevention Conference* (Syracuse, NY: Law Offices of Irwin Birnbaum), p. 203.

[25] Jerome Lederer, Guggenheim School of Aeronautics, "Infusion of Safety into Aeronautical Engineering Curricula," *Third International Conference, Royal Aeronautical Society,* Brighton, England, Sept. 3–14, 1951.

[26] "Operator Safety," in *Engineering,* May 1974, pp. 358–363.

[27] W. T. Fine, *A Management Approach to Accident Prevention,* White Oak TR 75–104 (AD A 014562) (Silver Springs, MD: Naval Surface Weapons Center, July 1975).

[28] Fred A. Manuele, *On the Practice of Safety* (New York: Van Nostrand Reinhold, 1993), p. 132.

[29] Aerojet Nuclear Company, *Human Factors in Design,* Idaho Falls, Idaho, Feb. 1976.

[30] Dan Cordtz, "Safety on the Job Becomes a Major Job for Management," *Fortune,* Nov. 1972, p. 112.

[31] *From Accident Prevention—A Worker's Education Manual,* International Labour Office, Geneva, Switzerland, 1961.

[32] I. Fredericks and D. McCallum, "International Standards for Environmental Management Systems: ISO 14000," *Canadian Environmental Protection,* Aug. 1995.

[33] Ronald P. Blake, *Industrial Safety* (New York: Prentice-Hall, 1943).

[34] Roger L. Brauer, "Educational Standards for Safety Professionals," *Professional Safety,* Sept. 1992.

[35] U. S. Department of Labor, Bureau of Labor Statistics, "Older Workers' Injuries Entail Lengthy Absences from Work," http://stats.bls.gov/osh, March 1998.

[36] Japan Industrial Safety and Health Association 1996 Report, www.jisha.or.jp.

[37] Donna Kotulak, "On the Job Older Is Safer," *Safety and Health,* Oct. 1990, pp. 29–33.

[38] Ronet Bachman, *National Crime Victimization Survey: Violence and Theft in the Workplace,* U.S. Department of Justice, Washington, D.C., 1994.

[39] OSHA, "Guidelines for Workplace Violence Prevention Programs for Night Retail Establishments," DRAFT, U.S. Department of Labor, Occupational Safety and Health Administration, June 28, 1996.

[40] OSHA, *Workplace Violence Awareness and Prevention,* U.S. Department of Labor, Occupational Safety and Health Administration, Washington, D.C., www.osha-slc.gov, Apr. 1998.

[41] "How to Pick Women Who Can Drive Cars," *Literary Digest,* April 5, 1924, p. 58.

[42] M. S. Schulzinger, *The Accident Syndrome* (Springfield, IL: Charles C. Thomas, Publisher 1956).

[43] Alkov, "The Life Change Unit and Accident Behavior," *Lifeline,* U.S. Naval Safety Center, Norfolk, VA, Sept./Oct. 1972.

[44] J. E. Davidson et al., "Intriguing Accident Patterns Plotted Against a Background of Natural Environment Features," SC-M-70–398, Sandia Laboratories, Albuquerque, NM, Aug. 1970.

[45] George Thommen, *Is This Your Day?* (New York: Crown Publishers, Inc., 1973).

[46] Willie Hammer, "Missile Base Disaster," *Heating, Piping, and Air Conditioning,* Dec. 1968.

[47] OSHA, *Bloodborne Pathogens Final Standard: Summary of Key Provisions,* U.S. Department of Labor, Occupational Safety & Health Administration, Fact Sheet: 92–46, 1992.

[48] *Industrial Safety & Hygiene News,* 14th Annual White Paper, Dec. 19, 1997, p. 14.

[49] E. Scott Geller, "How to Motivate Behavior for Lasting Results," www.safetyonline.net/ishn/9603/behavior.html, and in "Total Safety Culture," *Professional Safety,* Sept. 1994, pp. 18–24.

[50] Frank Rushmore, *Fire Aboard* (London: The Technical Press Ltd., 1961), p. 26.

[51] W. E. Tarrants, *Utilizing the Critical Incident Technique as a Method of Identifying Potential Accident Causes,* U.S. Department of Labor, Washington, D.C.

[52] National Institute of Occupational Safety and Health, "Self Evaluation of Safety and Health Programs," U.S. Government Printing Office, Washington, D.C.

[53] "Accident Facts," National Safety Council, Chicago, IL.

[54] Chauncey Starr, "Social Benefit versus Technological Risk," *Science,* vol. 164, Sept. 19, 1969, pp. 1232–1238.

[55] National Research Council, *Improving Risk Communication* (Washington, D.C.: National Academy Press, 1989).

[55a] Willie Hammer, *Product Safety Engineering and Management* (Englewood Cliffs, NJ: Prentice-Hall, 1980), p. 113.

[56] American National Standards Institute, *ANSI Z535.2–1998 Standard for Environmental and Facility Safety Signs,* New York.

[57] W. G. Johnson, "The Management Oversight and Risk Tree" in *Accident/Incident Investigation Manual,* prepared for U.S. Energy Research and Development, ERDA-76-20, Aug. 1, 1975.

[58] J. Ranill, "Cooperative Fire Fighting," *Chemical Engineering,* May 31, 1971.

[59] System Safety Society, *System Safety Analysis Handbook,* July 1993.

[60] Electric Power Research Institute and Science Applications International Corporation, *CAFTA for Windows Fault Tree Analysis System User's Manual,* July 1995.

[61] W. H. Muto, B. A. Caines, and D. L. Price, *System Safety Analysis of an Underwater Marine Tool,* VPISPO/NAVSWC-78-1, 91 pp., 1978.

[62] D. Conger and K. Elsea, *MORT User's Manual,* 1998, Conger & Elsea, Inc., 9870 Highway 92, Ste 300, Woodstock, GA 30188.

[63] 29 Code of Federal Regulations 1910, 119, "Process Safety Management of Highly Hazardous Chemicals," *Federal Register,* Vol. 57, No. 36, Feb. 24, 1991.

[64] C. Bullock, F. Mitchell, and B. Skelton, "Developments in the Use of the Hazard and Operability Study Technique," *Professional Safety,* Aug. 1991, pp. 33–40.

[65] Arthur D. Little, *HAZOPtimizer: Documentation and Reporting Software,* 1991.

[66] E. M. Roth, W. G. Teichner, and R. L. Craig, *Compendium of Human Responses to the Aerospace Environment* (Albuquerque, NM: Lovelace Foundation for Medical Education and Research for the National Aeronautics and Space Administration, NASA CR-1205).

[67] C. S. White and I. G. Bower, *Comparative Effects Data of Biological Interest* (Albuquerque, NM: Lovelace Foundation for Medical Education and Research, Apr. 1959).

[68] U. S. Department of Labor, Occupational Safety and Health Administration, Remarks of Joseph A. Dear before the American Psychological Association Conference "Work, Stress, & Health '95," Sept.14, 1995.

[69] E. R. Tichauer, "Occupational Biomechanics," Rehabilitation Monograph No. 51, The Center for Safety, School of Continuing Education and Extension Services, New York University, New York, NY, 10003, 1975.

[70] NIOSH, "Work-Related Musculoskeletal Disorders," NIOSH FACTS, NIOSH Publications Office, May 1997.

[71] National Institute of Occupational Safety and Health, *Musculoskeletal Disorders (MSDs) and Workplace Factors,* ed. B. P. Bernard, NIOSH Publications Dissemination, Cincinnati, Ohio 45226–1998.

[72] K. Kroemer, H. Kroemer, and K. Kroemer-Elbert, *Ergonomics* (Englewood Cliffs, NJ: Prentice-Hall, 1994), 766 pp.

[73] NIOSH, "Carpal Tunnel Syndrome," NIOSH FACTS, NIOSH Publications Office, June 1997.

[74] Thomas J. Armstrong, "An Ergonomics Guide to Carpal Tunnel Syndrome," American Industrial Hygiene Association, 1983.

[75] NIOSH, "Workplace Use of Backbelts, Review and Recommendations," DHHS [NIOSH] Publication No. 94-122, NIOSH Publications Office, 1994.

[76] J. F. Kraus et al., "Reduction of Acute Lower Back Injuries by Use of Back Supports," *International Journal of Occupational and Environmental Health,* Vol. 2, pp. 264–273, 1996.

[77] NIOSH, "Elements of Ergonomics Programs: A Primer Based on Workplace Evaluations of Musculoskeletal Disorders," DHHS (NIOSH) Publication No. 97-117, March 1997.

[76] T. G. Blocker, Jr., *Studies on Burns and Wound Heating* (Austin: University of Texas, 1965).

[77] J. B. Perkins et al., *The Relationship of Time and Intensity of Applied Thermal Energy to the Severity of Burns* (Rochester, NY: University of Rochester, 1952).

[78] U. W. Turnbull et al., *Crash Survival Design Guide,* Flight Safety Foundation, U.S. Army Aviation Material Laboratories, Technical Report 67-22, July 1967.

[79] J. F. Wing, *A Review of the Effects of High Ambient Temperature on Mental Performance,* Aerospace Medical Research Laboratories, AD-624-144, Sept. 1965.

[80] F. P. Ellis, *Personnel Research in the Royal Navy, 1939–1945,* Royal Navy Research Committee, London, March 1950.

[81] W. V. MacFarlane, "General Physiology Mechanisms of Acclimatization," in *Medical Biometeorology,* ed. S. W. Troup (New York: Elsevier Publishing Co., 1963), pp. 372–417.

[82] C. Brian Malley, "Cold Stress Revisited," *Professional Safety,* American Society of Safety Engineers, 1992, pp. 21–23.

[83] J. S. O'Connor II and Kim Querrey, "Heat Stress and Chemical Workers: Minimizing the Risk," *Professional Safety,* American Society of Safety Engineers, 1993, pp. 35–38.

[84] *Life Line,* U.S. Navy Safety Center, Norfolk, Va., Nov./Dec. 1973.

[85] D. Hunter, *The Diseases of Occupations* (Boston: Little, Brown, and Company).

[86] *Fire in the United States*, U.S. Fire Administration, U.S. Dept. of Commerce, Washington, D.C., Dec. 1978.

[87] U.S. Air Force, *Handling and Storage of Liquid Propellants,* AFM 160-39.

[88] M. J. Miller, "Risk Management and Reliability," *Third International System Safety Conference,* Washington, D.C., Oct. 1977.

[89] Naval Air Systems Command Manual 06-30-501, *Technical Manual of Oxygen/Nitrogen Cryogenic Systems,* July 1, 1968.

[90] F. S. Smith, *Air Quality Criteria for Propellants and Other Chemical Substances,* U.S. Air Force, Dec. 1965.

[91] R. J. Lewis, *Sax's Dangerous Properties of Industrial Materials,* Tenth Edition, New York: John Wiley, 1999.

[92] Robert Finch, "Technological versus Natural Destruction," *Technology Review,* Oct./Nov. 1977, pp. 16 ff.

[93] National Safety Council, *Reporting on Climate Change: Understanding the Science,* Chapter 1, Environmental Health Center: A Division of the National Safety Council, 1025 Connecticut Avenue, NW, Suite 1200, Washington, D.C., July 1997.

[94] National Aeronautics and Space Administration, "December 1997 is Coldest Month on Record in the Stratosphere," http://science.msfc.nasa.gov, Jan. 20, 1998.

[95] R. A. McFarland, *Human Factors in Air Transportation* (New York: McGraw-Hill Book Company, 1953).

[96] James S. Arnold, "Computer-Calculated Environments," *Machine Design,* Apr. 25, 1963.

[97] Derek Lawrence, "Nature's Artillery," *Engineering,* Aug. 1972, p. 764.

[98] R. K. Lemmler, "Protect Your Drive Systems from Hostile Environments," *Automation,* Aug. 1972, p. 42.

[99] T. E. Graedel and B. R. Allenby, *Industrial Ecology* (Englewood Cliffs, NJ: Prentice Hall, 1995), 412 pages.

[100] A. J. Suruda, D. N. Castillo, J. C. Helmkamp, and T. A. Pettit, "Epidemiology of Confined-Space-Related Fatalities," in *Worker Deaths in Confined Spaces,* National Institute for Occupational Safety and Health, Jan. 1994.

[101] National Institute for Occupational Safety and Health, *Criteria for a Recommended Standard, Working in Confined Spaces,* DHHS (NIOSH), 80–106, 1979.

[102] National Institute for Occupational Safety and Health, *Request for Assistance in Preventing Ocupational Fatalities in Confined Spaces,* DHHS (NIOSH), 86–110, 1986.

[103] T. A. Pettit and H. E. Linn, eds.,. *A Guide to Safety in Confined Spaces,* National Institute for Occupational Safety and Health, DHHS(NIOSH), 87–113, 1987.

[104] J. C. Manwaring and C. Conroy, "Occupational Confined-Space-Related Fatalities: Surveillance and Prevention," *Journal of Safety Research,* Vol. 21, pp. 157–165, 1990.

[105] A. Suruda and J. Agnew, "Deaths from Asphyxiation and Poisoning at Work in the United States 1984–6," *British Journal of Industrial Medicine,* Vol. 46, pp. 541–546, 1989.

[106] NIOSH, *Worker Deaths in Confined Spaces: A Summary of NIOSH Surveillance and Investigative Findings,* National Institute for Occupational Safety and Health, 1994, 273 pp.

[107] Code of Federal Regulations, 29 CFR 1910.146 [1993] *Permit-Required Confined Spaces,* U.S. Government Printing Office, Washington, D.C.

[108] American National Standards Institute, *Safety Requirements for Confined Spaces,* American National Standard, Z-117.1, 1977, New York.

[109] National Safety Council, *Confined Spaces: Training the Team,* Chicago, IL, 1991.

[110] Mr. Polly Story, President, the American Society of Radiologic Technologists, to the United States Senate Committee on Commerce, March 8, 1973.

[111] M. N. Gupta and H. Singh, *Ocular Effects and Visual Performance in Welders,* Directorate General Factory Advice Service and Labour Institutes, General Labour Institute, Sion-Bombay, India, 1968.

[112] *Slawinski v J. H. Williams Company,* 208 N.Y. 546 (1948).

[113] J. G. Casali, and G. S. Robinson, "Noise in Industry: Auditory Effects, Measurement, Regulations, and Management," in W. Karwowski, and W. Marras, eds., *Handbook of Occupational Ergonomics* (Boca Raton, FL: CRC Press, 1998).

[114] W. M. Nowikas, "The Noise of Sound," Special Report 652, *American Machinist,* Dec. 11, 1972.

[115] J. J. Crocker and A. J. Price, *Noise and Noise Control,* Vol. I (Cleveland, Ohio: CRC Press, 1975).

[116] American College of Occupational and Environmental Medicine Noise and Hearing Conservation Committee, "Guidelines for the Conduct of an Occupational Hearing Conservation Program," *Journal of Occupational Medicine,* Vol. 29, pp. 981–989, 1987.

[117] J. G. Casali and E. H. Berger, "Technology Advancements in Hearing Protection circa 1995: Active Noise Reduction, Frequency/Amplitude-Sensitivity, and Uniform Attenuation," *American Industrial Hygiene Association Journal,* Vol. 57, pp. 175–185, 1996.

[118] J. G. Casali and M-Y. Park, "Laboratory versus Field Attenuation of Selected Hearing Protectors," *Sound and Vibration Magazine,* Vol. 25, No. 10, pp. 28–38.

[119] Charles L. Piechota, "The Twilight Zone," Professional Safety, Jan. 1992, pp. 32–35.

[120] L. M. Ippolito and D. R. Wallace, "A Study on Hazard Analysis in High Integrity Software Standards and Guidelines," U. S. Department of Commerce, National Institute of Standards and Technology, NISTIR 5589, Jan. 1995.

[121] Michael L. Brown, "Tailoring MIL-STD-882B 300 Series Tasks," *Proceedings of the Ninth International System Safety Conference,* July 1989.

Index

F

G

H

I

M

Q

R

S

T

U

V

X

Y

Z